ALLE·ZEIT·WACH
SJ
1842

K. Shimoji T. Kurokawa T. Tamaki
W. D. Willis Jr. (Eds.)

Spinal Cord Monitoring and Electrodiagnosis

With 240 Figures and 70 Tables

Springer-Verlag
Berlin Heidelberg New York
London Paris Tokyo
Hong Kong Barcelona

KOKI SHIMOJI, M.D., Professor and Chairman
Department of Anesthesiology, Niigata University School of Medicine
Asahi-Machi 1, Niigata 951, Japan

TAKAHIDE KUROKAWA, M.D., Professor and Director
Department of Orthopaedic Surgery, Faculty of Medicine
University of Tokyo, 7-3-1 Hongo, Bunkyo-ku
Tokyo 113, Japan

TETSUYA TAMAKI, M.D., Professor and Chairman
Department of Orthopaedics, Wakayama Medical College
Wakayama, Japan

WILLIAM D. WILLIS Jr., M.D., Ph.D., Director
Marine Biomedical Institute Ashbel Smith,
Department of Anatomy and Neurosciences
The University of Texas Medical Branch at Galveston
200 University Boulevard, Galveston, Texas 77550-2772, USA

ISBN-13:978-3-642-75746-4 e-ISBN-13:978-3-642-75744-0
DOI: 10.1007/978-3-642-75744-0

Library of Congress Cataloging-in-Publication Data. Spinal cord monitoring and electrodiagnosis/ K. Shimoji ... [et al.] (eds.). p. cm. Based on the papers presented at the Fourth International Symposium on Spinal Cord Monitoring and Electrodiagnosis held in Niigata, Japan, June 26–29, 1989. Includes bibliographical references. Includes index. ISBN-13:978-3-642-75746-4
1. Evoked potentials (Electrophysiology — Congresses. 2. Spinal cord — Diseases — Diagnosis — Congresses. 3. Electrodiagnosis — Congresses. 4. Patient monitoring — Congresses. I. Shimoji, Kōki, 1935– . II. International Symposium on Spinal Cord Monitoring and Electrodiagnosis (4th: 1989: Niigata-shi, Japan) [DNLM: 1. Electrodiagnosis — congresses. 2. Evoked Potentials — congresses. 3. Monitoring, Physiology — congresses. 4. Spinal Cord — physiopathology — congresses. WL 400 S75778 1989] RC402.2.E94S67 1990 616.07'547 — dc20 DNLM/DLC 90-10106

Softcover reprint of the hardcover 1st edition 1991

Typesetting: Best-set Typesetter Ltd., Hong Kong

11/3130-543210 — Printed on acid-free paper

Preface

The spinal cord has a characteristic structure and functions that are distinct from those of the brain. Its functions are tremendously important since it modulates the peripheral sensory inputs to the dorsal horn, and it gives rise to the ascending pathways transmitting peripheral afferent inputs to the brain, and conveys the descending pathways from the brain both to the lower motor neurons, the final common pathway, and to dorsal horn sensory neurons. In spite of these vital functions, the spinal cord constitutes only a small percentage of the mass of the human central nervous system and is located far from the skin surface, which has obstructed the recording of its electrical activity. Recently, however, important advances have been made in several recording techniques, including epidural recording or averaging methods, allowing both sensory and motor evoked spinal cord potentials in man to be recorded.

This volume is based on the papers presented at the Fourth International Symposium on Spinal Cord Monitoring and Electrodiagnosis. Each of these international symposia has brought together many of the specialists involved in this research, with an important increase in the number of participants since the first symposium was held in Toyko in 1981. At the past symposia several attempts were made to standardize data, techniques, and clinical applications and to integrate the new findings into patient care.

The purpose of publication of this book is to provide an overview of recent advances in the field of spinal cord monitoring and electrodiagnosis, including data from basic physiological and pharmacological studies. It should thus serve as a forum for the prevention of spinal cord damage during operations and for the electrodiagnosis of spinal cord diseases.

We would like to thank Mitsuru Ebe, from the Japan Society of EEG and EMG, for his welcoming comments at the symposium.

Niigata,	K. Shimoji
Tokyo,	T. Kurokawa
Wakayama,	T. Tamaki
Galveston,	W.D. Willis Jr.

Contents

I. Basic Considerations

A. Animal Studies

1. Neurophysiology of Spinal Cord

B. Human Studies

1. Electrical Stimulations

B. Electrodiagnosis

List of Contributors

Addresses are given at the beginning of the respective contribution.

I. Basic Considerations

A. Animal Studies

The Spinothalamic Tract and Other Ascending Nociceptive Pathways of the Spinal Cord

W.D. WILLIS Jr. [1]

Spinal Cord Ascending Nociceptive Pathways

The sensation of pain can result from the activation of nociceptors (nociceptive pain) or it can be a consequence of damage to nervous tissue (neuropathic and central pain). Nociceptive pain in humans depends primarily upon pathways that ascend in the anterolateral quadrant of the spinal cord on the side of the body contralateral to that stimulated by a noxious stimulus. Evidence that one or more tracts in the anterolateral quadrant are necessary for pain sensation comes from the fact that an anterolateral cordotomy can relieve severe nociceptive pain in patients [7, 9]. A demonstration that the anterolateral quadrant is sufficient for pain was a human subject who had all of the spinal cord transected except for an anterolateral quadrant and who retained the ability to feel pain below the level of the lesion [8]. However, tracts ascending in other sectors of the human spinal cord have the potential to contribute to pain sensation, since pain recurs in about 50% of cases months to years after an initially successful cordotomy [9]. White and Sweet [9] argue that this recurrence of pain is not due to regeneration of fibers across the cordotomy. Instead, the pain may be conveyed by ipsilateral tracts in the anterolateral quadrant or by pathways ascending in the posterior part of the spinal cord. It is unclear if the tracts responsible for the return of pain after cordotomy normally contribute to pain sensation or if their capability for signalling pain after cordotomy results from some plastic change secondary to the surgical lesion. It should be noted that the anterolateral quadrant is also responsible for thermal sensation and participates in tactile sensation [7–9].

Potentially nociceptive somatosensory pathways that ascend to the brain from the spinal cord in the anterolateral quadrant include the spinothalamic, spinomesencephalic, and spinoreticular tracts, and those that are found in the posterior part of the spinal cord are the postsynaptic dorsal column pathways and the spinocervical tract [10]. Recently, Hodge's group has demonstrated

[1] Department of Anatomy and Neurosciences and Marine Biomedical Institute, University of Texas Medical Branch, Galveston, TX 77550-2772 USA

K. Shimoji et al. (Eds.)
Spinal Cord Monitoring and Electrodiagnosis

that the part of the spinothalamic tract originating from neurons in the marginal zone actually ascends in the posterolateral quadrant, at least in several animal forms, including two species of monkeys [1].

Nociceptive Pathways in the Anterolateral Quadrant

The anterolateral pathway that is most likely to convey pain sensation is the spinothalamic tract, since this pathway terminates directly in the main somatosensory nucleus of the thalamus (the ventral posterior lateral nucleus), and most of the neurons of this pathway in animals are nociceptive (see below). The spinothalamic tract also includes nocicpetive neurons which project to medial thalamic nuclei that are unlikely to be involved in the sensory discriminative aspects of pain. Nociceptive neurons have also been found in other pathways of the anterolateral quadrant, including the spinomesencephalic and spinoreticular tracts [10]. Many of these nociceptive neurons have large and complex receptive fields that make them unsuited for signalling the sensory discriminative aspects of pain. Furthermore, they end in nuclear regions that have at best a crude somatotopic organization. Projections from the midbrain and rhombencephalic reticular formation are chiefly to nuclei in the medial thalamus that are more likely to be involved in the motivational and affective components of pain than in the sensory discriminative aspects. In addition to actions at higher levels of the brain, the midbrain and pontomedullary reticular formation contribute to the descending control of nociception ("endogenous analgesia system") and to autonomic and somatic reflex reactions to pain.

Properties of Primate Spinothalamic Neurons

The functional properties of nociceptive pathways are best studied in animal experiments. A considerable amount of work has been done on the spinothalamic tract in the rat, cat, and monkey [10]. Emphasis here will be on the electrophysiological properties of primate spinothalamic tract cells.

Spinothalamic tract cells can be identified in experiments on anesthetized monkeys by antidromically activating their axons by electrical stimulation in the thalamic target nucleus. Spinothalamic tract cells that project to the ventral posterior lateral nucleus has been studied extensively, and some of those that project to the central lateral nucleus have also been examined [reviewed in 10].

While recording from an identified spinothalamic tract cell, it is easy to determine whether or not these cells are activated by volleys in fine nerve fibers in the peripheral nerves that supply the cutaneous receptive field.

Almost all of the spinothalamic cells that have been examined could be excited by volleys in both cutaneous A δ and C fibers, and most could also be excited, at least weakly, by volleys in A β fibers as well. However, some spinothalamic tract cells, especially some located in the marginal zone of the spinal cord dorsal horn, are excited by A δ but not by A β fibers. Although most cutaneous nociceptive afferent fibers have small diameters, in the A δ and C range, several other types of cutaneous receptors also have fine afferent fibers. For this reason, experiments demonstrating that volleys in fine afferent fibers activate a spinal cord neuron cannot show definitively whether or not the neuron is nociceptive.

A more direct demonstration that spinothalamic tract cells are nociceptive depends upon activation of the receptive field by "natural" forms of stimuli, such as intense mechanical stimulation of the skin or changing the temperature of the skin to levels that cause pain in humans. Of course, the responses of the same neuron to weak stimuli, such as tactile, vibratory, warm, or cool stimuli, must be compared with the responses to noxious stimuli; the responses to noxious stimuli must exceed those to innocuous stimuli before a neuron can be regarded as nociceptive.

Figure 1A shows the responses of several different spinothalamic neurons in the monkey to graded intensities of mechanical stimulation of the skin. The responses were to a series of stimuli, each lasting 10 s. The first stimulus (*brush*) consisted of brushing the receptive field; the second (*pressure*) was application of an arterial clip to the skin (near pain threshold in humans); the third (*pinch*) was application of a different arterial clip whose bite was stronger (distinctly painful in humans); and the fourth (*squeeze*) was squeezing the skin with serrated forceps (a damaging stimulus). The first cell responded best to *brush*. Its responses to the other stimuli were rapidly adapting and probably also due to activation of mechanoreceptors. Presumably, this spinothalamic tract cell conveyed tactile information to the thalamus. Other spinothalamic cells of this type often had graded responses to the compressive stimuli, as well to *brush*. The second cell had only weak responses to either the *brush* or the *pressure* stimuli, but its response to *pinch* was apparently its maximal response to cutaneous mechanical stimulation, since *squeeze* resulted in a comparable response. Evidently, the stimulus–response function for compressive stimulation of the skin of such neurons was quite steep. The third cell also showed only small responses to *brush* and *pressure* stimuli, but the response to *pinch* was not much larger. The maximum response required the *squeeze* stimulus. The stimulus–response function of this type of neuron had a shallow slope for lesser intensities of mechanical stimulation, but the curve became steep when the stimulus intensity was strong. Both the second and the third cells were clearly nociceptive, but they could have somewhat different functions. For example, spinothalamic cells with steep stimulus–response curves could be more important for sensory discrimination, whereas those with shallow stimulus–response curves might signal the presence of damage.

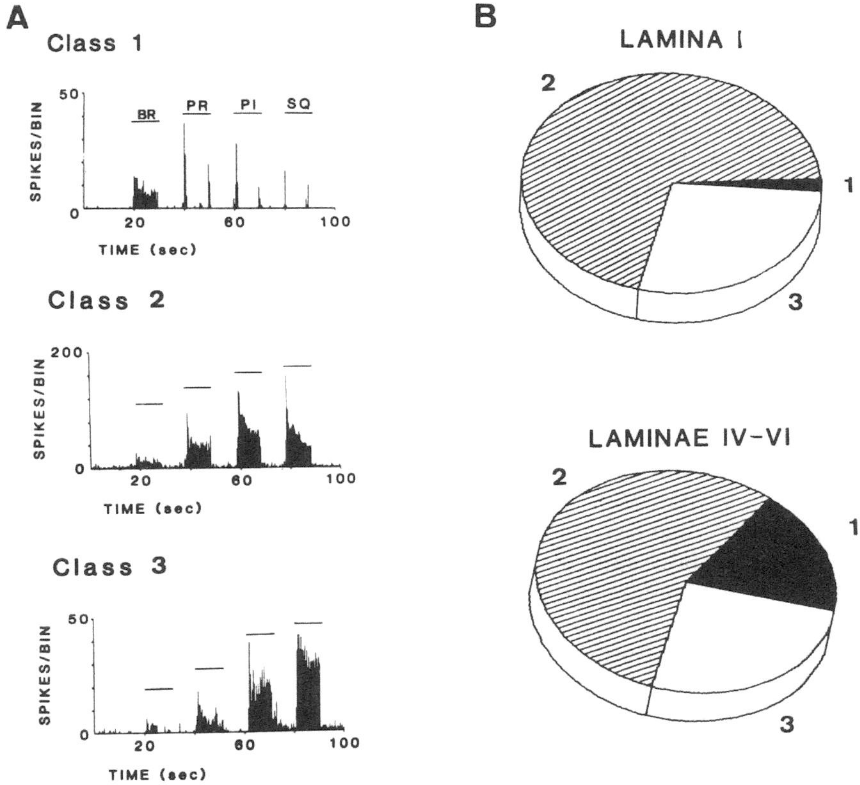

Fig. 1. A Peristimulus time histograms showing the responses of three different primate spinothalamic tract cells to graded intensities of mechanical stimulation of the skin. The cells belonged to classes *1*, *2*, and *3*, respectively (BR, *brush*; PR, *pressure*; PI, *pinch*; SQ, *squeeze*). **B** The proportions of spinothalamic neurons in lamina I and in laminae IV–VI belonging to classes 1–3

These cells are representative of three classes of spinothalamic tract cells that can be distinguished statistically [6], based on their responses to graded strengths of mechanical stimulation of the skin, using a multivariate statistical analysis (*k* means cluster analysis). Of a population of 318 spinothalamic tract cells, about 15% were of the tactile variety, 59% had steep stimulus–response curves, and 26% had shallow stimulus–response curves. However, the relative numbers of these cell types were different when spinothalamic tract cells in different laminae were compared (Fig. 1B). Only one spinothalamic cell in lamina I was found to belong to class 1; the remaining cells were either of class 2 (70%) or 3 (28%). On the other hand, 18% of spinothalamic cells in laminae IV-VI were in class 1, 56% in class 2, and 26% in class 3. This observation suggests that the tactile functions of the spinothalamic tract are more likely to be mediated by cells in the neck of the dorsal horn than by cells in the marginal zone [7].

Spinothalamic tract cells can also be activated by noxious thermal stimulation of the skin [10]. Graded noxious heat pulses produce graded responses of spinothalamic tract cells, with a threshold near 45°C, which is the pain threshold in human subjects. Many spinothalamic cells will also respond to noxious cold stimuli. The responses are graded for temperatures below about 15°C. The threshold for cold pain in humans is about 18°C [10].

Innocuous changes in skin temperature should also excite spinothalamic tract cells, since the spinothalamic tract is thought to be responsible for temperature sensation [9]. The spinothalamic cells that respond to weak thermal stimuli appear to be concentrated in lamina I. Some of these cells respond rather specifically to thermal stimuli [3], but others also respond to noxious mechanical and thermal stimuli [3, 6]. Furthermore, most of the specific thermoreceptive spinothalamic cells appear to project to the medial rather than to the lateral part of the thalamus, at least in the cat [3]. More experiments are needed to clarify the role of the spinothalamic tract in signaling warm and cold sensations.

A limited population of spinothalamic cells that project just to the central lateral nucleus of the intralaminar complex has been studied [reviewed in 10]. The properties of these neurons contrast with those of spinothalamic neurons that project to the ventral posterior lateral nucleus (or that project both medially and laterally). The spinal neurons that project to the central lateral nucleus often have very large receptive fields that may include all of the body and face bilaterally. Much of the receptive field depends upon a spino-bulbospinal loop, since the receptive field becomes limited to the ipsilateral hindlimb following transection of the spinal cord at an upper cervical level. These neurons are clearly nociceptive, but their receptive field characteristics suggest that they play a role in some other aspect of pain than sensory discrimination.

Nociceptive Neurons Projecting in Cord Sectors Other than the Anterolateral Quadrant

The axons of most lamina I spinothalamic tract neurons in the monkey appear to project in the posterior part of the lateral funiculus [1]. Barring a species difference, this implies that some nociceptive functions may be mediated by spinothalamic axons that are not interrupted in the classical anterolateral cordotomy. However, it is not clear if lamina I spinothalamic axons in the human ascend in the same position as in the monkey. There are also uncrossed spinothalamic axons in the anterolateral quadrant; these constitute about 5–10% of the primate spinothalamic tract [1, 10]. Thus, spinothalamic axons in either the contralateral posterolateral funiculus or in the ipsilateral

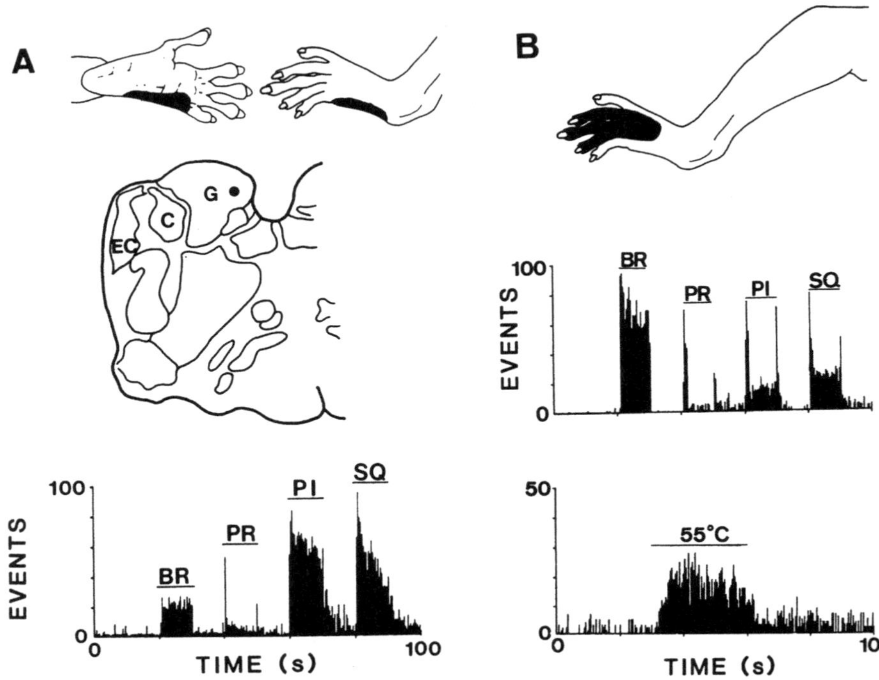

Fig. 2. A, B Nociceptive responses of two neurons in the nucleus gracilis of the monkey and the receptive fields of the neurons. The responses of both neurons to graded intensities of mechanical stimuli were tested. The first neuron responded best to noxious mechanical stimuli and less strongly to innocuous brushing. The second neuron responded best to brushing, but had a graded response to graded noxious mechanical stimuli. In addition, it responded to noxious heating to 55°C.

lateral funiculus may be responsible for pain sensation that recurs following an initially successful cordotomy.

In addition, nociceptive neurons belonging to other somatosensory pathways may contribute to pain sensation, or other aspects of the pain reaction, before or after cordotomy. Nociceptive responses have been observed in recordings from neurons of the postsynaptic dorsal column pathway and of the spinocervical tract of the cat [reviewed in 10]. These pathways ascend ipsilaterally in the posterior funiculus and posterior part of the lateral funiculus, respectively. However, it has been unclear if the nociceptive activity of these second order neurons was relayed on to the ventral posterior lateral thalamus. Nor was it known if similar responses occurred in the primate. Recently, however, our laboratory has recorded from neurons in the lateral cervical nucleus and the nucleus gracilis of the monkey that were activated antidromically from the ventral posterior lateral nucleus [4, 5]. At least 55% of the sample of neurons in the lateral cervical nucleus and 6% of that in the nucleus gracilis in the monkey could be considered nociceptive, based on the responses of these cells to graded intensities of mechanical stimuli. An even larger proportion of these cells responded to noxious thermal stimuli. Figure 2A shows the responses of a neuron in the nucleus gracilis to graded mechanical stimulation of the skin in its receptive field, and Fig. 2B shows the responses of another gracile neuron to a noxious heat stimulus.

Conclusions

Several of the somatosensory pathways of the spinal cord contain neurons that signal noxious stimuli. In humans, pain sensation is likely to depend primarily on the nociceptive responses of spinothalamic neurons, but it is possible that several other pathways also contribute. This is especially evident in cordotomy cases in which pain recurs months after a successful operation to relieve pain. Such recurrence of pain may depend upon components of the spinothalamic tract that ascend in the posterior part of the lateral funiculus or on the side of the cord ipsilateral to the source of pain. Alternatively, nociceptive spinal cord neurons whose axons ascend ipsilaterally to the dorsal column nuclei or to the lateral cervical nucleus may be able to signal pain under special circumstances. In addition, several somatosensory pathways contribute to pain reactions other than sensory experience or the activation of descending analgesia systems.

References

1. Apkarian AV, Hodge, CJ (1989) The primate spinothalamic pathways: II. The cells of origin of the dorsolateral and ventral spinothalamic pathways. J Comp Neurol 288:474–492
2. Chung JM, Surmeier DJ, Lee KH, Sorkin LS, Honda CN, Tsong Y, Willis WD (1986) Classification of primate spinothalamic and somatosensory thalamic neurons based on cluster analysis. J Neurophysiol 56:308–327.
3. Craig AD, Kniffki KD (1985) Spinothalamic lumbosacral lamina I cells responsive to skin and muscle stimulation in the cat. J Physiol 365:197–221
4. Downie JW, Ferrington DG, Sorkin LS, Willis WD (1988) The primate spinocervicothalamic pathway: responses of cells of the lateral cervical nucleus and spinocervical tract to innocuous and noxious stimuli. J Neurophysiol 59:861–885
5. Ferrington DG, Downie JW, Willis WD (1988) Primate nucleus gracilis neurons: responses to innocuous and noxious stimuli. J Neurophysiol 59:886–907
6. Ferrington DG, Sorkin LS, Willis WD (1987) Responses of spinothalamic tract cells in the superficial dorsal horn of the primate lumbar spinal cord. J Physiol 388:681–703
7. Kuru, M (1949) Sensory paths in the spinal cord and brain stem of man. Sogensya, Tokyo
8. Noordenbos W, Wall PD (1976) Diverse sensory functions with an almost totally divided spinal cord. A case of spinal cord transection with preservation of part of one anterolateral quadrant. Pain 2:185–195
9. White JC, Sweet WH (1969) Pain and the neurosurgeon. Thomas, Springfield
10. Willis WD (1985) The pain system. Karger, Basel

Motor and Sensory Tract Activation Following Epidural Electrical Stimulation

E. TRANSFELDT[1], G. NIZNIK, F. SHICHIJO, T. OHSHIMA, and B. POMERANZ

There is a need for a more precise method of spinal cord monitoring to assess entire spinal cord function including motor and sensory tracts and to do this with a single simultaneous test.

Somatosensory evoked potentials have been shown to measure function predominently in the dorsal columns. Significant loss of motor function may occur in the face of a normal evoked potential.

Motor evoked potentials popularized by Levy can be generated by electrical or magnetic stimulation transcranially and measure the function of the spinal motor tracts.

This study examines the technique of electrical stimulation of the spinal cord, a technique first popularized by Tamaki and Shimoji to see if both sensory and motor tracts could be activated. The study also seeks to define the origins of the evoked potentials.

Methods

The experiments were performed on 30 mongrel cats. Anesthesia was induced with ketamine and maintained with enflurane and nitrous oxide. Bipolar epidural electrodes were used for stimulation and recording.

A small laminectomy was performed at T13 and the epidural electrodes passed on the T7 and L3 levels. A hemi-laminectomy was made from L5-Sl on the right exposing the L7 nerve root. Bipolar needle electrodes were placed in the dorsal and ventral roots respectively.

The left and right sciatic nerves were exposed and bipolar needle electrodes placed into them. A similar bipolar recording electrode was also placed in the gastrocnemious muscle. The main stimulus sites were transcranial at the vertex. T7 epidural, and the sciatic nerve.

[1] Twin Cities Scoliosis Center, 920 East 28th Street Suite 40, Minneapolis, MN 55407, USA

K. Shimoji et al. (Eds.)
Spinal Cord Monitoring and Electrodiagnosis

Spinal cord lesions were made at the T12, T13 laminectomy levels. There were four lesion types. First, a dorsal column lesion alone was produced and after the electrophysiologic studies, a complete dorsal hemisection was performed just distal to the first lesion and the studies once again repeated. In the next group of cats a lesion of the ventral half of the spinal cord was made and after doing the appropriate physiologic studies, the remainder of the spinal cord, excluding the dorsal columns, was incised just distal to the first. After completion of the experiment, the spinal cord was removed for histological studies to verify the extent of each lesion. If the histology did not correspond to the intended lesion, animals were excluded from the experiments.

Results

Tracings of representative wave forms following right epidural stimulation at T7 were recorded. Recordings are from L3, the L7 ventral root, the L7 dorsal root, the sciatic nerves, and from the right gastrocnemius. The epidural response in particular consists of three distinctive peaks.

Recordings were made following the two types of dorsal spinal cord lesions. With a dorsal column lesion alone the N3 component of the L3 epidural response was abolished. N2 decreased slightly but not significantly in amplitude and N1 remained normal. There was no response at the L7 dorsal root and at the ventral root the late components were all abolished, but N1 remained present. No significant further changes were observed after total dorsal hemi-section.

Following a ventral hemi-section, the L3 response showed a decrease in amplitude of N1 and N2 (N3 being preserved) and an increased latency of the N1 component. With lesioning of the remainder of the spinal cord with preservation of the dorsal columns, the N1 component was totally abolished and there was considerable decrease in the N2, but N3 is preserved. Following a ventral hemi-section, the N1 and N2 components were abolished from the L7 ventral root, but N3 remained normal. There was no further change with further lesioning. The L7 dorsal root response in both lesions was unchanged except for a small, early negative component similar in latency to N1 of the ventral root response. There was no change in the right sciatic nerve recording following either lesion.

The T7 and L3 somatosensory evoked potentials following sciatic nerve stimulation are shown in Fig. 1. The T7 potential decreased in amplitude following a dorsal column lesion but was not obliterated. There was no change in the L3 recording. Following dorsal hemi-section, however, the T7 response was totally abolished and there was no significant change of the L3 component. Spinal evoked potentials at both T7 and L3 remain constant, however, following ventral lesions.

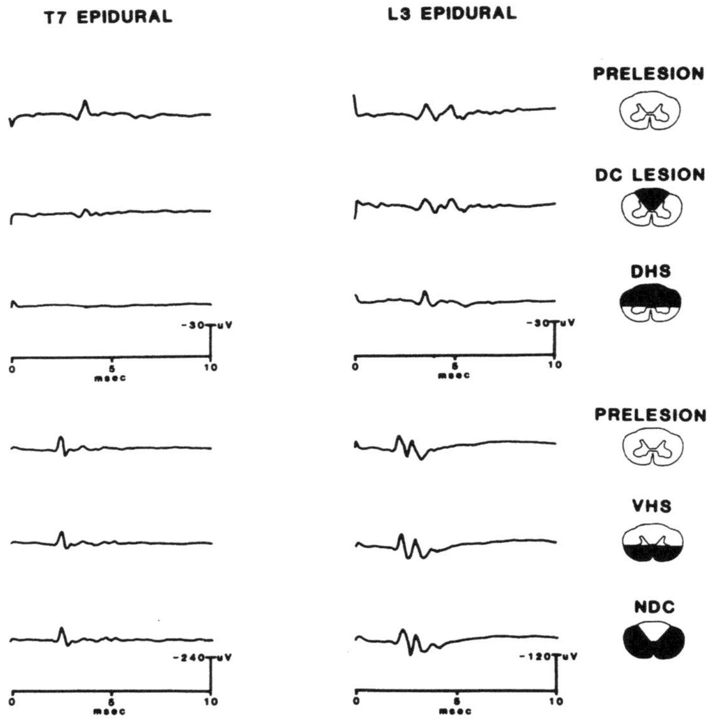

Fig. 1. Representative tracings of epidural spinal evoked potentials recorded from *T7* and *L3* following stimulation of the right sciatic nerve. Recordings were made prior to and following each type of spinal cord lesion. The T7 potential was affected by a *DC* lesion and was obliterated following a *DHS*. No consistant changes were observed in any other recording site or with any other type of lesion

The T7 and L3 epidural recordings following motor evoked potentials from transcranial stimulation are shown in Fig. 2. Following dorsal spinal cord lesions there was no change in the evoked response. However, following a ventral hemi-section, the N1 potential at L3 was obliterated while the late polyphasic peaks remained present. Following a non-dorsal column lesion, the motor evoked potential at L3 was totally obliterated. No changes occurred at T7 above the lesion.

Conclusions

1. Epidural electrical stimulation produced simultaneous motor and sensory tract activation.
2. Sciatic nerve potential reflects activity in the dorsal columns.

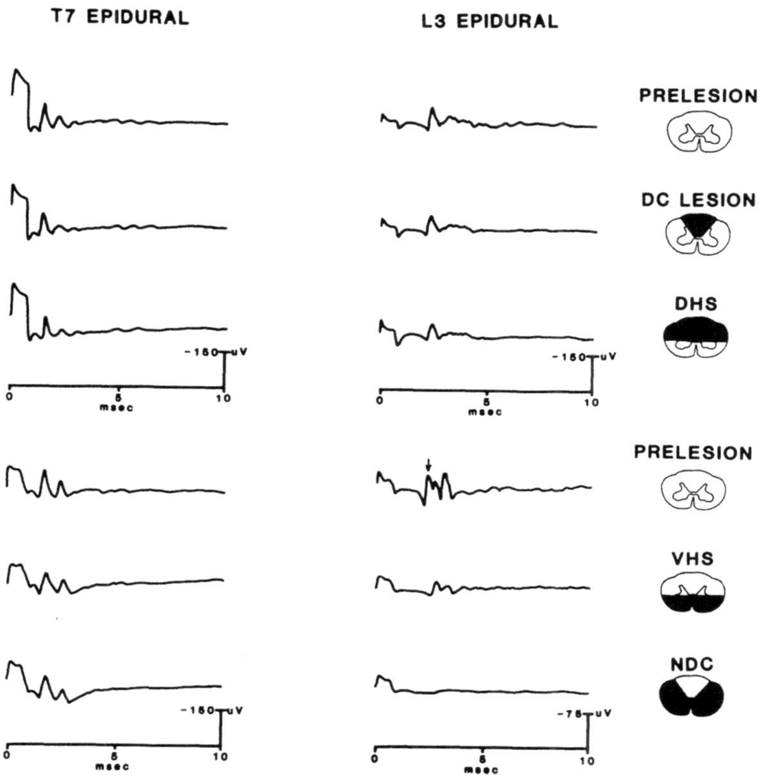

Fig. 2. Tracings of representative MEPs recorded from the *T7* and *L3* epidural electrodes after anodal transcranial stimulation at the vertex. Recordings were made prior to and following each type of spinal cord lesion. The early negative spike wave of *L3* was lost after a VHS (*arrow*) and the rest of the waveform lost after a NDC. No consistant changes were observed in any other recording site or with any other type of lesion

3. Epidural potentials measure activity in the dorsal columns, the ventral motor tracts, and the dorsal lateral tracts.
4. Global spinal cord function is measured by epidural stimulation. Gastrocnemius and ventral root recordings following epidural stimulation result from spinal motor neuron activation. Dorsal root recordings reflect antidromic activation of the dorsal columns and right sciatic nerve recordings reflect a peripheral activation of the dorsal roots.

References

Berry MS, Penthreath VM (1976) Criteria for distinguishing between monosynaptic and polysynaptic transmission. Brain Res 105:1–20

Ertekin C (1976) Studies on the human evoked electrospinogram – the conduction velocity along the dorsal funiculus. Acta Neurol Scandinav 53:21–38

Ginsburg HH, Shetter AG, Raudzens P (1985) Post-operative paraplegia with preserved intraoperative somatosensory evoked potentials. A case report. J Neurosurg 63:296–300

Grundy B (1982) Monitoring of sensory evoked potentials during neurosurgical operations: methods and applications. Neurosurg 11:556–575

Grundy B (1984) Evoked potentials in the operating room. Mt Sinai J Med 51:585–591

Guru K, Mailis A, Ashby P, Vanderlinden G (1987) Postsynaptic potentials in motoneurons caused by spinal cord stimulation in humans. EEG Clin Neurophysiol 66:275–280

Imai T (1976) Human electrospinogram evoked by direct stimulation on the spinal cord through epidural space. J Jap Orthop Ass 50:1037–1056

Jones SJ, Edgar MA, Ransford AO, Thomas NP (1983) A system for the electrophysiological monitoring of the spinal cord during operations for scoliosis. JBJS. 65:134–139

Kotani H, Katasuhiko S, Yamasaki H, Hattori S, Shinya K, Otome K (1986) Evaluation of cervical cord function in cervical myelopathy and/or radiculopathy using both segmental conductive spinal-evoked potentials (SEP). Spine 11:185–190

Kurokowa T (1972) Spinal cord action potential evoked by epidural stimulation of cord – report of human and animal records. Jpn J Electroencephalogr Electromyogr 1:64–66

Lance JW (1954) Pyramidal tract in spinal cord of cat. J Neurophysiol 17:253–270

Lesser RP, Raudzens P, Lueders H, Nuwer MR, Goldie WD, Morris HH, Dinner DS, Klem G, Hahn JF, Shetter AG, Ginsburg HH, Gurd AR (1986) Postoperative neurological deficits may occur despite unchanged intraoperative somatosensory evoked potentials. Ann Neurol 19:22–25

Levy WJ, McCaffrey M, Goldman D, York M (1986) Nonpyramidal motor activation produced by stimulation of the cerebellum, direct or transcranial: a cerebellar evoked potential. Neurosurgery 19:163–176

Levy WJ, York DH, McCaffrey M, Tanzer F (1984a) Motor evoked potentials from trascranial stimulation of the motor cortex in humans. Neurosurgery 15:287–302

Levy WJ McCaffrey M, York DH, Tanzer F (1984b) Motor evoked potentials from transcranial stimulation of the motor cortex in cats. Neurosurgery 15:214–227

Machida M, Weinstein SL, Yamada T, Kimura J (1985) Spinal cord monitoring: electrophysiological measures of sensory and motor function during spinal surgery. Spine 10:407–413

Mills KR, and Murray NMF (1986) Electrical stimulation over the human vertebral column: which neural elements are excited? EEG Clin Neurophysiol 63:382–589

Ryan TP, Britt RH (1986) Spinal and cortical somatosensory evoked potential monitoring during corrective spinal surgery with 108 patients. Spine 11:352–361

Shimoji K, Higashi H, Kano T (1971) Epidural recording of spinal electrogram in man. EEG Clin Neurophysiol 30:236–239

Tamaki T, Noguchi T, Takano H, Tsuji H, Nakagawa T, Imai K, Inoue S (1984) Spinal cord monitoring as a clinical utilization of the spinal evoked potential. Clin Orthop 184:58–64

Tamaki T, Tsuji H, Inoue S, Kobayashi H (1981) The prevention of iatrogenic spinal cord injury utilizing the evoked spinal cord potential. Int Orthop 4:313–317

Tsuyama N, Tsuzuki N, Kurokawa T, Imai T (1978) Clinical application of spinal cord action potential monitoring. Int Orthop 2:39–46

Effects of Electric Current Application on the Evoked Spinal Cord Potentials in Dogs

T. KANO[1] and A. HASHIGUCHI

At present, electric stimulation therapy is widely accepted as a useful means of pain management, because it is free from adverse pharmacological effects and neurological deficits. We have been studying electrically induced analgesia or anesthesia since 1969. In 1971, Shimoji et al. developed a noninvasive method of cord dorsum stimulation for the relief of intractable pain. A fine stainless steel wire which was used as a stimulation electrode was introduced through a polyethylene tube into the epidural space using the continuous epidural anesthesia technique [6]. In 1975, Shimoji and Kano reported 47 cases of successful clinical electroanesthesia in which a high frequency square pulse wave current was applied to the head through front-occipital electrodes [7]. In 1978, Kano succeeded in producing analgesia in the hand, solely by transcutaneous current application to the main nerve trunks of the arm, which was accompanied by strong vibratory sensation and muscle contraction [4].

In the present animal experiment, a high frequency pulse wave current, used for the production of electroanalgesia or electroanesthesia (EA), was applied to various parts of the body and its influences on somatosensory evoked spinal potentials were observed.

Materials and Methods

Experiments were carried out on 18 mongrel dogs, each weighing 9.5–14.3 kg. Following intravenous injection of pentobarbital (20 mg/kg), the trachea was intubated. The animals were paralyzed with gallamine and mechanically ventilated to maintain $PaCO_2$ around 40 mmHg. The femoral vein and artery were catheterized for infusion and for blood pressure monitoring and blood sampling, respectively. Esophageal temperature was also monitored and maintained at 37°C–38°C by using a heating pad. During the subsequent surgical preparations and electrode settings, nitrous oxide (50%)-oxygen-halothane (1%) anesthesia was used.

[1] Surgical Center, Kumamoto University Hospital, 1-1-1 Honjo, Kumamoto, 860 Japan

K. Shimoji et al. (Eds.)
Spinal Cord Monitoring and Electrodiagnosis

Laminectomies were performed at the L3, L4, and L5 spinal levels. After the spinal cord had been exposed, the ventral roots on one side were cut from L5 to L7. Recording electrodes were then set. To record the cord dorsum potentials (CDPs), a Teflon-coated wire electrode was placed on the dorsal surface of the spinal cord. The reference electrode was placed on the surrounding back muscle. To record the dorsal root potentials (DRPs), a filament of the dorsal root was cut approximately 10 mm away from the spinal cord and the central stump was mounted on a pair of J-shaped wire electrodes.

The tibial nerve was exposed on the side of the cut ventral roots and was mounted on a pair of J-shaped platinum electrodes. After infiltration of 0.5% lidocaine into the wound, halothane administration was reduced to 0.5%. The exposed spinal cord and the tibial nerve were immersed in a liquid paraffin pool heated to 36°C.

The CDPs or the DRPs were evoked by stimulating the tibial nerve with a square pulse of 0.2 ms at a frequency of 1 pulse per second (pps). A constant current stimulator (Nihon Kohden MSE-3) with an isolation unit was used. Stimulus intensity was adjusted to supramaximum for the P wave of the CDPs or the Vth wave of the DRPs. The bandpath of the recording system was 1 Hz–3 kHz. The evoked potentials were superimposed on a polaroid photograph, or averaged 10–30 times in an averaging system (Nihon Kohden ATAC 250), then drawn on a X-Y plotter. The CDPs were composed of a primary afferent volley, an intermediary N wave, and a long-lasting P wave. In the DRPs, potential polarities were reversed during the same event. The P wave of the CDPs and the Vth wave of the DRPs were a mirror image of each other.

In seven of the 18 dogs, partial laminectomy at the C1 spinal level was also performed for cord dorsum stimulation and/or for spinal cord transection. A twisted Teflon-coated wire electrode was inserted caudally under the dura mater. Four dogs received transection of the upper cervical spinal cord. In dogs with the spinal cord intact, EA current of high frequency bidirectional square pulse wave was applied under light halothane anesthesia. However, in dogs with the spinal cord transected, the EA current was applied after washing out the anesthetic gases. The EA current was delivered through a constant current generator (Neurodyne Dempsey Model 1660).

The animals were killed with overdoses of pentobarbital at the end of the experiment.

Results

Prior to EA current application, some fundamental features of the tibial nerve induced CDPs (TN-CDPs) or DRPs (TN-DRPs) were studied in dogs with an intact spinal cord. Upon increasing the stimulus frequency to 20 pps, the amplitudes of both N and P waves of the CDPs decreased gradually, while

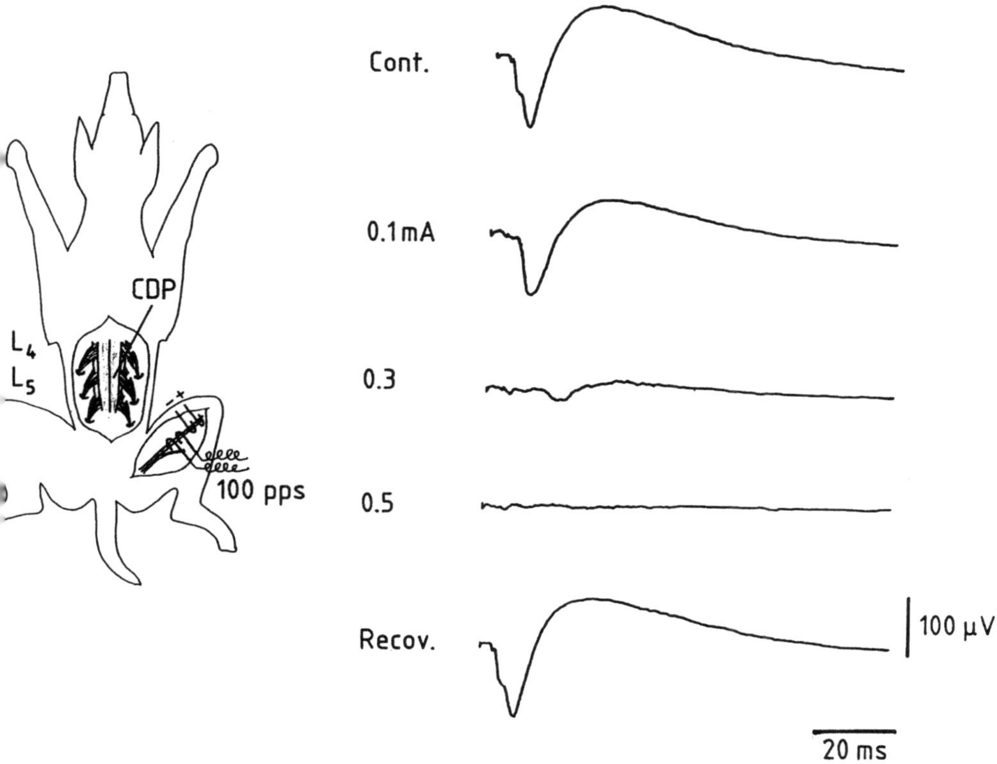

Fig. 1. Application of an EA current to the tibial nerve between the stimulation and the recording electrodes. The EA current applied was a bidirectional square pulse with a frequency of 100 pps and a duty ratio of 20%. On increasing the stimulus intensity of the EA current up to 0.5 mA, changes in the tibial nerve induced (TN)-DRPs were observed in dogs with an intact spinal cord under 60% nitrous oxide-0.5% halothane anesthesia. Upward deflections are positive in this and the following illustrations

the primary afferent volley showed little change. Lidocaine 0.5% or direct current, applied to the tibial nerve between the stimulation and the recording electrodes, depressed the TN-CDPs dose-dependently. All components of the TN-CDPs were eliminated by application of a 5 V direct current.

A high frequency EA current producing bidirectional pulses with a duty ratio of 20% was applied to various parts of the body and the influences on the TN-CDPs or the TN-DRPs were observed. An EA current with a frequency of 100 pps, which was applied to the tibial nerve between the stimulation and the recording electrodes, depressed the TN-CDPs dose-dependently. The EA current at an intensity of 0.5 mA flattened the curve of the TN-CDPs (Fig. 1). Application of a 500 pps EA current to the head through bitemporal safety pins, to the exposed superficial radial nerve through J-shaped platinum electrodes, or to the foreleg through safety pins did not affect the TN-DRPs of the dogs with an intact spinal cord. However, applica-

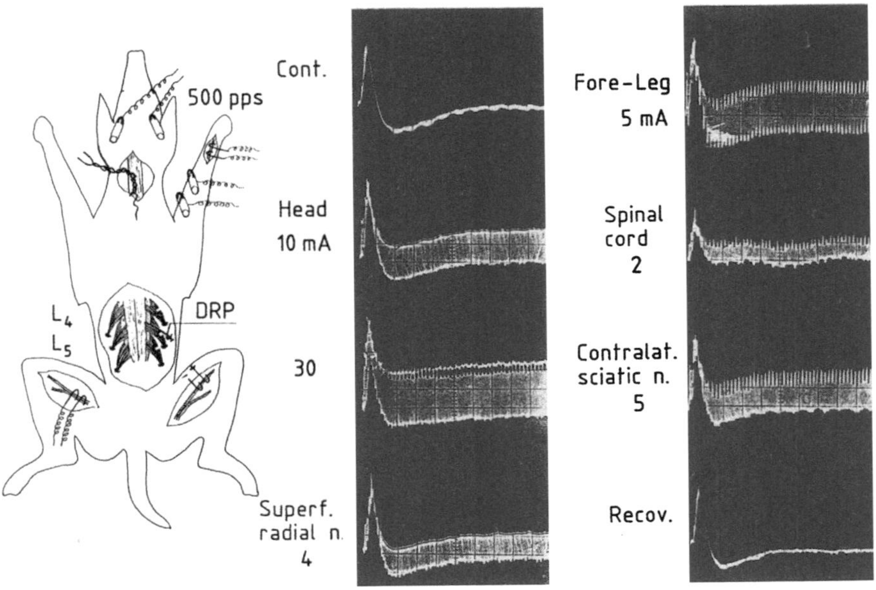

Fig. 2. Application of an EA current to the head, to the superficial radial nerve, to the foreleg, to the cervical spinal cord, or to the contralateral sciatic nerve. Changes in the TN-DRPs were observed during application of 500 pps EA current to intact dogs under light nitrous oxide-halothane anesthesia

tion of the EA current to the dorsum of the cervical spinal cord through a subdural twisted wire electrode markedly depressed the TN-DRPs. Its application to the contralateral sciatic nerve also caused slight depression of the TN-DRPs (Fig. 2).

Even in dogs with the spinal cord transected at the C1 spinal level, application of 500 pps EA current to the cord dorsum depressed the TN-CDPs in a dose-dependent manner (Fig. 3). An EA current was applied to the ipsilateral fibular nerve or to the cervical spinal cord in spinal dogs with the current intensity kept constant. As the pulse frequency of the EA current was reduced from 5000 pps to 500 pps, the depressive effect on the TN-DRPs became more marked (Fig. 4).

Discussion

An EA current of 100 pps, applied to the tibial nerve proximal to the stimulation electrode, depressed the TN-CDPs in the same way as a local anesthetic or a direct current do. Application of such a high frequency pulse

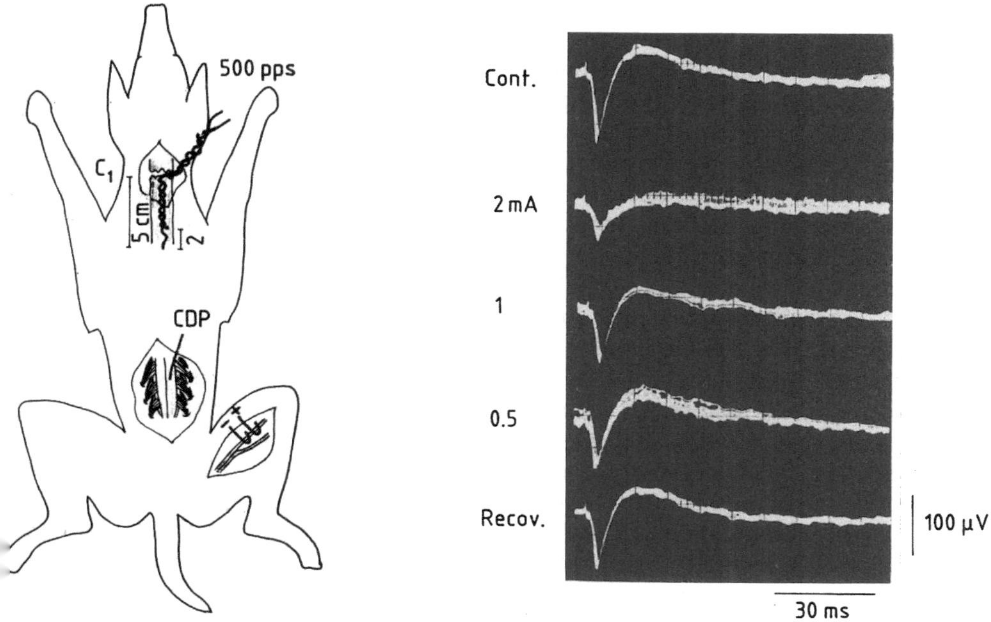

Fig. 3. Application of an EA current to the cervical spinal cord. Through a subdural electrode, 500 pps EA current was applied to the cord dorsum and the descending influences on the TN-DRPs were observed in dogs with the spinal cord transected at a higher level

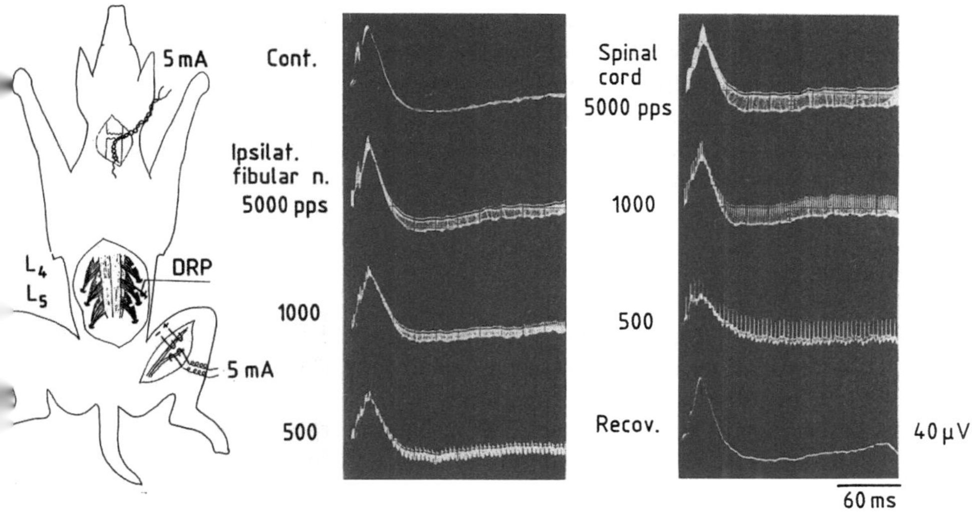

Fig. 4. Application of an EA current to the ipsilateral fibular nerve or to the cervical spinal cord. Changes in the TN-DRPs were observed during application of 500 pps, 1000 pps, or 5000 pps EA current in spinal dogs

wave develops a prolonged depolarization of the nerve and blocks afferent impulses from the periphery [1]. However, the depolarization itself produces new afferent and efferent impulses. The efferent impulses may contribute to the conduction block at a peripheral level through collision with the impulses from the periphery. Considering that afferent relays in the spinal cord were easily depressed by increasing the stimulus frequency, the afferent impulses also seem to contribute to the production of EA at the spinal level. Handwerker et al. demonstrated in spinal animals that the neuronal activities in the dorsal horn responding to nociceptive stimuli were postsynaptically suppressed by electrical stimulation of the peripheral Aβ fibers [3].

EA current application to the rostral cord dorsum was very effective in depressing the TN-CDPs and the TN-DRPs. Cord dorsum stimulation induces inpulses in the spinal cord that spread bidirectionally, cephalad and caudad. The descending impulses seem to interfere with the transmission of afferent impulses at the dorsal horn. There are many reports showing that a descending sensory-regulating system works in a tonic fashion and that its activation depresses afferent synaptic transmission in the spinal cord [2, 3, 9]. Electrical stimulation of the descending spinal tracts has also been reported to produce evoked potentials that resemble those following stimulation of primary afferent fibers [8, 10]. The ascending impulses may also contribute to activating a descending sensory inhibition via a feedback system.

The TN-DRPs were also slightly depressed by EA current application to the ipsilateral fibular nerve or to the contralateral sciatic nerve. This suggests that some unspecified common interneurons are segmentally distributed in the cord dorsum and are involved in the production of the TN-DRPs [2].

Application of EA current to such a remote region as the head, the foreleg, or the superficial radial nerve was not effective in inducing a descending inhibition of the TN-CDPs. Otherwise the activation of the descending inhibitory system might be depressed by anesthesia.

Application of current to the head, which is accompanied by loss of consciousness, is thought to exert its anesthetic action by a different physiological mechanism [5].

Summary

1. When an EA current (100 pps) was applied to the tibial nerve between the stimulation and the recording electrodes, all the components of the TN-CDPs were depressed completely.
2. Application of an EA current (500 pps) to the rostral spinal cord markedly depressed the TN-CDPs and the TN-DRPs, even after the spinal cord was transected at a higher level.
3. Application of an EA current (500 pps) to the ipsilateral fibular nerve or to the contralateral sciatic nerve also caused slight depression of the TN-DRPs.

4. Application of an EA current (500 pps) to the head, to the foreleg or to the superficial radial nerve did not affect the TN-DRPs.

References

1. Campbell J, Taub A (1973) Local analgesia from percutaneous electrical stimulation. Arch Neurol 28:347–350
2. Hagbarth KE, Kerr DIB (1954) Central influences on spinal afferent conduction. J Neurophysiol 17:295–307
3. Handwerker HO, Iggo A, Zimmermann M (1975) Segmental and supraspinal actions on dorsal horn neurons responding to noxious and non-noxious skin stimuli. Pain 1:147–167
4. Kano T (1978) Local electroanalgesia. I. Percutaneous current application to the human forehand to produce local analgesia. Jpn J Anesth 27:495–501
5. Kano T, Cowan GSM, Smith RH (1974) The role of the somatosensory system in electroanesthesia. Anesth Analg 53:667–671
6. Shimoji K, Higashi H, Kano T, et al. (1971) Electrical management of intractable pain. Jpn J Anesth 20:444–447
7. Shimoji K, Kano T (1975) Clinical studies in Japan. In: Sances A Jr, Larson SJ (eds) Electroanesthesia. Academic Press, New York
8. Shimoji K, Shimizu H, Maruyama Y, et al. (1982) Dorsal column stimulation in man: facilitation of primary afferent depolarization. Anesth Analg 61:410–413
9. Willis WD (1982) Progress in sensory physiology 3; Control of nociceptive transmission in the spinal cord. Springer, New York Berlin Heidelberg
10. Willis WD (1984) Evoked spinal cord potentials in the cat and monkey: Use in the analysis of spinal cord function. In: Homma S, Tamaki T (eds) Fundamentals and clinical application of spinal cord monitoring. Saikon Publishing, Tokyo

The Significance of Motor Evoked Spinal Cord Potentials Recorded from the Ventral Root in Cats

H. Kawamura[1], K. Yokogushi, G. Katahira, and M. Matsumoto

Introduction

The conventional methods of monitoring spinal cord potentials have been used experimentally and clinically; however these techniques do not seem to reflect directly the motor function of the spinal cord. It is probable that these potentials represent sensory evoked spinal cord potentials. We previously reported that the ventral root potentials (VRP) could be recorded from the lower lumbar ventral root by stimulating the dorsolateral surface of the upper thoracic spinal cord in cats [1–3]. VRPs had two negative peaks of several hundred microvolts in amplitude in many cases (Fig. 1a). The onset latency was about 8 ms and the peak latencies were about 10 ms for peak 1 and about 13 ms for peak 2 (Fig. 1b) [1–3].

The purpose of this study is to make clear that VRPs directly reflect the motor function of the spinal cord.

Materials and Methods

Thirty-two cats were anesthetized with a halothane-oxygen mixture, immobilized by pancuronium bromide under artificial respiration, and then mounted on a stereotaxic spinal apparatus [1, 3]. Their spinal cords were transected at the upper side of C1 level. VRPs were recorded at a lower lumbar ventral root through a silver hook electrodes in warm (37°C) paraffin oil by stimulating the dorsolateral surface of the upper thoracic spinal cord with a bipolar electrodes (0.2 Hz supramaximal stimulation). The rectal temperature of the cats was kept between 36°C and 38°C by means of external heat and their blood pressure was kept above 90 mmHg throughout the

[1] Department of Orthopedic Surgery, Sapporo Medical College, Minami-1 Nishi-17, Chuou-ku, Sapporo, 060 Japan

K. Shimoji et al. (Eds.)
Spinal Cord Monitoring and Electrodiagnosis

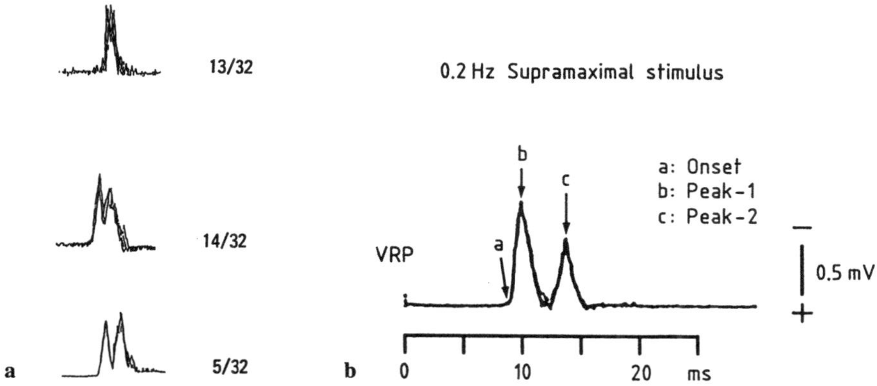

Fig. 1a,b. Wave form of VRPs recorded from the lower lumbar ventral root by stimulating the dorsolateral surface of the thoracic spinal cord in cats ($n = 32$; supramaximal stimulus, 0.2 Hz). **a** In many cases VRPs had two negative peaks of several hundred microvolts in amplitude, **b** the onset latency was about 8 ms (*a*) and the peak latencies were about 10 ms for peak 1(*b*) and about 13 ms for peak 2(*c*)

experiments. An electrophysiological study of VRPs (experiment 1) and an investigation of the origin of VRPs (experiments 2 and 3) were done.

Results

In experiment 1, we studied the following electrophysiological characteristics of VRPs and obtained the following results:

1. The relation between VRP amplitude and stimulus frequency (Fig. 2a). VRP amplitude was suppressed at higher stimulus frequencies [1–3].
2. The effect of stimulus frequency on peak 1 and peak 2 (Fig. 2b). Peak 2 amplitude decreased earlier than peak 1 amplitude, and was significantly suppressed at a stimulus frequency below 5 Hz [1–3].

In experiment 2, we investigated the origin of VRPs. In order to do so, we performed lesion experiments and electrophysiological studies on the VRPs of normal cats and of those with degenerated spinal cords.

1. Lesion experiment (Fig. 3a). In cases of dorsolateral column lesions, VRPs were markedly suppressed or diminished [2, 4].
2. Electrophysiological study. First, the left dorsolateral side of the upper spinal cord of the cat was cut using a thin surgical blade under general anesthesia. After 6 weeks, VRPs were recorded using the same method as described before. Immediately after recording the VRPs, the spinal cord was excised and then fixed in 10% formalin for 3 weeks. The lesion was confirmed

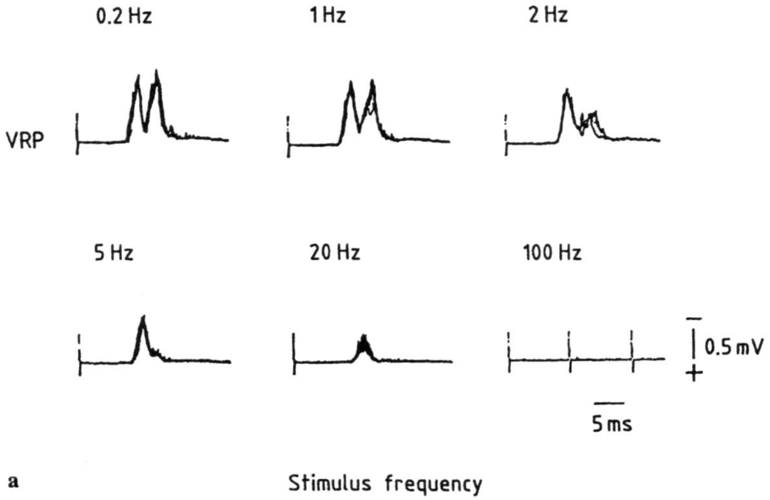

a

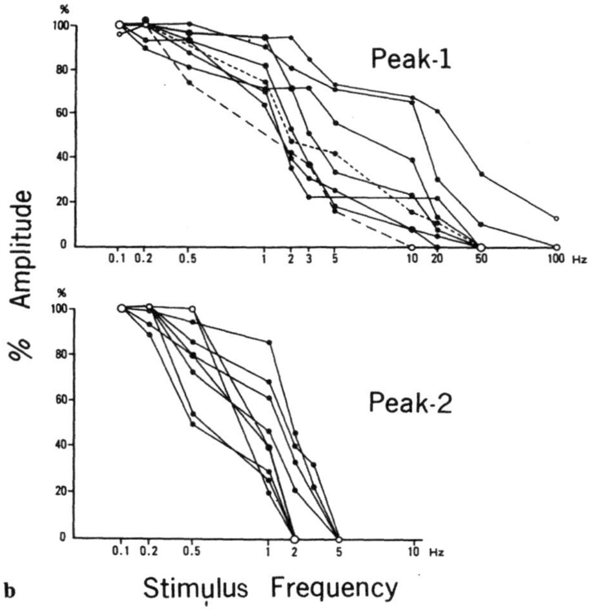

b

▲ **Fig. 2a,b.** Study of the electrophysiological characteristics of VRPs showing **a** the relation between the VRP amplitude and stimulus frequency, **b** the effect of stimulus frequency on peak 1 and peak 2

► **Fig. 3a–c.** Study of the origin of VRPs using lesion experiments and electrophysiological studies on the VRPs of normal cats and those with degenerated spinal cords. **a** The lesion experiment showed that VRP was markedly suppressed or diminished in cases of dorsolateral column lesions. **b** The electrophysiological study showed that in cats with degenerated spinal cords, the VRP amplitude was depressed and the latency prolonged. **c** The VRPs of cats with degenerated spinal cords had a higher threshold than normal cats

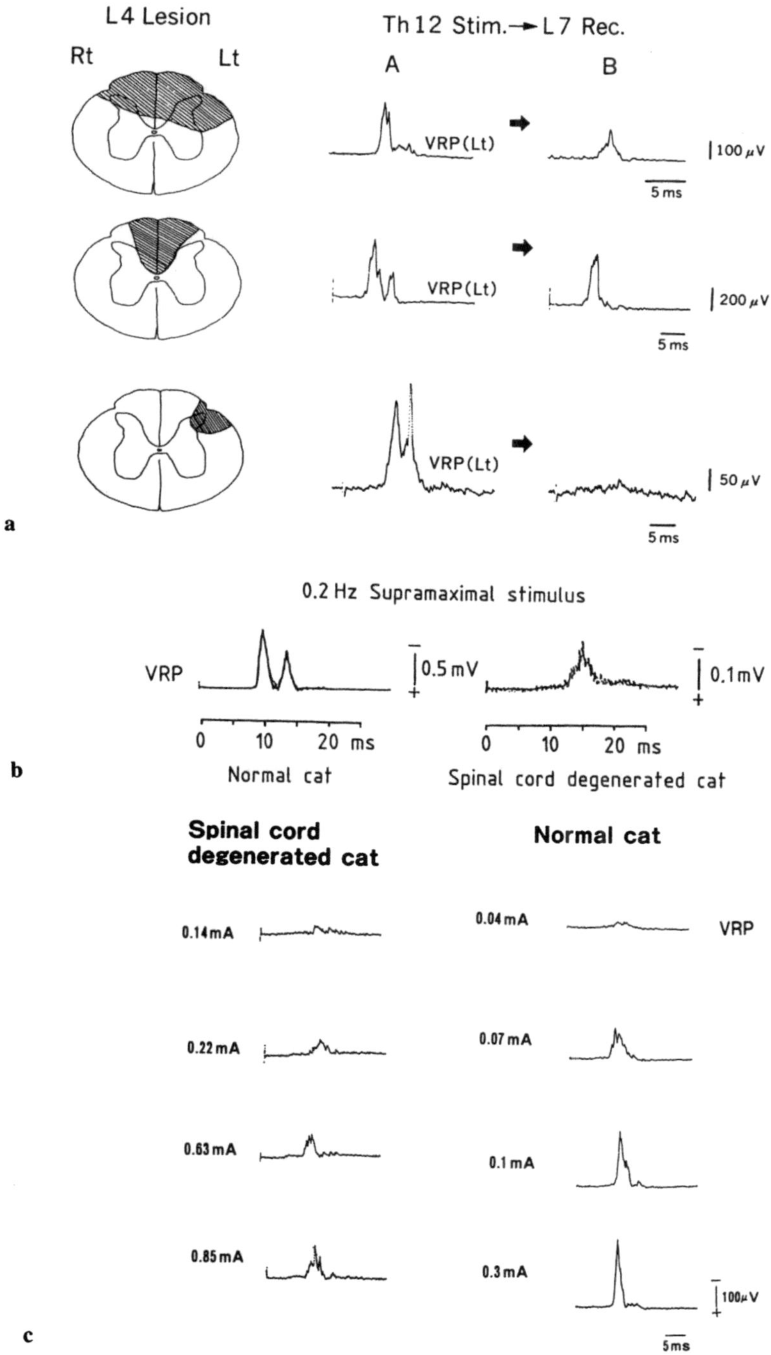
L4 Lesion
Rt
Lt
Th 12 Stim. → L 7 Rec.
A
B
VRP (Lt)
100 μV
5 ms
VRP (Lt)
200 μV
5 ms
VRP (Lt)
50 μV
5 ms
a
0.2 Hz Supramaximal stimulus
VRP
0.5 mV
0 10 20 ms
Normal cat
0.1 mV
0 10 20 ms
Spinal cord degenerated cat
b
Spinal cord degenerated cat
Normal cat
0.14 mA
0.22 mA
0.63 mA
0.85 mA
0.04 mA
0.07 mA
0.1 mA
0.3 mA
VRP
100μV
5ms
c

microscopically in serial sections stained with hematoxylineosin. Degenerated parts were thought to correspond to the descending pathways of the corticospinal tract and the rubrospinal tract [4]. The VRP amplitude was suppressed and latency was prolonged (Fig. 3b). As to threshold, VRPs of cats with degenerated spinal cords had a higher threshold than normal cats did (Fig. 3c).

In the final investigation, experiment 3, VRPs evoked by microstimulation of the crus cerebri were examined. VRP evoked by stimulation of the crus cerebri with four-pulse trains of 0.2 ms duration using a tungsten microelectrode was recorded from a ventral root at a lower lumber level in cats, which had been anesthetized using chloralose and urethan. Finally, the location of the stimulating electrode was confirmed by histological examination. At the end of the experiment, a 10% formalin solution was injected into the carotid artery. The site of the stimulating electrode was coagulated by passing currents of 20 μA for 20 s through the electrode.

The typical wave form of VRP was recorded by stimulation with four-pulse trains at 0.2 Hz and 3 mA (Fig. 4a). As the frequency of stimuli was increased, the size of VRP decreased gradually (Fig. 4b). Four pulse trains produced a higher amplitude of VRPs than single pulses (Fig. 4c). This result pointed out a summation effect on VRP. The electric lesion spot was found in the crus cerebri.

Discussion

As the frequency of the stimuli was increased, the size of VRPs decreased gradually, Peak 2 amplitude decreased earlier than peak 1 amplitude. Peak 2 diminished at a stimulus frequency below 5 Hz. On the basis of the recorded differences in latency and the characteristic response to stimulus frequency, it was assumed that peak 2 originated from relatively thin fibers of the descending pathway and had more synaptics in the spinal cord than peak 1 (experiment 1) [1–3]. Experiment 2 showed that the VRP did not represent antidromic potentials through Ia fibers [4].

In addition to experiments 1 and 2, it was thought that VRPs represented the action potentials of motor neurons evoked via synapses by fibers descending through the dorsolateral column (corresponding to the corticospinal and rubrospinal tracts) [2, 4]. VRP evoked by microstimulation of the crus cerebri was similar to VRP described before in term of wave form and characteristic frequency. Summation effect on VRP suggested that the origin of VRP might be the corticospinal tract.

Summary

VRPs were considered to be an evoked spinal cord potential directly reflecting the motor functions of the spinal cord.

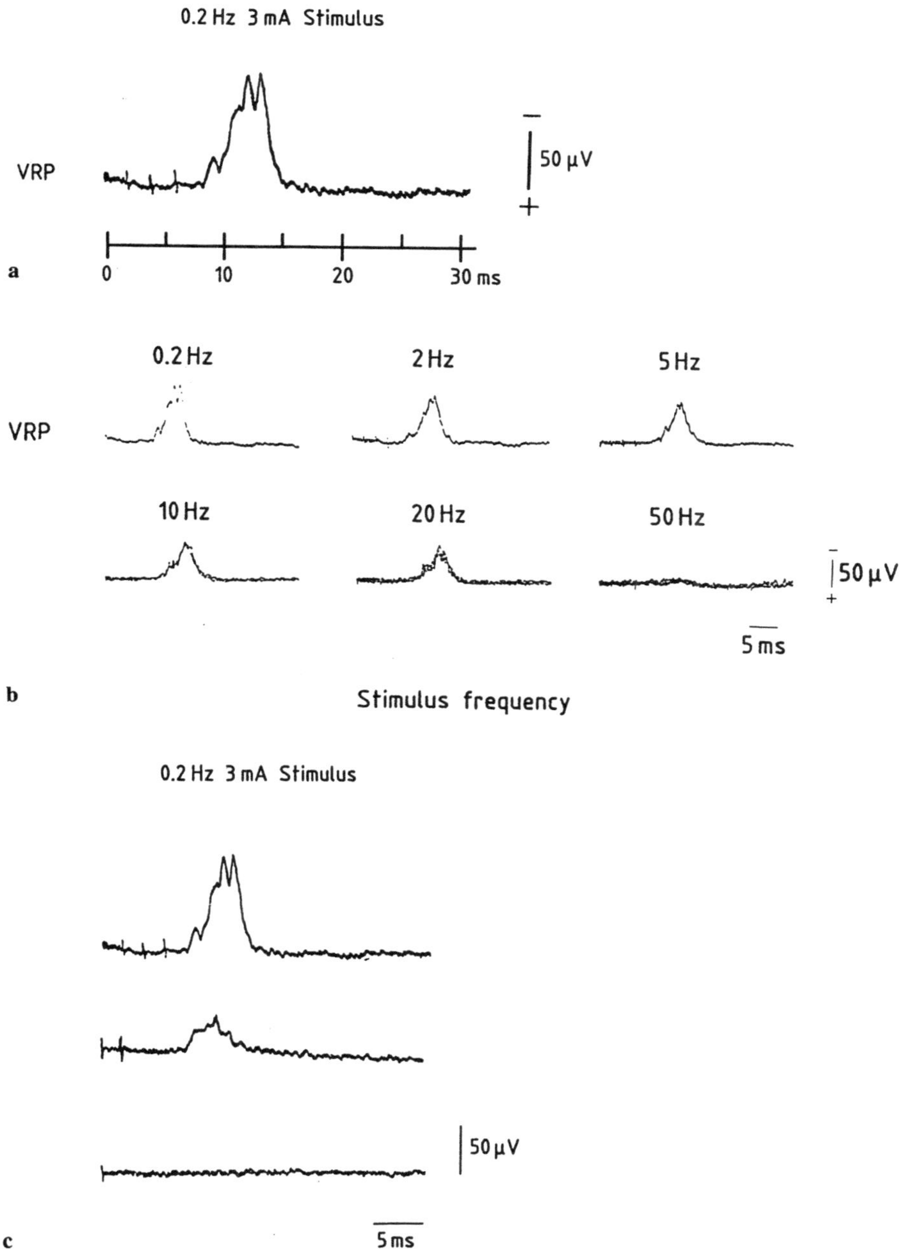

Fig. 4a–c. VRPs evoked by simulation of the crus cerebri with four-pulse trains of 0.2 ms duration using tungsten microelectrodes and recorded at the lower lumbar ventral root in cats. **a** the typical wave form of VRPs, **b** the effects of increasing stimulus frequency, **c** the difference in effects on VRP between four-pulse train stimulation and single pulse stimulation

References

1. Kawamura H (1987) Significance of the motor spinal evoked potentials recorded from the ventral root in cats. J Jpn Orthop Assoc 61(3):54
2. Kawamura H (1987) Significance of the motor spinal evoked potentials recorded from the ventral root in cats. The Journal of Electrodiagnosis of the Spinal Cord 10(1):23–26
3. Kawamura H (1988) Electrophysiological study on spinal cord monitoring in case of posterior compression of the spinal cord. J Jpn Orthop Assoc 62(2):33
4. Kawamura H (1988) Studies on ventral root potentials of spinal cord degenerated cats. J Jpn Orthop Assoc 62(7):686

Monitoring Motor Function of the Spinal Cord: The Descending Segmental Evoked Spinal Cord Potential

M. YOKOYAMA[1], K. SHINOMIYA, H. SATO, H. KOMORI, and K. FURUYA

Introduction

Conductive and segmental evoked spinal cord potentials (ESCPs) have been employed since 1982 as a monitor during cervical anterior decompression surgery [3]. Conductive ESCPs, showing the function of the lower extremities, are recorded from the cervical epidural space following stimulation of the thoracic spinnal cord, and segmental ESCPs are elicited by stimulating the median nerve to evaluate the function of the upper extremities. However, it is difficult to monitor motor dysfunction of the upper extremities using these ESCPs during surgery. We confirmed that a potential that descends the spinal cord and passes through the anterior roots is contained in the potential recorded from the median nerve after stimulating the upper cervical spinal cord. Calling this potential the descending segmental evoked spinal cord potential (DS-ESCP), we conducted a study on identification of the origin and conduction route of this potential and its usefulness as a monitor for motor function of the spinal cord.

Materials and Methods

The sixty adult cats used in this study were anesthetized by intramuscular injection of ketamine (30 mg/kg) and atropine (0.1 mg/kg) [2]. After tracheal intubation, the left femoral artery was cannulated in order to measure blood pressure and the left femoral vein was cannulated in order to allow infusion. The animals were immobilized with pancuronium bromide (0.1 mg/kg h) and placed under controlled ventilation. A stimulating bipolar electrode, consisting of two coils of copper wire set 10 mm apart at the end of a polyethylene catheter, was inserted into the epidural space at C1–2. For recording, a

[1]Department of Orthopedic Surgery, Tokyo Medical and Dental University, 1-5-45 Yushima, Bunkyo-ku, Tokyo, 113 Japan

K. Shimoji et al. (Eds.)
Spinal Cord Monitoring and Electrodiagnosis

bipolar electrode was hooked onto the median nerve in the forelimb.

Using a Disa 1500 system or an Ebomatic 8000 (Dantec, Denmark) system, the bandwidth of the recording system (2 Hz to 2kHz) was kept constant during recording. Generally, 50–100 evoked signals were amplified and averaged. Stimulation was applied with an intensity of 7–10 mA, a frequency of 5Hz, and a duration of 0.2 ms. After the control potential had been recorded in response to stimulation, as described previously, the following studies were performed:

1. Investigation of waveform characteristics including:
   a) Stimulation at high frequency
   b) Decerebration
   c) Effects of hypotension
2. Transection of the spinal cord
3. Transection of the nerve roots (rostrally and caudally)
   a) Posterior roots
   b) Anterior roots after transection of posterior roots
4. Compression of the spinal cord

Results and Discussion

Investigation of the Waveform Characteristics

The control DS-ESCP showed a negative spike with two peaks. The latency of the first peak was 3–5 ms. The inter-peak latency was approximately 0.9 ms. The conduction velocity of the first peak was 52.4 ± 6.8 m/s (Fig. 1).

With high frequency stimulation at 50 Hz, the first peak decreased in amplitude by 14%, but the second peak was marked attenuated. The first

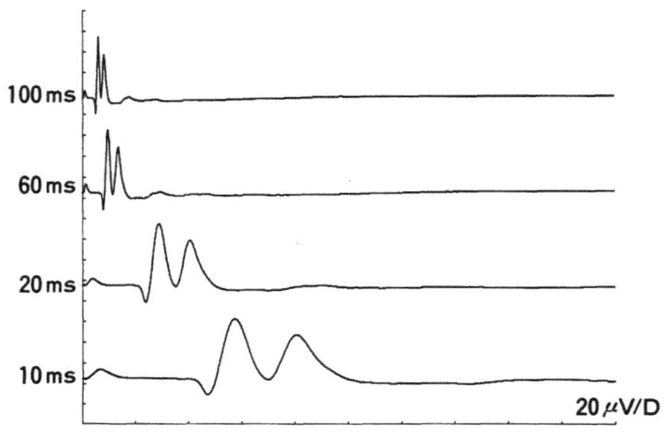

Fig. 1. Changes of analysis time. An early component of the DS-ESCP was recorded as a negative potential having two peaks. The analysis times were 10 ms, 20 ms, 60 ms, and 100 ms

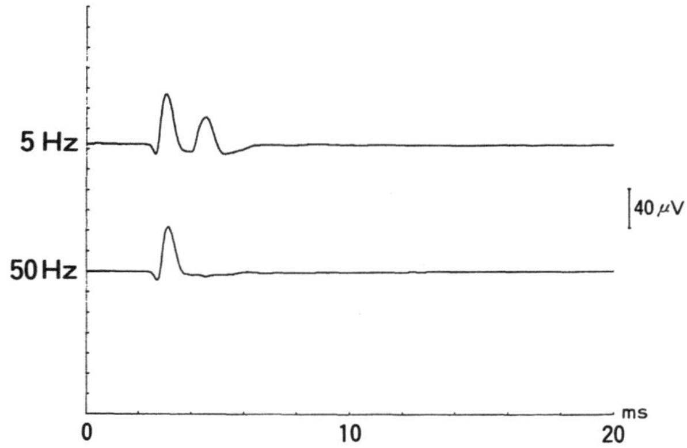

Fig. 2. The second peak of the DS-ESCP was remarkably diminished by high frequency stimulation at 50 Hz

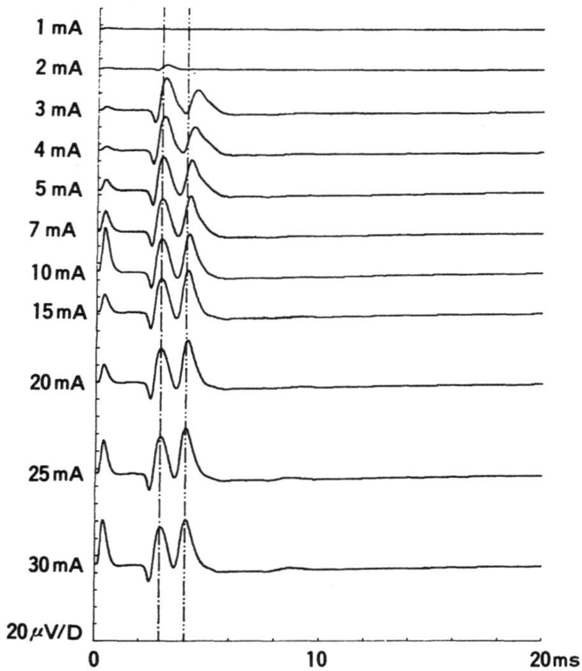

Fig. 3. The effect of graded stimulation on the DS-ESCP. The second peak showed an increase in amplitude and shortening of the latency period in response to an increase in stimulus intensity up to 15 mA

peak was thought to reflect a mainly nonsynaptic potential, while the second peak might reflect a synaptically relayed potential (Fig. 2).

When changes in the pattern according to the intensity of stimulation were examined, the first peak appeared at 2 mA and reached supramaximum at five times the threshold strength ($5 \times T$). The second peak appeared at 3 mA, showing an increase in amplitude and shortening of the latency, with an increase in the intensity of stimulation up to 10 mA. Over 15 mA the latency of the second peak showed no change. This is probably because descending fibers causing the second peak are present on the ventral side of the spinal cord (Fig. 3).

In cases where decerebration was performed, the amplitude of the second peak decreased temporarily as changes in blood pressure occurred immediately after transection of the spinal cord, but returned to its original level when blood pressure was stabilized; neither peak showed any change in latency or amplitude. When the effect of hypotension on the potential was examined with trimetaphan, the second peak responded sharply to a fall in blood pressure and decreased in amplitude.

Transection of the Spinal Cord

After the dorsal half of the dorsal column had been transected at the C4–5 level, the first peak decreased in amplitude by 27%, but the second deflection

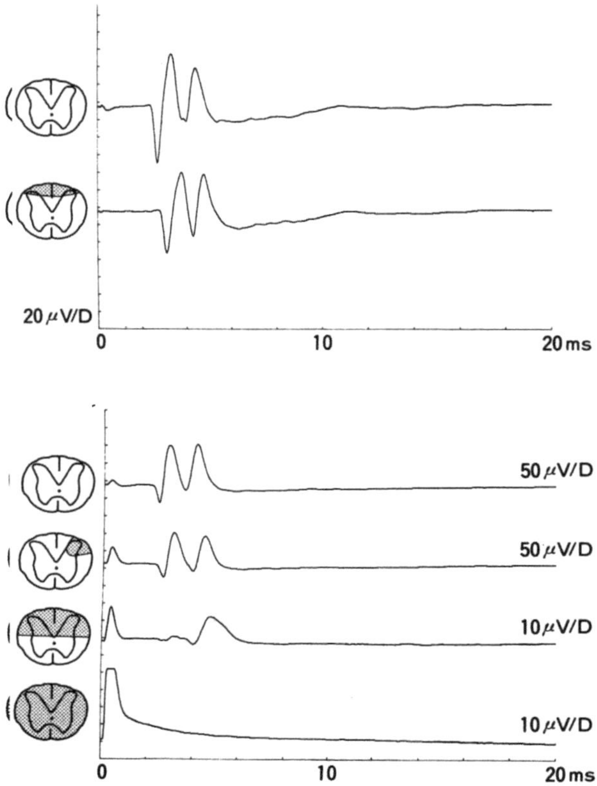

Fig. 4. Transection of the spinal cord at C4–5 level. When transecting the dorsal half of the spinal cord, the first peak almost disappeared, while the second peak remained 13% in amplitude. With total transection of the spinal cord, both peaks disappeared

showed little change. After transection of the dorsolateral column, both peaks decreased in amplitude by 24%.

When the dorsal half of the spinal cord had been transected, the first deflection almost disappeared, while the second remained at 13% in amplitude. When the entire spinal cord had been transected, both deflections disappeared (Fig. 4). The result of the spinal cord transection showed that both deflections reflect mainly potentials conducting through the dorsal half of the spinal cord, but some of the second conducting through the ventral side of the spinal cord is also present [1, 4].

Transection of the Nerve Roots

The posterior nerve roots were severed one by one from the C4 posterior root caudally. The potentials did not change following transection of the posterior roots of C4, C5, and C6. When the C7 posterior root was additionally cut however, the first deflection decreased in amplitude by 38%, and peak latency was delayed by 0.4 ms. With the additional severance of the C8 posterior root, the first deflection disappeared, while the second showed no change in peak latency or amplitude. Inversely, the posterior roots were severed one by one from the T1 posterior root rostrally. With the T1 posterior root cut, the first peak showed a 5% increase in amplitude. By additionally cutting the C8 posterior root, the amplitude of the first peak decreased by 58%. With the C7 posterior root severed additionally, the first peak was markedly attenuated, and peak latency was delayed by 0.3 ms. After the C6 posterior root had been cut, the first peak disappeared. At this point, the amplitude of the second peak did not change, but peak latency was delayed by 0.2 ms. The above results suggest that the first peak is a potential that passes chiefly through the C7 and C8 posterior roots (Fig. 5).

After transection of the posterior roots from C4 to T1, the anterior roots were severed one by one from the C4 anterior root caudally for observation of changes in the remaining peak. With the C4, C5, and C6 anterior roots transected, the amplitude decreased by degrees and the latency of the second peak was delayed gradually.

When the C7 anterior root was severed in addition, the amplitude decreased to 62%, and the peak latency was delayed by 0.8 ms. With additional severing of the C8 anterior root, the amplitude decreased to 5%, and the peak latency was delayed by 1.0 ms. With the T1 anterior root cut, the potential disappeared.

Inversely, the anterior roots were severed one by one from T1 rostrally. Transection the T1 anterior root resulted in an 8% increase in the amplitude. With the C8 anterior root severed additionally, the amplitude decreased to 17%, and the peak latency was delayed by 1.0 ms. With the C7 anterior root also severed, the potential disappeared (Fig. 6). The above results suggest that the second peak is a potential that passes chiefly through the C8 anterior roots [5].

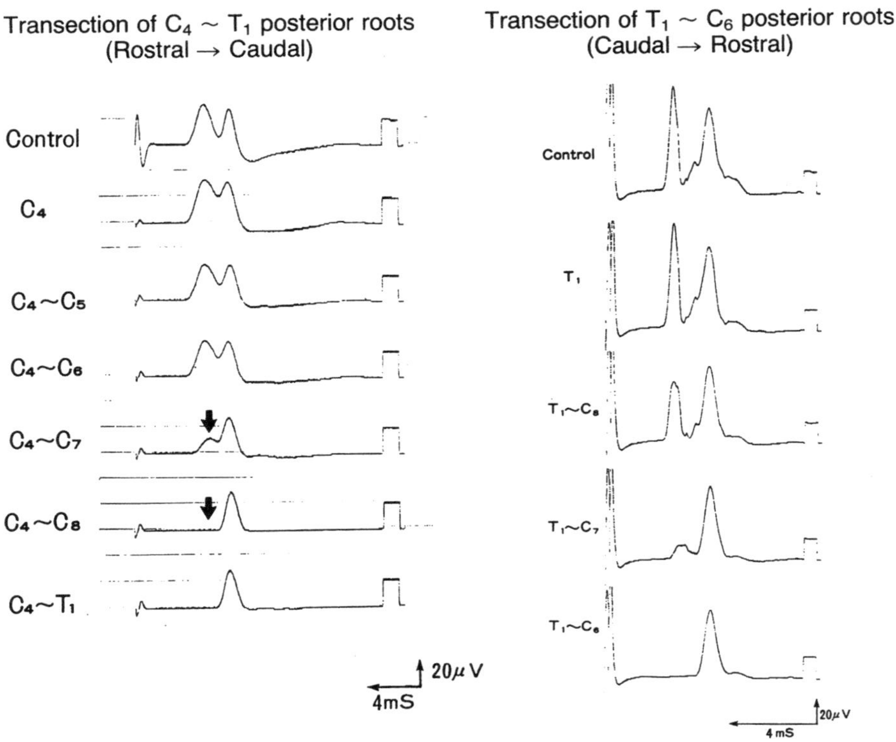

Fig. 5. Transection of the posterior roots in order from the C4 posterior root caudally and from the T1 posterior root rostrally. Approximately 90% of the amplitude of the first peak was a potential that went through the C7 and C8 posterior roots

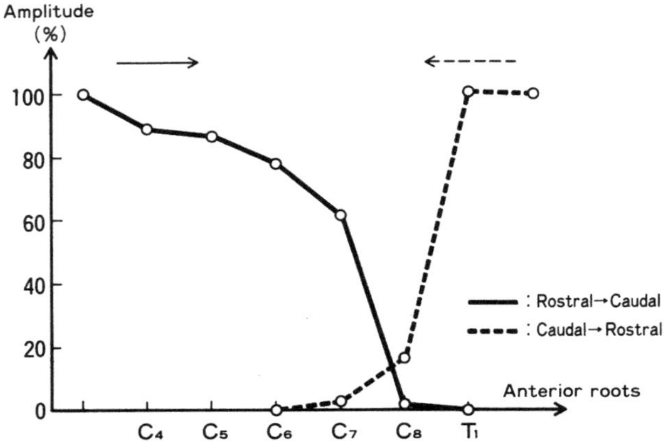

Fig. 6. Transection of the anterior roots in order from the C4 anterior root caudally and from the T1 anterior root rostrally after transection of the posterior roots. Changes in amplitude of the second peak were expressed as a percentage of the control value

Compression of the Spinal Cord

For anterior compression of the spinal cord, a screw 7 mm in diameter, which advances 1 mm per rotation at the C5–6 disc level, was used. Neither peak showed any change in latency and amplitude until the screw for compression protruded 2.0 mm from the posterior edge of the body into the spinal canal. At 2.0 mm of protrusion, the latency of the second peak was delayed by 0.2 ms. and the amplitude increased by 7% in the first peak and by 17% in the second peak. At 3.0 mm of protrusion, the latency of the first peak began to delay slightly. At 3.5 mm of protrusion, the amplitude decreased by 90% in the first peak and by 75% in the second peak. The screw was fixed at 3.5 mm, but the potential change in the course of time. The second peak disappeared in 10 minutes. In the first peak, the latency was delayed by 1.2 ms. and amplitude decreased by 56% (Fig. 7). Anterior compression of the spinal cord produced a delay in the second peak, a delay in the first peak, a decrease and a disappearance of the second peak, and a decrease in the first peak, in that order. This shows that the second peak reflects spinal damage more sharply.

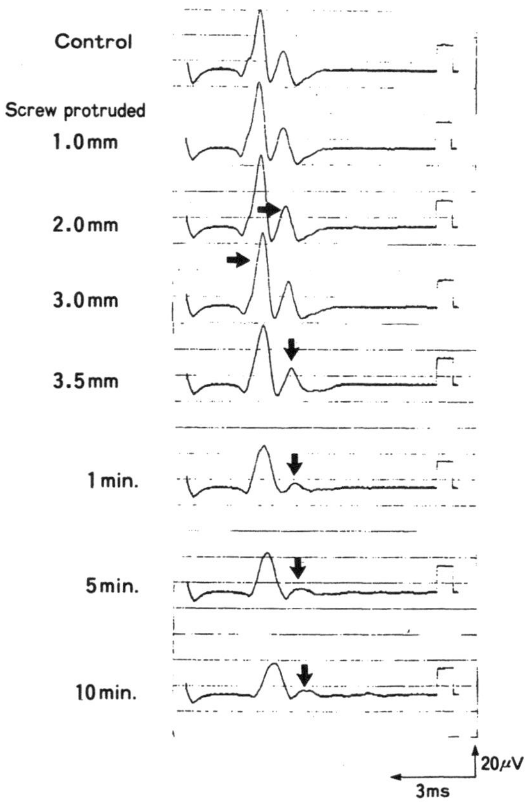

Fig. 7. Anterior compression of the spinal cord at C5–6 disc level by using a screw. The second peak gradually decreased in amplitude and finally disappeared entirely at 3.5 mm of protrusion. The first peak remained, although decreased in amplitude by 56%

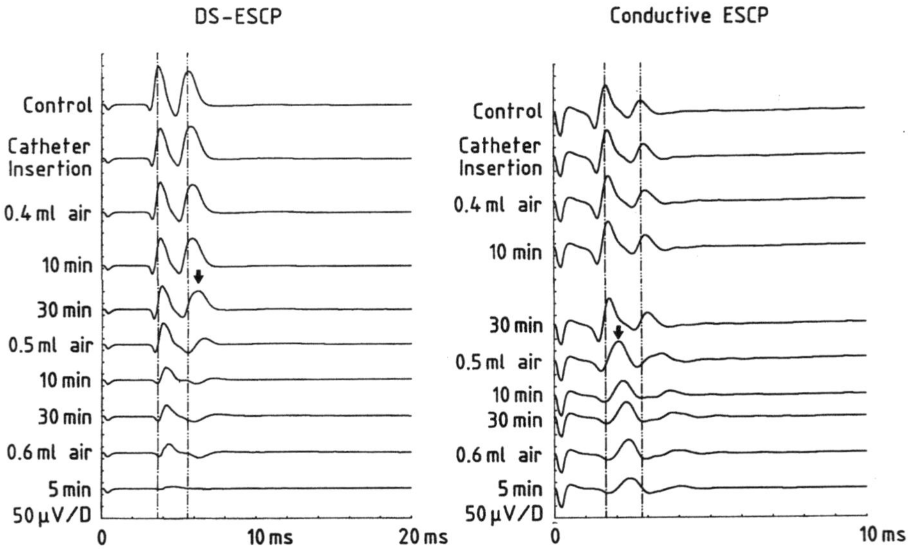

Fig. 8. Posterior compression of the spinal cord using a balloon. The amplitude of the second peak of the DS-ESCP began to decrease with injection of 0.4 ml of air. At injection of 0.5 ml, the first peak of the conductive ESCP began to decrease in amplitude and the second peak of the DS-ESCP disappeared

For posterior compression of the spinal cord, a balloon at the tip of a 4 French catheter for thrombectomy was used. Changes in the potential were observed by injecting air in 0.1 ml doses. When 0.4 ml air had been injected, the amplitude of the second peak began to decrease. After injection of 0.5 ml, the second peak disappeared and the amplitude of the first peak began to decrease. After injection of 0.6 ml had been injected both peaks had disappeared.

Comparison was made between the second peak of the DS-ESCP and the first peak of the conductive ESCP as to changes in the potentials with posterior compression of the spinal cord. The first peak of the conductive ESCP showed no change, even when the amplitude of the second peak of the DS-ESCP decreased. The delay in the latency and the decrease in the amplitude occurred at the first peak of the conductive ESCP when the second peak of the DS-ESCP disappeared. The first peak of the conductive ESCP remained even when the potential disappeared with the DS-ESCP (Fig. 8). These findings show that changes in the second peak with the DS-ESCP can act as a monitor of motor dysfunction of the spinal cord more sharply than the conductive ESCP.

Summary

1. The DS-ESCP consists of a negative potential with two peaks recorded in the median nerve after stimulation at the C1-2 epidural space.
2. The first peak was mainly derived from the C7 and C8 posterior roots. The second peak was a synaptically relayed potential mainly derived from the C8 anterior root.
3. The second peak of the DS-ESCP sharply reflected spinal dysfunction due to compression of the spinal cord before the conductive ESCP did.
4. The DS-ESCP was found to be useful as a monitor of the function of the spinal cord.

References

1. Harada Y, Takemitsu Y, Atsuta Y, Imai M (1984) Determination of the pathways of ascending and descending conductive spinal cord evoked potentials (SCEP). In: Homma S, Tamaki T (eds) Fundamentals and clinical application of spinal cord monitoring. Saikon Publishing, Tokyo, pp 33–43
2. Matsumoto M, Namiki A, Kawamura H (1988) The effects of intravenous anesthetics on the VRP (Ventral Root Potentials). The Journal of Electrodiagnosis of the Spinal Cord 11:25–28
3. Shinomiya K (1983) Monitoring of the spinal cord function using evoked spinal cord potentials during surgery of the cervical spine. Orthop Traum Surg 26:1961–1968
4. Toyoda A, Kanda K (1984) Origins of spinal cord potentials evoked by stimulation of the cat spinal cord. In: Homma S, Tamaki T (eds) Fundamentals And Clinical Application of Spinal Cord Monitoring. Saikon Publishing, Tokyo, pp 99–111
5. Yokoyama M, Shinomiya K, Sato H, Komori H (1987) The experimental study of nerve action potentials recording from the median nerve by stimulating the cervical spinal cord. The Journal of Electrodiagnosis of the Spinal Cord 10:27–30

Experimental Study of the Origin of Transcranially Evoked Descending Spinal Cord Potentials

T. YAMAMOTO[1], Y. KATAYAMA, T. TSUBOKAWA, S. MAEJIMA, T. HIRAYAMA, and J. XING

Introduction

It has been demonstrated that a corticospinal response to direct stimulation of the motor cortex can be recorded from the lateral column of the spinal cord or the spinal epidural space [4, 5, 9, 10]. It has also been reported that identical responses can be recorded in cats and humans using transcranial brain stimulation [6, 7]. A reliable technique of transcranial brain stimulation would be of great value for monitoring pyramidal tract function, since this method does not require any operation to be performed in order to stimulate the motor cortex directly. However, there is some doubt as to whether spinal cord responses to transcranial brain stimulation really represent impulses mediated by pyramidal neurons. We previously compared the corticospinal D response to stimulation of exposed motor cortex and the spinal cord responses to transcranial brain stimulation in cats [4], and we found that spinal cord responses to transcranial brain stimulation are easily confused with responses other than the corticospinal D response.

In the present study, spinal cord responses to direct stimulation of the motor cortex and the cerebellum are compared with responses to transcranial brain stimulation, and the origin of the transcranially evoked spinal cord potentials is discussed.

Methods

Experiments were performed on 30 adult cats weighing between 2.4 kg and 3.6 kg. These animals were initially anesthetized with ketamine (10 mg/kg, i.m., supplemented if necessary). The femoral vein and artery of one leg were cannulated for drug and fluid infusion and for recording the arterial blood

[1] Department of Neurological Surgery, Nihon University School of Medicine, 30-1 Oyaguchikami-machi, Itabashi-ku, Tokyo, 173 Japan

K. Shimoji et al. (Eds.)
Spinal Cord Monitoring and Electrodiagnosis

pressure, respectively. Anesthesia was subsequently maintained by periodic administration of various doses of sodium pentobarbital (5–40 mg/kg, i.v.). The electroencephalographic pattern, recorded from stainless steel screws implanted in the skull, was used to ensure that each cat was adequately anesthetized throughout the recording session.

A laminectomy which exposed the dorsal surface of the dura of the lumbar (L1–2) segments was performed. A flexible platinum-wire electrode, insulated except at its tip (Medtronic Co. M-8483), was inserted into the epidural space from the area of laminectomy, and the tip of the electrode was advanced up to the upper cervical spinal segments. Descending spinal cord evoked potentials were recorded from the whole length of the spinal cord on pulling out the flexible wire electrode. The end-tidal CO_2 concentration was measured and maintained at 4.0%–4.5% while the animals were paralyzed and artificially ventilated during the recording session. The rectal temperature was monitored and maintained at between 37°C and 39°C with the aid of a water circulating heat pad.

For recording cortically evoked spinal cord potentials in response to direct stimulation of the motor cortex (CESPs-M), the motor cortex (anterior sigmoid gyrus) and other cortical areas were stimulated through the exposed overlying dura with a bipolar silver ball electrode (interpolar distance, 2.5 mm) by means of monophasic square pulses of 0.2 ms duration delivered at a frequency of 2 Hz. Cortically evoked spinal cord potentials (CESPs-C) in response to direct stimulation of the cerebellar cortex were recorded in response to direct stimulation with a bipolar silver ball electrode under the same conditions used in recording CESPs-M. CESPs-C were also recorded after stimulating the dentate nucleus with a bipolar concentric electrode. For recording transcranially evoked spinal cord potentials (TESPs), a stainless steel screw was inserted into the lamina externa of the skull. The skull was not penetrated by the screw. Square pulses of constant phases were applied between a screw electrode located at the vertex and a needle electrode placed in the hard palate while applying an anodal pulse to the skull. Pulses of 0.5 ms were delivered at a frequency of 2 Hz. The intensity of stimulation was adjusted to supramaximal levels for the investigated potentials.

Signals from each recording electrode were fed into an amplifier (Nihon-Koden VC10) with a band-pass of 5 Hz–5 kHz and then into a signal averaging processor (San-ei Sokki 2TOB).

Results

CESPs-M recorded at the epidural space of the upper cervical segments in response to stimulation of the motor cortex (anterior sigmoid gyrus) had an initial negative wave (N1) followed by polyphasic waves. N1 was often preceded by a positive wave (P1), which covaried with N1. Although as the

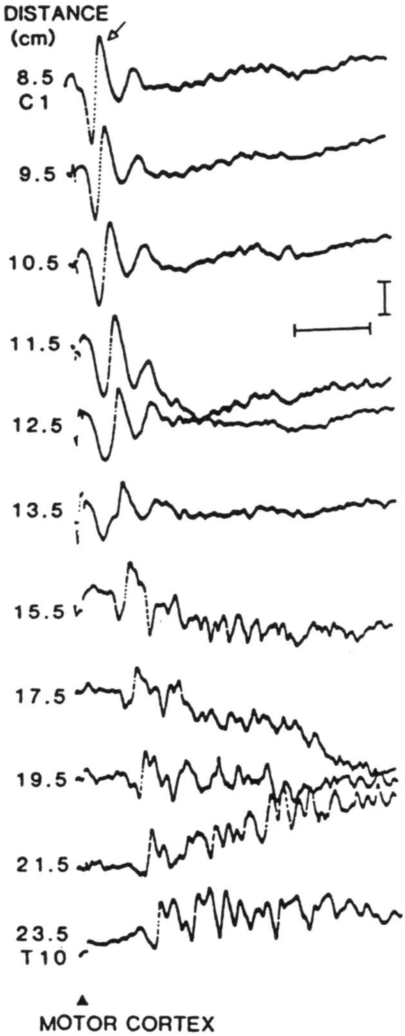

Fig. 1. Representative examples of cortically evoked spinal cord potentials in response to direct stimulation of the motor cortex (CESPs-M). The CESPs-M were recorded at various levels of the spinal cord between C1 and T10. The distances between the stimulation and recording electrodes are shown on the left (*arrow*, N1 of the CESPs-M; time scale, 5 ms; amplitude, 2 μV for the distance range 8.5–13.5 cm and 1 μV for the distance range 15.5–23.5 cm

recording electrodes were moved caudally. P1 and N1 became progressively decreased in amplitude and prolonged in latency. P1 and N1 were clearly traced down to T10 (Fig. 1). The conduction velocity calculated from the differences in latency between several recording sites was 68.3 ± 14.7 m/s

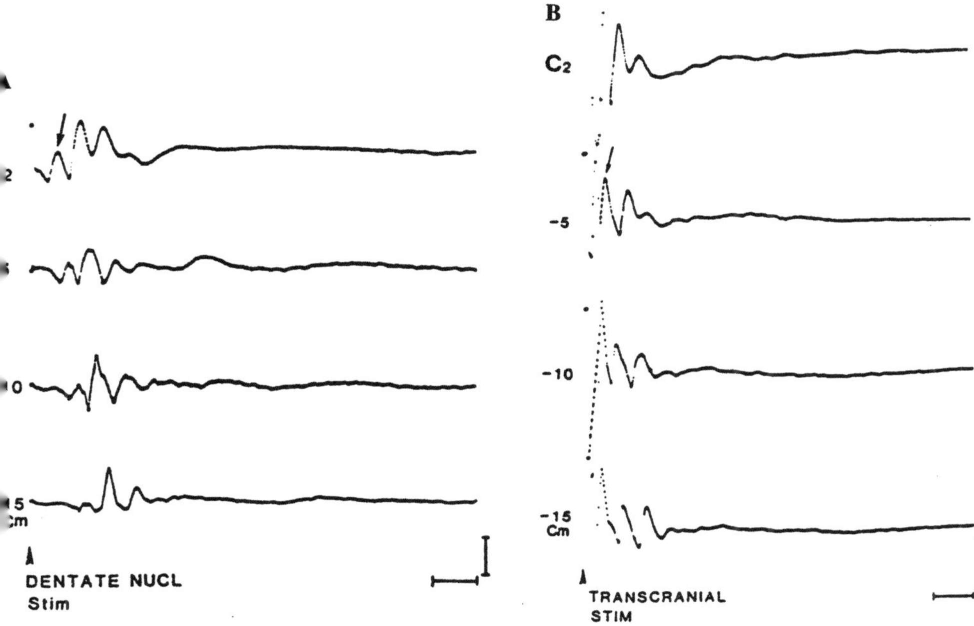

Fig. 2A,B. Representative examples of (**A**) cortically evoked spinal cord potentials in response to direct stimulation of the cerebellar cortex (CESPs-C). These recordings relate to direct stimulation of the dentate nucleus. The distances between the C2 segment and the recording electrodes are shown on the left (*arrow*, N1 of CESPs-C; time scale, 2 ms; amplitude, 5 μV), and (**B**) transcranially evoked spinal cord potentials (TESPs). The distances between the C2 segment and the recording electrodes are shown on the left (*arrow*, N1 of TESPs; time scale, 2 ms; amplitude, 5 μV)

(mean ± SD, $n = 5$) for N1. P1 and N1 and the following polyphasic waves were all abolished by intercollicular transection (Fig. 4).

TESPs recorded from the epidural space of the lower cervical segments consisted of an initial negative wave (N1) followed by polyphasic waves. Responses to transcranial stimulation recorded at the upper cervical segment contained large shock artifacts. Thus, recordings of TESPs were started from the lower cervical segments. As the recording electrodes were moved caudally, P1, which preceded N1, appeared, following shock artifacts. N1 and the following polyphasic waves became progressively decreased in amplitude and prolonged in latency. The conduction velocity calculated from the differences in latency between several recording sites was 122.3 ± 16.3 m/s ($n = 5$) for N1 (Fig. 2B). N1 did not disappear, but the following polyphasic waves were abolished by intercollicular transection (Fig. 4).

CESPs-C recorded at the epidural space of the lower cervical segments in response to stimulation of the region of the cerebellar cortex showed an initial negative wave (N1) and polyphasic waves. N1 was preceded by a positive

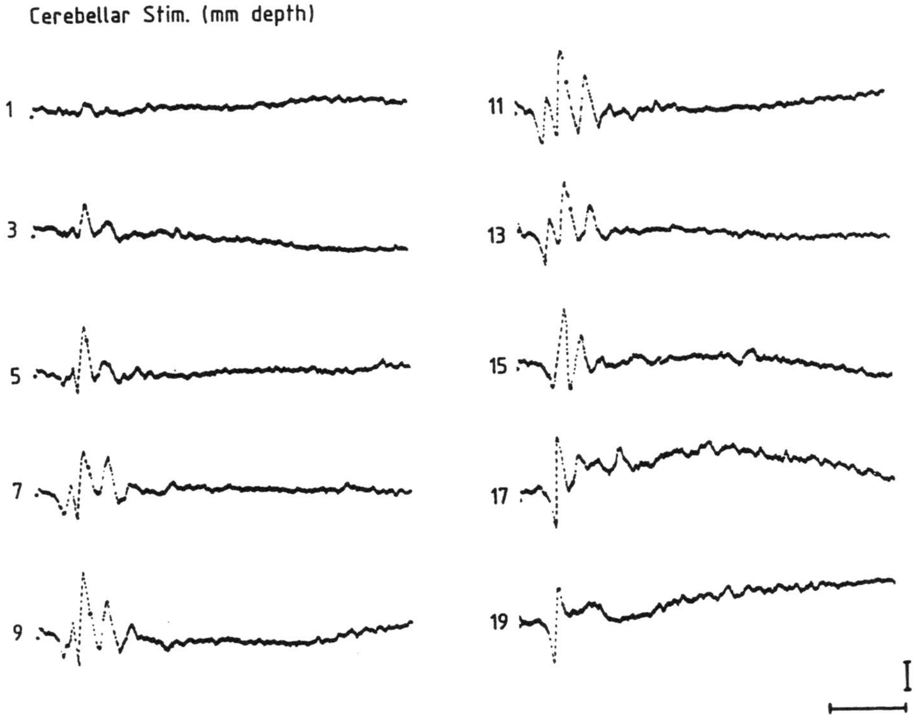

Fig. 3. Recording of CESPs-C. A bipolar concentric stimulation electrode was advanced from the cerebellar cortex towards the dentate nucleus. The distances between the cerebellar cortex and the tip of the stimulating electrode are shown on the *left* (time scale, 2 ms; amplitude, 5 μV)

wave (P1), which covaried with N1. These waves were recorded at highest amplitude and at shortest latency on stimulation around the lower part of the junction between the cerebellar vermis and the cerebellar hemisphere. A bipolar, concentric, stimulating electrode was inserted from the cerebellar surface towards the dentate nucleus when recording the descending spinal cord potentials from the lower cervical segments. (Fig. 3) As the stimulating electrode advanced, the waves were recorded at highest amplitude and at short latency at a depth from 9–13 mm. Histological examinations showed these areas to be consistent with the region of the dentate nucleus. As the recording electrodes were moved caudally, P1 and N1 becamė progresively decreased in amplitude and prolonged in latency. The conduction velocity calculated from the difference in latency between several recording sites was 120.2 ± 16.2 m/s (n = 5) for N1 (Fig. 2A). N1 did not disappear, but the amplitude of the following waves was markedly decreased by intercollicular transection (Fig. 4).

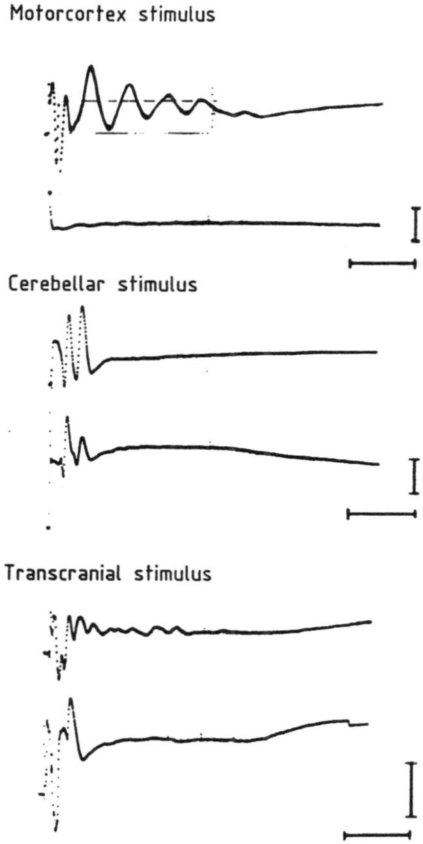

Fig. 4. Effect of intercollicular transection. Each of the upper traces represents a recording made before intercollicular transection, and each of the lower traces represents a recording made after intercollicular transection (time scale, 2 ms; amplitude, 5 μV)

Discussion

N1 of CESPs-M is resistant to surgical doses of sodium pentobarbital and has the ability to follow high frequency stimulation of more than 500 Hz [4, 10]. Furthermore, the sites which evoke these components are restricted to the vicinity of the motor cortex. N1 of CESPs-M has thus been employed for intraoperative monitoring of the functions in the case of intracranial and intra-spinal lesions in humans [5, 9]. While recordings of CESPs-M involve invasive procedures for stimulation, it has recently been claimed that spinal cord potentials identical to CESPs-M can be recorded after transcranial brain stimulation through the intact skull in humans [7] as well as cats [6]. These previous studies, however, did not provide direct evidence that the motor cortex was actually stimulated by that technique. We previously compared CESPs-M and TESPs in the cat, and demonstrated that TESPs are easily confused with responses other than CESPs-M [4].

The N1 of CESPs-M recorded in the present study showed conduction velocities of 68.3 ± 14.7 m/s. This value agrees with those calculated previously for the pyramidal tract. The N1 of TESPs recorded in the present study showed conduction velocities of 122.3 ± 16.3 m/s. Although the values of the conduction velocities of TESPs were quite different from those of CESPs-M, the values of the conduction velocities of CESPs-C were almost the same as those of TESPs. The presence of fast conducting fibers and the fact that many reticulospinal neurons activated monosynaptically or polysynaptically from the cerebellum [1, 8] have fast (90–150 m/s) conduction velocities [2, 3], suggest that reticulospinal, vestibulospinal, or other tracts may be included in TESPs. Furthermore, the CESPs-M disappeared completely after intercollicular transection, but the CESPs-C and TESPs remained. These results suggest the possibility that N1 of TESPs may arise from the same origin as CESPs-C. In addition, the fact that the polyphasic waves following N1 disappeared after intercollicular transection also indicates that CESPs-M may be involved in these polyphasic waves of the TESPs.

The above results imply that transcranial activation of the brain, especially when the method of stimulation between the skull and hard palate is used, stimulates not only the pyramidal tract but also extrapyramidal tracts, and that TESPs have their origins in both the pyramidal and extrapyramidal tracts.

References

1. Bantli H, Bloedel JR (1975) Monosynaptic activation of a direct reticulospinal pathway by the dentate nucleus. Pflugers Arch 357:237–242
2. Eccles JC, Nicoll RA, Schwartz DWF, et al. (1974) Cerebello spinal pathway via the fastigial nucleus and the medial reticular nucleus. 66:525–530
3. Ito M, Udo M, Mano N, et al. (1970) Synaptic action of fastigiobulbar impulses upon neurons in the medullary reticular formation and vestibular nuclei. Exp Brain Res 11:29–47
4. Katayama Y, Tsubokawa T, Yamamoto T (1986) Assessment of spinal cord injury with multimodality evoked spinal cord potentials Part 1, localization of lesions in experimental spinal cord injury. Neuro-Orthopedics 1:130–141
5. Katayama Y, Tsubokawa T, Maejima S, et al. (1988) Corticospinal direct response in humans; identification of the motor cortex during intracranial surgery under general anesthegia. J Neurol Neurosurg Psychiatry 51:50–59
6. Levy WJ, McCaffrey M, York DH, Tanzer F (1984) Motor evoked potentials from transcranial stimulation of motor cortex in cats. Neurosurgery 15:214–227
7. Levy WJ, York DH, McCaffrey M, Tanzer F (1984) Motor evoked potentials from transcranial stimulation of the motor cortex in humans. Neurosurgery 15:287–302
8. Orioli FL (1961) Descending limb of brachium conjunctivum in cat. An electrophysiological study. Neurophysiol 24:583–594
9. Tsubokawa T, Katayama Y, Maejima S (1987) Assessment of spinal cord injury with multimodality evoked spinal cord potentials Part 2, Correlation with neurological outcome in clinical spinal cord injury. Neuro-Orthopedics 3:82–89
10. Yamamoto T, Katayama Y, Tsubokawa T, et al. (1985) Non-invasive monitoring of corticospinal evoked responses from the spinal epidural space in the cat. Nihon Univ J Med 27:17–21

The Use of the Dog as a Model for Spinal Cord Monitoring Research

S. Uchiyama[1], R.H. Brown, and C.L. Nash

Introduction

Spinal cord monitoring (SCM) research has and is being conducted using a wide variety of methods and several different animal models. There are many factors which affect monitoring such as stimulus rates and intensities, bandwidth settings, electrode montage and physiologic parameters such as hypovolemia, hypocarbia and anesthetic variations. Recently, many investigators have raised doubts regarding the appropriateness of using the cat model for this purpose. This study was undertaken for two reasons: (1) to establish a standard anesthetic protocol for the dog research model, and (2) to test the dog as a suitable model for SCM experimentation.

Materials and Experiments

First, 22 mongrel dogs weighing 15–20 kg were studied to delineate a stable, standard anesthetic technique. This anesthetic regimen was then used to study the effects of varying several monitoring parameters on the patterns of recordings from both the cervical and cortical areas.

In the anesthetic study, premedication in 19 of the 22 dogs consisted of secobarbital, 5 mg/kg, given intramuscularly 15–30 min before induction. In three dogs no premedication was given. Induction was accomplished with an initial bolus of 5 μg/kg fentanyl followed by 10 mg/kg Na-Pentothal (Na-thiopenthal) in all animals. Additional Na-Pentothal was given as required, depending on the condition of the animal. Maintenance anesthesia was studied in two groups. Group 1 had seven dogs in which continuous infusion of Na-Pentothal with fentanyl under 100% O_2 was used. Group 2 had 15 dogs in which continuous infusion of fentanyl was used with or without Ethrane

[1] Department of Orthopedic Surgery, Niigata University School of Medicine, Asahi-machi, Niigata City, 951 Japan

K. Shimoji et al. (Eds.)
Spinal Cord Monitoring and Electrodiagnosis

(enflurane) under 50% N_2O and 50% O_2. All dogs were paralyzed with d-Tc (d-tubocurarine chloride) or pancronium bromide and ventilated artificially. Blood pressure was monitored through a femoral artery, and blood gas analysis was done every 30–60 min to control PCO_2 from 30–40 mmHg. Animals were considered to be in steady state when their BP and HR were stable and HR showed no changes or an increase by no more than 10 bpm to painful stimulation. Dosage of each anesthetic was changed to evaluate the requirement to reach the steady state.

The second part of this study looked at SCM and the effects of changing parameters. This part of the study was done under fentanyl anesthesia with or without Ethrane under 50/50 N_2O and O_2 as described above. Stimulation was presented to the posterior tibial nerve at the ankle with 20 mA constant current, and the cortical and cervical potentials were recorded separately. Stimulus rate was 0.8 or 1.2 Hz for cortical potentials and 14–17 Hz for cervical potentials. Cortical potentials were recorded by subcutaneous platinum needle electrodes. Analysis time was 125 ms and 64 averages were done. The input filter settings were 0.2–100 Hz. Using these settings, the optimum locus for the cortical electrode montage as well as the effects of Ethrane were studied. Cervical recording was done with bipolar epidural electrodes placed in the upper cervical area. The reference electrode was inserted into paraspinal muscle 2 cm lateral at the level of recording. Analysis time was 50 ms and 1000 averages were done with an input filter setting of from 10 to 1000 Hz. After basic wave patterns were obtained, the effects on both cortical and cervical potentials of various filter settings were studied.

Results

Looking first at the anesthetic portion of the study, in the seven dogs in Group 1, a steady state could not be reached in any dogs using 3.5 mg/kg per

Table 1. Results of anesthesia

Group 1 (n = 7 dogs): continuous infusion under 100% O_2			
Na-Pentothal (mg/kg per hour)	+	Fentanyl (μg/kg per hour)	Steady state achieved
3.5		3.5	0/7
5		5	1/4
Group 2 (n = 15 dogs): continuous infusion under 50% N_2O and 50% O_2			
Fentanyl (μg/kg per hour)	+	Ethrane (%)	Steady state achieved
8		0	0/5
5		0.5	2/15
5		1.2	9/11

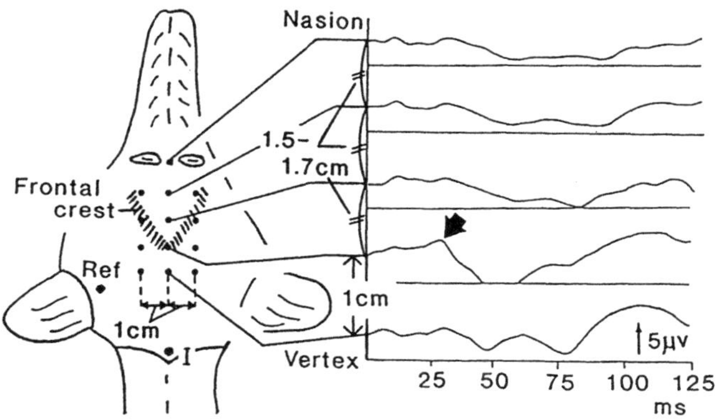

Fig. 1. Mapping study of cervical potentials. Recording was done through all these *points* with reference at the base of an ear and indifferent (*I*) at inion. Most stable potentials (*arrow*) were obtained from over the junction of frontal crests

hour of Na-Pentothal and 3.5 µg/kg per hour of fentanyl. Only one of four dogs reached a steady state when the dosages were increased to 5 mg/kg per hour of Na-Pentothal and 5 µg/kg per hour of fentanyl. Accordingly, this method was not considered to be acceptable.

In Group 2, without Ethrane a steady state could not be reached in five dogs even when the fentanyl was raised up to 8 µg/kg per hour. When 0.5% Ethrane was added, only 2 of 15 dogs achieved a steady state, compared with 9 of 11 who did so with 1.2% Ethrane (Table 1).

In the second portion of the study, a number of parameters were studied. In part 1, a mapping study of cortical potentials (Fig. 1) in which recording was done from the scalp with the indifferent located at the inion and the reference at the base of an ear, the most stable potentials were recorded from over the junction of the frontal crest which is about 1 cm anterior to the vertex. Potentials consisted of 3–5 peaks, of which the first positive peak was the most consistent, both in each individual animal and in the group. Its latency ranged from 25 to 30 ms and its amplitude was 2–5 µV.

In part 2 in which the effects of Ethrane on cortical potentials (Fig. 2) were studied, the use of Ethrane caused depression of all peaks other than the first one. In the steady state with 5 µg/kg per hour fentanyl and 1.2% Ethrane, the amplitude of the first peak was maintained with a latency prolongation of only 2–3 ms while other potentials almost disappeared.

In part 3, looking at cervical potentials after posterior tibial nerve stimulation (Fig. 3D), polyphasic potentials consisting of 6–10 peaks with an amplitude of only 0.2–1.0 µV were recorded. The latency of the first peak was 12–16 ms and the duration of each potential was 0.8–2.0 ms. Although some variabilities existed between animals, the first or second peak consistently had the largest amplitude. The first 5–6 peaks were consistent within

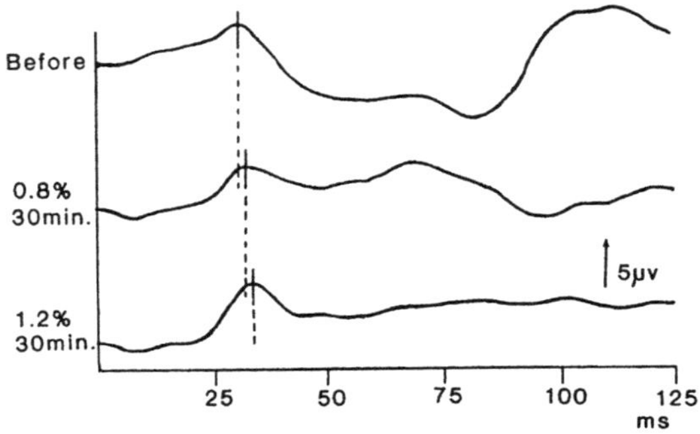

Fig. 2. Effects of Ethrane on cortical potentials. Potentials were recorded from over the junction of frontal crests. First peak was maintained even when later components almost disappeared with the use of Ethrane

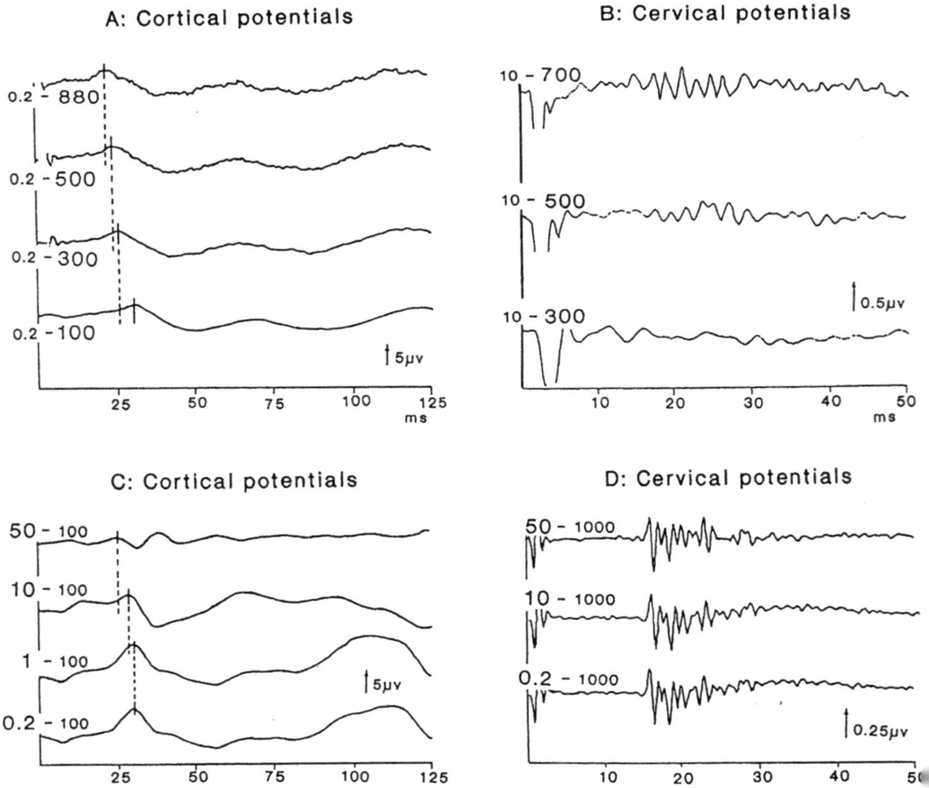

Fig. 3A–D. Filter effects on recording patterns. Bandwidth settings are indicated at the *left margin* of each recording

each animal so that these peaks were considered to be useful for SCM study. Later peaks were smaller and inconsistent even in the same animal.

Finally in part 4, the study of filter effects (Fig. 3), filter settings affected both amplitude and latency. An increase in the high cutoff level resulted in a decrease in the latency and sharpening of the wave form (Fig. 3A). Cervical potentials were not recorded with a high cut at 300 Hz or less. At 500 Hz they could be detected, and at 700 Hz or more they were sharp (Fig. 3B). Increasing the low cutoff setting above 50 Hz caused a 3- to 4-ms decrease in latency and a significant decrease in the amplitude in the cortical potentials (Fig. 3C). On the other hand, cervical potentials were not affected by low cutoff levels from 0.2–50 Hz (Fig. 3D).

Discussions

This study was undertaken because we felt that the dog had several distinct advantages for SCM research. Adult dogs have a spine of appropriate size for almost any kind of surgical procedures being done on humans. They are also easy to obtain and handle. The ability to achieve steady state anesthesia, which is extremely important as the basis for obtaining reproducible data, could also be achieved. Initially a steady state condition could not be obtained with a continuous infusion of Na-Pentothal and fentanyl under 100% O_2 even with 5 mg/kg per hour of Na-Pentothal and 5 μg/kg per hour of fentanyl. However, with a fentanyl infusion and Ethrane under 50% O_2 and 50% N_2O, steady state anesthesia was achieved although some interanimal variability existed in the required dosage of the drugs. In the final analysis, 5–7 μg/kg per hour fentanyl infusion with 1%–1.5% Ethrane seemed to be the minimal requirement for spinal surgery in dogs. As a whole, dogs seemed to be less sensitive to anesthetics than humans in whom steady condition can usually be obtained with balanced N_2O and narcotic technique in spinal surgery [2].

In addition to the variety of research animals and anesthetic techniques noted in the literature, basic wave patterns of cortical potentials in dogs were also noted to be different among investigators [3, 4]. The differences are considered to come from the positions of electrodes and experimental parameters such as filter settings. In our mapping study, the most stable recordings were obtained by locating the active electrode over the junction of frontal crests with the indifferent electrode at the inion and reference at the base of an ear. Using this montage and no Ethrane, five peaks were recognized, with the first peak being the most consistent within a given animal and among different animals. The sensory cortex of dogs is considered to exist mainly in the parietal portion [9], and our mapping study suggests that the first peak comes from the sensory cortex.

The results of human studies of the effects of volatile anesthetics on cortical potentials are different among investigators [5, 7, 8]. In our dog studies with Ethrane, the first peak showed little or no change in amplitude and only a 2- to 3-ms increase of latency even when later peaks almost disappeared. This result shows that cortical potentials evoked by posterior tibial nerve stimulation can be used for SCM studies even if higher doses of Ethrane are required to reach a steady state.

The value of the cervical potentials generated from peripheral nerve stimulation is controversial. Satomi et al. [10] noted that in cats the potentials in the upper thoracic to cervical epidural space after sciatic nerve stimulation were very small in amplitude and not reproducible between animals. In our dog experiments, these potentials were very small and polyphasic with inter-animal variations in the wave pattern. However, since the first 5–7 peaks were consistent within an animal, these potentials are considered to be useful for SCM studies in a particular dog. Changes in filter settings produced changes both in the amplitude and latency, which is the case in human studies [1, 6]. These results emphasize the fact that filter settings should never be changed during intraoperative monitoring because clinically caused changes in the potentials may be masked by the effects of the filter.

Summary

The dog is a suitable animal model for studying SCM if close attention is paid to the anesthetic methodology presented in this study. Also, the importance of standardizing parameters such as filter settings and anesthetic agents has been demonstrated. Finally, investigators are encouraged to follow standard protocols so that data can be compared and/or transferred among centers.

References

1. Brown RH (1988) Technical factors influencing perioperative evoked response recording. Semin Hearing 9:89–98
2. Brown RH, Nash CL (1988) Standardization of evoked potential recording. In: Ducker TB, Brown RH (eds) Neurophysiology and standards of spinal cord monitoring. Springer, Berlin Heidelberg New York, pp 1–10
3. Gaines R, York DH, Watts C (1984) Identification of spinal cord pathways responsible for the peroneal-evoked response in the dogs. Spine 9:810–814
4. Hukada S, Mochizuki T, Ogata M, Fujii S, Shichikawa K (1985) Therapeutic trial of hypertension and hyperoxia combined with hypercarbia on experimental acute spinal cord injury in the dog. In: Schramm J, Jones SJ (eds) Spinal cord monitoring. Springer, Berlin Heidelberg New York, pp 143–153
5. Mc Pherson RW, Mahla M, Johnson R, Traystman RJ (1985) Effects of enflurane, isoflurane and nitrous oxide on somatosensory evoked potentials during fentanyl anesthesia. Anesthesiology 62:626–633
6. Nuwer MR, Dawson EC (1983) Intraoperative evoked potential monitoring of the spinal cord — a restrictive filter, scalp method during Harrington instrumentation for scoliosis. Clin Orthop 183:42–50

7. Paterson DO, Drummond JC, Todd MM (1986) Effects of halothane, enflurane, isoflurane and nitrous oxide on somatosensory evoked potentials in humans. Anesthesiology 65:35–40
8. Pathak KS, Ammadio M, Kalamchi A, Scoles PV, Shaffer JW, Mackay W (1987) Effects of halothane, enflurane and isolfurane on somatosensory evoked potentials during nitrous oxide anesthesia. Anesthesiology 66:753–757
9. Redding RW (1971) Anatomy and physiology. In: Hoerlein BF (ed) Canine neurology — diagnosis and treatment, 2nd edn. Saunders, Philadelphia, pp 9–52
10. Satomi K, Nishimoto GI (1985) Comparison of evoked spinal potentials by stimulation of the sciatic nerve and the spinal cord. Spine 10:884–890

Neurotransmitter Receptors and Voltage-Operated Channels Expressed by Rat Spinal Cord mRNA in *Xenopus* Oocytes*

H. Akagi[1,2,3] and R. Miledi[1]

Spinal cord function involves complex interactions between spinal neurons, segmental input from primary afferent fibers, and suprasegmental input from descending pathways. Various neurotransmitters are thought to be involved in spinal cord function, and many studies have shown that the distribution of neurotransmitters in the spinal cord differs from that in other parts of the central nervous system. Perhaps there is also a difference in the types and densities of the neurotransmitter receptors; but in this respect, only limited information is available, and it derives mainly from receptor binding studies with radiolabeled ligands. Here we describe an alternative approach, using oocytes of the South African clawed frog, *Xenopus laevis* [9, 10].

As the fully grown oocyte is quite large (1.0–1.5 mm in diameter), it can be easily injected with "foreign substances" such as messenger ribonucleic acid (mRNA) extracted from mammalian neuronal cells. The oocyte injected with mRNA, in turn, can translate the foreign message to synthesize various proteins. Furthermore, some of these products are processed, assembled and inserted into the oocyte membrane where they form functional ion channels and neurotransmitter receptors. Measuring the amplitude of ionic currents that flow through "transplanted" membrane proteins allows us to estimate the number of the receptor and ion channels expressed. We have used this approach to compare the "expressional potencies" of mRNAs extracted from the spinal cord and cerebral cortex as well as from the neonatal spinal cord.

* This work was supported by grant R01-NS23284 from the U.S. public Health Service

[1] Laboratory of Cellular and Molecular Neurobiology, Department of Psychobiology, University of California, Irvine CA92717, USA

[2] Department of Neuroscience, Mitsubishi Kasei Institute of Life Sciences, 11 Minamiooya, Machida-shi, Tokyo 194, Japan

[3] *Present address and correspondence*: Department of Pharmacology, Gunma University, School of Medicine, Maebashi-shi, Gunma 371, Japan

K. Shimoji et al. (Eds.)
Spinal Cord Monitoring and Electrodiagnosis

Methods

Mature male rats (Sprague-Dawley, 12–14 weeks old) were anesthetized with pentobarbital (200–250 mg/kg, i.p.), and the cerebral cortex and whole spinal cord were dissected. The spinal cord was also dissected from 3- to 4-day-old rats which were anesthetized with ether. The collected tissues were immediately frozen in liquid nitrogen and stored at −70°C until used. Isolation and purification of mRNA was essentially as used previously [8]. The mRNA was dissolved in water to concentrations of about 1 mg/ml, and 50 nl were injected into oocytes. Injected and noninjected (native) oocytes were then cultured at 16°C for 3–8 days in modified Barth's medium with antibiotics [5]. Most of the oocytes were treated with collagenase (Sigma, Type I, 0.5–1.0 mg/ml for about 1–2 h) on the 2nd or 3rd day after injection to remove follicular and other enveloping cells [6]. Electrophysiological recordings were carried out at room temperature (20°–23°C). To study currents elicited by activation of voltage-operated channels the membrane potential was clamped at −100 mV; while for currents induced by receptor activation it was clamped at −60 mV, so as to be away from the equilibrium potentials for Na^+, K^+, Cl^- and Ca^{2+} [6]. Receptor agonists were applied by bath perfusion in normal frog Ringer (pH 7.0) at concentrations which gave nearly maximal responses.

Results

The yields of total RNA and mRNA in different preparations are shown in Table 1. Similar amounts of mRNA per gram wet weight were obtained from adult cortex and spinal cord. In contrast, the concentration of mRNA in neonatal spinal cord was more than doubled although, due to its small size, each neonatal cord had only about one fifth of the mRNA contained in an adult spinal cord.

As with mRNAs from other sources, different preparations of spinal cord mRNA varied considerably in their ability to express receptors in *Xenopus*

Table 1. Yields of RNAs from rat cerebral cortex and spinal cord

Tissue	Wet weight (g)	Total RNA (mg)	Poly(A)$^+$ RNA (μg)	Poly(A)$^+$/ Total RNA (%)	Poly(A)$^+$RNA/ g wet weight (μg)
Adult rat cortex	9.8	9.3	345	3.7	35.2
Adult rat spinal cord (A)	7.3	7.1	245	3.5	35.6
Adult rat spinal cord (B)	11.2	11.0	390	3.5	34.8
Adult rat spinal cord (C)	9.2	–	239	–	26.0
Neonatal rat spinal cord	4.5	9.6	376	3.9	83.6

Table 2. Agonist-activated membrane currents in oocytes injected with mRNA from adult rat spinal cord

Preparation	GABA (1 m*M*)	β-Alanine (1 m*M*)	Glycine (1 m*M*)	Kainate (100 μ*M*)	Serotonin (10 μ*M*)
A	21.3 ± 4.2 (32)	360 ± 47 (36)	1004 ± 110 (36)	182 ± 16 (31)	624 ± 142 (26)
B	26.9 ± 6.3 (8)	558 ± 97 (9)	1616 ± 246 (9)	162 ± 20 (9)	935 ± 340 (9)
C	15.7 ± 4.8 (11)	418 ± 64 (11)	894 ± 95 (11)	174 ± 30 (10)	1064 ± 248 (11)

Xenopus oocytes were voltage-clamped at −60 m*M*. Each value represents peak amplitude of the current (mean ± S.E. of the mean, in nA) obtained in oocytes (number in parenthesis) derived from two to seven donors

oocytes. Furthermore, the number of receptors expressed by a given mRNA preparation also varied greatly in oocytes from different donors, as well as in oocytes from the same donor. Nevertheless, the results obtained from the three preparations of spinal cord mRNA were essentially the same (Table 2). For the comparisons presented below we used the spinal cord mRNA preparation labeled A in Table 1.

Expression of Voltage-Operated Channels

In addition to passive outward currents, depolarizing pulses applied to oocytes injected with spinal cord mRNA elicited three main types of currents. These voltage-dependent currents were like those described previously for oocytes injected with adult rat cortex mRNA [4, 10] and consisted of a tetrodotoxin-sensitive inward current carried by Na^+ and two transient outward currents, one carried by K^+ and the other by Cl^- (Fig. 1). For simplicity we will refer to oocytes injected with adult spinal cord mRNA as "adult cord oocytes," and so on.

The tetrodotoxin-sensitive Na^+ currents expressed in adult or neonatal spinal cord oocytes were similar to the Na^+ current elicitable after injection of adult cortex mRNA. However, the amplitude of the Na^+ current generated depended on the source of the mRNA in such a way that the expressional potency of the mRNAs followed the sequence: adult cortex > adult spinal cord > neonatal spinal cord (Fig. 1A). Oocytes from some donors possess native tetrodotoxin-sensitive Na^+ currents; but the seven donors used for the present experiments yield oocytes in which the tetrodotoxin-sensitive currents, if any, were less than 10 nA.

It has been shown recently that rat brain Na^+ channels are immunologically different from Na^+ channels in the spinal cord [3]. However the gross properties of the channels expressed in the oocytes after injection of either spinal cord or cortex mRNAs were similar. For example, they all inactivate rapidly and are blocked by tetrodotoxin. Nevertheless, it is possible that a more detailed physiological and pharmacological analysis will disclose

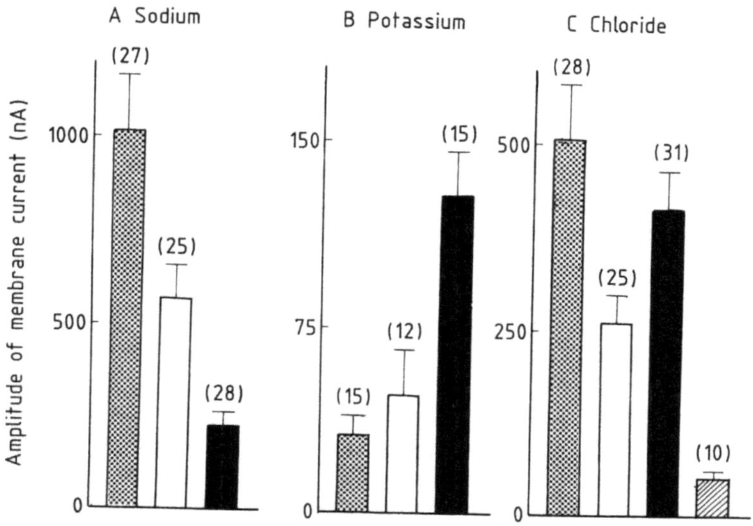

Fig. 1A–C. Peak amplitudes of voltage-gated currents expressed in *Xenopus* oocytes by mRNA derived from adult rat cerebral cortex (*stippled bars*), adult rat spinal cord (*open bars*) and neonate rat spinal cord (*solid bars*). Each oocyte was injected with mRNA (55–60 ng/50 nl) and currents were measured 3–8 days later. **A** Peak sodium current elicited by a pulse from −100 to −20 mV, **B** transient potassium currents evoked by a pulse to −30 mV, **C** calcium-dependent chloride current evoked by a pulse to 0 mV. Peak currents were measured by subtracting the size of the current induced by same pulse applied in the presence of 0.3 μ*M* tetrodotoxin in **A** or by subtracting the amplitude of the current recorded at the end of the applied pulse (150 ms in **B** and 3 s in **C**). The *hatched bar* in **C** shows the amplitude of the current observed in native (noninjected) oocytes. *Each bar* represents the average value with S.E. of the mean obtained in oocytes (*number in parenthesis*) derived from six different donors

more subtle differences that correlate with the different structures of the Na^+ channel subtypes.

The two transient outward currents expressed by spinal cord mRNA were also similar to those expressed by cerebral cortex mRNA [4, 10]. One of these currents begins to be activated at about −50 mV, increases progressively with depolarization, is carried mainly by K ions and is rapidly inactivated if the membrane potential is held at about −60 mV. A transient K^+ is occasionally seen also in native oocytes [9], but this was undetectable in the noninjected oocytes used for these experiments. Interestingly, for the K^+ current the expressional potency of the mRNAs was: neonatal spinal cord > adult spinal cord∼adult cerebral cortex; that is the opposite sequence of what is observed for the Na^+ current (Fig. 1A, B).

The other outward current is also transient and is frequently observed in native oocytes. It is caused by Cl ions flowing through channels that are opened following an influx of Ca^{2+} through native [7] and induced voltage-activated Ca^{2+} channels [10]. The time course of this Ca^{2+}-dependent Cl^- current is slower than that of the transient K^+ current. It begins to be activated at about −30 mV, reaches a peak at about −20 mV and disappears at about +40 mV. As in the case of the Na^+ current, the size of this Cl^- current in adult cord oocytes was about one half of that in adult cortex oocytes (Fig. 1C). However, while the Na^+ current in the adult cord oocytes was larger than in neonatal oocytes, the Ca^{2+}-dependent Cl^- current was larger in the neonatal cord oocytes, perhaps because this mRNA is richer in message encoding the voltage-operated Ca^{2+} channels.

Expression of Receptors That Active Smooth Inward Currents

Messenger RNA from both adult and neonatal spinal cord induced the oocytes to acquire the ability to generate "smooth" inward currents in response to amino acid receptor agonists, γ-aminobutyrate (GABA), β-alanine, glycine and kainate [1]. The properties of these currents were similar to those already reported for oocytes injected with mRNAs from rat, chick or

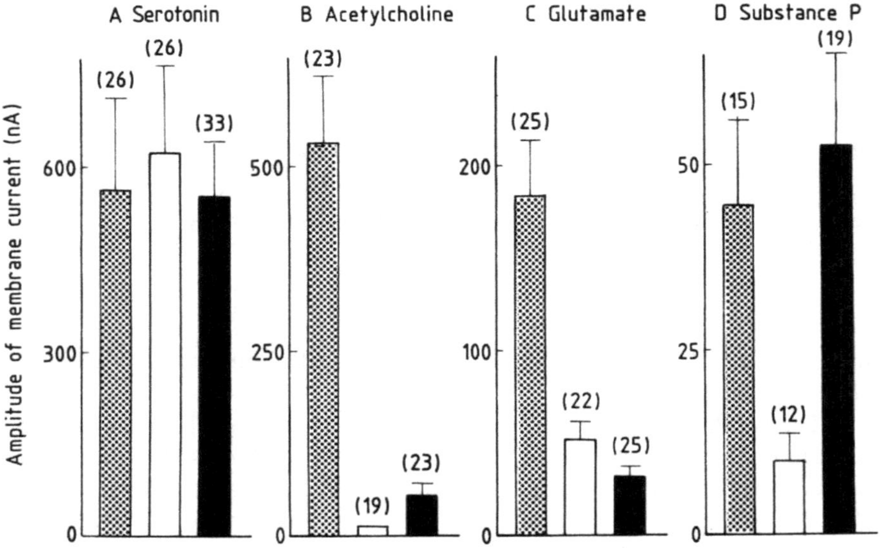

Fig. 2A–D. Peak amplitudes of oscillatory inward currents induced by agonists in oocytes injected with mRNAs from adult rat cerebral cortex (*stippled bars*), adult rat spinal cord (*open bars*) and neonate spinal cord (*solid bars*). The holding membrane potential was −60 mV and concentrations were: serotonin 10 μ*M* (**A**), acetylcholine (ACh) 100 μ*M* (**B**), Glutamate 1 mM (**C**) and substance P 1 μ*M* (**D**). Each *bar* represents mean ± S.E. of the mean obtained from oocytes (*number in parenthesis*) from four to seven donors. Control oocytes from two donors gave ACh current greater than 50 nA and were not included in **B**. Control oocytes from the other donors did not respond to ACh or gave currents of less than 10 nA

human brains [9, 10]. For example, the currents elicited by GABA, β-alanine and glycine were all carried mainly by Cl ions and they reversed direction at about −20 mV, which corresponds to the equilibrium potential for Cl^-. Pharmacological studies showed that the responses to 1 mM glycine and β-alanine were greatly depressed by strychnine (0.5 μ*M*) and picrotoxin (20 μ*M*). The cerebral cortex and spinal cord mRNAs expressed the various responses with different potencies. For instance, kainate and GABA-induced currents, which are usually quite large after injection of adult rat cortex mRNA, were rather small in adult and neonatal cord oocytes. In contrast, the glycine-induced current, which was relatively small in cerebral cortex oocytes, was the most prominent current in adult and neonatal spinal cord oocytes.

The spinal cord and cerebral cortex mRNA encoding glycine receptors were analyzed further after separation by sucrose density gradient sedimentation. The detailed results were already reported elsewhere [2, 11]. Briefly, after centrifugation, the glycine receptor mRNAs isolated from both sources were separated in two peaks; one is large (an estimated size of >8 kb) and the other is small (~2 kb). The sedimentation profile revealed that adult cerebral cortex has mainly the small size glycine receptor mRNA, while adult spinal cord is richer in the larger mRNA. The neonatal rat spinal cord is rich in glycine receptor mRNA, with an expressional potency comparable to that of the adult cord. However, the profile of the currents elicited by glycine in oocytes injected with fractions of neonatal spinal cord mRNA differed from that of adult spinal cord and rather resembled that obtained with adult cerebral cortex mRNA. The properties of the glycine receptors encoded by the large mRNA differed from those of the receptor encoded by the small mRNA, with respect to dose-response relationship, time course of desensitization and some pharmacological action [2, 11].

Expression of Receptors That Activate Oscillatory Inward Currents

Serotonin, acetylcholine, glutamate and substance P, applied to oocytes injected with spinal cord mRNA, all elicited oscillatory inward currents like those evoked in rat cortex oocytes [9]. The responses to serotonin were blocked by methysergide, a serotonin antagonist, while that to acetylcholine was suppressed by atropine, but not by D-tubocurarine. Acetylcholine elicited some oscillatory currents in noninjected oocytes from some donors, but the other transmitter substances were practically ineffective in the native oocytes used for the present experiments.

When comparing the ability of the mRNAs to express oscillatory currents to the different agonists, the most striking feature is that while the serotonin currents was more or less equally expressed with the three types of mRNAs, the responses to acetylcholine, glutamate and substance P were markedly smaller in the spinal cord oocytes (Fig. 2). Furthermore, while serotonin and glutamate generated currents of similar amplitudes in oocytes injected with adult and neonatal spinal cord mRNA, the responses to acetylcholine and substance P were four to five times larger in the oocytes injected with neonatal spinal cord mRNA (Fig. 2).

Discussion

The present results show that mRNA from the rat spinal cord induces the oocytes to acquire functional transmitter receptors and voltage-operated membrane channels that are essentially like those induced by cerebral cortex mRNA. However, cerebral cortex and spinal cord mRNAs have greatly different expressional potencies.

Although more detailed studies are required we may assume, for the time being, that the single channels expressed by cerebral cortex and spinal cord mRNAs have similar characteristics. Then, the peak amplitude of the evoked currents provides a measure of the number of receptors expressed and allows us to assay the relative amounts of mRNAs encoding the different neurotransmitter receptors, and voltage-operated channels, in the cerebral cortex and the spinal cord. Judging from the size of the currents elicited, and provided that the messages are expressed with equal efficiency, it appears that the mRNA encoding the Na^+ channels in the spinal cord increases from neonate to adult. However, this increase is comparatively smaller than the 10-fold increase in the wet weight of the spinal cord. This difference may be due to a slowing down in the production of Na^+ channel mRNA during the development as well as to neuronal death and a proportionally larger increase in the number of nonneuronal cells that do not carry many Na^+ channels.

In contrast, the evoked K^+ current was much larger in the oocytes injected with the neonatal spinal cord mRNA; and, again, several possibilities could account for these results. For example, it may be that neonatal and adult K^+-channel mRNA are different; but perhaps a more likely explanation is that in the adult cord the K^+-channel mRNA becomes less abundant.

Another interesting outcome of the present studies concerns the response to glycine. Comparing the currents elicited by glycine suggest that both adult and neonatal spinal cords are much richer in glycine receptor mRNA than adult cerebral cortex. Size fractionation of mRNA by sucrose gradient centrifugation revealed that there are two types of glycine receptor mRNAs in the rat central nervous system; one is large and the other is small. The adult spinal cord has predominantly the large glycine receptor mRNA while the adult cerebral cortex has mainly the small one. Interestingly, the small size mRNA encoding glycine receptors is abundant in neonatal spinal cord, whereas the large mRNA is comparatively scant. Thus, it seems that the production of two types of glycine receptor mRNA is regulated not only in a tissue-dependent, but also in an age-dependent manner.

In conclusion, *Xenopus* oocytes provide a convenient and powerful tool for studies of neurotransmitter and voltage-operated channels in general, developmental, comparative, and also clinical neurobiology.

References

1. Akagi H, Miledi R (1988) Expression of glycine and other amino acid receptors by rat spinal cord mRNA in *Xenopus* oocytes. Neurosci Lett 95:262–268
2. Akagi H, Miledi R (1988) Heterogeneity of glycine receptors and their messenger RNAs in rat brain and spinal cord. Science 242:270–273
3. Gordon D, Merrick D, Auld V, Dunn R, Goldin AL, Davidson N, Catterall WA (1987) Tissue-specific expression of the R_I and R_{II} sodium channel subtypes. Proc Natl Acad Sci USA 84:8682–8686
4. Gundersen CB, Miledi R, Parker I (1983) Voltage-operated channels induced by foreign messenger RNA in *Xenopus* oocytes. Proc R Soc Lond B 220:131–140
5. Gundersen CB, Miledi R, Parker I (1984) Glutamate and kainate receptors induced by rat brain messenger RNA in *Xenopus* oocytes. Proc R Soc Lond B 221:127–143
6. Kusano K, Miledi R, Stinnakre J (1982) Cholinergic and cathecolaminergic receptors in the *Xenopus* oocyte membrane. J Physiol (Lond) 328:143–170
7. Miledi R (1982) A calcium-dependent transient outward current in *Xenopus laevis* oocytes. Proc R Soc Lond B 215:491–497
8. Miledi R, Sumikawa K (1982) Synthesis of cat muscle acetylcholine receptors by *Xenopus* oocytes. Biomed Res 3:390–399
9. Miledi R, Parker I, Sumikawa K (1989) Transplanting receptors from brains into oocytes. In: Smith J (ed) Fidia award lecture series. Raven, New York pp. 57–90
10. Sumikawa K, Parker I, Miledi R (1984) Partial purification and functional expression of brain mRNAs coding for neurotransmitter receptors and voltage-operated channels. Proc Natl Acad Sci USA 81:7994–7998
11. Akagi H, Miledi R (1989) Discrimination of heterogenous mRNAs encoding strychnine-sensitive glycine receptors in Xenopus oocytes by antisense oligonucleotides. Proc Natl Acad Sci USA 86:8103–8107

Prototypical Features of the Inhibitory Synapses in the Frog Spinal Cord as Revealed Pharmacologically

Y. KUDO[1] and E. AKIYOSHI

Introduction

Inhibitory synapses of the spinal cord modulate excitatory synaptic transmission and control the tones of agonistic and antagonistic muscles to ensure smooth movements of limbs and fingers. Thus terrestrial vertebrates which are destined to move with their limbs under high gravity are expected to have well organized inhibitory synapses as well as excitatory ones. During studies on the functions of inhibitory synapses in the spinal cord of frog, which is the most primitive terrestrial vertebrate, we encountered many prototypical features. For instance bicuculline and strychnine, well-known antagonists for GABA and glycine in mammalian central nervous system, respectively, showed little antagonism of the GABA- and glycine-induced membrane potential changes in amphibians [6]. Interestingly, we found that these drugs block the effect of taurine which exerts biphasic effects on primary afferent terminals [6]. We have also found that diazepam did not influence GABA binding on synaptic membranes prepared from frog spinal cord [7]. Taurine seems to be a major inhibitory neurotransmitter in the spinal cord of this species. These results give the impression that the inhibitory systems in frog spinal cord are prototypical as compared with those in mammalian central nervous system.

In the present study we examined the effects of inhibitory amino acids on ventral and dorsal root potentials and their interactions with diazepam and pentobarbital and obtained further evidence for the primitiveness of the inhibitory system in the frog spinal cord.

[1] Department of Neuroscience, Mitsubishi Kasei Institute of Life Sciences, Minamiooya 11, Machidashi, Tokyo, 194 Japan

K. Shimoji et al. (Eds.)
Spinal Cord Monitoring and Electrodiagnosis

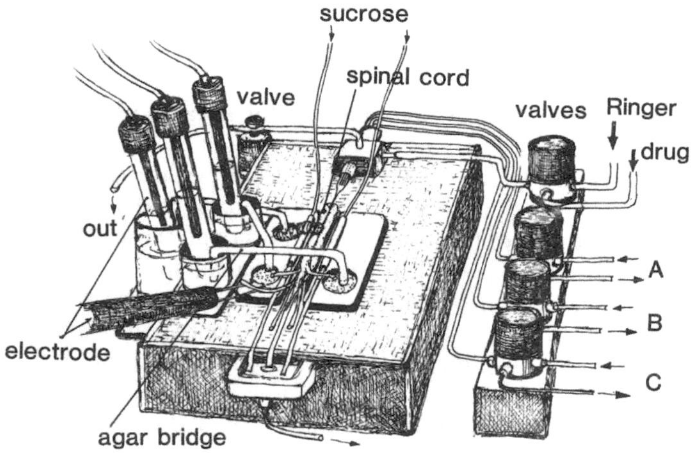

Fig. 1. Experimental arrangement of a sucrose-gap method applied on the ventral and dorsal roots of the isolated, intraarterially perfused spinal cord of the frog

Method

Isolated intraarterially perfused spinal cords of the bullfrog (*Rana catesbiana*, 100–150 g) were used in this study (Fig. 1). An arterial cannula (about 200 μm in diameter) was inserted into the ventral spinal artery and the spinal cord was perfused at the rate of approx. 0.3 ml/min with Ringer solution composed of (m*M*): NaCl 117, KCl 2.7, $CaCl_2$ 1.8, $MgCl_2$ 0.1, glucose 5.5, NaH_2PO_4 0.25, with pH adjusted to 7.6 ± 0.1 by addition of $NaHCO_3$. In some experiments, the concentrations of $CaCl_2$ and $MgCl_2$ were changed as indicated in the text. The temperature of the recording chamber was kept at 20°C ± 1°C with a thermomodule temperature control unit (Diamedical System, DTC-100). The potential differences between the spinal cord and the peripheral stumps of the 9th dorsal root (dorsal root potential; DRP) and the 10th ventral root (ventral root potential; VRP) were assessed by means of the sucrose-gap method. The 10th dorsal root was stimulated via a bipolar electrode to evoke the dorsal root potential (DR-DRP) and the corresponding ventral root potential (DR-VRP). Taurine, GABA or glycine (Wako Pure Chem.) were dissolved in Ringer solution at concentrations of 1–10 m*M* and perfused for 1–10 s at a rate of 0.04 ml/s (0.1–4 μmol) through a fine polyethylene tubing placed in the glass cannula for perfusion with the aid of a peristaltic microtube pump (LKB2155) and timer-operated three-way electric valves (General valve, USCC1T-3C-12D). Diazepam (Sumitomo Chem.) and sodium pentobarbital (Tokyo Kasei) were dissolved in Ringer solution at 10 μ*M*. These drugs were administered by replacing the perfusing medium with the drug-containing Ringer solutions.

Results

Effects of Inhibitory Amino Acids on the DRP and VRP

As shown in Fig. 2A, GABA (0.1–1 m*M*) caused a marked depolarization in the dorsal root (primary afferent terminals) and a marked hyperpolarization in the ventral root (motoneurons). Although these effects of GABA were the same as those expected in the mammalian spinal cord [3], a difference was seen when evoked potentials were recorded in the presence of GABA. The amplitudes of the DR-VRPs were only weakly affected by the amino acid. A concentration as high as 10 m*M* was required to abolish the DR-VRP(Fig. 2B). On the other hand, taurine as low as 0.01 m*M* showed biphasic effects (a fast-onset hyperpolarization followed by a slow-onset depolarization) in the dorsal root and in the ventral root during the administration (Fig. 2A). The effects were accompanied by a marked inhibition of the amplitude of the evoked response (DR-VRP). Almost complete inhibition of DR-VRP was obtained by 1 m*M* of taurine (Fig. 2B). Glycine had various effects on the DRP and VRP; depolarization and/or hyperpolarization, depending on specimens. The inhibitory effect of glycine on DR-VRP was not obvious even at a concentration of 10 m*M* (Fig. 2B).

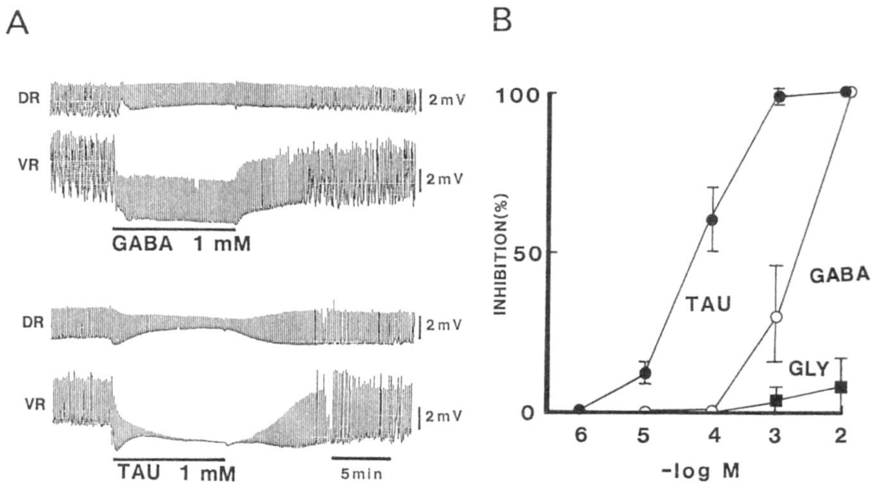

Fig. 2A,B. Effects of inhibitory amino acids on the root potentials. **A** Simultaneous recording of dorsal and ventral root potentials. *Upward deflections* indicate the evoked dorsal root potentials (DR-DRP; *upper tracings*) and ventral root potentials (DR-VRP; *lower tracings*). **B** Dose-response curves for inhibitory amino acids on the amplitude of DR-VRP. Means are shown ($n = 4$) with standard errors (S.E.)

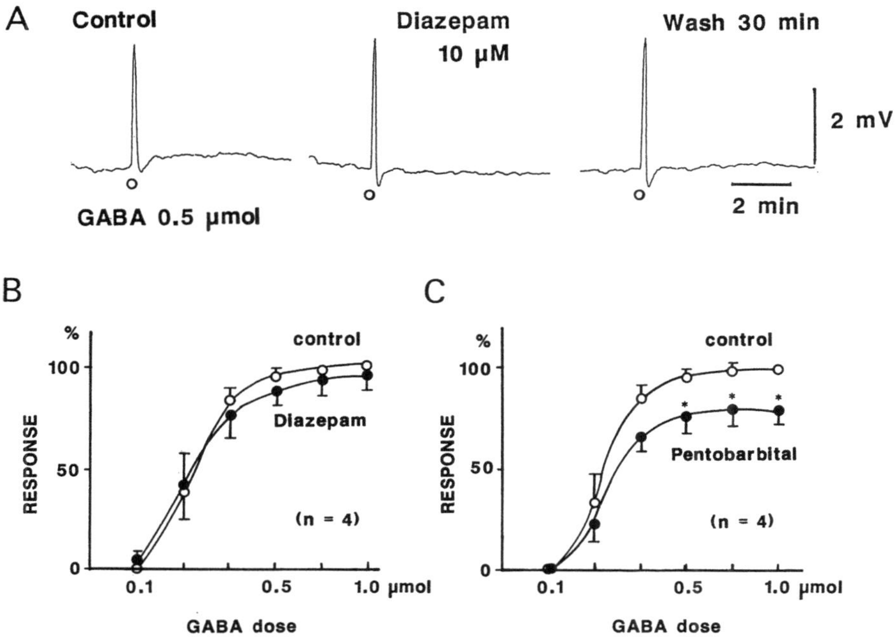

Fig. 3A–C. Effects of diazepam and pentobarbital on GABA-induced depolarization of the dorsal root. **A** Representative recording of depolarization in the dorsal root induced by GABA and the effect of diazepam (10 μ*M*, for 10 min) on it. **B** Dose-response curves for GABA before (*open circles*) and during (*solid circles*) the administration of diazepam (10 μ*M*). **C** Effect of pentobarbital on the dose-response curves for GABA. *Asterisks* indicate a statistically significant inhibition, $p < 0.05$. In **B** and **C**, the maximum response to the highest amount of GABA in the control experiment was taken as 100%

Effects of Diazepam and Pentobarbital on the Depolarization of DRP Induced by GABA

When a high dose of GABA (10 m*M*) was infused for 5–10 s at a rate of 0.3 ml/min, a marked depolarization was induced dependent upon the total amount of GABA applied in DRP (primary afferent terminal potential; Fig. 3A). Diazepam at a dose (10 μ*M*) supramaximal in mammalian specimens had no facilitatory effect on the GABA-induced depolarization of the primary afferent terminal (Fig. 3A). When pentobarbital coexisted during the administration of GABA, the depolarization in DRP induced by the amino acid was significantly reduced. Two sets of dose-response curves for GABA were obtained in four preparations before and after the treatment (for more than 10 min) with diazepam (10 μ*M*) or pentobarbital (10 μ*M*), shown in Fig. 3B, C.

Effects of Diazepam and Pentobarbital on the Depolarization Induced by Taurine on the Primary Afferent Terminal

As shown in Fig. 4A, taurine showed biphasic effects on the primary afferent terminal (DRP) in the preparation perfused with a medium containing 0.1 m*M* of Ca^{2+} and 9.0 m*M* of Mg^{2+} as has been reported elsewhere [6]. In the present study, the effects of diazepam and pentobarbital were evaluated during the depolarization phase. As shown in Fig. 4A and 4B, diazepam (10 μ*M*), which had no effect on the DRP-depolarization induced by GABA, significantly augmented the maximum effect of taurine. This was also the case for pentobarbital, which augmented the maximum depolarization induced by taurine. These effects of diazepam and pentobarbital on the response to taurine are quite similar to those reported on the GABA-induced DRP-depolarization in mammalian spinal cord [1, 8].

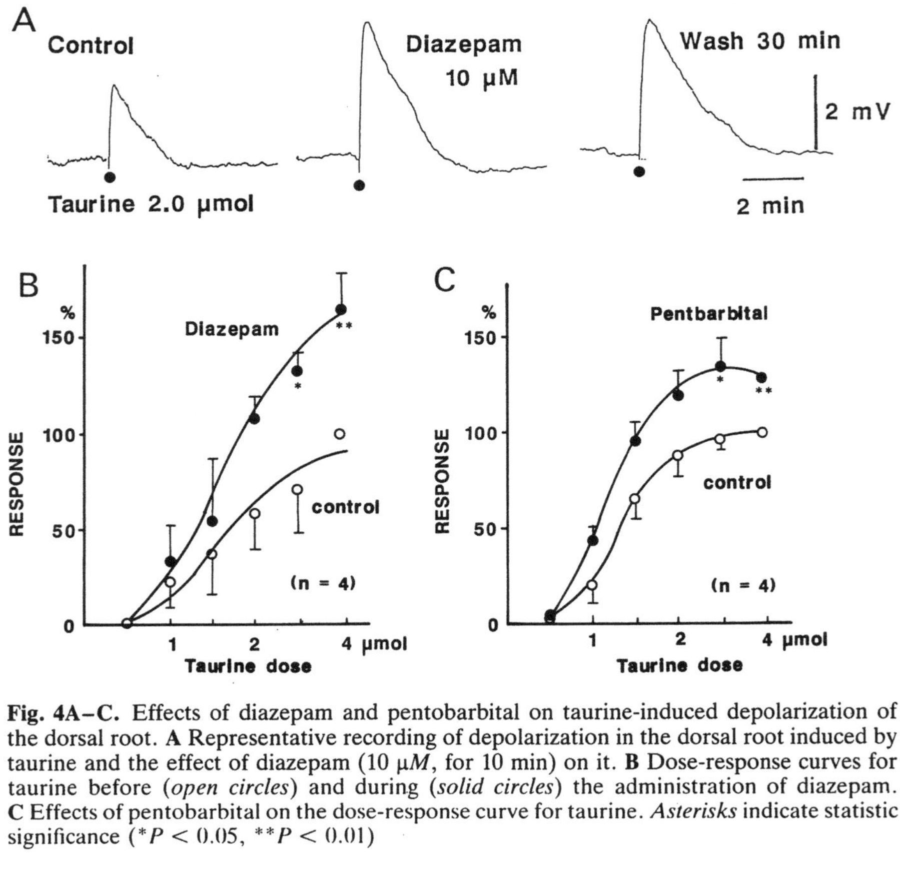

Fig. 4A–C. Effects of diazepam and pentobarbital on taurine-induced depolarization of the dorsal root. **A** Representative recording of depolarization in the dorsal root induced by taurine and the effect of diazepam (10 μ*M*, for 10 min) on it. **B** Dose-response curves for taurine before (*open circles*) and during (*solid circles*) the administration of diazepam. **C** Effects of pentobarbital on the dose-response curve for taurine. *Asterisks* indicate statistic significance ($*P < 0.05$, $**P < 0.01$)

Discussion

We have previously reported that the inhibitory actions of GABA and glycine are not obvious in the frog spinal cord and that the interactions with respective antagonists bicuculline and strychnine are insignificant [4–6]. Additionally, our previous study has shown that the interaction between binding sites for GABA and diazepam has features different from those in synaptic membranes prepared from mammalian spinal cord [7]. We also reported recently that taurine exerts biphasic effects on DRP which are discriminated by raising extracellular Mg^{2+} concentration or by administering bicuculline or strychnine [6]. Those results indicate the inhibitory systems in the frog spinal cord are different from those of mammals. Taurine seems to participate in the inhibitory system of the frog spinal cord rather than GABA and/or glycine. The present study provides new evidence which further supports the participation of taurine as a presynaptic inhibitory transmitter at the primary afferent terminal. The effect of taurine on the primary afferent terminal was modified by diazepam and pentobarbital in a manner analogous to GABA modulated by these drugs in the mammalian spinal cord.

Our preliminary experiments on dissected preparations of dorsal root ganglia, a piece of dorsal root or a portion of dorsal horn (primary afferent terminal) with a dorsal root using a sucrose-gap method revealed that GABA showed depolarizations not only on the dorsal root ganglia and primary afferent terminal, but also on the dorsal root of the frog spinal cord. On the other hand, taurine had no effects on the dissected dorsal root ganglia and dorsal root, but did induce an obvious hyperpolarization and depolarization of the dissected dorsal horn. Thus the receptor for taurine seemed to be of a more specific nature as a functional receptor than that for GABA on the primary afferent terminals in the frog spinal cord.

Since taurine has been shown to be rich in the central nervous system of mammalian embryos and early infants and since the level of taurine goes down during maturation, in contrast to the increase of the level of GABA [9], taurine is suggested to play a role of a transient inhibitory neurotransmitter in immature or underdeveloped central nervous systems.

Early comparative histological studies on the vertebrate spinal cord have shown that the primary afferent terminals of amphibians end in the dorsal horn and connect monosynaptically to motoneuronal dendrites invading the dorsal horn [2]. In contrast, the primary afferent terminals of mammals invade toward the ventral horn in the course of development and make monosynaptic connections onto motoneuronal somas.

In addition to biochemical and histological evidence, the present findings provide further clues to elucidate the evolution of the inhibitory system of the vertebrate. In the frog spinal cord, GABA receptors do exist on the primary afferent terminal, but they appear to be poorly functional, so to say "prefunctional", in this system.

Acknowledgement. We thank Dr. Akihiko Ogura and Dr. Hiroyuki Akagi for their valuable criticism and corrections on our manuscript.

References

1. Braestrup C, Honore T, Nielsen M, Petersen EN, Jensen LH (1983) Benzodiazepine receptor ligands with negative efficacy chloride channel coupling. Ad Biochem Psychpharmacol 38:29–36
2. Joseph BS, Whitelock DG (1968) The morphology of spinal afferent-efferent relationships in vertebrates. Brain Behav Evol 1:2–18
3. Krnjevic K (1979) Inhibitory action of GABA and GABA-mimetics on vertebrate neurons. In: Roberts E, Chase TN, Tower DB (eds) GABA in nervous system function. Raven Press New York pp 269–281
4. Kudo Y, Oka J-I, Yamada K (1981) Anisatin, a potent GABA antagonist, isolated from *Illicium anisatum*. Neurosci Lett 25:83–88
5. Kudo Y, Tanaka A, Yamada K (1983) Dendrobine, an antagonist of β-alanine, taurine and of presynaptic inhibition in the frog spinal cord. Br J Pharmacol 78:709–715
6. Kudo Y, Akiyoshi E, Akagi H (1988) Identification of two taurine receptor subtypes on the primary afferent terminal of frog spinal cord. Br J Pharmacol 94:1051–1056
7. Oka J-I, Fukuda H, Kudo Y (1981) The immaturity of interactions between GABA- and benzodiazepine-binding sites in the frog spinal cord. Gen Pharmacol 12:385–389
8. Polc P, Mohler H, Haeffely W (1974) The effect of diazepam on spinal cord activities: Possible sites and mechanisms of action. Naunyn-Schmiedeberg's Arch Pharmacol 284:319–337
9. Sturman JA, Gaull GE (1976) Taurine in the brain and liver of the development human and Rhesus monkey. In: Huxtable R, Barbeau A (eds) Taurine. Raven Press, New York pp 73–84

Metabolic Depression of the Spinal Cord and Brain with Epidural Bupivacaine in Rats

Y. Kuroda[1], T. Sakabe[2], K. Nakakimura[2], T. Maekawa[2], T. Ishikawa[2], and H. Takeshita[2]

Introduction

Despite the wide use of epidural anesthesia, little is known about the effects of epidural local anesthetics on central nervous system (CNS) metabolism. We examined the effects of epidural bupivacaine on CNS glucose metabolism as a reflection of CNS function in conscious rats. To determine whether or not the metabolic effects are related to the systemic effects of bupivacaine and deafferentation, we also examined the effects of intramuscular bupivacaine and spinal cord transection on CNS metabolism.

Materials and Methods

The experiments were performed in 39 adult male Wistar rats. In 10 rats, given bupivacaine epidurally ($n = 5$) and intramuscularly ($n = 5$), the duration of analgesia (by tail flick and hot plate tests) and the plasma concentrations of bupivacaine were measured. In another 29 rats, local spinal cord glucose utilization (LSGU) and local cerebral glucose utilization (LCGU) were measured. The rats were randomly divided into four groups: epidural bupivacaine group ($n = 7$), intramuscular bupivacaine group ($n = 7$), spinal cord transection group ($n = 7$), and control group ($n = 8$). The control group consisted of eight rats in total: four rats received epidural saline and four rats received intramuscular saline. In the epidural bupivacaine group, the drug was given via the catheter inserted through the lumbar (L-5) epidural space as described previously [1]. In the intramuscular bupivacaine

[1] Department of Anesthesiology, Kokura Memorial Hospital, 1-1 Kifune-machi, Kokura-kita-ku, Kitakyushu, Fukuoka, 802 Japan
[2] Departments of Anesthesiology-Resuscitology, Yamaguchi University Hospital, 1144 Kogushi, Ube, Yamaguchi 755, Japan

K. Shimoji et al. (Eds.)
Spinal Cord Monitoring and Electrodiagnosis

group, the drug was injected into the major gluteal muscle. In all groups given bupivacaine, 300 μg bupivacaine hydrochloride dissolved in 40 μl saline was injected. In the spinal cord transection group, the spinal cord between T1-2 vertebra was transected, and the mean arterial blood pressure (MABP) was maintained above 75 mmHg with a continuous infusion of phenylephrine.

Glucose utilization was measured using the 2-[^{14}C] deoxyglucose autoradiographic method [4]. The method was identical as reported in our previous paper [1].

The statistical significance of the differences of the plasma bupivacaine concentrations between the epidural and intramuscular groups was analysed with an unpaired *t* test. For the comparison of the glucose utilization between four groups, one way analysis of variance for repeated measures, with the least significant difference test for multiple comparisons, was used. $P < 0.05$ was considered statistically significant.

Results

The duration of analgesia with epidural bupivacaine was 40–50 min. Intramuscular bupivacaine did not produce analgesia. There were no significant differences of bupivacaine concentrations between the epidural and intramuscular groups at 10 min (epidural, 0.39 ± 0.06 vs. intramuscular, 0.46 ±

Table 1. Local spinal cord glucose utilization

Rexed's laminae	Control ($n = 8$)	Epidural bupivacaine ($n = 7$)	Intramuscular bupivacaine ($n = 7$)	Spinal cord transection ($n = 7$)
Cervical				
I-II	31 ± 1	28 ± 1	32 ± 1	29 ± 2
III	35 ± 1	31 ± 1	35 ± 1	34 ± 2
IV-VI	41 ± 2	36 ± 2	42 ± 1	42 ± 2
VII-IX	38 ± 2	35 ± 2	41 ± 1	41 ± 1
Thoracic				
I-II	32 ± 1	24 ± 1^a	27 ± 2^a	25 ± 1^a
III	35 ± 1	27 ± 1^a	32 ± 2^b	29 ± 1^a
IV-V	36 ± 2	28 ± 2^a	34 ± 1^b	31 ± 1^a
VII-IX	35 ± 2	28 ± 2^a	34 ± 1^b	35 ± 2^b
Lumbar				
I-II	35 ± 1	26 ± 1^a	32 ± 2^b	29 ± 2^a
III	39 ± 1	29 ± 1^a	36 ± 2^b	33 ± 2^a
IV-VI	44 ± 2	34 ± 2^a	43 ± 2^b	41 ± 2^b
VII-IX	44 ± 3	34 ± 1^a	43 ± 2^b	41 ± 2^b

Values are mean ± SE (μmol $100g^{-1}$ min^{-1})
[a] Significantly different from control group ($P < 0.05$)
[b] Significantly different from epidural bupivacaine group ($P < 0.05$)

Table 2. Local cerebral glucose utilization

Structures	Control ($n = 8$)	Epidural bupivacaine ($n = 7$)	Intramuscular bupivacaine ($n = 7$)	Spinal cord transection ($n = 7$)
Somatosensory cortex	88 ± 3	68 ± 3[a]	75 ± 3[a]	83 ± 4[b]
Thalamus (VB complex)	81 ± 3	65 ± 3[a]	71 ± 3	81 ± 4[b]
Reticular formation	53 ± 2	40 ± 2[a]	44 ± 1[a]	50 ± 2[b]
Cuneate nu.	80 ± 6	61 ± 3[a]	67 ± 4	63 ± 4[a]
Gracile nu.	56 ± 2	43 ± 2[a]	53 ± 4[b]	49 ± 3
Periaqueductal gray	61 ± 2	45 ± 3[a]	49 ± 2[a]	56 ± 3[b]
Dorsal raphe nu.	66 ± 3	53 ± 3[a]	59 ± 2	66 ± 4[b]
Median raphe nu.	73 ± 3	56 ± 3[a]	63 ± 2[a]	72 ± 4[c]
Gigantocel. ret. nu.	51 ± 2	40 ± 2[a]	42 ± 1[a]	47 ± 2[b]
Paragigant. ret. nu.	53 ± 3	42 ± 3	48 ± 1	50 ± 3
Raphe magnus nu.	43 ± 3	34 ± 3	36 ± 1	41 ± 3

Values are mean ± SE (μmol 100 g^{-1} min^{-1})
[a] Significantly different from control group ($P < 0.05$)
[b] Significantly different from epidural bupivacaine group ($P < 0.05$)
[c] Significantly different from intramuscular bupivacaine group ($P < 0.05$)

0.05 μg/ml) and 30 min (epidural, 0.17 ± 0.03 vs. intramuscular, 0.24 ± 0.02 μg/ml) after injection (mean ± SE).

LSGU and LCGU values are shown in Tables 1 and 2. Epidural bupivacaine decreased glucose utilization in the thoracic (by 17%–25%), lumbar (by 23%–26%) spinal cord, and most of the brain structures examined (by 20%–26%). Evans blue injected through the epidural catheter at the end of the experiment was distributed to the epidural space of the thoracic and lumbar spinal cord. With intramuscular bupivacaine, LSGU did not change significantly except in one structure, while LCGU decreased (by 14%–20%) in about half of the structures examined. With the transection of the upper thoracic spinal cord, LCGU did not decrease significantly in most of the structures, while LSGU decreased significantly in laminae I-V (by 14%–22%) in the lower thoracic region and laminae I-III (by 15%–17%) in the lumbar region.

Discussion

The present study demonstrated that epidural bupivacaine in a dose sufficient to produce analgesia decreased glucose utilization in the spinal cord and brain. With epidural bupivacaine, the decrease in LSGU was observed only in the thoracic and lumbar regions where the drug was assumed to be distributed (judged by the Evans blue distribution.) There were no changes in the cervical spinal cord metabolism. Therefore, the decreased LCGU with epidural bupivacaine could be attributed to reduced sensory input to the spinal cord and the direct drug effect on the spinal cord. It is unlikely that bupivacaine absorbed systemically has affected spinal cord metabolism because intramuscular bupivacaine had no effect on LSGU in spite of the comparable plasma concentrations of bupivacaine obtained.

In regard to the effects of other local anesthetics given epidurally, Lin et al. (abstract) [2] reported minimum depression of the resting LSGU with epidural lidocaine in rats anesthetized with halothane. Though the experimental designs are different between their study and ours, it appears that epidural local anesthetics, when given in a sufficient dose to produce analgesia, depress the spinal cord glucose metabolism.

LCGU in many brain structures was decreased with intramuscular bupivacaine as observed with epidural administration. The plasma concentrations of bupivacaine obtained in both groups were comparable. Almost the same magnitude of metabolic depression by intramuscular and epidural bupivacaine in many brain structures would indicate that the drug systemically absorbed has affected brain glucose metabolism. Furthermore, a transection of the spinal cord did not significantly affect the cerebral metabolism. These results indicate that the decreased somatosensory input from periphery to the brain may not be the major contributing factor for the cerebral metabolic depression with epidural bupivacaine.

Little is known about the effect of spinal cord transection on CNS metabolism: The only report we are aware of is that by Schwartzman et al. [3], who demonstrated in monkeys that 24 h after spinal cord transection at T-10 level, LSGU decreased in laminae IV-IX and increased in lamina I in the lumbar region. They attributed the increase in LSGU in lamina I to loss of descending inhibitory influences. They also speculated that the decrease in glucose utilization in the ventral horn might be caused by loss of descending facilitatory influences. In the present study, metabolic depression by transection was prominent in laminae I-V in the lower thoracic and laminae I-III in the lumbar regions. We have no explanation for the discrepancies between Schwartzman's study and ours. It could be due to the differences in the species and the time of measurement after spinal cord transection. In any event, it can be said that normal baseline spinal cord metabolism is modified not only by the input to the spinal cord from the periphery, but also by the neuronal activity of descending modulatory systems in the brain and upper part of the spinal cord above the transection.

In summary, epidural bupivacaine depresses spinal cord glucose metabolism, and the metabolic depression may be due to the reduced sensory input to the spinal cord and the direct effect of bupivacaine on the spinal cord. Cerebral metabolic depression with epidural bupivacaine is mainly due to the effects of bupivacaine absorbed systemically. The decrease of the glucose metabolism in the spinal cord after transection may indicate that descending neuronal regulatory mechanism also affects normal spinal cord metabolism.

References

1. Kuroda Y, Nakakimura K, Sakabe T, Maekawa T, Takeshita H (1987) Analgesic doses of epidural morphine do not affect local glucose utilization in the spinal cord in rats. Anesth Analg 66:1175–1179
2. Lin DM, Shapiro HM, Shipko EM (1985) Comparison of epidural lidocaine and fentanyl on spinal cord metabolism during sensory stimulation. Anesthesiology 63:A232
3. Schwartzman RJ, Eidelberg E, Alexander GM, Yu J (1983) Regional metabolic changes in the spinal cord related to spinal shock and later hyperreflexia in monkeys. Ann Neurol 14:33–37
4. Sokoloff L, Reivich M, Kennedy C et al. (1977) The [^{14}C]deoxyglucose method for the measurement of local cerebral glucose utilization: Theory, procedure, and normal values in the conscious and anesthetized albino rat. J Neurochem 28:897–916

The Effects of Sevoflurane on Heterosegmental Slow Positive Cord Potentials in the Rat

M. TOHYAMA[1], S. DENDA, Y. SATO, N. FUJIWARA, and K. SHIMOJI

Abstract

To define the effects of sevoflurane on brain and spinal cord functions, simultaneous recordings of cortical evoked potentials as well as segmental and heterosegmental spinal cord potentials (SCPs), in response to fore- and hindpaw stimulations, were made in the rat. The heterosegmental slow positive potentials, suggested to be a primary afferent depolarization exerted by the descending impulses via the supraspinal structures, were most vulnerable to the anesthetic. The second component of the segmental slow positive wave (P_2s) of the segmental SCP, also thought to be produced by a feed-back loop, was more strongly suppressed by sevoflurane than the first component (P_2f). The negative wave (N_1) of segmental SCP, believed to be a synchronized activity of dorsal horn interneurons, was not significantly affected in either peak latency or amplitude, but was prolonged in duration by sevoflurane. No differential effects of sevoflurane on spinal cord potentials produced by fore- and hindpaw stimulations were found.

Thus, the results suggest that clinical concentration of sevoflurane has a minimum direct effect on the synaptic transmission from the first to second order neurons in the spinal dorsal horn, but a profound blocking effect on primary afferent depolarization particularly activated by a feed-back loop via supraspinal structures.

Introduction

Somatosensory stimulation of the forepaw (FP) produced a segmental spinal cord potential (SCP) in the cervical enlargement and at the same time evoked a heterosegmental slow positive potential (HSP) in the lumbar enlargement.

[1] Department of Anesthesiology, Niigata University School of Medicine, Asahimachi 1-757, Niigata, 951 Japan

K. Shimoji et al. (Eds.)
Spinal Cord Monitoring and Electrodiagnosis

Likewise, hindpaw (HP) stimulation produced a segmental SCP in the lumbar enlargement and an HSP in the cervical enlargement [6, 11, 12]. The origin of the HSP is suggested to be an inhibitory potential exerted by the impulses feeding back via the supraspinal structures [6, 12]. Sevoflurane, fluoromethyl-2,2,2-trifluoro-1-(trifluoromethyl)-ethylether, is a new volatile anesthetic with a low blood/gas partition coefficient of 0.59 [1, 3]. Anesthetic induction and recovery with sevoflurane is therefore rapid and easily controllable [1, 3].

For evaluation of the effect of an anesthetic drug on the somatosensory system, recordings of evoked potentials not only from the brain but also from the spinal cord seem more valuable. Thus to define the effects of sevoflurane on brain and spinal cord function, simultaneous recording of electroencephalogram (EEG), cortical evoked potential (CEP), and segmental and heterosegmental SCPs was made in the rat.

Material and Methods

Twenty five male wistar rats weighing 382.5 ± 8.5 g (mean ± SE) served as our subjects. Following an intraperitoneal injection of 125 mg/kg ketamine with 0.05 mg/kg atropine, controlled ventilation was carried out through a tracheotomized orifice under muscle relaxation with intravenous injection of pancuronium, 0.1 mg/kg/h, through the femoral vein. The vein was also used for administration of ketamine, 30 mg/kg/h and other drugs. The femoral artery was cannulated for measurement of blood pressure and samplings of arterial gases. Ventilation volume was adjusted to maintain endtidal $Paco_2$ at 35–40 mmHg. For monitoring urine volume, a cannula was inserted into the urinary bladder. The rectal temperature of the rats was kept constant between 37°–38°C using a water heating blanket (Fig. 1).

The electroencephalogram and cortical evoked potential were recorded with nonpolarizable needle electrodes from the scalp fixed at the sensorimotor area. The heterosegmental slow positive potential and segmental spinal cord potential were recorded with nonpolarizable needle electrodes (200 μm in diameter) from the supraspinous ligaments of the cervical and lumbar enlargements. Pairs of stimulus electrodes were inserted 10 mm apart into both the fore- and hindpaws. Electrical stimuli were delivered through an isolation unit and changes in evoked potentials were recorded on an X-Y plotter before, during, and after 1%, 2%, 3%, and 4% sevoflurane anesthesia. The electroencephalogram, electrocardiogram, and arterial pressure were continuously monitored and recorded on polygraph.

All evoked responses were averaged (N = 50). The averaged evoked responses were recorded consecutively 2–4 times during the control and at each stepwise increased concentration of sevoflurane. Arterial blood concentrations of sevoflurane were measured by a gaschromatograph (Shimadzu GC-6AM) equipped with a blood gas sampler (Shimadzu BGS-1A) 10–20

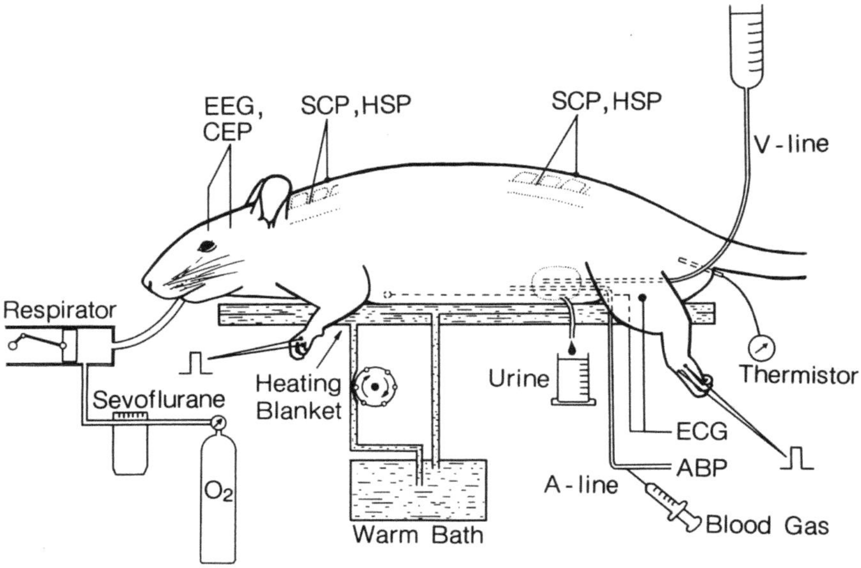

Fig. 1. Schematic illustration of the experimental setup. *EEG*, electroencephalogram; *CEP*, cortical evoked potential; *SCP*, spinal cord potential; *HSP*, heterosegmental slow positive wave; *V-line*, venous line; *A-line*, arterial line; *ECG*, electrocardiogram; *ABP*, arterial blood pressure

min following the inhalation of 1.0%–4.0%. Percent changes in latencies, amplitude, and duration observed 10–20 min after inhalation of each concentration of sevoflurane were measured. A two-way analysis of variance was conducted to identify significant differences in amplitude, latency, and duration changes in cervical and lumbar spinal cord potentials following sevoflurane administration. The duration of sevoflurane inhalation lasted 80 min.

Results

Figure 2 shows an example of the measured parameters for cortical evoked potential, segmental spinal cord potential, and heterosegmental slow positive potential in response to forepaw stimulation under ketamine anesthesia before inhalation of sevoflurane. Forepaw stimulation produced a cortical evoked potential as well as segmental spinal cord potential consisting of a sharp negative (N_1) and slow positive (P_2) wave on the dorsal surface of cervical enlargement. The P_2 wave was usually divided by a dip (N_2) into the first (P_2f) and second (P_2s) components (Fig. 2). At the same time a slow positive wave, heterosegmental slow positive potential, was also evoked

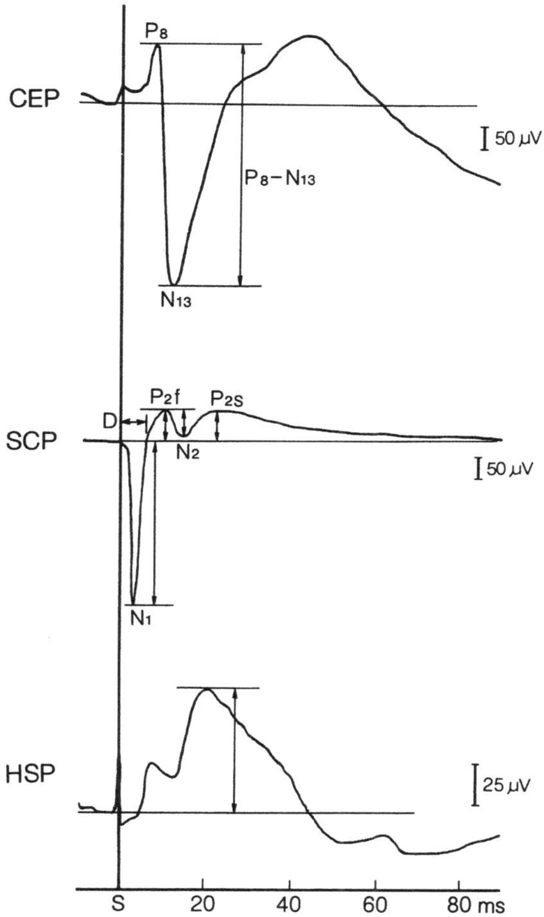

Fig. 2. Measured parameters for cortical evoked potential (*CEP*), segmental spinal cord potential (*SCP*) and heterosegmental slow positive potential (*HSP*) in response to forepaw stimulation, under ketamine anesthesia before inhalation

in the lumbar enlargement. Hindpaw stimulation also produced segmental potentials on the dorsal surface of lumbar enlargement and a heterosegmental slow positive potential on the cervical enlargement.

The most profound effect of sevoflurane was a decrease in amplitude of the dip (N_2) imposed on the P_2 wave of the segmental spinal cord potential (Table 1). Heterosegmental slow positive potentials were also extremely vulnerable to sevoflurane (Table 1). The amplitudes of cortical evoked potential (P_8-N_{13} peak-to-peak), and the P_2 wave of the segmental spinal cord potential were also decreased significantly (Table 1). There was no significant change in N_1 latency or amplitude but a significant prolongation of its duration (Table 1). Latencies of cortical evoked potentials (P_8, N_{13}) and the heterosegmental slow positive wave were significantly prolonged at 2% of the drug (Table 1). With 3% sevoflurane, the heterosegmental slow positive potential disappeared

Table 1. Effects of sevoflurane on cortical evoked potential (CEP), segmental spinal cord potential (N_1, N_2, P_2) and heterosegmental slow positive cord potential (HSP) in response to forepaw stimulation

		Before	1[a] (0.26 ± 2)	2 (0.58 ± 3.6)	3 (0.90 ± 4.9)	4 (1.17 ± 7.3)	Rec. 10 min	30 min (0.17 ± 11)	60 min (0.04 ± 0.5)
CEP									
P_8-N_{13}	A	100	66.3 ± 8.6*	27.5 ± 8.4**	7.3 ± 4.7**	0	96.6 ± 7.0**	100.2 ± 10	100.1 ± 5.1
P_8	L	100	108.2 ± 3.2	172.4 ± 12.2***	187.6 ± 24.5	–	120.9 ± 3.0**	108.6 ± 3.6	106.1 ± 2.8
		(10.3 ± 0.1)	(11.0 ± 0.3)	(17.0 ± 1.4)			(12.4 ± 0.4)	(11.1 ± 0.4)	(10.8 ± 0.3)
N_{13}	L	100	104.8 ± 2.4	211.9 ± 18.9**	–	–	118 ± 3.4**	109.9 ± 2.4**	106.5 ± 2.8
		(15.8 ± 0.2)	(16.6 ± 0.4)	(33.3 ± 2.9)			(18.7 ± 0.6)	(17.4 ± 0.4)	(16.9 ± 0.5)
N_1	A	100	100.7 ± 2.8	102.3 ± 4.5	101.4 ± 6.6	103.1 ± 7.2	97.3 ± 5.1	96.3 ± 5.1	103.3 ± 6.4
	L	100	103 ± 1.8	107.7 ± 1.8**	107.2 ± 2.7*	110.7 ± 1.9**	108.2 ± 2.2**	103.3 ± 3.0	102.5 ± 1.9
	D	100	118.4 ± 5.7*	181.5 ± 17.4**	220.7 ± 22.4**	244.5 ± 26.9**	131.2 ± 10.1	103.5 ± 5.5	98.1 ± 2.8
P_sf	A	100	64.1 ± 7.6*	63.8 ± 5.8	50.2 ± 4.5	40.3 ± 3.5	69.7 ± 11.4*	90.5 ± 9.3	100.1 ± 6.8
	L	100	114.1 ± 2.8	116.4 ± 5.9*	130.3 ± 4.8**	141.1 ± 3.9**	126.4 ± 2.4**	106.6 ± 2.5	106.1 ± 2.0*
P_2S	A	100	92.7 ± 4.3	–[b]	–	–	83 ± 4.1**	98 ± 5.8	95.2 ± 4.7
	L	100	103.5 ± 1.5	–	–	–	115 ± 1.6**	106.8 ± 2.8	103.5 ± 3.1
N_2	A	100	28.9 ± 9.2**	–	–	–	36.8 ± 13.3**	90.7 ± 9.7	114.9 ± 9.1
	L	100	97.4 ± 2	–	–	–	109.3 ± 2.9*	103 ± 1.8	102.6 ± 0.7*
HSP	A	100	60.6 ± 9.6*	22.4 ± 6.5***	0***	0***	39.8 ± 6.7**	68.7 ± 4.7**	76.7 ± 8.4
	L	100	117.5 ± 4.8	142 ± 7.2*	–	–	149.6 ± 8.2**	113.4 ± 4.4*	109.7 ± 3.6

$n = 12$; mean ± SE. *$P < 0.05$; **$P < 0.01$; ***$P < 0.001$. All values are expressed as percentage of the control values at 10–20 min after the administration of 1%–4% sevoflurane. *A*, peak amplitude, *L*, peak latency, *D*, duration

[a] Sevoflurane concentration by dial setting (%) and in arterial blood (mM) (parentheses)

[b] A negative notch (N_2) dividing P_2f and P_2s disappeared after 2% sevoflurane, forming one peak thereafter. Therefore, the P_2s component could not be discriminated from the P_2f component at sevoflurane concentration greater than 2%

(Table 1). With 4% sevoflurane, the cortical evoked potentials were also completely suppressed, and P_2 was decreased in amplitude to 40% of the control value with a prolongation in latency, while the amplitude of N_1 was still not significantly changed (Table 1). The latency and duration of N_1 was increased at 4% of the drug (Table 1). No differential effects of sevoflurane on spinal cord potentials in response to fore- or hindpaw stimulations were found (data not shown). Then minutes after the end of sevoflurane inhalation, the amplitude of the cortical evoked potential (P_8-N_{13} peak-to-peak) was the first to recover (Table 1). Within 30 min after the end of sevoflurane inhalation, all evoked potentials recorded were recovered but a significant suppression of the cortical N_{13} and heterosegmental slow positive potential existed (Table 1). Sixty minutes after termination, all evoked potential values became insignificantly different from the control values (Table 1).

Discussion

The present results have demonstrated that sevoflurane has profound suppressive effects on the negative dip (N_2), the second component of P_2 of the segmental spinal cord potential, heterosegmental slow positive potential and cortical N_{13} component. By contrast, N_1 amplitude remained unchanged even with 4% sevoflurane, though its duration was prolonged more than two times with a slight increase in peak latency compared to that before inhalation. The first component of P_2(P_2f) was also suppressed in amplitude with prolongation in peak latency, but still kept its configuration even during 4% sevoflurane.

Initially a positive small spike (P_1) exists before N_1, but it is sometimes not visible, being masked by the negative-going deflection of N_1. P_1 is thought to be incoming afferent volleys along the roots [7, 8].

A slow positive wave (P_2) has been believed to be a primary afferent depolarization which is an index of presynaptic inhibition [9]. The P_2 wave usually contains two components, the first (P_2f) and the second (P_2s) waves, interposed by a negative dip (N_2). The second component (P_2s) has been suggested to reflect a primary afferent depolarization produced by a feedback loop via supraspinal structures [10, 14], because it disappears following spinal transection at the rostral level or under deep anesthesia [11]. We have previously demonstrated that heterosegmental slow positive potentials can be recorded from the cervical enlargement in response to hindpaw stimulation and from the lumbar enlargement in response to forepaw stimulation [11]. These potentials disappear following spinal transection as the P_2s wave does. Thus, it also is suggested to be a primary afferent depolarization exerted by descending impulses via the supraspinal structures.

The P_8 and N_{13} components of CEP in the rat have been variously interpreted with regard to their origins [5]. Although several origins may be involved in those components, P_8 and N_{13} may largely reflect a synchronized activity of the lemniscal and also reticular systems, respectively, judging from the positive-negative configurations and latencies [4].

The most profound suppressive effects of sevoflurane on the second component of P_2 with a negative dip (N_2) and HSP may indicate that primary afferent depolarization, a reflection of presynaptic inhibition, exerted by feed-back loops via supraspinal structures on the same or heterosegments, is vulnerable to sevoflurane. Similar results were also demonstrated in the effects of isoflurane on the spinal cord potentials in the rat [2, 13]. Similar profound suppression by the anesthetic of the cortical N_{13}, demonstrated in the present experiments, may indicate that both the specific and nonspecific cerebral activities are also sensitive to sevoflurane. Taken together, it is suggested that suppression by sevoflurane on cerebral as well as spinal presynaptic feed-back inhibitory activities progresses at a similar step.

A profound decrease in amplitude of the first component of the P_2 suggests that primary afferent depolarization activated segmentally is also suppressed substantially even during the light surgical stage of sevoflurane. On the other hand, the insignificant effect of sevoflurane on the N_1 component, particularly in its early phase, may indicate that segmentally activated second-order or interneuronal activities in the dorsal horn are minimally affected during the surgical stage of sevoflurane anesthesia. Thus, a minimal effect on segmentally activated interneuronal activities and profound effects on primary afferent depolarization may lead to excitation at the spinal level in terms of somatosensory transmission. However, widening of the N_1 wave in its later phase does not fully support the above thesis. The later phase of N_1 may reflect activities of smaller neurons, slower synaptic delays or smaller afferent fibers, which are probably suppressed sustantially even in the surgical stage of sevoflurane anesthesia.

A decrease in mean arterial blood pressure and heart rate to 48.5 ± 2.4 mmHg (mean $\pm$ SE) and 310.5 ± 11.1 bpm, respectively, was noticed during 4% sevoflurane. Our previous experiment [13], however, has shown that the decrease of mean arterial blood pressure to less than 40% of the control value does not have any significant effect on the evoked potentials. Therefore, the suppressive effects of sevoflurane on the evoked potentials as demonstrated in the present experiments are thought to be not due to hemodynamic changes. Some additive or potentiating affects of hypotension, however, should be considered in the deeper stage of sevoflurane anesthesia.

References

1. Inada Y, Ikeda K, Mori K, Morio M, Oyama T, Iwatski N, Suzuki H, Nagano M, Tanaka R, Fujita M, Miyake T, Shimoji K, Sato A (1987) Clinical evaluation of sevoflurane vs enflurane – a multicenter well controlled study. Jpn J Anesthesiol 36:875–889

2. Maruyama Y, Fujioka H, Sato Y, Taga K, Nishimura Y, Shimoji K (1987) The effect of isoflurane on evoked potentials from the brain and spinal cord in the rat. Jpn J Anesthesiol 36:1964–1970
3. Scheller MS, Tateishi A, Drummond JC, Zornow MH (1988) The effects of sevoflurane on cerebral blood flow, cerebral metabolic rate for oxygen, intracranial pressure, and the electroencephalogram are similar to those of isoflurane in the rabbit. Anesthesiology 68:548–551
4. Shaw NA (1985) Effect of electroconvulsive shock on the somatosensory evoked potential in the rat. Exp Neurology 90:566–579
5. Shaw NA, Cant BR (1981) The effect of pentobarbital on central somatosensory conduction time in the rat. Electroenceph Clin Neurophysiol 51:674–677
6. Shimoji K (1988) Human spinal cord potentials (SCPs): Ascending recording variations – an uptake. In: Ducker TB, Brown RH (eds) Neurophysiology and standards of spinal cord monitoring. Springer Berlin Heidelberg New York, pp 19–28
7. Shimoji K, Higashi H, Kano T (1971) Epidural recording of spinal electogram in man. Electroenceph Clin Neurophysiol 30:236–239
8. Shimoji K, Matsuki M, Shimizu H (1971) Wave-form characteristics and spatial distribution of evoked spinal electrogram in man. J Neurosurg 46:304–313
9. Shimoji K, Ito Y, Ohama K, Sawa T, Ikezono E (1975) Presynaptic inhibition in man during anesthesia and sleep. Anesthesiology 43:388–391
10. Shimoji K, Matsuki M, Ito Y, Masuko K, Maruyama M, Iwane T, Aida S (1976) Interactions of human cord dorsum potential. J. Appl Physiol 40:79–84
11. Shimoji K, Maruyama Y, Shimizu H, Fujioka H, Taga K (1985) Spinal cord monitoring – A review of current techniques and knowledge. In: Schramm J, Jones SJ (eds) Spinal cord monitoring. Springer, Berlin Heidelberg New York, pp 16–28
12. Shimoji K, Fujioka H, Maruyama Y, Shimizu H, Hokari T, Takada T (1988) Spinal cord potentials (SCPs) produced by descending volleys in the rat. In: Duker TB, Brown RH (eds) Neurophysiology and standards of spinal cord monitoring. Springer Berlin Heidelberg, New York, pp 73–81
13. Shimoji K, Fujiwara N, Fukuda S, Denda S, Takada T, Maruyama Y (1990) Effects of isoflurane on spinal inhibitory potentials. Anesthesiology 72:851–857
14. Tang AH (1969) Dorsal root potentials in the chloralose-anesthetized cat. Exp Neurol 25:393–400

The Effects of Anesthetics on the Ventral Root Potential in Spinal Cats

M. MATSUMOTO[1], H. KAWAMURA, and A. NAMIKI

Intraoperative monitoring of spinal cord function by means of spinal cord potentials was considered to be a simple and reliable technique. Some studies, however, pointed out that neither spinal evoked potential (SEP) nor somatosensory evoked potentials could detect a perioperative dysfunction of spinal motor activity. The ventral root potential (VRP) was introduced experimentally as a monitor of spinal injury during surgical procedures [1] because it is believed to be a more sensitive indicator of spinal motor activity than the other techniqus, including SEP. As the VRP is so sensitive that it is reasonable to think some anesthetics will considerably affect monitoring of VRP, we studied the effects of clinically used anesthetics on the VRP in cats.

Methods

Thirty-nine cats, body weight ranging from 3.0–4.0 kg, were used in this experiment (Fig. 1). Tracheostomy was done under halothane anesthesia, and the animal was immobilized by pancuronium bromide. End-tidal concentration of carbon dioxide was maintained from 4.5%–5.5% by means of artificial ventilation. A cannula was inserted into the femoral vein for intravenous administration of drugs. Blood pressure was continually monitored via a cannula inserted into one femoral artery and attached to a pressure transducer.

Body temperature was maintained between 37°C and 38°C by means of a thermostatically controlled heating blanket. Urine output was maintained within normal limits by controlling the volume of infusion. Bilateral pneumothorax was made so that respiratory movements of the spinal cord should be minimal.

Following a laminectomy extending from C1–C4, T4–T7, and L4–S1 vertebrae, the spinal cord was transected at C1. The exposed spinal cord was

[1] Department of Anesthesiology, Sapporo Medical College, South-1, West-16, Sapporo, 060 Japan

K. Shimoji et al. (Eds.)
Spinal Cord Monitoring and Electrodiagnosis

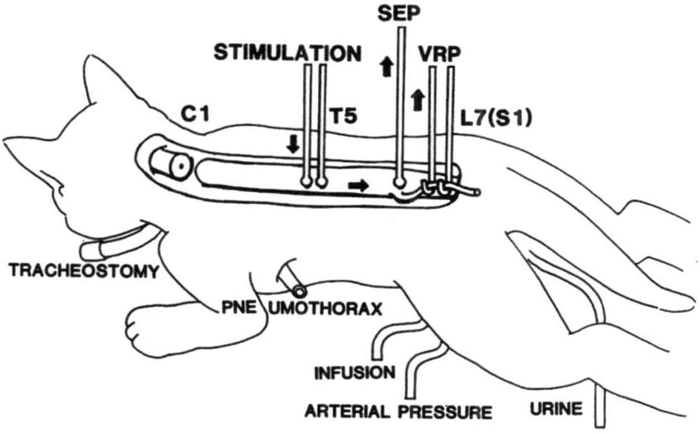

Fig. 1. Preparation for the recording of VRP and SEP. All surgical procedures were done under halothane anesthesia

covered with a pool of paraffin oil maintained at 36°C–37°C with the aid of an infrared lamp.

Measurement of the VRP was started at least 4 hours after the end of operation and anesthesia. The spinal cord was stimulated by means of silver ball bipolar electrodes (electrode distance: 8 mm) placed longitudinally at the T5 dorsal root entry zone with supramaximal square wave stimuli of 0.2-ms duration. Ventral root S1 or L7 was cut ipsilaterally, and the central end of S1 or L7 was mounted up on bipolar platinum electrodes for recording of the VRP. The SEP was also recorded with a silver monopolar electorode placed on the 7th lumbar or 1st sacral dorsal root entry zone. Recording of both VRP and SEP was done with Sanei Medelec M S-6 (Nihondenki) system. Control evoked potentials were recorded for at least 15 min to ensure stability of the response before drugs were given.

Inhalational Anesthetics Study

One of the following halogenated inhalational anesthetics was given for a duration of 30 min. Spontaneous recovery was also observed for more than 30 min after the end of anesthesia. The VRP and SEP were measured every 5 min for at least 1 h.

Halothane	0.5%	(n = 3)	1.0%	(n = 3)
Enflurane	0.25%	(n = 3)	0.5%	(n = 3)
Isoflurane	0.5%	(n = 3)	1.0%	(n = 3)
Sevoflurane	0.5%	(n = 3)	1.0%	(n = 3)

Each cat was given only one concentration and one kind of anesthetic.

Intravenous Anesthetics Study

One of the three following drugs was given intravenously into the femoral vein. The VRP and SEP were recorded 5 min after the injection and every 5 min for 30 min. The last recording of the VRP and SEP was done 60 min after the injection to confirm a spontaneous recovery.

Fentanyl	10 mcg/kg	($n = 3$)	25 mg/kg	($n = 4$)
Ketamine	1 mg/kg	($n = 2$)	10 mg/kg	($n = 2$)
Thiamylal	1 mg/kg	($n = 2$)	10 mg/kg	($n = 2$)

Results

Figure 2 shows the VRP and SEP with the peak latency and amplitude that we used arbitrarily in this study.

Inhalational Anesthetics Study

No significant change was seen in the amplitude and peak latency of the SEP. Peak latency of the VRP also showed no change by administration of inhalational anesthetics. An example of a suppression of the amplitude of VRP induced by inhalation of halothane, 1%, is shown in Fig. 3. Within 5 min after inhalation of halothane the amplitude of the VRP began to decrease. A suppression of the amplitude of the VRP continued more than 30 min after the end of inhalation of halothane, and it took more than 30 min to recover completely. The percent suppression of the amplitude of VRP induced by inhalation of anesthetics is summarized in Fig. 4. The suppressive effect of halogenated anesthetics on the amplitude of VRP seemed to be concentration related. In all anesthetics (i.e., halothane, enflurane, isoflurane, and sevoflurane), percent amplitude of VRP began to decrease within 5 min of inhalation. The spontaneous recovery is delayed when the concentration of anesthetics increased.

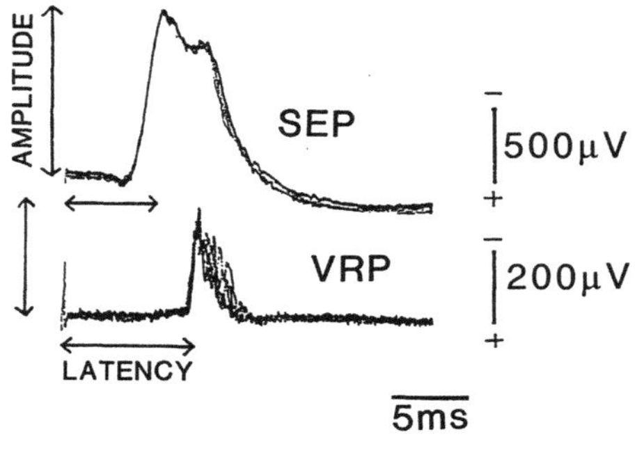

Fig. 2. Peak latency and amplitude of SEP and VRP. SEP and VRP were recorded simultaneously

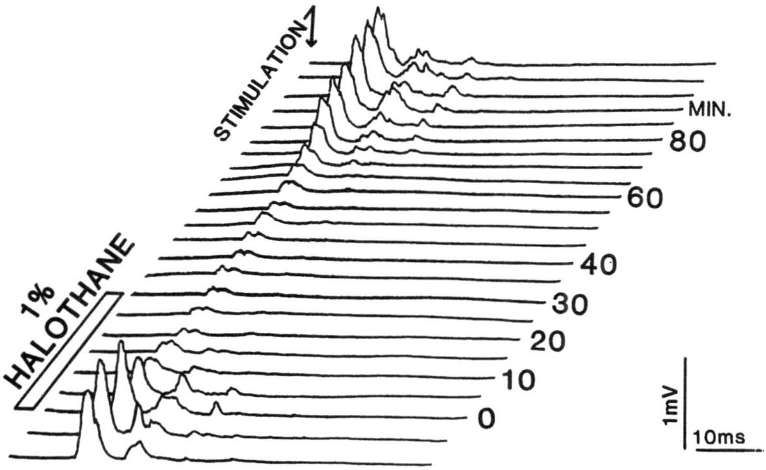

Fig. 3. Suppression of VRP by inhalation of halothane, 1%, for 30 min. Within 5 min of inhalation, the amplitude of VRP began to decrease. Spontaneous recovery was seen about 30 min after anesthesia

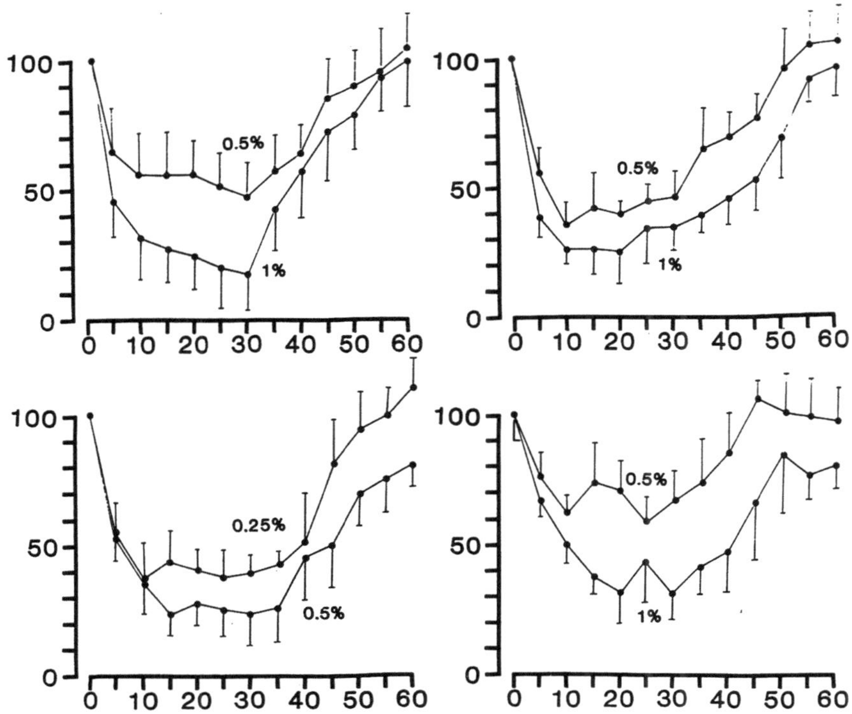

Fig. 4. Suppression (percent amplitude; mean + SD) of VRP induced by inhalation of halogenated anesthetics (halothane, *top left*; enflurane, *bottom left*; isoflurane, *top right*; sevoflurane, *bottom right*; *X-axis*, Time (min.); *Y-axis*, % amplitude of VRP. Anesthetics were given for 30 min

Arterial blood pressure was slightly decreased with inhalation of any concentrations of halogenated anesthetics. None of the subjects, however, showed a mean blood pressure under 80 mmHg.

Intravenous Anesthetics Study

In this study the SEP and the peak latency of the VRP did not show any significant changes by the administration of intravenous anesthetics. Fentanyl and thiamylal in both concentrations tested decreased the amplitude of the VRP in all cats. The suppression induced by fentanyl is shown in Fig. 5 and the results are summarized in Fig. 6. In the fentanyl study, a significant suppression of the amplitude of the VRP was observed 5 min after the administration of fentanyl, and almost complete spontaneous recovery is also seen within 30 min.

Discussion

While the SEP is believed to represent the activities of spinal cord dorsal roots [3], the origin of the VRP is not fully understood. Kawamura et al [1] reported that as the frequency of stimuli increased from 0.1 Hz to 100 Hz, the VRP elicited by the stimuli gradually decreased and was almost completely suppressed at a frequency of 10 Hz. They also reported that resection of the lateral or dorsolateral column of spinal cord made the amplitude of VRP decrease. These data suggest that the VRP includes synaptic activities and probably originates from the spinal motor system. As the synapse is now known to be the place that anesthetics exert their suppressive effects,

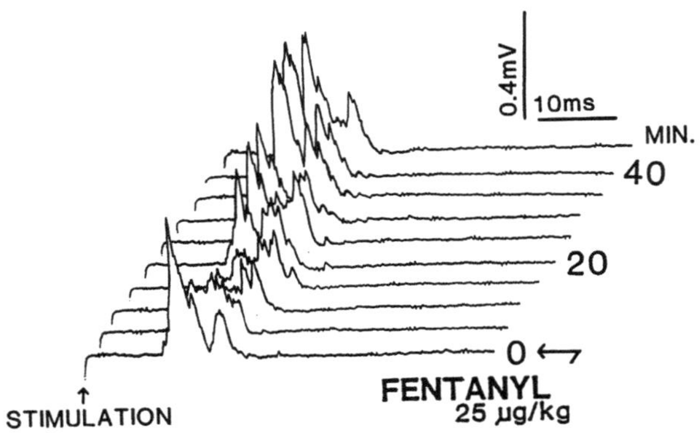

Fig. 5. Suppression of VRP induced by the intravenous administration of fentanyl, 25 mcg/kg. Spontaneous recovery was observed within 10 min after the injection of the drug

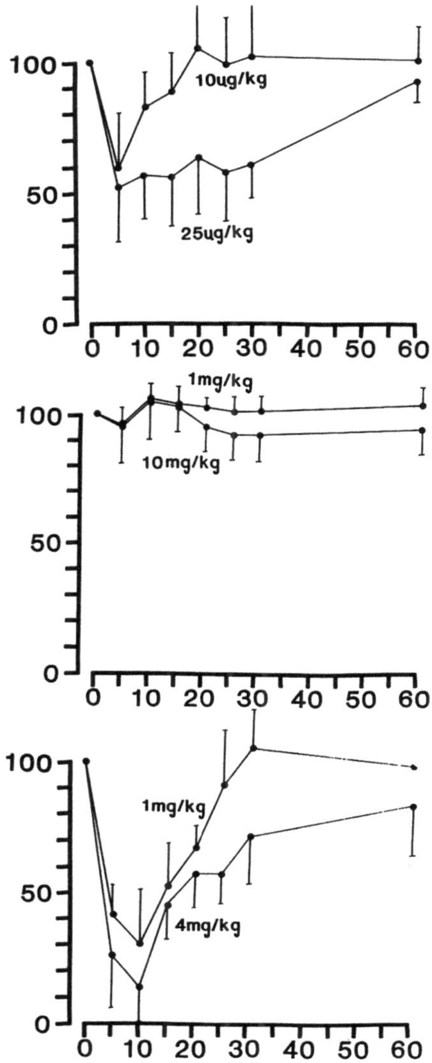

Fig. 6. The effect of intravenously administered fentanyl (*top*), ketamine (*middle*), and thiamylal (*bottom*) on percent amplitude of VRP (mean + SD). *X-axis*, Time (min); *Y-axis*, % amplitude of VRP. Ketamine, even 10 mg/kg, did not suppress the amplitude of VRP

it is easily understood that the VRP is suppressed by the administration of anesthetics. The present study demonstrates that some kinds of anesthetics produce a suppressive effect on the VRP.

Many anesthetics drugs, such as halogenated anesthetics, are known to lower the blood pressure. In this study suppression of the VRP is not due to decreased blood pressure because we previously reported that the amplitude of the VRP is not suppressed until the mean blood pressure is under 60 mmHg [2].

Thus, in conclusion, all halogenated volatile anesthetics and some kinds of intravenous anesthetics depress the VRP, which is believed to represent spinal motor functions. It is suggested that the selective action of inhalation anesthetics in the spinal cord probably contributes to its muscle relaxing activity.

References

1. Kawamura H, Matsumoto M (1987) The origin of ventral root potentials in cats. J Electrodiag Spinal Cord 10:23–26
2. Matsumoto M, Kawamura H (1987) The effects of halothane and enflurane on the SEP (spinal evoked potentials) and VRP (ventral root potentials). J Electrodiag Spinal Cord 10:39–42
3. Szilagyi DE, Hageman JH, Smith RF, et al. (1978) Spinal cord damage in surgery of the abdominal aorta. Surgery 83:38–56

The Effects of Anesthetics on Spinal Cord Nociceptive Neural Activities

K. SHINGU[1], M. OSAWA, and K. MORI

Abstract

The effects of several anesthetics on spinal cord nociceptive neural mechanisms and their interactions with naloxone were studied in acute spinal cord transected cats. Intraarterial injection of bradykinin was used to stimulate chemical nociceptors. Spontaneous activity and the bradykinin-induced responses were recorded by the multiunit activity technique in the lateral funiculus. Thiamylal, nitrous oxide, halothane and diethylether suppressed the bradykinin-induced response considerably, but it was not antagonized by naloxone, 0.1–2.0 mg/kg iv. Enflurane had little effect on the bradykinin-induced response. The effects of these anesthetics on the spontaneous activity was divergent: nitrous oxide enhanced while the other anesthetics suppressed it to various degrees. Fentanyl, 30 μg/kg iv, suppressed both the bradykinin-induced response and the spontaneous firing. These effects were antagonized completely by naloxone, 0.1 mg/kg iv. All these findings indicate that anesthetics, except enflurane, suppress nociceptive neural mechanisms in the spinal cord and that their actions are not exerted through the opioid receptor-endorphin system.

Introduction

Since the discovery of the opioid receptors and endorphins, the interaction of general anesthetics with the opioid receptor-endorphin system has become one of the major interests in the neurophysiology of anesthesia. The conclusions, however, are controversial. Naloxone reversal has been reported in the case of EEG changes and behavioral depression induced by cyclopropane, halothane and enflurane [2]. In contrast, this was not the case with

[1] Department of Anesthesia, Faculty of Medicine, Kyoto University, 54 Kawahara-cho, Shogoin, Sakyo-ku, Kyoto, 606 Japan

K. Shimoji et al. (Eds.)
Spinal Cord Monitoring and Electrodiagnosis

the minimum alveolar concentration (MAC) of halothane [4] and the loss of righting reflex induced by nitrous oxide or halothane. Further, stereospecific binding of ketamine with opioid receptors in the rat brain has been reported, while Fratta et al. [3] could not confirm these findings.

The evaluation of the analgesic action in the studies cited above was based mainly on behavioral observations. However, the analgesic action is only one component, various other CNS actions involved would be expected to modify the behavioral expressions of nociception during anesthesia, e.g., depression of the motor system would decrease the motor response to noxious stimulation even with little reduction in the afferent activity related to pain. In the present study, we stimulated nociceptors by intraarterial injection of bradykinin and measured spinal cord nociceptive cell responses. We obtained conclusive evidence of the absence of interaction of anesthetics with the opioid receptor-endorphin system. Most parts of this study have been published in a previous report [8].

Materials and Methods

Thirty-eight cats of either sex, weighing 2.4–4.7 kg, were used. Details of the surgical procedures, recording techniques of electrical activities and other preparations were described previously [7].

Following intravenous administration of thiamylal, 10 mg/kg iv, and alcuronium, 10 mg iv, the cats were intubated, anesthetized with a mixture of 1% halothane and 75% nitrous oxide in oxygen. A polyethylene cannula, 0.5 mm in diameter, was introduced into the left femoral artery and fixed so that the tip was just distal to the bifurcation. The animals were decerebrated by electrocoagulation, laminectomy was performed at C1–2 and L1–4, and the dura was opened at L1–4 to expose the spinal cord. Cervical cord transection was performed with a bipolar electrocoagulator at C1, following an additional administration of 10 mg/kg thiamylal. Norepinephrine was then infused continuously to maintain the arterial blood pressure at near the control level. A sharp tapered elgiloy electrode, insulated with glass except for 0.5–1.0 mm at the tip, was inserted into the left lateral funiculus, L1–3, and then the exposed spinal cord was covered with warm liquid paraffin at 36°C–38°C. The reference electrode was placed at the dorsal spinal process of T12. Firing of axons in the lateral funiculus was measured using the multiunit activity technique described previously [7].

After completion of all surgical preparations, halothane-nitrous oxide anesthesia was terminated and the experiment was begun 2 h later when the multiunit recording was stable and a constant magnitude of bradykinin-induced response had been obtained by repeated bradykinin injections. Bradykinin was dissolved in saline (100 μg/ml) and 0.1 ml of the solution was injected rapidly into the femoral artery through the cannula. The injection of bradykinin was performed at 10-min intervals. Following three to four recordings of control bradykinin-induced responses, the drug effects were examined. The level of spontaneous activity was measured by the level of multiunit recording 2 min prior to intraarterial injection of bradykinin, and the magnitude of bradykinin-induced response was measured by subtracting the level of spontaneous activity from the peak of excitation induced by bradykinin injection.

The animals were divided into six groups, according to drugs studied, as follows:

Group 1 (fentanyl, eight cats)
Fentanyl, 30 μg/kg iv, was administered initially. After complete recovery of spontaneous activity and bradykinin-induced response, naloxone, 0.1 mg/kg iv, was administered and an identical drug study was done.

Group 2 (thiamylal, five cats)
The same study as for fentanyl was done in five cats. The dose of thiamylal was 5.0 mg/kg iv.

Group 3 (nitrous oxide, seven cats)
Nitrous oxide, 75% in oxygen, was administered for 30 min initially. After recovery of the spinal cord activities to the control levels, naloxone, 2.0 mg/kg iv, was administered and an identical drug study was done.

Group 4 (halothane, five cats)
Halothane, 1.0% in oxygen, was administered, and naloxone, 0.5 mg/kg iv, was administered at 65 min during the halothane administration. The effects of naloxone were evaluated by comparing the spinal cord activities before and after naloxone administration.

Group 5 (ether, six cats)
Ether, 10% in oxygen, was administered, and naloxone, 0.5 mg/kg iv, was administered at 95 min during ether inhalation. The effects of naloxone were evaluated using the same method as in group 4.

Group 6 (enflurane, seven cats)
Euflurane, 2% in oxygen, was administered, and naloxone, 1 mg/kg iv, was administered at 65 min during enflurane inhalation. The effects of naloxone were evaluated using the same methods as in group 4.

All gaseous anesthetics were administered by the nonrebreathing method and with a mechanical ventilator (Acoma AR 100). Statistical analysis was performed using repeated measures analysis of variance for the effects of anesthetic agents and the paired Student's *t* test for the effects of naloxone.

Table 1. Anesthetic action on the activities of spinal cord lateral funiculus and their modification by naloxone[a]

	Spontaneous activity Mean ± S.E.M. (%)	p	p*	Bradykinin-induced response Mean ± S.E.M. (%)	p	p*
Fentanyl 30 μg/kg	−31.3 ± 3.3 − 6.7 ± 3.9	0.001 n.s.	0.01	−76.7 ± 8.3 + 5.8 ± 7.5	0.001 n.s	0.001
Thiamylal 5mg/kg	−41.5 ± 4.1 −33.8 ± 2.2	0.001 n.s.	n.s.	−42.7 ± 7.4 +39.7 ± 7.9	0.01 0.01	n.s.
Nitrous oxide 75%	−10.4 ± 2.3 + 6.6 ± 3.8	0.01 n.s.	n.s.	−18.4 ± 2.6 −15.1 ± 3.3	0.001 0.01	n.s.
Halothane 1%	−18.1 ± 2.8 − 9.4 ± 7.0	0.01 n.s.	n.s.	−44.5 ± 7.9 −43.8 ± 7.2	0.01 0.01	n.s.
Ether 10%	−34.4 ± 11.0 −22.3 ± 14.6	0.05 n.s.	n.s.	−54.8 ± 4.9 −60.6 ± 5.3	0.001 0.001	n.s.
Enflurane 2%	−36.6 ± 6.1 −17.7 ± 8.0	0.001 n.s.	n.s.	−10.3 ± 4.7 − 8.4 ± 6.2	n.s. n.s.	n.s.

* P values for the significance of naloxone effect
[a] Percent changes (− decrease, + increase) compared with control values of spontaneous activity and bradykinin-induced response. In each agents, the drug effect only (*upper*) and that following naloxone administration (*lower*) are presented

Results

The results are summarized in Table 1.

Effects of Naloxone in the Control Study

The effects of naloxone were obtained from 17 cats in groups 1–3. Different doses of naloxone did not produce apparently significant differences, and the data were pooled. Naloxone, 0.1–2.0 mg/kg iv, induced activation of spontaneous activity by 24.9% ± 5.0% of control (mean ± S.E.M., $p < 0.001$) and depression of bradykinin-induced response by 6.6% ± 3.1% of control at 5 min after administration. The effects of naloxone on the bradykinin-induced response were not statistically significant.

Effects of Fentanyl and Naloxone (Group 1)

Fentanyl, 30 μg/kg iv, depressed both spontaneous activity ($p < 0.001$) and the bradykinin-induced response ($p < 0.001$) at 3 min after administration. After pretreatment with naloxone, the changes became negligible.

Effects of Thiamylal and Naloxone (Group 2)

Thiamylal, 5 mg/kg iv, depressed spontaneous activity ($p < 0.001$) and the bradykinin-induced response ($p < 0.01$) at 3 min after administration. After pretreatment with naloxone, the same dose of thiamylal depressed spontaneous activity ($p < 0.001$) and the bradykinin-induced response ($p < 0.01$). The differences in depressive actions before and after the naloxone pretreatment were not statistically significant.

Effects of Nitrous Oxide and Naloxone (Group 3)

Administration of 75% nitrous oxide for 30 min activated spontaneous activity ($p < 0.01$), but depressed the bradykinin-induced response ($p < 0.001$). Pretreatment with naloxone did not change the time course of the effects of nitrous oxide statistically. Spontaneous activity was activated (n.s.) but the bradykinin-induced reponse was depressed ($p < 0.01$) at 30 min of inhalation. Changes in the effects of nitrous oxide induced by naloxone pretreatment were not statistically significant.

Effects of Halothane and Naloxone (Group 4)

Administration of halothane, 1% in oxygen for 60 min, depressed spontaneous activity ($p < 0.01$) and the bradykinin-induced response ($p < 0.01$). Naloxone reversed spontaneous activity partially, but had little effect on the bradykinin-induced response. The changes in halothane actions induced by naloxone were not statistically significant.

Effects of Ether and Naloxone (Group 5)

Administration of ether, 10% in oxygen for 90 min, depressed spontaneous activity ($p < 0.05$) and the bradykinin-induced reponse ($p < 0.001$). Naloxone reversed spontaneous activity, but had little effect on the bradykinin-induced response. Changes induced by naloxone in the depressant actions of ether were not statistically significant.

Effects of Enflurane and Naloxone (Group 6)

Administration of enflurane, 2% in oxygen for 60 min, depressed spontaneous activity ($p < 0.001$), but induced no change of the bradykinin-induced response. Naloxone reversed the depression of spontaneous activity, but had little effect on the bradykinin-induced reponse. The changes in enflurane actions induced by naloxone were not statistically significant.

Discussion

The present study confirmed the depression of bradykinin-induced spinal cord response with all agents except enflurane. Further, a complete reversal of depressant actions by naloxone was confirmed only in the case of fentanyl, and reversals in cases of anesthetic agents were not statistically significant.

If the depressions of bradykinin-induced responses in the present study were mediated by a direct interaction of anesthetics with the opioid receptor-endorphin system, the depression should be antagonized by the opioid antagonist naloxone. The failure of this antagonism indicates the fundamental difference in pharmacological mechanism between opioids and anesthetics. In a previous report [5], nitrous oxide was shown to have a dual mode of antinociceptive action, similar to that seen with opioids. However, as was shown, a direct interaction with the opioid receptor-endorphin system was ruled out in the present study. The possible absence of interaction of ketamine with the opioid receptor-endorphin system was discussed in a previous report [9] and was not examined in the present study.

We found that naloxone activates some CNS structures not directly related to nociception: this compound activated spontaneous activity in the lateral funiculus, yet had no effect on the bradykinin-induced response either in the control state or following administration of anesthetics. Harper et al. [4] reported that a high dose of naloxone (100 mg/kg iv) induced convulsions in untreated rats. There is a possibility that these CNS excitatory actions not related directly to the nociceptive neural mechanisms lighten the level of anesthesia and are mistaken for reversal of analgesic actions of anesthetics.

One of us has reported that one group of anesthetics, such as ketamine, nitrous oxide, ether and cyclopropane, induces activation of the neuronal firing of the brain-stem reticular core [6]. Although the present study ruled out the direct interaction of these agents with the opioid receptor-endorphin system, there still remains the possibility that the CNS excitatory actions of these agents involve this antinociceptive system trans-synaptically, as in the case of stimulation-produced analgesia, in which stimulation of the midbrain periaqueductal gray produced analgesia. Akil et al. [1] reported a total or partial reversal of this stimulation-produced analgesia by intravenous administration of naloxone, but Yaksh et al. [10] reported the absence of the reversal. These findings indicate that in stimulation-produced analgesia, the contribution of the opioid receptor-endorphin system is only partial and that other neural systems are probably activated electrically or trans-synaptically. If anesthetics are indeed related to this system, then the effects would be indirect, nonspecific and trans-synaptic.

References

1. Akil H, Mayer DJ, Liebeskind JC (1976) Antagonism of stimulation-induced analgesia by naloxone, a narcotic anatagonist. Science 191:961–962
2. Finck AD, Ngai SH, Berkowitz BA (1977) Antagonism of general anesthesia by naloxone in the rat. Anesthesiology 46:241–245
3. Fratta W, Casu M, Balestrieri M, Loviselli A, Biggio G, Gessa GL (1980) Failure of ketamine to interact with opiate receptors. Eur J Pharmacol 61:389–391
4. Harper MH, Winter PM, Johnson BH, Eger EI (1978) Naloxone does not antagonize general anesthesia in the rat. Anesthesiology 49:3–5
5. Komatsu T, Shingu K, Tomemori N, Urabe N, Mori K (1981) Nitrous oxide activates the supraspinal pain inhibition system. Acta Anaesth Scand 25:519–522
6. Mori K, Winters WD (1975) Neural background of sleep and anesthesia. Int Anesth Clin 13:67–108
7. Mori K, Komatsu T, Tomemori N, Shingu K, Urabe N, Seo N, Hatano Y (1981) Pentobarbital-anesthetized and decerebrate cats reveal different neurophysiological responses in anesthetic-induced analgesia. Acta Anaesth Scand 25:349–354
8. Shingu K, Osawa M, Omatsu Y, Komatsu T, Urabe N, Mori K (1981) Naloxone does not antagonize the anesthetic-induced depression of nociceptor-driven spinal cord response in spinal cats. Acta Anaesth Scand 25:526–532
9. Tomemori N, Komatsu T, Shingu K, Urabe N, Seo N, Mori K (1981) Activation of the supraspinal pain inhibition system by ketamine hydrochloride. Acta Anaesth Scand 25:355–359
10. Yakah TL, Yeung JC, Rudy TA (1976) An inability to antagonize with naloxone the elevated nociceptive thresholds resulting from electrical stimulation of the mesencephalic central gray. Life Sci 18:1193–1198

Monitoring the Expression of the C-*fos* Proto-Oncogene to Study the Spinal Circuits that Contribute to Pain and its Control*

A.I. Basbaum[1], R. Presley, S.-I. Chi, K.R. Gogas, and J.D. Levine

There are several traditional approaches to analyzing the organization of spinal cord neurons that contribute to the processing of nociceptive information and to the control of pain. Electrophysiological studies have established that dorsal horn neurons in laminae I, II and V contain nociresponsive cells, and iontophoretic studies have demonstrated that various neurotransmitter substances can either excite (e.g., substance P), or inhibit, (e.g., enkephalin and GABA) the firing of nociresponsive neurons [5]. Such studies, however, are rarely performed under conditions in which the behavior of the animal can be monitored and correlated; in fact, most of these studies are performed under anesthesia. Thus, it is hard to determine whether the firing of any of these cells is necessary for the generation and ultimate appreciation of pain. Although pharmacological studies provide the necessary correlative behavioral data, such studies do not provide information about the neural circuits through which the different drug effects are mediated.

Recently, a novel approach to the problem has been developed. The method beautifully complements the information provided in electrophysiological and pharmacological studies. Specifically, Hunt et al. [6] reported that noxious stimulation of the periphery increases the expression of the c-*fos* proto-oncogene in spinal cord neurons. C-*fos* is the cellular homologue of a viral oncogene, v-*fos*, the expression of which results in the development of osteosarcoma in the FBJ mouse. Induction of the cellular form of the gene leads to the expression of a 55-Da nuclear phosphoprotein, *fos*, that can be localized in neurons by immunocytochemistry with antisera directed against the *fos* protein. An increase in *fos* messenger RNA can be detected within 15 min of stimulation; the protein can be detected within 30 min. C-*fos* is an early response gene whose induction is presumed to regulate the induction of numerous genes, most of which are as yet undefined. Although we do not know the significance of finding neurons which express *fos*, its use as a marker

* This work was supported by NS 21445 and NS 14627

[1] Department of Anatomy, University of California San Francisco, San Francisco, CA 94143, USA

K. Shimoji et al. (Eds.)
Spinal Cord Monitoring and Electrodiagnosis

of neuronal activity is unequaled by existing techniques. In this report, I will review our recent studies of the organization of nociceptive transmission systems in the spinal cord of rats and the regulation of transmission by opiate analgesics using *fos* immunocytochemistry to monitor neural activity.

We used several different noxious peripheral stimulation protocols in rats: hindpaw injection of Freund's adjuvant to produce a delayed inflammation of the paw; noxious visceral stimulation, with dilute acetic acid injection; formalin injection into the paw. The latter is of particular value because it evokes a characteristic and quantifiable behavioral syndrome that is presumed to be indicative of pain. The experimental procedure is to stimulate the animals and after different times perfuse the animal with fixative and prepare the tissue for *fos* immunocytochemistry. With the exception of the Freund's adjuvant studies, the rats were killed 2 h after presentation of the noxious stimulus.

Our initial studies [8] emphasized the large spatial extent of activity generated by a relatively restricted peripheral noxious stimulus to the hindpaw of the rat (Fig. 1). As expected, the majority of cells were labeled in the L4–5 segment of the cord, which receives its innervation from the hindpaw. Many neurons, however, could be detected throughout the lumbar enlargement and into the sacral cord. Neurons in laminae V through VIII had the most extensive rostrocaudal spread; neurons in laminae I and II were restricted to the L3–6 segments. This suggests that the brain "receives" a pattern of activity made of up large populations of neurons extensively distributed. Importantly, we also demonstrated that there is a significant effect of general anesthetic on the distribution of *fos* positive neurons. Barbiturate anesthetic profoundly depressed the labeling of neurons in the deeper laminae of the dorsal horn and in laminae VII and VIII of the ventral horn. These latter studies emphasized the importance of performing these experiments in awake animals. Our next studies asked whether the pattern of labeling differs with different noxious stimuli. Using noxious visceral stimulation, we, in fact, found that the labeling in the superficial dorsal horn is considerably more confined to lamina I; very few cells were found in the substantia gelatinosa (as they were with the adjuvant, or with formalin injection). The rostral-caudal pattern was also considerably different; the greatest spread was found in lamina I, not in the ventral horn.

Clearly very large numbers of cells are activated by relatively restricted noxious peripheral stimuli. Are these interneurons, projection neurons, or both? To address this question, we used a double-labeling technique in which the projection neurons of the spinal cord, including spinoreticular and spinothalamic tract neurons, were labeled with a retrograde marker recently developed in our laboratory [2]. Specifically, 2 weeks before the noxious stimulation experiments began, we injected a protein-gold retrograde tracer complex, namely colloidal gold-labeled wheatgerm agluttinin conjugated to enzymatically inactive horseradish peroxidase (WGAapoHRP-Au) into various targets in the brain, including the thalamus, parabrachial region/

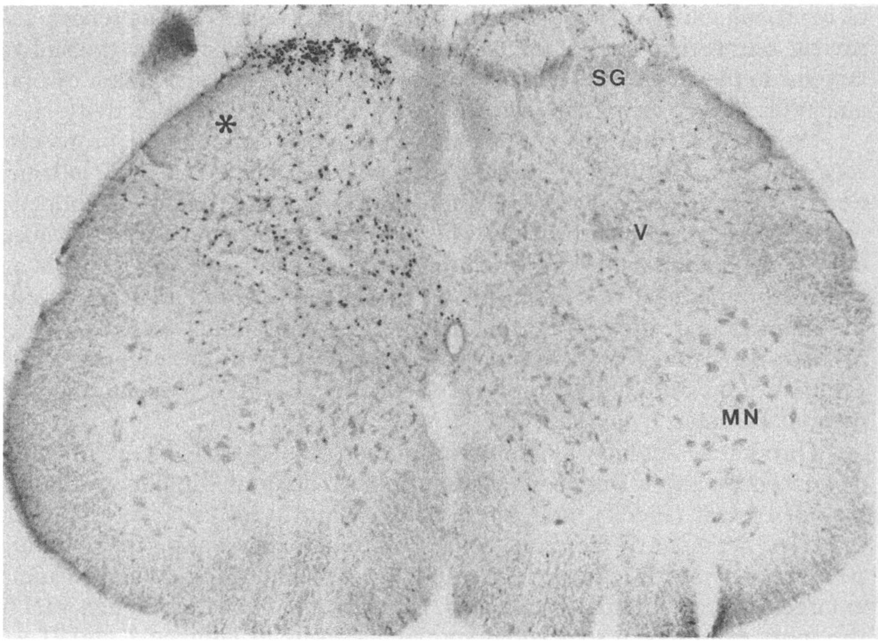

Fig. 1. Pattern of labeling in the lumbar spinal cord after plantar injection of formalin into the hindfoot of the rat. There is very dense staining of *fos*-immunoreactive neurons in the superficial dorsal horn, in the neck of the dorsal horn (lamina V) and in the ventral horn. Note that with the exception of medial ventral horn, there is minimal labeling contralateral to the stimulated foot. Also note the absence of motoneuron labeling. *Asterisk* indicates lateral region of substantia gelatinosa, which is not labelled; this is consistent with the topographic projection of the hindpaw to the medial SG. *SG*, substantia gelatinosa; *V*, lamina V of the dorsal horn; *MN*, motoneurons of the ventral horn

nucleus cuneiformis, and medullary reticular formation, i.e., the major targets of spinal projection neurons. The colloidal gold in the tracer is identified by silver intensifying sections from perfusion-fixed animals. This produces a dense silver precipitate in the cytoplasm of neurons which retrogradely transported the marker intraxonally. The same sections can then undergo immunostaining to determine whether they also expressed *fos* in response to the noxious stimulation. Double-labeled cells contain silver precipitate in the cytoplasm and a *fos*-immunoreactive nucleus (Fig. 2). Thus 2 weeks after the tracer was injected, we stimulated the rats with Freund's adjuvant, allowed the animals to survive for the appropriate time (in this case 16 h) and then prepared tissue for silver intensification, followed by *fos* immunocytochemistry. Our mapping studies revealed that the vast majority of *fos* positive neurons do *not* project to the brain; however, in each of the laminae different proportions of projection neuron could be demonstrated. The largest percentage was found in lamina I.

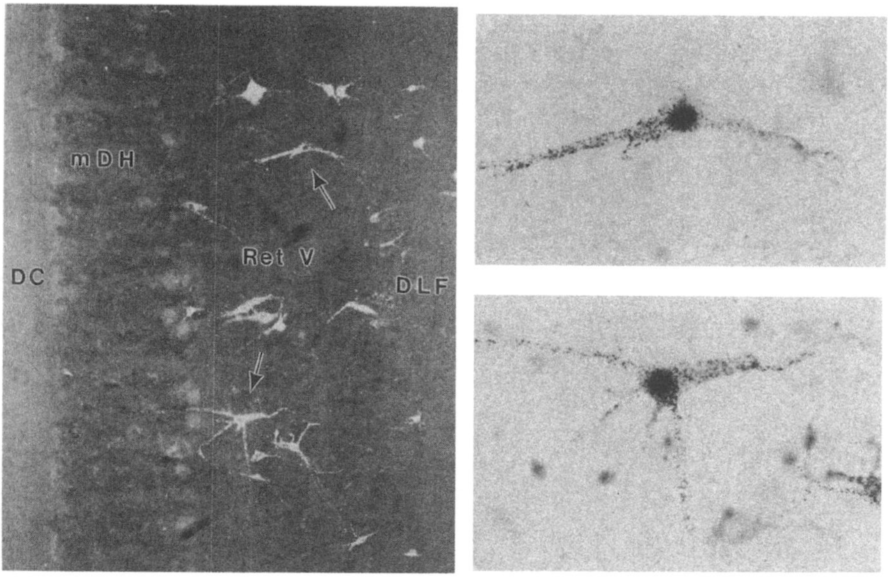

Fig. 2. *Fos*-immunoreactive projection cells. On the *left* is a dark field photograph of a horizontal section through the reticular part of the neck of the dorsal horn (*Ret V*). The section contains many retrogradely labeled cells. The retrograde tracer (WGAapoHRP-Au) was injected into various brainstem and thalamic sites. (*mDH*, medial part of the dorsal horn; *DC*, dorsal column white matter; *DLF*, dorsolateral funiculus white matter). The *arrows* identify the two double-labeled cells which are illustrated in bright field on the *right*. The silver precipitate is densely distributed through the cytoplasm of the retrogradely labeled cells. These cells also expressed the *fos* protein in response to peripheral noxious stimulation. The *fos*-immunoreactive nucleus is heavily labeled and is readily distinguished from the cytoplasmic silver label. (From Menétrey et al. [8])

Our next series of studies addressed the regulation of *fos* expression by different analgesic drugs and different routes of drug administration. We first examined the effect of systemic administration of morphine in the formalin test. Briefly rats received a subcutaneous injection of morphine 20 min prior to receiving a subcutaneous injection of dilute formalin in the plantar surface of the hindpaw. As described above, the formalin stimulus evokes a characteristic behavior pattern that is presumed to result from both acute and persistent noxious stimulation. In fact, a great advantage of this test is that it is considered to be a test of tonic pain. Such tests are necessary if one is to try to model pain that is more common in clinical situations. We found that morphine produced a dose-dependent suppression of *fos* expression in the spinal cord. The pattern of suppression, however, was not uniform. Thus, we found almost complete suppression of activity in neurons of the ventral horn at the lowest doses which produced analgesia. At these same doses labeled neurons always persisted in the superficial dorsal horn. Even at 10 mg/kg sc, a dose which completely blocked the formalin-evoked behavior,

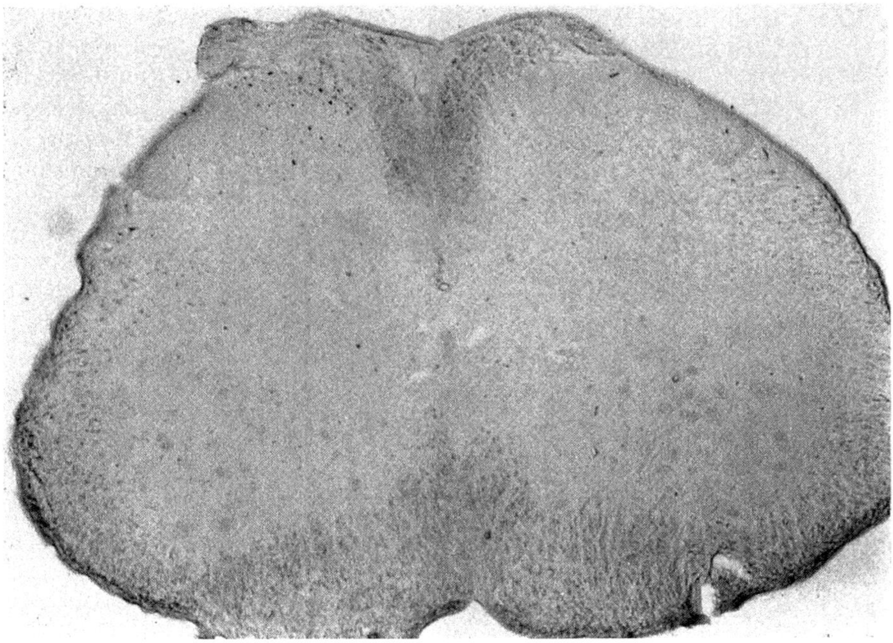

Fig. 3. Lumbar cord of a rat which received morphine (10 mg/kg; sc) 20 min before being injected with the formalin. Note the dramatic loss of label in all regions of the cord except for the superficial dorsal horn where some staining persists. Despite the persistence of staining in laminae I and II, this animal was demonstrably analgesic

labeled neurons could still be detected in laminae I and II (Fig. 3). Analgesic doses of morphine not only reduced the expression of the c-*fos* gene in the L4–5 segment, but eliminated the extensive rostral caudal spread of neurons in laminae V through X. This indicates that under conditions in which the animal is rendered analgesic, the brain receives a much more limited input from the spinal cord. Taken together with the fact that activity in deeper cells only appeared after 2 h of stimulation, these data suggest that the persistent noxious input (which is provided by the formalin stimulus and which presumably is essential for the generation of persistent, or chronic pain) results from activity in neurons of the deeper parts of the dorsal horn and of the ventral horn. In contrast, activity in the more superficial layers may be more relevant to the transmission of information related to more localized, acute noxious stimuli.

The mechanism(s) through which systemic narcotics exert their analgesic effect and suppress *fos* staining a spinal neurons remains to be determined. There is considerable evidence that systemic opiate drugs activate at least two major systems [1]. That there is a direct action of opiates on spinal cord nociceptive neurons is well established. Iontophoretic injection of opiates inhibits the firing of spinal cord neurons and intrathecal administration of morphine evokes a powerful and prolonged analgesia [10]. On the other hand, we have demonstrated that bilateral lesions of the dorsolateral funiculi of the spinal cord interrupt the analgesic effect of systemic morphine injection [3]. That result indicated that systemic narcotics also produce analgesia by activating descending inhibitory controls that originate in the brain stem. Specifically, it has been proposed that morphine binds to opiate receptors in the midbrain periaquaductal gray, the region which is considered to be at the origin of the supraspinal controls. The periaqueductal gray neurons initiate descending control via projections to neurons of the medullary nucleus raphe magnus, many of which are serotonin containing. Neurons from the latter regions project to the dorsal horn via the dorsolateral funiculus of the spinal cord and inhibit the firing of dorsal horn nociceptors. Consistent with this formulation, electrical stimulation in either the periaqueductal grey or the nucleus raphe magnus inhibits the firing of spinal neurons and produces a profound analgesia, in animals and in humans.

Our studies using systemic administration of morphine could not distinguish between the contribution of spinal and supraspinal sites of action in the suppression of *fos*. Furthermore, although morphine is a better mu than delta opioid ligand, it poorly discriminates between those different opiate receptors. More selective ligands that permit one to address the differential contribution of the three major classes of opioid peptides and opioid receptor (mu, delta and kappa) are now available. Thus, with a view to discerning the mechanism of action of systemic narcotics, we have extended our analysis to the effects of injecting opioids at different sites, in spinal cord and brain.

As described above there is evidence for both spinal and supraspinal targets through which morphine can exert its effect. We began our studies by examining the effect of intracerebroventricular injection of the mu opioid agonist, DAMGO (D-Ala^2, MePhe^4, Gly-ol^5, enkephalin). The paradigm was comparable to that which we used with systemic morphine injection. Animals were implanted with an indwelling intracerebroventricular cannula 7 days prior to the plantar injection of the formalin. Some animals also had bilateral lesions of the dorsolateral funiculi made at midthoracic levels 7 days earlier. This permitted us to determine whether any descending effect generated by microinjection of DAMGO was exerted via the descending pathways through which supraspinal opiates and electrical brain stimulation had been previously shown to operate. Our results provided the strongest evidence to date demonstrating that injection of opioids at supraspinal sites activates a powerful descending inhibitory control system. We found that DAMGO produced a dose-dependent inhibition of *fos*-immunoreactive neurons in the spinal cord. Again we found that the first neurons to be blocked were in the deep dorsal horn (laminae V and VI) and in the ventral horn (laminae VII, VIII and X). Of particular interest was quantitative analysis of the ED_{50} for inhibition of *fos* immunoreactivity and the ED_{50} for inhibition of the pain-related behavior produced by formalin injection into the paw. For the neurons of the ventral horn and laminae V, the ED_{50}s for suppression of *fos* immunoreactivity and the behavior were almost identical. In contrast, the ED_{50} for inhibition of *fos* immunoreactivity in neurons in the superficial dorsal horn was considerably higher. Consistent with our results after systemic injection of morphine, these data suggest that the behavioral responses indicative of pain are probably generated by activity within neurons of laminae V and the ventral horn, rather than by activity in the superficial dorsal horn cells.

To what extent the residual *fos* staining in the superficial dorsal horn is associated with a preserved perception of the stimulus is unknown. Rendering the animal analgesic does not mean that the stimulus was not perceived at all. One important possibility is that the animal can still recognize and localize the stimulus, but the diffuse, noxious qualities of the stimulus are lost. That view would be consistent with the fact that in patients one typically finds that although narcotics can control clinical pains, they do not alter the ability of the patient to recognize acute noxious stimuli, such as pin prick.

Importantly, we found that the inhibition of *fos* expression by intracerebroventricular administration of DAMGO, and the concomitant inhibition of the formalin-evoked behavior, were both antagonized in the animals which received bilateral DLF lesions in the thoracic spinal cord. In light of the strong similarity in the pattern of *fos* suppression by systemic morphine and by intracerebroventricular injection of DAMGO, these data suggest that the inhibition of *fos* staining by systemic morphine was indeed mediated by activation of supraspinal descending inhibitory systems. We believe that the spinal antinociceptive opiate sensitive circuits may only come into play at higher systemic does of narcotics. Interestingly, since opiate receptors and opioid peptides are most concentrated in the superficial dorsal horn [1, 5], it is possible that activation of opioid circuits via direct action of opiates at the spinal level may have very different functional consequences from those which result when opiates act at brain-stem sites. It would be of interest, in fact, to determine whether spinal injection of narcotics can block the firing of neurons in the ventral horn, laminae VII and VIII, where opiate receptors are found in much lower concentrations.

It is not clear how the cells of the deeper dorsal horn and ventral horn are inhibited when the periaqueductal gray-nucleus raphe magnus descending control system is activated. Note that the raphe neurons do not project to the ventral horn and thus a direct raphe-mediated inhibition of nociresponsive neurons in these regions is unlikely. One possibility is that a reduction in the output of superficial dorsal horn neurons (and possibly neurons in lamina V) is sufficient to eliminate the response of the deep cells. This could result from one of two different circuits. Activity generated in cells of the superficial dorsal horn neurons (driven by small diameter inputs from the periphery) could cascade ventrally, to activate the more ventrally located neurons. Alternatively, output of the lamina I projection cells may activate neurons in the brainstem, which in turn *facilitate* the firing of cells in the ventral part of the cord. This formulation suggests that there is a positive feedback system operating which maintains firing in the ventral horn as a result of activity in the superficial dorsal horn. There is, in fact, evidence that marginal neurons contribute to such supraspinal loops [4]. We are presently investigating this question by attempting to block selectively the ascending transmission of marginal neurons, by making lesions of the DLF, the pathway through which the marginal neurons reach the brainstem [7].

Of particular interest is the possibility that activity in the superficial dorsal horn may persist independent of activity in the periphery; this could, in turn maintain firing in other regions of the spinal cord. In fact, many chronic pain states that begin with a defined peripheral injury may become independent of the injury site. Because of this, pain therapies solely directed at the periphery often fail. For example, denervation by dorsal rhizotomy may only produce transient pain relief. Based on such results, it has been hypothesized that circuits in the spinal cord can sustain activity in the absence of peripheral stimulation; this activity, in fact, may underlie pains that are very difficult to control.

Because of the importance of this question to clinical pain problems, we are studying the changes in *fos* expression in a rat model of neuropathic pain [9]. Specifically, we are studying the effect of creating a neuroma in the sciatic nerve of the rat. Our preliminary studies established that there are waves of *fos* expression that appear in the spinal cord at different times after the injury. Not surprisingly, acute nerve section resulted in very intense staining in the cord; within 12 h most of the *fos* staining provoked by the nerve section had disappeared. Two days after injury, however, *fos* staining reappeared in the cord. Most importantly, the *fos* staining seen at 2 days could be blocked by peripheral injection of local anesthetics. This indicated that the expression of *fos* at 2 days resulted from hyperactivity arising in the neuroma that had formed on the cut nerve. Within 7 days of the nerve section, the *fos* levels had decreased in the cord, only to increase again by 2 weeks. Local anesthetic injection of the cut nerve at 2 weeks, however, did not completely reduce the central expression of *fos*. This indicates that activity generated either from the dorsal root ganglia attached to the cut nerve (i.e., proximal to the nerve block) or from intrinsic circuits in the spinal cord may be sufficient to maintain the expression of *fos*. This result is of particular interest because it provides new insights into the contribution of spinal cord circuits to the maintenance of neuropathic pains. With a view to developing novel approaches to the treatment of neuropathic pain problems, we have initiated studies of the effects of different analgesics on the pattern of *fos* immunoreactivity that appear 2 weeks after peripheral nerve injury. Hopefully, those studies will provide new information on the controls that can be generated in situations in which activity of spinal neurons may become independent of peripheral inputs and may contribute to various clinical pain states.

In summary, with this new approach to monitoring the activity of spinal neurons under conditions in which the behavior of the animal can be assessed and quantified, it is now possible to provide a much better global picture of the patterns of activity generated by peripheral noxious stimuli. This new approach offers a very powerful way to evaluate not only the mechanisms of action of different analgesic agents, but also a new way to assess novel approaches to blocking the transmission of nociceptive information over spinal pathways.

References

1. Basbaum AI, Fields HL (1984) Endogenous pain control system: brainstem pathways and endorphin circuitry. Ann Rev Neurosci 7:309–338
2. Basbaum AI, Menétrey D (1987) Wheat germ agglutinin-apoHRP gold: a new retrograde tracer for light- and electron microscopic single- and double label studies. J Comp Neurol 261:306–318
3. Basbaum AI, Marley N, O'Keefe J, Clanton CH (1977) Reversal of morphine and stimulus-produced analgesia by subtotal spinal cord lesions. Pain 3:43–56
4. Cervero F, Wolstencroft JH (1984) A positive feedback loop between spinal cord nociceptive pathways and antinociceptive area of the cat's brain stem. Pain 20:125–138
5. Fields HL, Basbaum AI (1978) Brainstem control of spinal pain transmission neurons. Ann Rev Physiol 40:193–221
6. Hunt SP, Pini A, Evan G (1987) Induction of the c-*fos*-like protein in spinal cord neurons following sensory stimulation. Nature 328:632–634
7. Jones MW, Apkarian AV, Stevens RT, Hodge Jr. CJ (1987) The spinothalamic tract: an examination of the cells of origin of the dorsolateral and ventral spinothalamic pathways in cats. J Comp Neurol 260:349–361
8. Menétrey D, Gannon A, Levine JD, Basbaun AI (1989) Expression of c-*fos* protein in interneurons and projection neurons of the rat spinal cord in response to noxious somatic, articular, and visceral stimulation. J Comp Neurol 285:177–195
9. Wall PD, Gutnick M (1974) Properties of afferent nerve impulses originating from a neuroma. Nature 248:740–743
10. Yaksh TL (1981) Spinal opiate analgesia: characteristics and principles of action. Pain 11: 293–346

3. Pathophysiology of Spinal Cord

Electrophysiological Differentiation Between Muscle and Sensory Nerve Funiculi for Peripheral Nerve Repair

N. NAKATA[1], H. MATSUDA, A. MIYAUCHI, M. SEKI and A. SHIMAZU

Summary

The purpose of this investigation was to differentiate the muscle funiculi from the sensory funiculi in the proximal stump of the peripheral nerve. Five cats anesthetized with sodium pentobarbital were used. An epidural electrode was inserted under the C7 lamina and needle electrodes were inserted at disc spaces C2–3, C3–4, C4–5 and C5–6. The amplitude of cortical somatosensory evoked potential (SEP) induced by stimulation of the superficial radial nerve, which has a sensory function, was usually larger than that of the deep radial nerve, which has a motor function. When a stimulus of less than 1.9 mA was applied through the needle electrode at the C5–6 disc space, a spike component suggesting orthodromic passage along the motor fibers was observed only from the muscle nerve. More intense stimulation (5.7 mA) under C7 lamina elicited two components from the sensory branch. The second component was almost completely abolished after intravenous injection of 0.5 mg/kg picrotoxin, hence it was considered to be the dorsal root reflex. Especially at C4–5 and C5–6 disc spaces, the amplitude of the positive component subsequent to the first negative component upon stimulation of the sensory nerve was higher than that following stimulation of the muscle nerve. These experiments may be clinically applicable to surgery of the funicular suture.

Introduction

Funicular suture has been recommended to obtain better functional recovery following peripheral nerve repair except at the level of a mixed nerve. Several methods have been used to distinguish between sensory and muscle funiculuIi

[1] Department of Orthopedic Surgery, Osaka City University Medical School, 1–5–7 Asahi-machi, Abeno-ku, Osaka, 545 Japan

K. Shimoji et al. (Eds.)
Spinal Cord Monitoring and Electrodiagnosis

[1, 2, 4–7]. It has been reported that the muscle afferents are little involved in scalp recorded cortical responses [3, 4, 6]. In addition, SEP were usually depressed with the duration of general anesthesia and the concentration of anesthetics [4, 6]. The present study was designed to differentiate electrophysiologically one from the other using a method that is less influenced by general anesthesia.

Materials and Methods

Five cats were used. A tracheal catheter for artificial ventilation and a venous catheter for drug administration were inserted under anesthesia with 30 mg/kg sodium pentobarbital, injected intraperitoneally. The radial nerve was exposed and divided near the elbow into two branches: the deep radial nerve, which has motor function, and the superficial radial nerve, which had sensory function. SEP was recorded from a screw electrode placed on the sensory cortex contralateral to the stimulated side. An epidural electrode was inserted under the C7 lamina and needle electrodes were inserted into the disc spaces at C2–3, C3–4, C4–5 and C5–6. These electrodes were used for both recording from and stimulation of the spinal cord. Usually 128 responses were averaged by the use of a DANTEC 1500 system with a bandpass of 20 Hz–2 kHz. The electric pulse used for the stimulation had a duration of 0.2 ms with 2–3 Hz.

Results

The amplitude of the SEP induced by the stimulation of the superficial radial nerve was usually larger than that of the deep radial nerve (Fig. 1). When a stimulus of 1.2 mA was applied through the epidural electrode under the C7 lamina, three components were observed from the muscle branch (Fig. 2). The first negative component was suggestive of a nonsynaptic action potential. Simultaneous recording from the sensory branch revealed only one component. However, with increasing intensity of epidural stimulation, another component was elicited from the sensory branch. The first component was considered to be a nonsynaptic action potential. The duration of the subsequent component was about 20 ms and its latency was 6.0 ms. The amplitude of this potential was almost completely abolished 30 min after administration of picrotoxin (Fig. 3).

When a stimulus of less than 1.9 mA was applied through the needle electrode in the C5–6 disc space, a component was clearly observed from the muscle branch (Fig. 4). By increasing the intensity to 2.7 mA, a component with an amplitude of 4 μv was detected in the sensory branch. Simultaneous

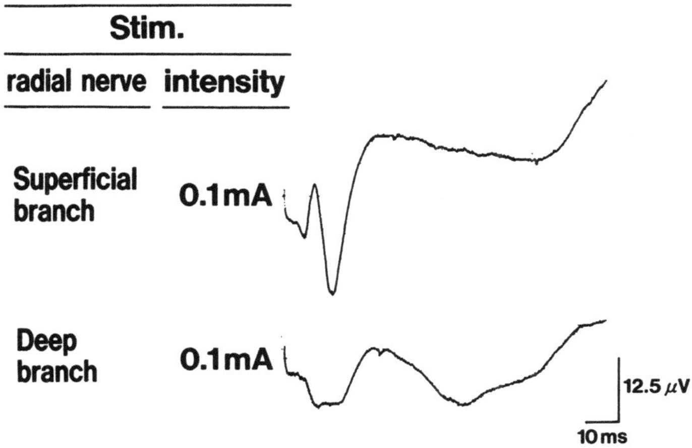

Fig. 1. Somatosensory evoked potential

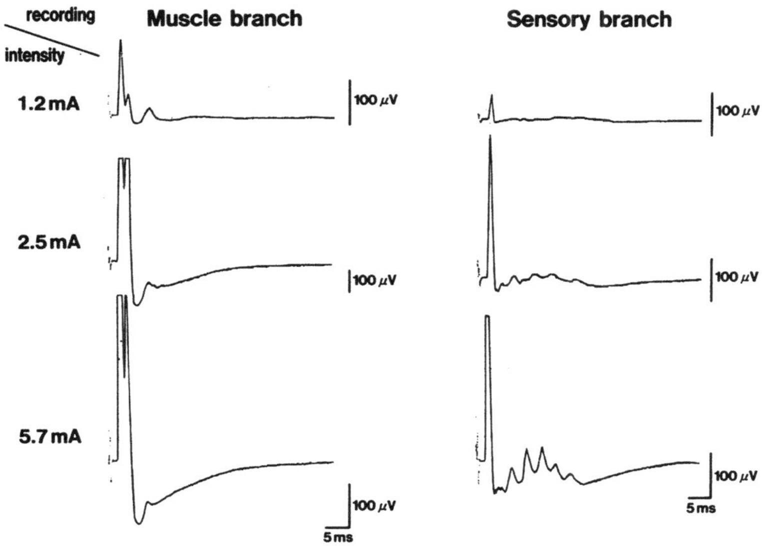

Fig. 2. Simultaneous recording from the muscle branch (*left*) and sensory branch (*right*) upon epidural stimulation under C7 lamina

recording from the muscle branch revealed that the amplitude of the first component was about 400 μv.

The amplitude of the positive component subsequent to the first negative component following the stimulation of the sensory nerve was higher than that following the stimulation of the muscle nerve, especially at C4–5 and

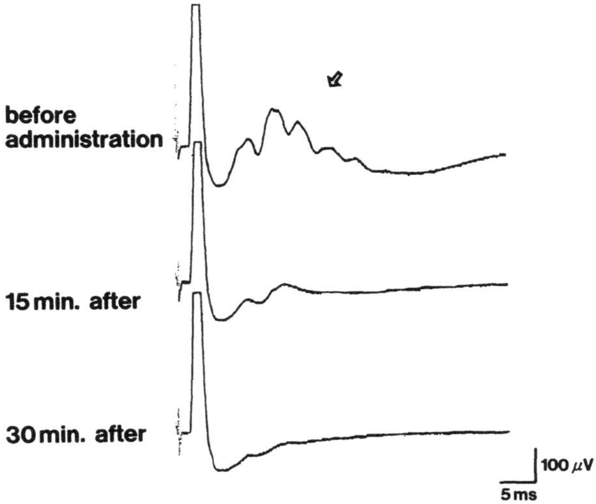

Fig. 3. Effects of picrotoxin on the potential (*arrow*) from the sensory branch upon C7 epidural stimulation. *Upper trace*; control response; *middle and lower traces*, responses recorded at 15 and 30 min after intravenous injection of 0.5 mg/kg, picrotoxin, respectively

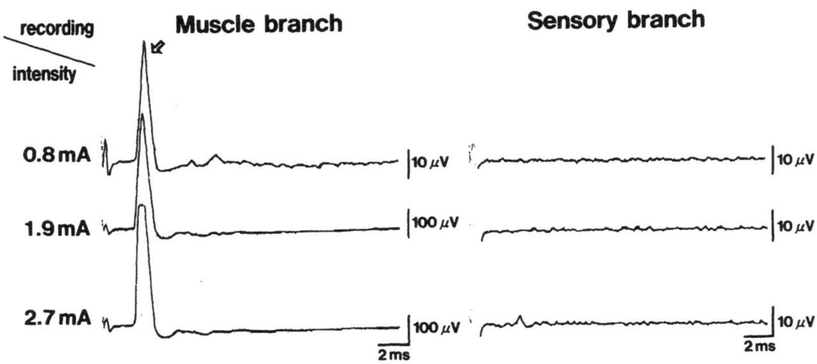

Fig. 4. Simultaneous recording from the muscle branch (*left*) and sensory branch (*right*) upon C5/6 disc stimulation

C5–6 (Fig. 5). After high frequency stimulation of 50 Hz for 20 s, the amplitude of the large positive component subsequent to the first negative component was about one half of that of the control, especially at C4–5 and C5–6.

Figure 6 shows the clinical application of these findings. The amplitude of the second positive components following stimulation of the digital nerve of the middle finger were larger than those following stimulation of the muscle branch of the triceps brachii, especially at C5–6 and C6–7.

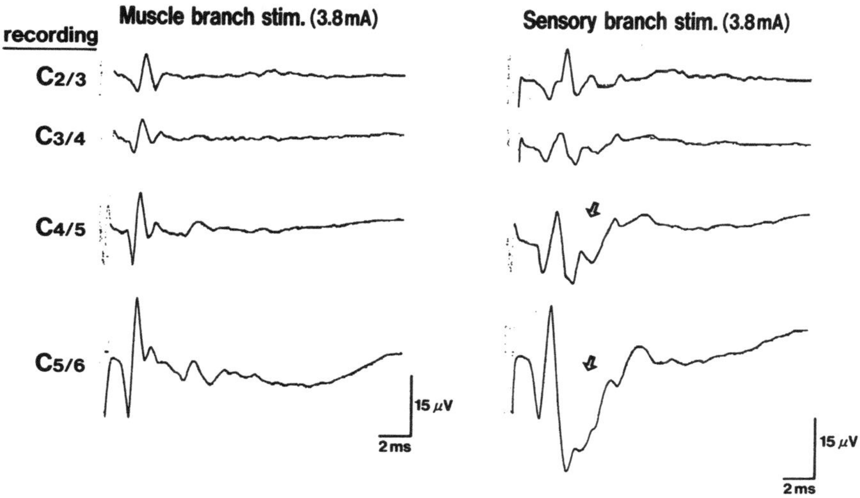

Fig. 5. Potentials recorded from the needle electrode in each disc space upon stimulation of the muscle branch (*left*) and the sensory branch (*right*)

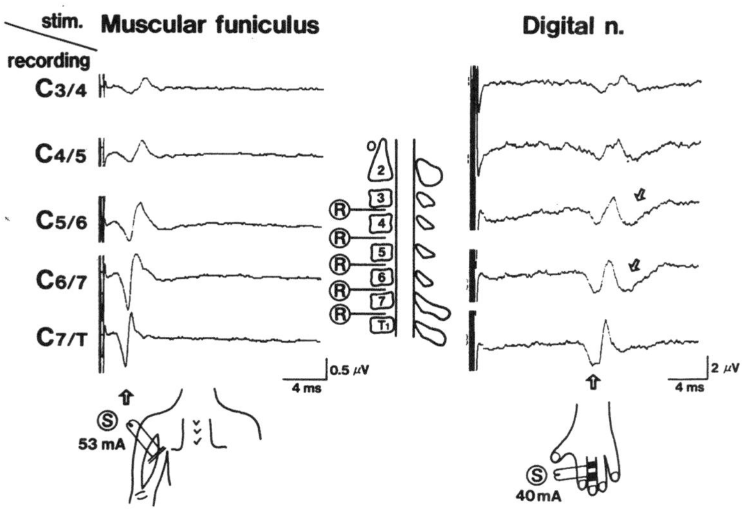

Fig. 6. Clinical application of a potential with upper type of brachial plexus injury during operation. *R*, recording; *S*, stimulation; *arrows*, amplitude of the second positive components

Discussion

More intense stimulation via the epidural electrode elicited two components from the sensory branch. The latter polyphasic potential was abolished after administration of picrotoxin, which is a GABA inhibitor. Thus, this potential was considered to be a dorsal root reflex originating from primary afferent depolarization.

Upon C5/6 disc stimulation, the threshold for the potential recorded from the muscle branch was lower than that from the sensory branch. Thus, the potential was considered to be an orthodromic nonsynaptic action potential passing along the motor fibers.

Upon stimulation of the sensory branch, the large positive component subsequent to the first negative component was especially observed at C4–5 and C5–6 levels. This component was abolished by after high frequency stimulation, and was considered to be a potential mediated by synapses. The radial nerve was innervated by C5–T1 roots, so this component may be a segmental potential. Further investigation on the origin of this component is necessary.

References

1. Gruber H et al. (1976) Identification of motor and sensory funiculi in cut nerves and their selective re-union. Br J Plast Surg 29:70–73
2. Hakstian RW (1968) Funicular orientation by direct stimulation. J Bone Joint Surg 50-A:1178–1186
3. Halonen JP, Jones S, Shawkat F (1988) Contribution of cutaneous and muscle afferent fibers to cortical SEPs following median and radial nerve stimulation in man. Electroencephalogr Clin Neurophysiol 71:331–335
4. Hashimoto T et al. (1980) Experimental study on the identification of sensory branch of the proximal stump at the site of nerve suture. Orthopedic Surgery 31:1371–1374 (Jpn)
5. Kondo N (1979) Clinical Study of Somatosensory Evoked Potentials in Orthopaedic Surgery. Int Orthop (SICOT) 1:9–15
6. Matsuda H et al (1981) Funicular identification of a muscle-branch for the nerve repair; clinical application of somatosensory evoked potential. Orthopedic Surgery 32:1594–1596 (Jpn)
7. Sunderland S (1953) Funicular suture and funicular exclusion in the repair of severed nerves. Br J Surg 40:580–587

Experimental Studies on the Spinal Cord Evoked Potentials in Cervical Spine Distraction Injuries

N. KAWAHARA[1], H. BABA, S. NAGATA, Y. KIKUCHI, K. TOMITA, S. NOMURA, and H. YUGAMI

Summary

Effects of stepwise distraction of the spinal cord on the spinal cord evoked potentials (SEP) were studied in cats. During the distraction between two adjacent cervical vertebrae, the SEP was examined upon single electric shocks to the lower spinal cord level (ascending SEP). The SEP was recorded from three points in the epidural space of the distracted spinal cord. Among the three sites the central site showed suppression of the ascending SEP earliest and most severely. The critical point for reversibility of spinal cord function should be set at a 30% amplitude decrease from the control SEP. Monitoring of the SEP at the central site of two cervical vertebrae seems to be best during corrective surgery to avoid possible damage to the spinal cord.

Introduction

It is possible to incur spinal cord damage during corrective spine surgery or reduction of the distracted spine. Spinal cord evoked potentials (SEP) have been widely utilized for prevention of spinal cord damage [3]. For utilization of SEP, it is very important to know how the SEP changes with various degrees of spinal distraction. Effects on SEP of stepwise distraction of the spinal cord were studied in cats. The reversibility of spinal cord function after release from distraction was also investigated.

Materials and Methods

Experiments were carried out on 30 adult mongrel cats (body weight, 3.0–4.5 kg). Anesthesia was induced with pentobarbital sodium and maintained

[1] Department of Orthopedic Surgery, Kanazawa University School of Medicine, 13-1 Takara-machi, Kanazawa, 920 Japan

K. Shimoji et al. (Eds.)
Spinal Cord Monitoring and Electrodiagnosis

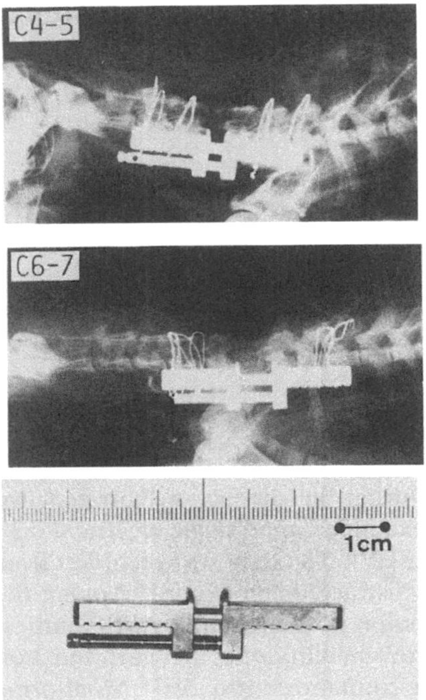

Fig. 1. Distraction device

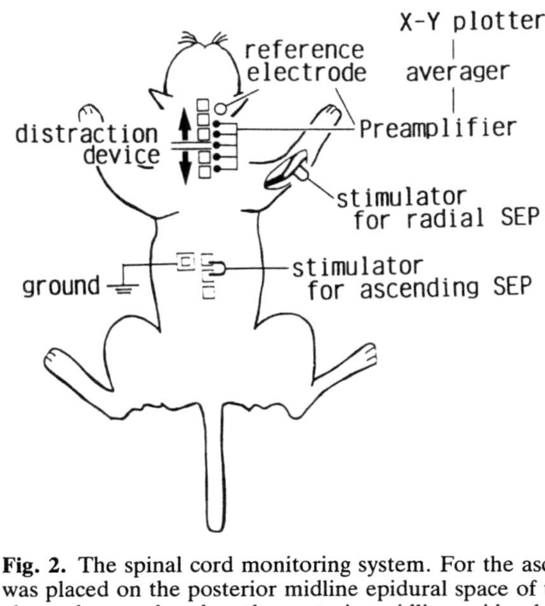

Fig. 2. The spinal cord monitoring system. For the ascending SEP, a recording electrode was placed on the posterior midline epidural space of the cervical spine and a stimulating electrode was placed on the posterior midline epidural space of the 12th thoracic vertebra. For the radial SEP, a recording electrode was placed on the same site as for the ascending SEP, while a stimulating electrode was placed on the radial nerve at the right elbow

with enflurane added to a constant flow of 3 l of oxygen per minute through an intubation tube. The distraction device was applied between the two adjacent cervical vertebrae (C4–5 or C6–7). Distraction was performed by turning the screw 1.4 mm every 3 min to increase the distance between the two adjacent cervical vertebrae (Fig. 1.).

During distraction the SEP was examined upon single electric shocks to the lower spinal cord level (ascending SEP) and through electrodes on the radial nerve at the right elbow (radial SEP) with stimulation parameters of 5 times signal threshold and 20 times per second for 0.2 ms each, and 64 responses were averaged by a DISA 1500 averager system. The SEP was recorded from three points at the posterior midline epidural surface of the distracted spinal cord; the central site of the two adjacent cervical vertebrae distracted, 10 mm rostral to the central site and 10 mm caudal to the central site (Fig. 2).

Results

Typical wave forms of the SEP recorded from the posterior midline epidural space of cervical spine are shown in Fig. 3. In evaluation of the ascending SEP, the peak amplitudes of the Pa1 component and the Na1 component from the baseline was measured, respectively. The radial SEP was evaluated in peak-to-peak amplitude from Pr1 to Nr1 (Fig. 3, Table 1).

During a series of stepwise distractions, the ascending SEP from the central site was earliest and most severely compromised. Then the SEP from the rostral site became abnormal and finally the SEP from the caudal site reversed abnormally (Figs. 4, 5).

In the cases in which the amplitude decrease remained less than 29% of the control level, the SEP completely recovered 120 min after the release of distraction. However, in the cases in which the decrease was 30%–49% of the control level, the potentials deteriorated down to 36.8%

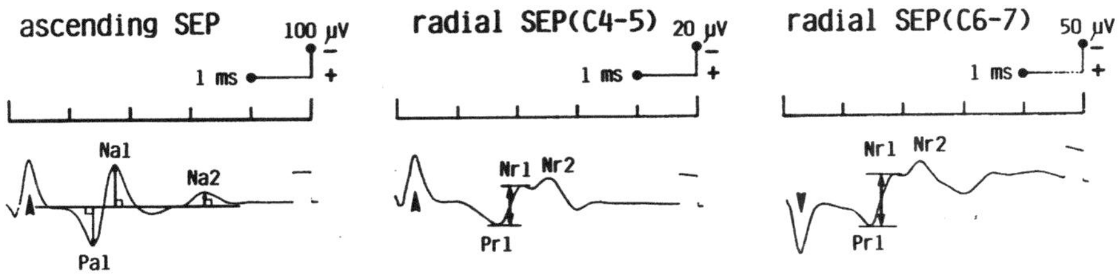

Fig. 3. Typical wave forms of the spinal cord evoked potentials recorded from the posterior midline epidural space of the cervical spine. *Left*, ascending SEP; *middle*, radial SEP recorded at level of C4–5; *right*, radial SEP recorded at the level of C6–7. The initial deflection of each trace was an artifact following the electrical stimulation (▲). In all traces negative is towards top

Table 1. Conduction velocities and amplitudes of spinal cord evoked potentials

Ascending SEP[a]

Component	Pa1	Na1	Na2
Conduction velocity (m/s)	114.6 ± 10.3	88.8 ± 7.5	43.6 ± 6.4
Amplitude (μV)	130.2 ± 18.5	153.7 ± 21.2	42.5 ± 6.8

Radial SEP[b]

	C4–5 (n = 20)			C6–7 (n = 10)		
Component	Pr1	Nr1	Nr2	Pr1	Nr1	Nr2
Conduction velocity (m/s)	87.7 ± 6.9	72.4 ± 4.1	54.7 ± 6.9	86.6 ± 10.9	72.1 ± 11.0	56.0 ± 7.6
Pr1-to-Nr1 peak-to-peak amplitude (μV)		42.8 ± 7.6			111.3 ± 29.8	

[a] Values are expressed in mean ± S.D. (n = 30)
[b] Values are expressed in mean ± S.D.

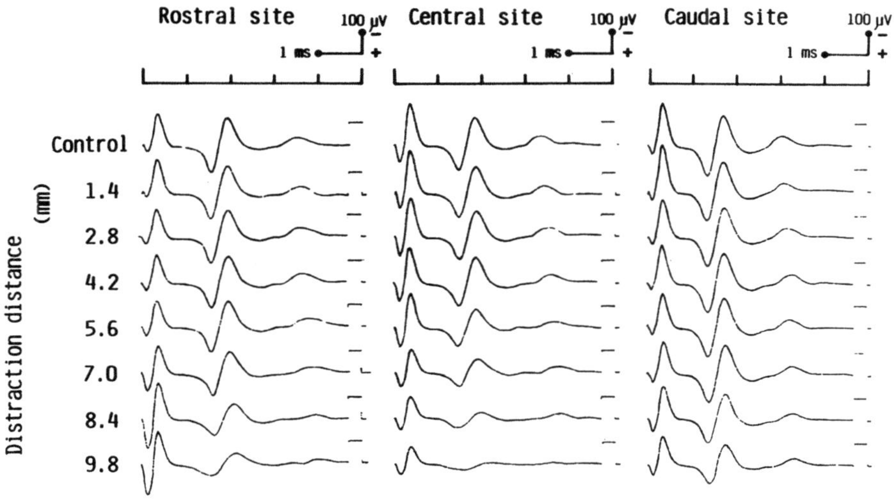

Fig. 4. Changes of the ascending SEP during stepwise distraction between the two adjacent cervical vertebrae. Recording electrodes were placed on the posterior midline epidural space at the level of the rostral, central and caudal sites, respectively. *Uppermost traces* (control) show responses with no distraction in this figure and Fig. 7

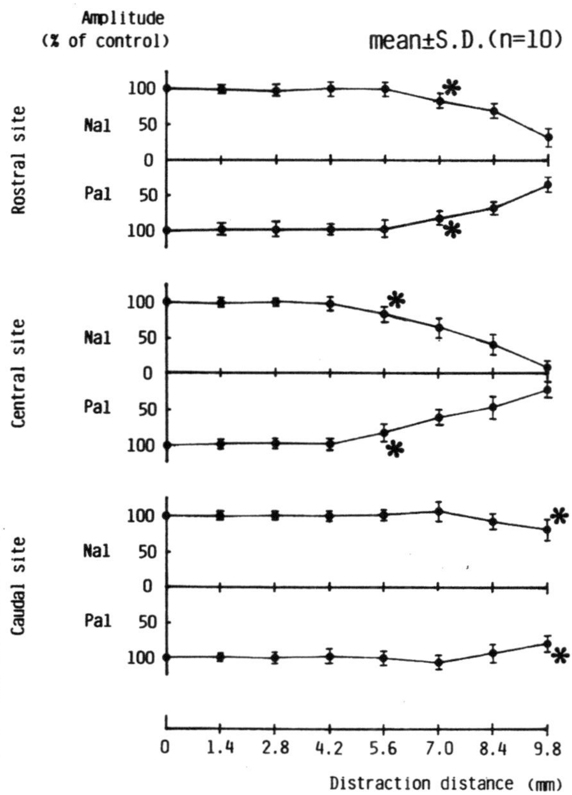

Fig. 5. Changes in the amplitude of ascending SEP components of Pa1 and Na1 (as percentage of the predistraction control values). The percent amplitude was plotted during stepwise distraction recorded at the rostral, central and caudal site. $^*p < 0.05$ vs control value by ANOVA followed by Dunnett's multiple comparison with control

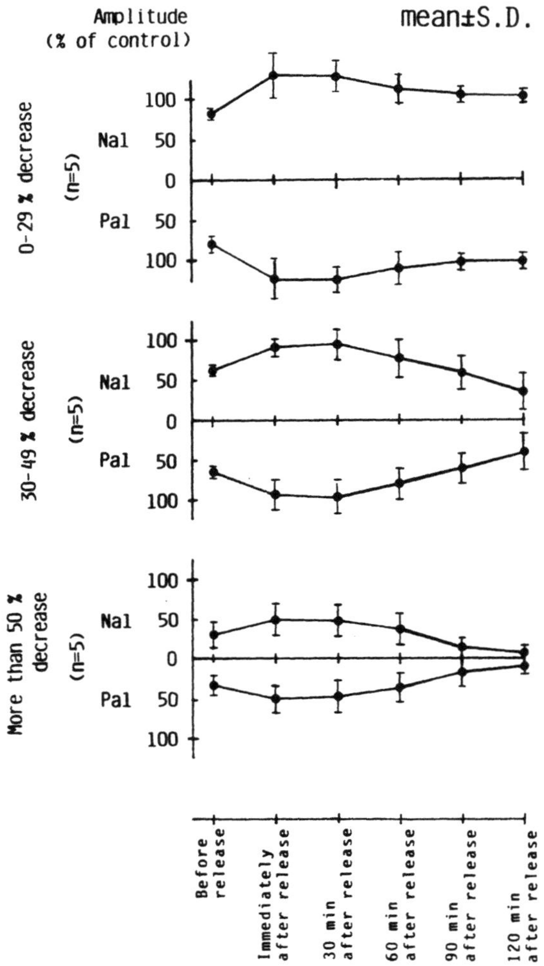

Fig. 6. Changes of the ascending SEP components of Pa1 and Na1 (as percentage of the predistraction control values) after distraction release in groups with various degrees of decrease of control values. *Top*, group with 0%–29% decrease of control amplitude; *middle*, group with 30%–49% decrease of control amplitude; *bottom*, group with more than 50% decrease of control amplitude

± 22.6% (mean ± S.D.; n = 5) 120 min after the release of distraction. In the cases in which the decrease was more than 50% of the control level, no recovery in the amplitude of the ascending SEP took place after the release of distraction (Fig. 6).

A high correlation was found between amplitude decrement of the radial SEP and distracting distance increment when the distraction was applied to the C4–5 vertebrae, whereas no correlation was observed between them in the radial SEP when the distraction was given to the C6–7 vertebrae (Figs. 7, 8).

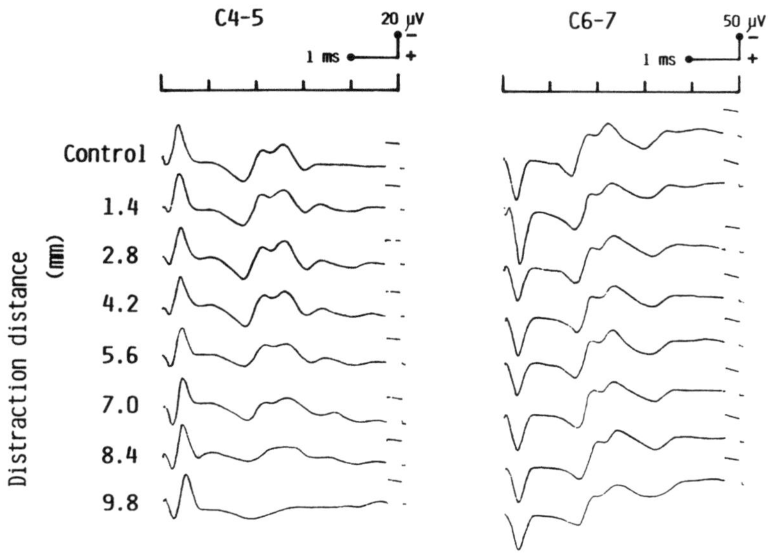

Fig. 7. Changes of the radial SEP during stepwise distraction between the two adjacent cervical vertebrae. Recording electrode was placed on the posterior midline epidural space at the level of the central site of the distraction. *Left*, C4–5 distraction; *right*, C6–7 distraction

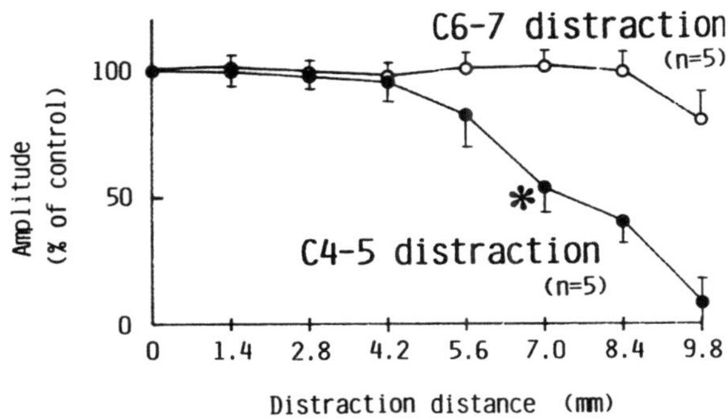

Fig. 8. Changes in the ampltidue of the radial SEP (Pr1-to-Nr1, peak-to-peak amplitude, was percentage of the predistraction control values) during distraction between the 4th and 5th, or the 6th and 7th cervical vertebrae, $^{*}p < 0.05$ vs control value by ANOVA followed by Dunnett's multiple comparison with control

Discussion

Spinal distraction is considered to be a major factor in spinal cord injury, in correction for scoliosis or the reduction of fracture dislocation and other spinal surgeries. There are a few discussions on experimental distraction injury of the spinal cord [2, 6]. There has been no study that reports the changes of the SEP recorded from the central site of the distraction. The present study clearly revealed that the ascending SEP recorded from the central site of the distraction was suppressed earliest and most severely [4]. Therefore the appropriate site for monitoring the ascending SEP to detect spinal cord dysfunction due to distraction is the central site between the two adjacent distracted cervical vertebrae. In this experiment on cervical distraction, positive-going potentials, which are important SEP parameters in compression spinal cord damage, were never observed. It seems that positive-going potentials do not appear because the spinal cord injury spreads longitudinally and gradually from the central site to both rostral and caudal sites in the distraction.

The SEP did not recover to normal when the initial amplitude reduction was 30%–49% at the central site of the distraction. It can be stated that the critical point for reversibility of spinal cord function should be set at the 30% amplitude decrease of the control SEP.

In the C4–5 distraction, the radial SEP changes are parallel to the ascending SEP at the central site of the distraction, whereas in the C6–7 distraction, there was no significant change in the radial SEP during distraction. It suggests that the radial SEP recorded at the level of C4–5 represents spinal electrical activity, whereas the radial SEP at the C6–7 represents the activity of both the cervical nerve root and spinal cord.

Conclusion

1. The monitoring of the SEP at the central site between the two adjacent cervical vertebrae is absolutely necessary during corrective surgery to avoid possible damage to the spinal cord.
2. The critical point for reversibility of spinal function should be set at a 30% amplitude decrease of the control SEP without distraction.
3. The radial SEP recorded at the level of C4–5 represents the spinal electrical activity, whereas the radial SEP at the C6–7 represents the activity of both the cervical nerve root and spinal cord.

References

1. Baba H (1986) Experimental studies of the spinal cord evoked potentials in cats. (in Japanese) J Jpn Orthop Ass 60:623–636
2. Dolan EJ, Transfeldt E, Tator CH, Simmons EH, Hughes KF (1980) The effect of spinal distractionon regional spinal cord blood flow. J Neurosurg 53:756–764
3. Jones SJ, Edgar MA, Ransford AO (1984) Epidural recording of spinal SEPs following posterior tibial nerve stimulation in man. Conduction properties of normal human spinal cord and changes during scoliosis correction. In: Homma S, Tamaki T (eds) Fundamentals and clinical application of spinal cord monitoring. Saikon Publishing, Tokyo, pp 223–236

4. Kawahara K (1988) Experimental studies on the spinal cord evoked potentials in the cervical spine distraction injury. (in Japanese) J Juzen Med Soc 97:672–687
5. Nagata S (1988) Experimental studies on the spinal cord evoked potentials in compression injury of the cervical spinal cord. J Juzen Med Soc 97:688–703 (in Japanese)
6. Nordwall A, Axelgaard J, Harada Y, Valencia P, Mcneal DR, Brown JC (1979) Spinal cord monitoring using evoked potentials recorded from feline vertebral bone. Spine 4:486–494

Experimental Studies on the Spinal Cord Evoked Potentials in Compression Injury of the Cervical Spinal Cord

S. NAGATA[1], H. BABA, U. KAWAHARA, Y. KIKUCHI, K. TOMITA, S. NOMURA, and H. YUGAMI

Summary

The waveform changes and their recovery processes were analyzed by recording the spinal cord evoked potential (SEP) before, during the after graded cervical ventral compression of the cat spinal cord. The SEP was recorded at three levels: compression site, and rostral and caudal sites, in response to single electric shocks at the level of the 8th thoracic vertebra. In the series of graded compression, the most significant changes were observed in the first negative component (Nc1) of the SEP. The magnitude of injury was classified into three groups according to the amplitude decrement of Nc1 component. After release of compression, the SEP showed no recovery in the group whose amplitude decrement was more than 50%.

The radial nerve evoked potential (radial NEP) was also induced by single electric shocks to the C1–2 spinal cord level during graded compression. A significant correlation was observed between amplitude decrements of SEP (Nc1 component) and the radial NEP when compression was applied to the spinal cord at the level of the C5 vertebra. For spinal surgery the most reliable monitoring technique is to record the SEP at the injury site; the distal peripheral nerve evoked potential is also useful when an appropriate nerve is selected for recording.

Introduction

There are many pathological or disease states in which spinal cord compression occurs; compression of vertebral bodies, discs, tumors etc. may occur during decompressive surgery. Now more precise evaluations using electrophysiological diagnostic methods are required for the diagnosis of

[1] Department of Orthopedic Surgery, Kanazawa University, School of Medicine, 13-1 Takara-machi, Kanazawa, 920 Japan

K. Shimoji et al. (Eds.)
Spinal Cord Monitoring and Electrodiagnosis

injury level and intraoperative spinal cord functional monitoring. But there is still much controversy concerning the correlation between the extent of injury and the waveform changes of spinal cord evoked potentials (SEP). The objectives of this study are: (1) to measure waveform changes in spinal cord injured cats, (2) to find the critical point of reversibility of SEP after the release of compression, and (3) to evaluate the clinical usefulness of peripheral NEP as a new method of monitoring spinal cord function after spinal cord compression injury.

Materials and Methods

Forty mongrel cats were used in this study, anesthetized with pentobarbital sodium and intubated. This study was divided into two parts, according to electrophysiologic measurement of ascending SEP and radial NEP.

Ascending SEP

Recording Method

A laminectomy was performed at the 12th thoracic vertebral level. A bipolar epidural electrode was inserted and set at the level of T8 as a stimulating electrode. Then after the partial laminectomy of T2, a five polar epidural electrodes (a distance between each electrode of 5 mm) were inserted and placed on the posterior midline surface of the cervical spinal cord. The third electrode (R3) was placed at the compression site using radiographic imaging. Measurements at the injury site, and at caudal and rostral sites were recorded. The reference electrode was placed in the back musculature. The DISA digital 1500 system was used for stimulation and recording. A rectangular pulse of 0.2-ms duration was used for stimulation. Frequency was 20 Hz. A bandpass filter from 10 Hz to 2000 Hz was used. The analysis times used were 5 ms and 10 ms. For each average 64 signals were used.

Spinal Cord Injury Model

After the cat was placed in supine position, a small hole with a diameter of 5 mm was made at the level of the 5th or the 7th cervical vertebra using an electrical minidrill without destroying the posterior longitudinal ligament. Then a dual screw (whose diameter of screw tip was 3 mm) was put into the hole and set. The graded compression injury was made by the inner screw of the dual screw with the speed of 0.125 mm per 3 min and the ascending SEP was recorded each time.

The compression ratio (X-ray measurement: depth of screw tip/AP spinal canal diameter × 100) is varied, producing different degrees of spinal cord injury. After the release of compression the ascending SEP was recorded every 30 min for 120 min. Histological evaluation followed these experiments.

Radial NEP

Recording Method

Three electrodes were used, the stimulating, recording, and reference electrode. Bipolar stimulating electrodes (5 mm distance between points) were inserted under radiographic guidance at the C1–2 level and placed on the posterior epidural midline surface. A rectangular pulse of 0.2 ms duration was used for stimulation. Frequency was 20 Hz and 3 Hz. Bandpass filter from 10Hz to 2000 Hz was used. Analysis times used were 5 ms and 10 ms. Some 64 repetitions were averaged. The recording needle electrode was inserted into the deep branch of the radial nerve 5 mm from bifurcation of the elbow. The reference electrode was placed in surrounding musculature.

Radial NEP on Cervical Ventral Compression

The cervical graded compression by the dual screw was performed at C5 (eight cats) and C7 (five cats). The same model was used as in the ascending SEP. The difference of waveform changes of radial NEP between the C5 and C7 groups was examined. Correlation of radial NEP and ascending SEP between the C5 and C7 groups was done.

Statistics

All the data of radial NEP and ascending SEP during graded compression were subjected to statistical analysis using ANOVA, following by Dunnett's multiple comparison with control. Three groups were classified according to amplitude decrease ratios of Nc1 component of the ascending SEP. The compression ratios were then analyzed statistically using Tukey's *t* test. The correlation between ascending SEP and radial NEP was examined by Pearson's correlation analysis. The data of changes in the amplitude of ascending SEP after decompression were also subjected to statistical analysis with Student's *t* test. Statements in this study indicate a significant difference with a *p* value 0.05 or less.

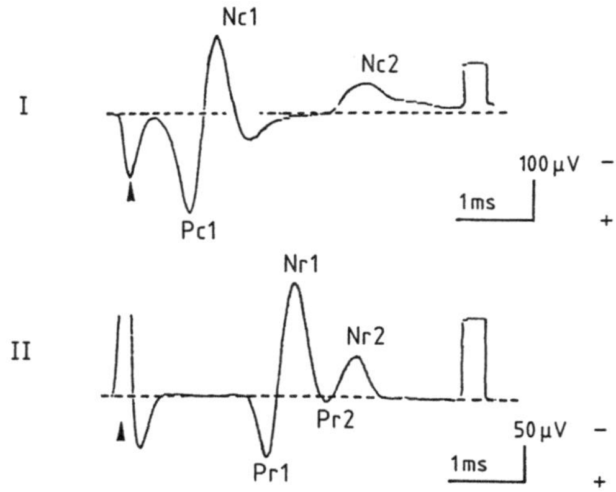

Fig. 1. Typical wave forms of evoked potentials *Upper trace* (*I*) shows the ascending SEP recorded on the posterior midline epidural surface. *Lower trace* (*II*) shows the radial NEP recorded on deep branch of the radial nerve. ▲, stimulation artifact; *top of frace*, negative potentials

Results

Ascending SEP

Normative Data of Ascending SEP

The ascending SEP consisted of two components, the positive-negative complex (Pc1, Nc1) and a slow-negative component (Nc2). The conduction velocity of Pc1 was 114.6 ± 10.3 m/s (mean ± SD): Nc1, 88.8 ± 7.5 m/s; Nc2, 43.6 ± 6.4 m/s (n = 30). Peak-to-peak amplitude of Pc1-Nc1 was from 150 to 320 μV. The amplitude of Pc1 and Nc1 from the baseline was calculated as the percentage of the precompression control value because of its constancy (Fig. 1).

Ascending SEP After Graded Ventral Compression

There was no significant change in amplitude at injury site, 1.0 cm at rostral site, and 1.0 cm at caudal site during graded compression until the tip of the inner screw depth achieved 2.5 mm. In the 3.0 mm compression the amplitude of Nc1 was markedly decreased to 53.6% ± 24.7% (n = 28) of the control value at the injury site ($p < 0.05$); at the 1.0-cm caudal site the amplitude of Nc1 was increased to 127.8% ± 24.5% of control value. At the 1.0-cm rostral site the amplitude of Nc1 was slightly decreased. In 3.5 mm compression, Nc1 of the injury site disappeared and the positive component

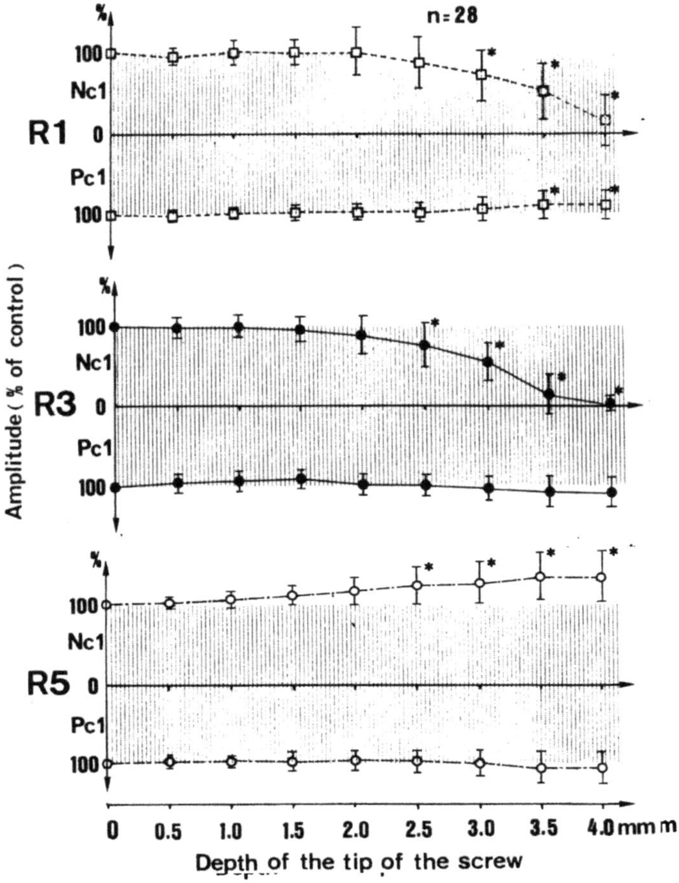

Fig. 2. Changes in the amplitude of the asecnding SEP during graded ventral compression. *R1* (□), 1.0 cm rostral to the compression site; *R3* (●), compression site; *R5* (○), 1.0 cm caudal to the compression site. Each value represents mean ± S.D. ($n = 28$). *, $p < 0.05$ by ANOVA, followed by Dunnett's multiple comparison with control

remained. The amplitude decrement of Nc1 at the 1.0 cm caudal site became more noticeable (130.8% ± 35.0%). The amplitude of Nc1 at the rostral site decreased to 52.1% ± 35.0%, but the negative component still remained. There were changes of Pc1 on all recording sites, the injury site demonstrated slight amplitude increment associated with the amplitude decrement of Nc1, the caudal site showed amplitude increment of Pc1 and the amplitude increment of Nc1. In contrast, at the rostral site the small amplitude decrement of Pc1 was exhibited with the amplitude decrement of Nc1. In the 4.0-mm compression injury, the Nc1 at the caudal site showed still greater amplitude increment and the Nc1 component at the rostral site showed only a positive component (Fig. 2).

Table 1. The compression ratios in three groups classified depending upon amplitude decrease ratios of Nc1 component of the ascending SEP

Group	Amplitude decrease ratio (%)	Compression ratio (%)
A (n=12)	0–30	10.7–55.6 (27.1±14.1)
B (n=10)	30–50	20.8–59.5 (36.5±11.6)
C (n=10)	50–100	24.3±63.5* (41.8±12.5)

The value in ( ) represents mean ±SD, *$p < 0.05$ compared to group A by Tukey's t test

Correlation Between the Amplitude Decrement of Nc1 Component and Compression Ratio

Three groups were arranged according to the amplitude decrement of Nc1. The first group, Group A, had a value 0%–30% precompression control value; Group B, 30%–50%; Group C, 50%–100%. The compression ratio for each group was 27.1% ± 14.1% in Group A (n = 12), 36.5% ± 11.6% in Group B (n = 10), 41.8% ± 12.5% in Group C (n = 10). There was no significant correlation between Group A and Group B, between Group B and Group C in regard to the compression ratio, but a significant correlation was recognized between Group A and Group C ($p < 0.05$); (Table 1).

Recovery Process of Ascending SEP After Release of Compression

The waveform changes of ascending SEP after release of compression were evaluated in three subset groups, Group A′ (five subjects), Group B′ (six subjects), and Group C′ (five subjects). In Group A′ there was complete recovery in the Nc1 and Pc1 amplitude immediately after release of compression at each of three recording sites. Even 120 min after the release of compression the amplitude of each component maintained the control value. In Group B′, 30 min after compression release the compression site amplitude of Nc1 showed the recovery of 88.3% ± 22.3% and at the rostral site 103.3% ± 33.8%. Nevertheless, the amplitude increment of Nc1 at the caudal site remained even 120 min after compression release ($p < 0.05$). In Group C′ at 30 min, the partial recovery of precompression control value occurred at 47.5% ± 33.7%, but by 120 min the value decreased to 31.8% ± 11.3% of control value (Fig. 3).

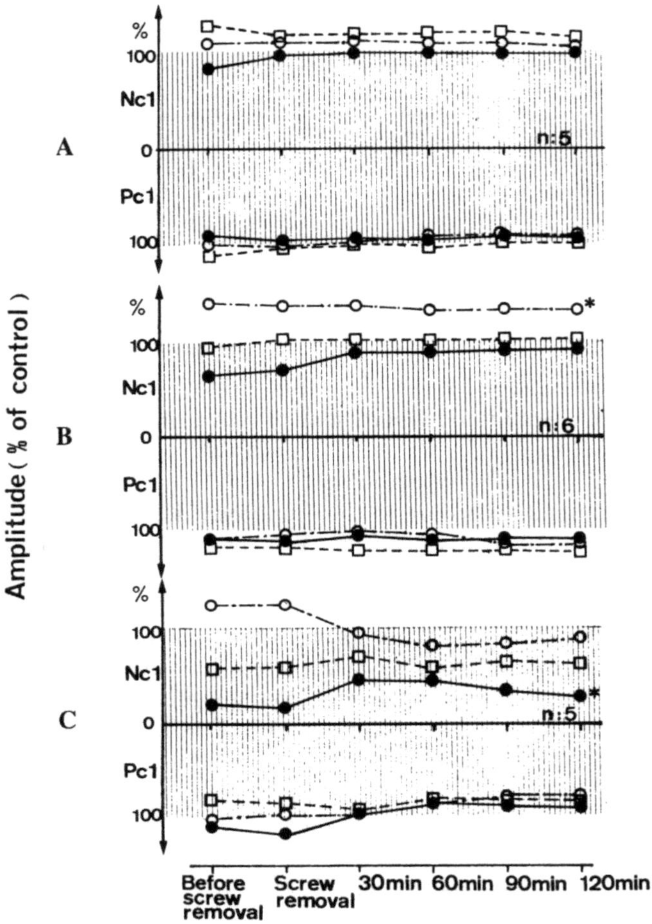

Fig. 3. Changes in the amplitude of the ascending SEP after removal of the inner screw. *A'*, group whose amplitude decrease of Nc1 was within 30% of the compression control value ($n = 5$); *B'* and *C'*, groups whose amplitude decrease was in 30%–50% ($n = 6$) and more than 50% ($n = 5$). □, 1.0 cm rostral site; •, compression site; ○, 1.0 cm caudal site. *, $p < 0.05$ vs. control value by Student's *t* test

Histological Evaluation

Transverse sections of the compression site were examined by HE staining. No significant pathological bleeding could be found in any of the groups.

Radial NEP

Normative Data of Radial NEP

The radial NEP consisted of basically two positive-negative complexes. The first complex consisted of the first positive component (Pr1) and the first negative component (Nr1). The second had mainly a slow negative (Nr2) and a positive component was not seen. The conduction velocities of Pr1 and Nr1 were 72.6 ± 23.3 m/s and 62.3 ± 5.3 m/s (n = 13), respectively. The conduction velocity of Nr2 was 55.6 ± 6.7 m/s (n = 13). The first complex of the radial NEP showed no changes in amplitude associated with the change of stimulus frequency. The amplitudes of Pr1 and Nr1 from the baseline were 64.7 ± 11.3 μV, 94.8 ± 25.7 μV (n = 13). The peak-to-peak amplitude of the second potential was 26.0 ± 6.8 μV during at stimulation 20 Hz. And during stimulation at 3Hz the amplitude of the second potential showed 67.5 ± 10.1 μV (n = 13; Fig. 1). The second potential of the radial NEP had a tendency to increase in amplitude with low frequency stimulation.

The Waveform Change of Radial NEP on Graded Cervical Ventral Compression and the Correlation Between Radial NEP and Ascending SEP

After the 5th cervical vertebral compression the waveform change of radial NEP showed the same response as the ascending SEP. At 2.5 mm compres-

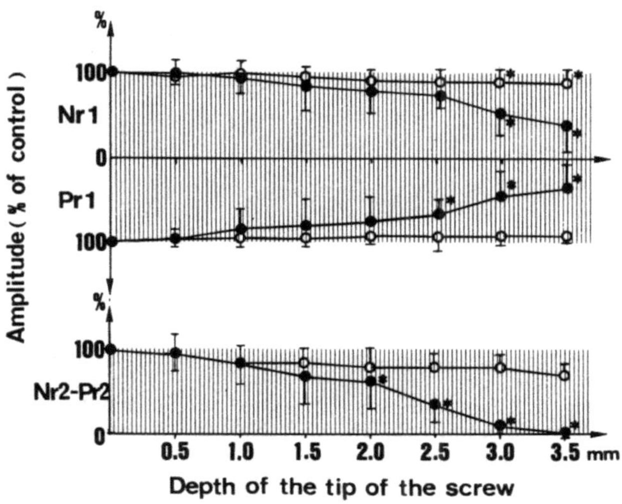

Fig. 4. Comparison between amplitude changes in the radial NEP during graded ventral compression at C5 and C7. ●, C5 compression (n = 8); ○, C7 compression (n = 5). Each value represents mean ± S.D. *, $p < 0.05$ by ANOVA, followed by Dunnett's multiple comparison with control

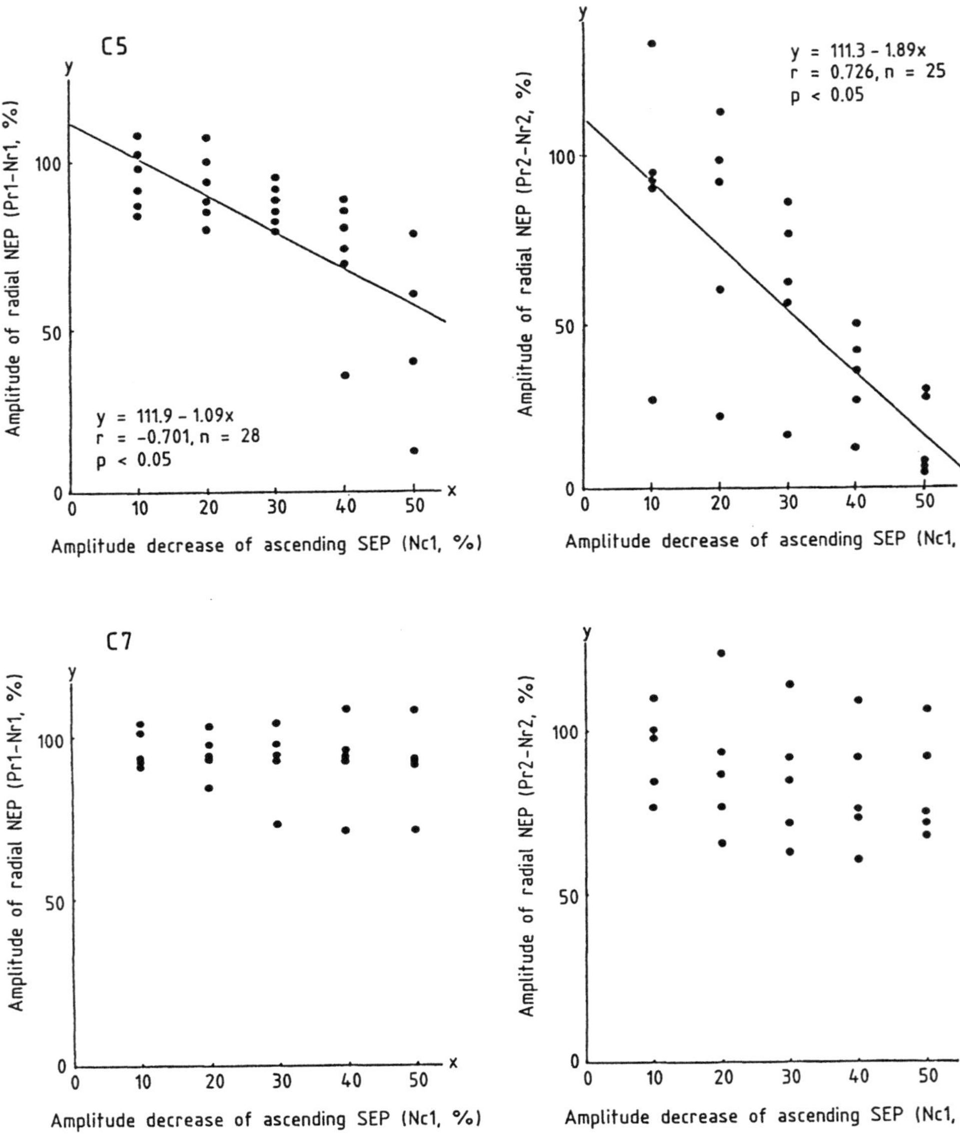

Fig. 5. Correlation between peak-to-peak amplitude of the radial NEP and the amplitude decrease in the ascending SEP (Nc1) in C5 compression and C7 compression. High correlation was observed between both components only at C5 compression. Regression line (*upper left*), Y = 111.9–1.09x; correlation coefficient (*upper left*), $r = -0.701$ ($p < 0.05$); regression line (*upper right*), Y = 11.3–1.89x; correlation coefficient (*upper right*), $r = -0.726$ ($p < 0.05$)

sion Pr1 and Nr1 of radial NEP decreased to 64.9% ± 12.1% and 75.4% ± 13.9% of precompression control value (n = 8). The peak-to-peak amplitude of Pr2–Nr2 also decreased to 36.4% ± 20.8% of control. At 3.5 mm compression the amplitude of Pr1 and Nr1 decreased to 40.1% ± 37.5% and 35.0% ± 27.5% of control and the Pr2–Nr2 complex disappeared.

The correlation between peak-to-peak amplitude of the radial NEP and the amplitude decrease in the ascending SEP, especially the Nc1 component, was examined. High correlation between both components at the 5th cervical vertebral compression was observed. The correlation of radial NEP and ascending SEP was −0.701 and between the second potential and ascending SEP was −0.726 (Figs. 4, 5).

After the 7th vertebral compression the waveform of radial NEP showed no significant change during graded compression, but slight amplitude decrement of Nr1 component at 3.0 mm compression. There was no sigificant correlation between the radial NEP and ascending SEP (Figs. 4, 5).

Discussion

Ever since the first recording of the spinal cord evoked potentials by Gasser and Graham (1933) [2], many studies concerning SEP in animals and humans have been reported [1, 4, 6, 7]. But there are differences between the results of each experiment because of the differences in injury and evoked potential recording methods.

In our experiment the changes of the ascending SEP on different recording sites during cervical-ventral graded compression and the point of reversibility of SEP was examined. The ascending SEP consisted of two components: the first positive-negative component (Pc1–Nc1), and the second slow negative component (Nc2). After cervical graded compression there was no significant waveform change at any of the recording sites, compression site, 1.0 cm rostral and 1.0 cm caudal, until 2.5 mm compression. It was concluded that there was an electrophysiological "silent zone" until 2.5 mm compression. At 3.0 mm of compression the amplitude of Nc1 at the compression site decreased to 53.7% ± 24.5% (n = 28) and the amplitude of Nc1 at 1.0 cm caudal site increased to 128.8% ± 24.5%. At the 1.0 cm rostral site the amplitude of Nc1 decreased slightly to 73.2% ± 35.6%. With 3.5 mm of compression the Nc1 component of compression site disappeared and the amplitude increment of Nc1 at caudal site increased more. Nevertheless, at the rostral site the amplitude decrement of Nc1 was 52.1% ± 35.0% and the negative component of the first potential remained. The amplitude of Pc1 showed no significant change, but a slight decrease at the rostral site with 3.5 mm compression.

These results clearly demonstrated that on ventral compression of the spinal cord the waveform changes of ascending SEP occurred mainly in the negative component Nc1. Initially, the Nc1 component of ascending SEP disappeared at the injury site, then the amplitude increment of Nc1 occurred at the caudal site, and at the rostral site the negative component gradually disappeared. It was concluded that in the ventral compression the earliest waveform change was found at the injury site in its amplitude decrement of Nc1 and the waveform change of the surrounding area occurred secondarily. The phenomenon that the negative component of the action potential disappears and the positive component remains demonstrates a nerve fiber state which could not conduct through the injury site, that is to say, there was no depolarization. Woodbury called this kind of potential "killed end potentials" [8] and Schramm reported it as "spinal cord evoked injury potentials" [5]. Clinically these potentials could be recorded sometimes at the damaged spinal cord level. It could be an excellent indicator of spinal conduction block and was considered extremely important for monitoring epidural spinal conduction. The first waveform change in the ascending SEP, the amplitude decrement of Nc1 component and injury potential occurred at the injury site and immediately after the amplitude increment of Nc1 occured at the caudal site.

Kojima et al. did chronic cervical ventral compression experiments using dogs and correlated the compression ratio and the recovery of spinal motor function [3]. They recorded the SEP from an electrode inserted into cerebral cistern by stimulation of sciatic nerve. They reported that the latency elongation of the evoked potential began at the point of 51% compression ratio and the wave of evoked potential disappeared at a 67% compression ratio. They concluded that the 67% compression ratio was an electrophysiologic maximum tolerable compression. In our experiment the correlation between the amplitude decrement of Nc1 component and the compression ratio was evaluated by the three groups (according to the amplitude decrement of NCI). The compression ratios were 27.1% ± 14.1% in Group A (n = 12), 36.5% ± 11.6% in Group B (n = 10), and 41.8% ± 12.5% in Group C (n = 10). It was suggested that the severe spinal cord damage occurred at the compression site even within the electrophysiologic safety limit (which Kojima et al. reported) because of the fact that the negative component of ascending SEP on injury site completely disappeared at the point before the wave at the rostral site disappeared. It is felt that more stringent criteria should be developed regarding the change of the first negative component of ascending SEP at the compression site.

The waveform change of ascending SEP after release of compression was observed on the compression site, 1.0-cm rostral and 1.0-cm caudal sites until 120 min. Group A′, whose amplitude decrease of Nc1 components at the injury site was within 30%, showed good recovery immediately after release of compression at each recording site. Group B', whose amplitude decrease of Nc1 component at injury site was 30%–50%, showed recovery 120 min after release of compression at the compression site the rostral site recordings. But the amplitude increment of Nc1 component at the caudal site persisted in the order of 131.0% ± 26.4% of precompression control value ($p < 0.05$). Group C', whose amplitude decrease of Nc1 component was more than 50%, showed partial recovery (47.5% ± 33.7% of control value) 30 min after release of compression, however, 120 min after release of compression the amplitude decreased again to 31.8% ± 11.3% of control value. This result indicates that it is less likely that recovery of SEP will occur if there has been more than 50% amplitude decrement of Nc1 component at the compression site. The increase in amplitude of Nc1 at the caudal site in Group B also suggests a possible spinal cord injury. Therefore it is concluded that an amplitude decrement of the Nc1 component of 30% is the relative critical point of the SEP reversibility and an amplitude decrement of Nc1 component to 50% is the absolute critical point of the SEP reversibility.

There were several problems during the epidural recording of SEP. Massive bleeding into epidural space caused an impedance change surrounding the recording electrodes and loss of the signal. The electrodes can be moved inadvertently during surgery and cause small changes in the signal. This should therefore be avoided. In order to monitor spinal cord function using peripheral NEP the recordings should be made from the nerve whose segmental supply is the same as spinal cord segments at risk, e.g., injury levels of C5 cord segments are best monitored by radial nerve recordings. A significant correlation was observed between amplitude decrements of SEP (Nc1 component) and the radial NEP when the compression was applied to the spinal cord at the level of the C5 vertebra, whereas no correlation was observed between them when the compression was applied at the C7 vertebral level. The reason for the difference was thought to be related to compression; in C5 vertebral compression the compression site was placed on the tracts before branching to the radial nerve, and in C7 compression it was placed after branching to the radial nerve.

It is concluded that for spine surgery the most reliable monitoring technique is to record the SEP at the injury site and that the distal peripheral NEP is also useful when an appropriate nerve is selected for recording.

References

1. Baba H (1986) Experimental study of the spinal cord evoked potential in cats. J Jpn Ortho Assoc 60:623–636
2. Gasser HS, Graham HT (1933) Potentials produced in the spinal cord by stimulation of dorsal roots. Am J Physiol 103:303–320
3. Kojima Y, Yamamoto T, Ogino H, Okada K, Ono K (1979) Evoked spinal potentials as a monitor of the spinal cord viability. Spine 4:651–657
4. Magladery JW, Porter WE, Park AM, Teasedall RD (1951) Electrophysiological studies of nerve and reflext activity in normal man. IV: The two-neuron reflex and identification of certain action potentials from spinal roots and cord. Bull Johns Hopkins Hosp 88:499–519
5. Schramm J, Krause R, Shigeno T, Brock M (1983) Experimental investigation on the spinal cord evoked injury potential. J Neurosurg 59:485–492
6. Shimoji K, Higashi H, Kano T (1971) Epidural recording of spinal electrogram in man. Electroencephalogr Clin Neurophysiol 30:2326–2329

7. Shinomiya K (1982), Spinal cord monitoring of spinal cord function using evoked spinal cord potentials. J Jpn Ortho Assoc 56:1551–1560
8. Woodbury JW (1965), Potentials in a volume conductor. In: Ruch TC, Patton JW, Woodbury JW (eds), Neurophysiology, 2nd edn., Saunders, Philadelphia, pp 93–91

Comparison Between Spinal Cord Evoked Potential and Cortical Evoked Potential by Experimental Dorsal Compression on the Thoracic Spinal Cord

S. SATO[1], S. UCHIYAMA, and K. TAKAHASHI

To prevent iatrogenic spinal cord injury during surgery, in Japan spinal cord evoked potential monitoring is frequently performed, but in Europe and America cortical evoked potential monitoring is mainly used. However, only a few references about the comparison of both potentials are found in the literature. The purpose of this report is to define differences in monitoring abilities between both potentials during the experimental dorsal compression of the spinal cord.

Materials and Method

The experiments were performed on nine cats. Anesthesia was maintained with intermittent administration of intravenous pentobarbital. The animals were paralyzed with pancuronium bromide and ventilated with a respirator.

Laminectomies were performed at the low lumbar, thoracic and cervical regions, leaving the dura intact. The skull was drilled with a dental drill, and the dura was exposed.

A catheter electrode was positioned on the dura at the lumbar spine, and the spinal cord was stimulated electrically. The intensity of the stimulation was supramaximal for the amplitude of 1st wave of the spinal cord evoked potential.

Compression was made on the thoracic cord. Recordings were made simultaneously at cervical and cortical levels in the same animal. Spinal cord potential was recorded with a catheter electrode at a cervical level, and cortical evoked potential was recorded with two silver ball electrodes placed over the frontal and parietal lobes.

The compression device (Fig. 1) was made of steel screws and plates, and it was fixed to the bone at the thoracic laminectomy site. The compression

[1] Department of Orthopedic Surgery, Niigata University School of Medicine, Asahimachi-dori 1, Niigata City, 951 Japan

K. Shimoji et al. (Eds.)
Spinal Cord Monitoring and Electrodiagnosis

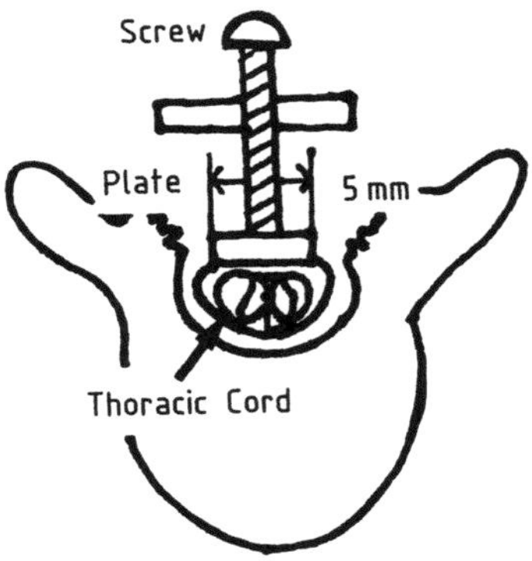

Fig. 1. Compression device

was done over the dura and gradually increased by rotating the screw. Plate contact with the dura was 5 mm in width.

A preliminary experiment suggested that administration of pentobarbital decreased the amplitude of the cortical evoked potential. So compression was started after the amplitude has finished decreasing, and pentobarbital was not given during compression period.

Results and Discussion

Basic patterns of both potentials and the measuring methods of latency and amplitude are shown in Fig. 2.

Spinal cord evoked potential appears about 3 ms after the stimulation, and peak cortical evoked potential more than 10 ms after the spinal cord potential. In a preliminary experiment the cortical evoked potential disappeared after a complete cut of the spinal cord at the C1 vertebral level, which was cranial to the recording site for spinal cord evoked potential, so we can say that this potential originates from brain.

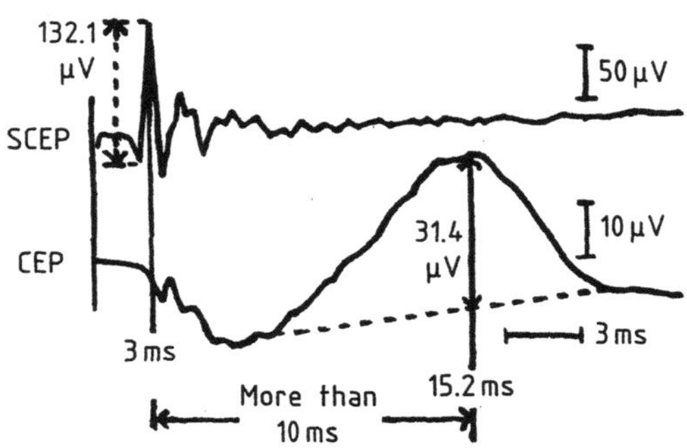

Fig. 2. Basic patterns of both potentials and the measuring methods of latency and amplitude

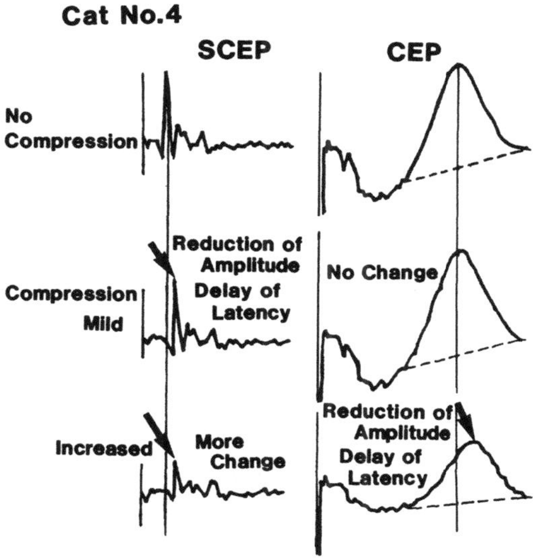

Fig. 3. Beginning of change

Beginning of Change

In cat No. 4 (Fig. 3), under the mild compression, the spinal cord evoked potential begins to reduce in amplitude and there is delay in latency, but the cortical evoked potential shows no change at that point. When the compression is increased, the cortical evoked potential begins to change. Four cats showed an earlier change in spinal cord evoked potential. Only one cat showed earlier change in cortical evoked potential. The remaining four cats showed a simultaneous change in both potentials. These data suggest that the spinal cord evoked potential is better when attention is paid to the beginning of a change.

Delay Times

Delay times of cortical evoked potential are constantly longer than those of spinal cord evoked potential. And the more the compression, the greater the difference between them becomes (Fig. 4). This tendency was observed in seven cats. In two other cats the peaks of cortical evoked potential were gently sloping, so their latencies were not clear. This result suggests that a delay of the cortical evoked potential can be seen more easily.

Reduction of Amplitude

The percentages of amplitude during the compression period were compared when there was 100% amplitude without compression. In six cats the spinal

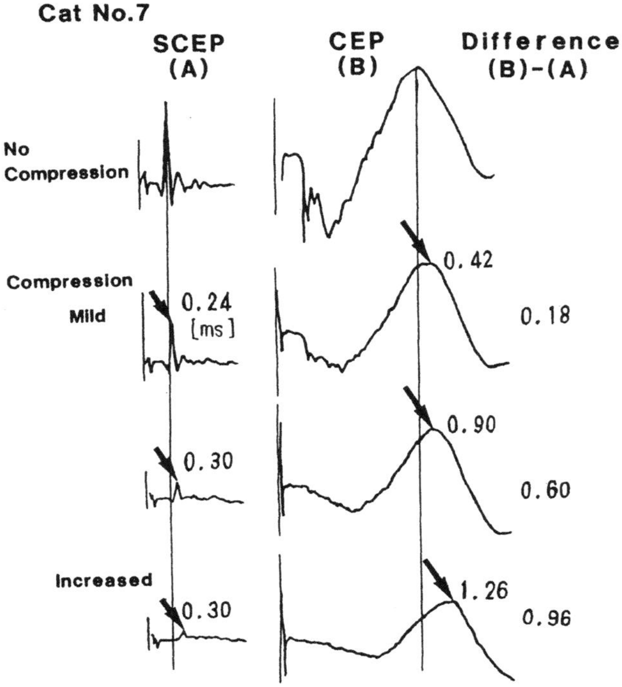

Fig. 4. Delay time of latency

cord evoked potential showed a greater reduction of amplitude than the cortical evoked potential. Only one cat (no. 2) showed a greater reduction of the cortical evoked potential, and in two other cats there was no particular relationship between the two potentials. So the reduction of amplitude of the spinal cord evoked potential can be seen more easily.

Critical Point

The "critical point" of spinal cord potential is generally considered to be about 50% of amplitude without compression. Critical means that dysfunction of spinal cord maybe become symptomatic. In six cats spinal cord evoked potential showed greater reduction of amplitude. Cortical evoked potential amplitudes was reduced to 84.5% of amplitude without compression on the average, when the amplitude of the spinal cord evoked potential was reduced to the critical point.

Conclusion

1. Monitoring the amplitude of spinal cord evoked potential was suitable for the early discovery of spinal cord injury.
2. Amplitude of spinal cord evoked potential and latency of cortical evoked potential were good indicators of spinal cord injury.
3. About 85% of amplitude without compression was a reasonable critical point for the cortical evoked potential monitored in experimental dorsal compression on spinal cord.

Spinal Cord Functions with Spinal Evoked Potentials and Spinal Blood Flow by Retracting the Spinal Cord

H. Kuchiwaki[1], S. Inao, K. Andoh, H. Ishiguri, and K. Sugita

Introduction

Careful manipulation is needed in spinal surgery to avoid postoperative impairment of spinal cord function. For the purpose of predicting these impairments there have been many studies which evaluate spinal cord function after various kinds of experimental models of trauma or ischemic insult. However, there are few reports of estimating the impairment of spinal cord function using a model of spinal cord retraction after laminectomy.

Our study was designed to detect the relationship between spinal cord function as determined by spinal motor evoked potentials (SpMEP), spinal somatosensory evoked potentials (SpSSEP), and spinal cord blood flow (SCBF) at the time of spinal cord retraction after laminectomies of the high cervical region.

Material and Methods

Four adult mongrel cats, weighing 2.5 to 4.6 kg, were anesthetized under an intramuscular injection of ketamine (50 mg/kg). A tracheal cannula was intubated and breathing was spontaneous. An arterial catheter was introduced via the femoral artery and connected to a pressure transducer (T-ICP II, Toyota Cent Res Inst., Aichi) for continuous measurement of systemic arterial blood pressure (SAP). One catheter was inserted in the femoral vein for injecting anesthetics. These agents were added, if necessary, using 1–2 ml of chloralose/urethane solutions (chloralose 10 mg/urethane 50 mg/kg). Expiratory CO_2 gas was monitored with a carbon dioxide analyzer (1H21, NEC-SANEI Co., Ltd., Tokyo). Blood gases were also analyzed in several experiments.

[1] Department of Neurosurgery, Nagoya University School of Medicine, 65 Tsurumai-cho, Showa-ku, Nagoya, 466 Japan

K. Shimoji et al. (Eds.)
Spinal Cord Monitoring and Electrodiagnosis

After an animal was fixed prone on the operating table, a midline incision was made from the nasion to the level of the third cervical spinal process. A laminectomy of the first cervical vertebra (Cl) and a partial laminectomy of the second (C2) were carried out. A small suboccipital craniectomy was also undertaken, followed by the construction of a small window at the interlaminal space (C2) and the third cervical lamina (C3) for the placement of electrodes.

A small craniectomy was undertaken to stimulate the motor cortex under the dura mater. A bipolar electrode (UH-1012, Unique Medical Co., Ltd., Osaka) was attatched to the dural surface. Stimulation for SpMEP was brought about using 50–70 V pulse for 0.1 ms at 4 Hz. A tungsten electrode (Clark Co., UK) with impedance of 12 MOhm, 0.5 mm tip diameter, and 75 mm in length was used for monitoring evoked potentials at the level of C1 (proximal portion) and the interlaminal space between C2 and C3 (peripheral portion). These electrodes were connected to an acoustic amplifier (3G26, Nec-Sanei Co., LTD., Tokyo). A low-pass filter (less than 1.5 Hz) and a high-pass filter (greater than 32 Hz) were used. A reference electrode was placed at the nasion when the left sciatic nerve was being stimulated and on the muscles when the motor area was being stimulated because of a large stimulation artifact. Electrical stimulation for the sciatic nerve was 1–2 V for 0.1 ms at 4 Hz (Electric Stimulator Model SEN-1101, Nihon Kohden, Tokyo). A computer system (7T17, NEC-SANEI Co., Ltd.) was used for averaging (200 times), displaying, and recording of both SpMEP and SpSSEP.

SCBF was measured using two H_2 clearance electrodes (World Medical Co., Nagoya), one of which was placed in the junctional area of the medulla and the cervical cord for comparison with values at the retracted portion of the spinal cord, where another electrode was inserted. Pure hydrogen gas was inhaled in 2 min.

A retractor with a 2.9-mm width was perpendicularly attached to the right side of the cord at the C1/C2 portion and was retracted medially by a micromanipulator (Sumitt Medical Co., Ltd., Tokyo) at a rate of 0.3 mm/s to one half of the cord width (Fig. 1). The retraction was continued for 60 min, followed by a 60-min observation of the electrophysiological recovery of the spinal cord function and SCBF. All the data during and after the retractions were compared with the control data.

Each peak of SpSSEP and SpMEP was respectively considered as follows: (1) SpSSEP: The first positive peak(P), which is clearly observable in the proximal portion, indicates a latency of 5.5 ms (P5.5) followed by a negative peak (N) of 6.1 ms (N6.1). Their peak to peak heights (P5.5/N6.1) were evaluated as voltages. P5.1 and N5.8 were also taken as peaks of SpSSEP at the peripheral portion. Their peak heights were also considered as voltages. The abnormal values are considered to be a delay time greater than 5.9 ms for P5.5 and a delay time greater than 6.5 for N6.1 at the proximal portion, respectively. The abnormal values at the peripheral portion are P5.1 $>$ 5.7

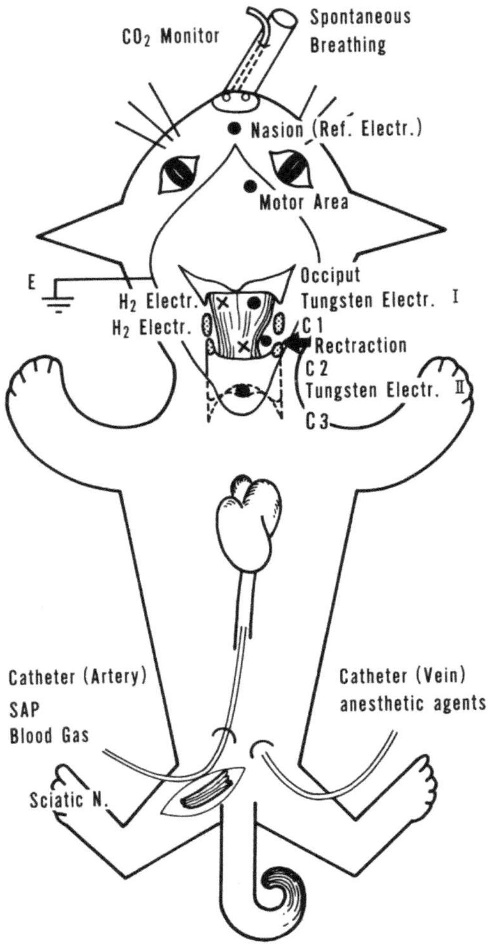

Fig. 1. Monitoring model. *Ref. Electr.* reference electrode; *N*, nerve; *C* cervical vertebra; *SAP*, systemic arterial blood pressure

and N5.8 > 6.4, respectively. These mean values and definition of the abnormal values from a large series will appear elsewhere. (2) SpMEP: SpMEPs at the proximal portion were P2.2 (2.4) and N3.1 (3.9). Those at the peripheral portion were evaluated at P1:P1.3 (1.5); N1, N2.0 (2.6); P2, P2.7 (3.7); and N2, N3.5 (4.3). The peak latencies and their abnormal values, which are greater than the value given in parentheses, from a large series will also appear elsewhere.

Results

General Condition of the Animals

The SAP ranged from 140 mmHg systolic to 66 mmHg diastolic pressure. No hypotensive state appeared in any experiment. States of blood gas were found within normal range. A retractor was carefully attached to the spinal cord so that respiratory movement did not become irregular. No animal died during these experimental procedures.

Changes in SCBF

SCBF ranged from 55.0 to 22.3 ml/100 g per minute. Two animals (cat no. 1 and 2) showed about 50% of the control SCBF during the retraction. By a release of the retraction, SCBF recovered up to a level of more than 60% of the control. These were grouped as "A", with SCBF decreasing less than 59% of the control by a retraction. The other two animals (cat no. 3 and 4) showed reduced SCBF that was more than 60% of the control SCBF during the retractions. After a retraction was released the SCBF showed about 90% of the control (cat no. 4). A recovery of SCBF of cat no. 3 was not obtained because of a nonfunctioning electrode. This was group "B". SCBF at a reference point, where a retraction was not involved, changed between 90% and 120% of the control.

Evaluation of SpSSEP

Changes of peak latency time in SpSSEP are shown in Table 1. The variations in P/N time ranged within 10% of control values in group B. However, changes in group A were greater than 10% of the control value. In contrast, the evoked potentials changed but showed no tendency to decrease or increase. Each abnormal datum is marked with an asterisk.

Evaluation of SpMEP

Changes of peak latency times in SpMEP are shown in Table 2. In cat no. 3 SpMEP was not evaluated. As shown in Table 2 SpMEP varied between animals. However, the changes in group B were slightly smaller than those in group A. Each abnormal datum is marked with an asterisk.

Discussion

There have been many experimental models of spinal cord trauma by dropping a weight from several centimeters above [5] or balloon compression [2, 3]. An ischemic insult was frequently introduced by lowering SAP. For protection against deterioration of the spinal cord, multimodality evoked potentials are frequently used. Recently a technique to monitor motor potentials was developed for clinical practice [4]. As the pyramidal tract

Table 1. Changes in SpSSEP

Cat no.			Control (ms) (μV)	After 60 min of retraction (ms) (μV)	60 min after release of retraction (ms) (μV)
Group A					
	Proximal portion	P	5.9	6.7*	6.8*
			]0.02	]0.23	]0.20
1		N	6.3	7.3*	7.6*
	Peripheral portion	P	5.7	5.7	5.7
			]0.14	]0.10	]0.20
		N	6.4	6.3	6.3
	Proximal portion	P	5.7	6.4*	5.9
			]0.15	]0.94	]0.11
2		N	6.3	7.3*	7.6*
	Peripheral portion	P	5.4	5.7*	6.0*
			]0.13	]0.25	]0.37
		N	6.0	6.7*	6.8*
Group B					
	Proximal portion	P	5.2	5.7	5.5
			]0.31	]0.04	]0.36
3		N	6.0	6.0	6.5
	Peripheral portion	P	5.0	5.2	5.3
			]0.14	]0.13	]0.27
		N	5.5	5.9	6.0
	Proximal portion	P	5.3	5.5	5.7
			]0.14	]0.14	]0.11
4		N	6.0	6.0	6.5
	Peripheral portion	P	5.0	5.2	5.2
			]0.16	]0.31	]0.7
		N	5.6	5.7	5.7

P, positive; *N*, negative; * abnormal value

descends in the anterior and lateral sides of the spinal cord, monitoring the motor functions seem to be useful in cases of a traction at this region.

Although our study consists of a small number of experimental animals, favorable data by SpMEP were not obtained for the monitoring of the functions. Variations of data in SpMEP were greater than those in SpSSEP. However, there is support for SpMEP from a weight drop injury in the literature [6].

From our data, SCBF seems to be an important factor for the impairment of cord function. There are reports concerning the mechanism of deterioration of the spinal cord due to ischemia [1, 3]. On the other hand, ischemia of the spinal cord was not the reason for the deterioration of spinal function [5, 7]. However, this point is controversial.

Dolan et al. [2] showed a model of a spinal injury with traction of the spinal canal. In this case, when the retraction of the spinal canal was increased, the cord was not only compressed, but also tracted. If this mechanism is involved in a spinal injury, it may act as another factor in the

Table 2. Changes in SpMEP

Cat no.			Control (ms) (μV)	After 60 min of retraction (ms) (μV)	60 min after release of retraction (ms) (μV)
Group A					
	Proximal portion	P	2.1	2.3	2.3
			]0.54	]1.00	]0.9
1		N	3.2	3.6	3.2
	Peripheral portion	P1	1.3	1.4	1.5
			]0.75	]1.00	]1.20
		N1	1.8	1.9	2.1
		P2	2.1	2.8	2.3
			]0.50	]0.12	]0.95
		N2	3.1	3.4	3.2
	Proximal portion	P	2.3	3.0*	3.0*
			]0.52	]0.66	]0.70
2		N	3.1	3.9*	3.9*
	Peripheral portion	P1	1.3	1.7*	1.7*
			]0.18	]0.20	]0.22
		N1	2.2	2.3	2.3
		P2	2.6	3.1	3.1
			]0.18	]0.04	]0.16
		N2	3.9	3.9	3.9
Group B					
	Proximal portion	P	2.1	2.1	2.2
			]1.50	]0.87	]0.84
4		N	2.5	2.8	3.0
	Peripheral portion	P1	1.3	1.4	2.2
			]0.88	]0.15	]0.84
		N1	1.8	1.8	3.0
		P2	2.6	2.6	3.1
			]0.17	]0.25	]0.24
		N2	3.1	3.4	3.4

P, positive; *N*, negative; * abnormal value

deterioration of the spinal function. In our retraction model, when it becomes at most near to one half of the spinal cord width, the peripheral side of the cord is tensed and lifted towards the proximal side.

In conclusion, with our model of stimulated spinal surgery, it is important to keep SCBF greater than 60% of the control by retraction of the spinal cord. It would be better to monitor SCBF and SpSSEP frequently rather than monitor SpMEP.

References

1. Bennett MH (1983) Effects of compression and ischemia on spinal cord evoked potentials. Exp Neurol 80:508–519

2. Dolan EJ, Transfeldt EF, Tator CH, Simmons EH, Hughes KF (1980) The effect of spinal distraction on regional spinal cord blood flow in cats. J Neurosurg 53:756–764
3. Grifiths IR, Trench JG, Crawford RA (1979) Spinal cord blood flow and conduction during experimental cord compression in normotensive and hypotensive dogs. J Neurosurg 50:353–360
4. Grundy BL (1982) Monitoring of sensory evoked potentials during neurosurgical operations: methods and applications. Neurosurgery 11:556–575
5. Kobrine AI, Evans DE, Rizzoli HV (1979) Experimental acute ballon compression of the spinal cord. J Neurosurg 51:841–845
6. Levy W, McCaffrey M, York D (1986) Motor evoked potential in cats with acute spinal cord injury. Neurosurgery 19:9–19
7. Tarlov IM (1972) Acute spinal cord compression paralysis. J Neurosurg 36:10–20

New and Sensitive Methods for Detecting Acute Axonal Dysfunction After Experimental Spinal Cord and Root Compression Injury

K. SAKATANI[1], Y. HATABU, W. YOUNG, J.A. GRUNER, H. IIZUKA, and K. TANIGUCHI

Introduction

Neural conduction in the central and peripheral nervous system is commonly evaluated by averaged evoked potentials. For example, somatosensory evoked potentials (SEP) are widely used for clinical diagnosis and monitoring of spinal cord injury. However, averaged evoked potentials have several disadvantages when used in the assessment of dysfunctional conduction. First, responses evoked by stimuli delivered at a constant frequency overlook the dynamic aspects of neuronal function. Injured neuronal responses are not stable but fluctuate depending on the frequency of stimulation. Second, the averaging process degrades the information content of evoked responses. For example, the latency and amplitude of averaged waveforms do not necessarily reflect the average latencies and amplitudes of individual responses. Third, averaged evoked potentials often may not reveal subtle changes in neuronal function following mild trauma.

To circumvent these limitations of conventional SEP, we developed a new stimulus method and several new signal analysis approaches. These methods were tested in rat spinal cord and root compression injury model. These methods are based on the measurement of axonal conduction characteristic over a range of stimulus frequencies.

Materials and Methods

Adult Long-Evans rats weighing 200–300 g were anesthetized with 40 mg/kg i.p. pentobarbital. A laminectomy was performed from T8 to L6 to expose the spinal cord and roots. The animals were mounted in a stereotaxic frame, and vertebral clamps were used to stabilize the spinal column. The spinal cord and spinal roots were exposed by laminectomy and covered with warm

[1] Department of Neurosurgery, New York University Medical Center, 550 First Avenue, New York, NY 10016, USA

K. Shimoji et al. (Eds.)
Spinal Cord Monitoring and Electrodiagnosis

(36°–37°C) paraffin oil. The animals were paralyzed with gallamine triethiodide (1 mg/kg i.v.) and respired with a rodent respirator. Pneumothoraces were created to reduce spinal cord movements due to respiration.

Stimulation and Recording Methods

L5 dorsal roots were cut and placed on bipolar hook stimulating electrodes (cathode proximal). The roots were stimulated supramaximally for myelinated fibers. Trains of double pulses with interpulse intervals 3, 4, 5, 8, 10, 20, 30, 50, and 80 ms were delivered. For the purpose of data presentation, the interpulse intervals were converted to a frequency term, i.e. 333, 250, 200, 125, 100, 50, 33, 20, and 12.5 Hz, which we call interpulse frequency. These double pulses were randomly delivered with a constant base frequency (5 Hz). Compound action potentials (CAP) were recorded above (10 mm rostral) and below (10 mm caudal) a spinal cord injury site with monopolar silver ball electrodes. The evoked responses were amplified with band-pass filters set at 0.08 and 5k Hz.

Injury Methods

To injure the spinal cord, we compressed the T10 spinal cord segment dorsoventrally with a plastic plate (3 × 5 mm) attached to a micromanipulator. Compression was increased every 5 min in stepwise increments of 0.5 mm until the dorsal column CAP at 1 Hz were absent. The spinal cord was then decompressed after 10 min. To injure the dorsal roots, the L5 dorsal roots were compressed with forceps until the dorsal root CAP were approximately 50% of preinjury amplitudes.

Data Analysis

The first stimulus of the double pulses is called "conditioning" pulse and the second the "test" pulse. We evaluated the evoked responses activated by the test (second) pulses in three ways. First, we plotted scatter diagrams in which the amplitude of a given response (n) was plotted against the amplitude of the next response (n + 1). Second, we assessed the effect of interpulse intervals on response. The amplitude of the test pulse responses were plotted against interpulse intervals. Third, to assess changes in waveforms, the evoked responses were displayed using a phase space diagram of amplitude vs. velocity (time derivative or change of amplitude at each time point of the waveform). The time derivative of each response, dV/dt, was calculated for each point of the waveform. Next, the derivatives of the point were plotted against the amplitude at that point, i.e., $[(dV/dt)_t$ vs. $V_t]$.

Results

Figure 1A and B, respectively, show the dorsal column and dorsal root CAP to the first 100 test stimuli before and after injury. Baseline shifts occurred when the interpulse frequency was high enough to overlap with the synaptic components evoked by the conditioning stimuli. Before injury, both CAP showed minimal changes in amplitude and latency. However, injured axons showed larger fluctuations of CAP amplitude and latency. The negative peak was particularly unstable, resulting in large changes in the slope from the positive to negative peak.

Figure 2A shows the changes of the dorsal column CAP amplitudes to the first 100 test stimuli before and after spinal cord injury. After spinal cord

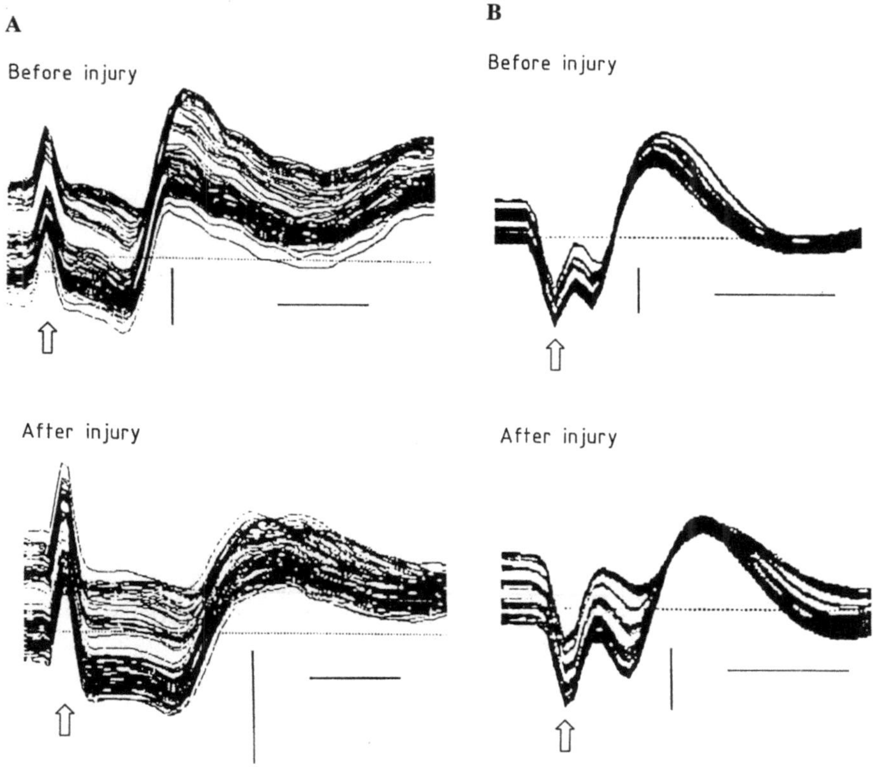

Fig. 1A,B. Dorsal column (**A**) and dorsal root (**B**) CAP to the first 100 test (second) stimuli before and after injury. Before injury, both CAP showed minimal changes in amplitude and latency. However, the CAP showed larger fluctuations after injury. Fluctuation of the negative peak was particularly prominent, resulting in a large fluctuation of the slope from the positive to negative peak. Baseline shifts occurred when the interpulse frequency was short enough to overlap with the synaptic components evoked by the conditioning stimuli. The *horizontal bar* indicates 0.5 ms and the *vertical bar* indicates 0.5 mV (**A**) and 1 mV (**B**). Positivity is downward. The *arrow* indicates stimulus artifact

A

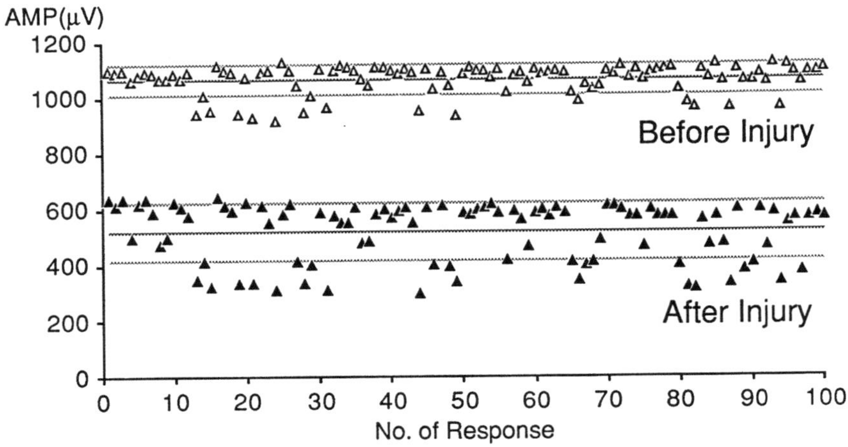

B

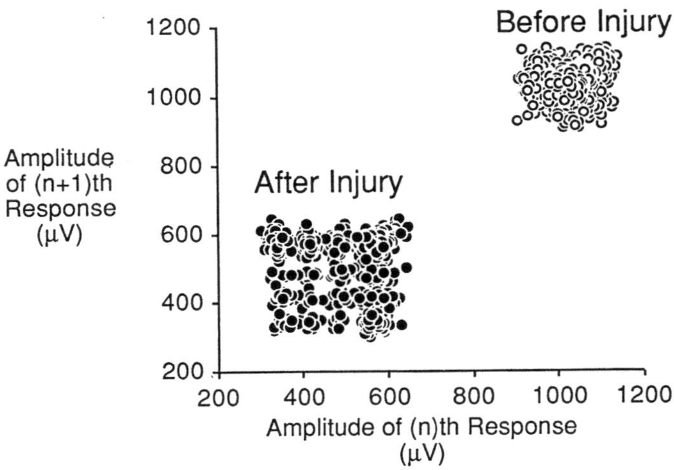

Fig. 2. **A** Changes of the dorsal column CAP amplitudes to the first 100 test stimuli before and after spinal cord injury. After spinal cord injury, the standard deviation of the amplitude increased from 5% of the mean amplitude to 20%, in spite of a decrease in mean amplitude of over 500 μV. The *dotted lines* indicate the mean amplitude of the first 100 responses and the standard deviation of amplitude. **B** Scatter diagram of the dorsal column CAP amplitude. The response amplitude (n) was plotted against the response amplitude (n + 1). Before injury, the dorsal column responses grouped around the mean amplitude. After injury, the responses scattered, forming several subgroups

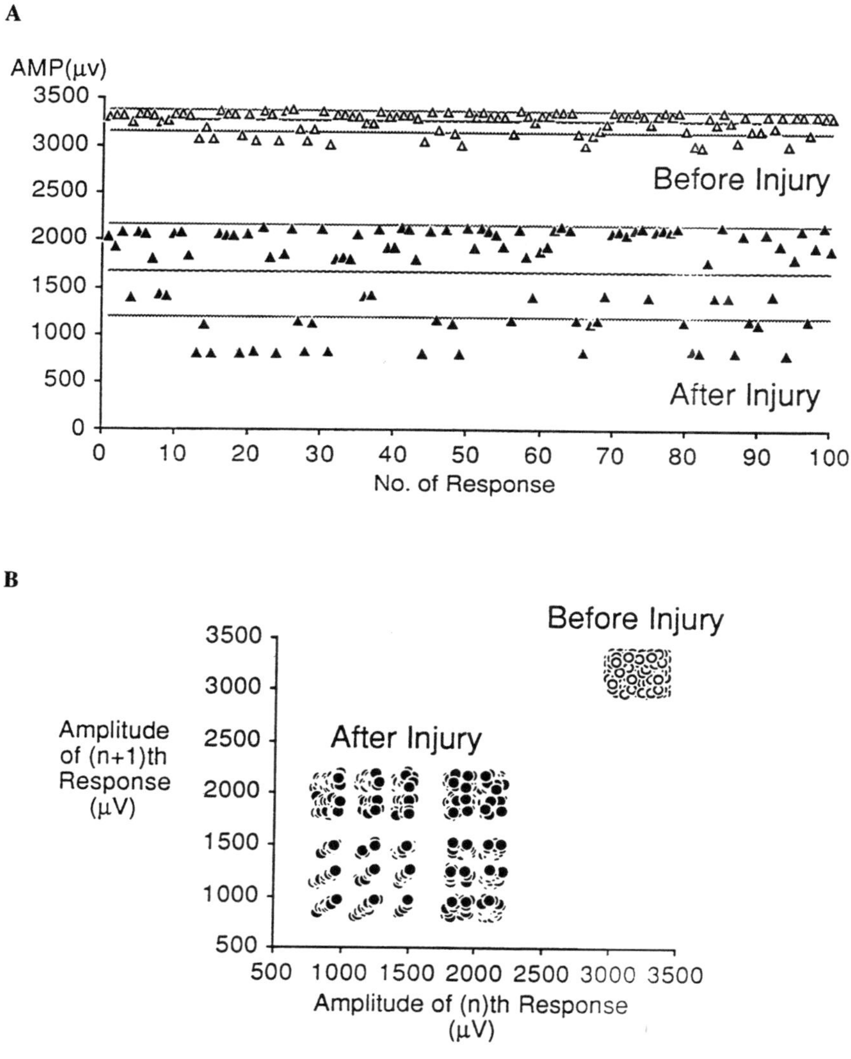

Fig. 3. **A** Changes of the dorsal root CAP amplitudes to the first 100 test stimuli before and after injury. The dorsal root CAP amplitude exhibited a relatively small fluctuation (3% of the mean amplitude) compared with the dorsal column CAP before injury. However, after injury, the standard deviation of the amplitude increased from 3% of the mean amplitude to 29% in spite of a decrease in mean amplitude of over 1500 μV, indicating a larger fluctuation than the injured dorsal column. The *dotted lines* indicate the mean amplitude of the first 100 responses and the standard deviation of amplitude. **B** Scatter diagram of the dorsal root CAP amplitude to the first 1000 stimuli before and after injury. The group formation before injury was smaller than the injured dorsal column, and the subgroup formation after injury was observed more prominently than the injured dorsal column

injury, the standard deviation of the amplitude increased from 5% of the mean amplitude to 20% in spite of a decrease in mean amplitude of over 500 μV. The scatter diagram clearly demonstrates amplitude fluctuations (Fig. 2B). Before injury, the dorsal column responses grouped around the mean amplitude. After injury, the responses scattered forming several subgroups.

Figure 3A shows the changes of the dorsal root CAP amplitudes to the first 100 test stimuli before and after injury. The dorsal root CAP amplitude exhibited a relatively small fluctuation (3% of the mean amplitude) compared with the dorsal column CAP before injury. However, after injury, the

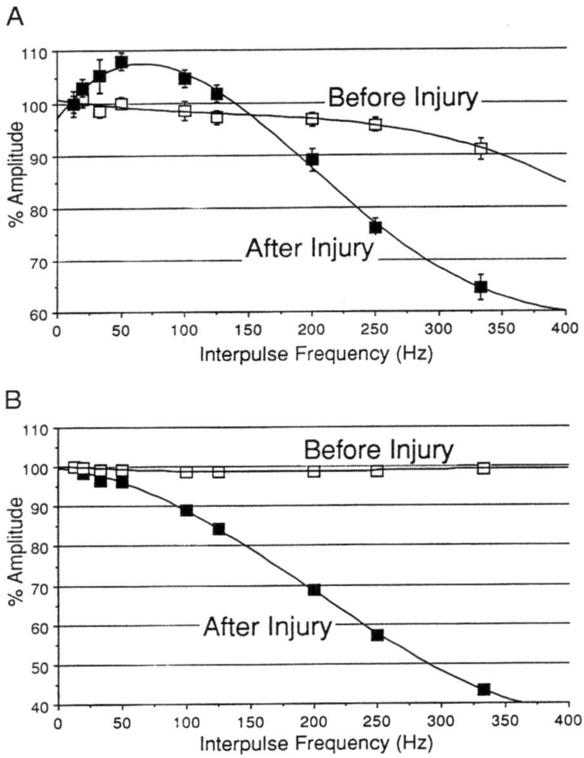

Fig. 4. A Dorsal column and dorsal root CAP amplitude as a function of interpulse frequency before and after spinal cord injury. The amplitude at each interpulse frequency was normalized to that at 12.5 Hz. Before spinal cord injury, the amplitude decreased slightly with increase in interpulse frequency. After spinal cord injury, the amplitude decreased progressively beyond 200 Hz. Furthermore, the amplitudes between 20–125 Hz exceeded those at 12.5 Hz. **B** The dorsal root responses showed a different pattern of frequency response. Normal dorsal root CAP amplitudes did not show the decreasing tendency with increase in interpulse frequency. After injury, a progressive decrease in amplitude was also observed at over 100 Hz; however, the magnitude of decrease in amplitude was more prominent than the dorsal column CAPs. The paradoxical augmentation between 20–125 Hz was not observed in dorsal root CAP

standard deviation of the amplitude increased from 3% of the mean amplitude to 29% in spite of a decrease in mean amplitude of over 1500 μV, indicating a larger fluctuation than the injured dorsal column. Figure 3B shows the scatter diagram of the dorsal root CAP amplitude to the first 1000 stimuli before and after injury. The scatter of the points before injury was smaller than the normal dorsal column. The scatter of the points after injury was observed more prominently than the injured dorsal column.

Figure 4A shows the dorsal column and dorsal root CAP amplitude as a function of interpulse frequency before and after spinal cord injury. The amplitude at each interpulse frequency was normalized to the amplitude of the response activated with interpulse frequency of 12.5 Hz. Before spinal cord injury, the amplitude decreased slightly with increased interpulse frequency. After spinal cord injury, the amplitude fell progressively with an increased interpulse frequency greater than 200 Hz. Note the increased amplitudes between 20–125 Hz. Such a paradoxical augmentation did not occur before spinal cord injury. The dorsal root responses showed a different pattern of frequency response (Fig. 4B). Normal dorsal root CAP amplitudes did not change with increased interpulse frequency. After injury, a progressive decrease in amplitude occurred at interpulse frequencies greater than 100 Hz. However, the decrease in amplitude was more prominent than the dorsal column CAP. Dorsal root CAP did not show the paradoxical augmentation between 20–125 Hz.

Figure 5 shows the dorsal column and dorsal root CAP in the phase space diagram before and after injury. Fluctuations of the negative peak of the dorsal column CAP were particularly prominent, resulting in a large variation of the slope between the positive to negative peaks (see Fig. 1). This portion of the curve is mapped in the phase space as the portion below the abscissa. The fluctuation of this region of the waveforms is evident. Thus, the phase space diagram allows direct visualization of the fluctuations in CAP velocity.

Discussion

Fluctuation of Dorsal Column CAP Before and After Spinal Cord Injury

Before injury, the dorsal column and dorsal root CAP to the test stimuli showed amplitude changes, suggesting that impulse conduction of the normal dorsal column and dorsal root was fluctuating. It is commonly assumed that action potentials along the axon travel at a constant velocity. However, several observations argue against this assumption. First, Derksen and Verveen [2] found that the membrane potentials of the myelinated axon are not absolutely constant. Second, Lass and Abeles [4] found fluctuation in the conduction velocity along the frog sciatic nerve. The fluctuations increased

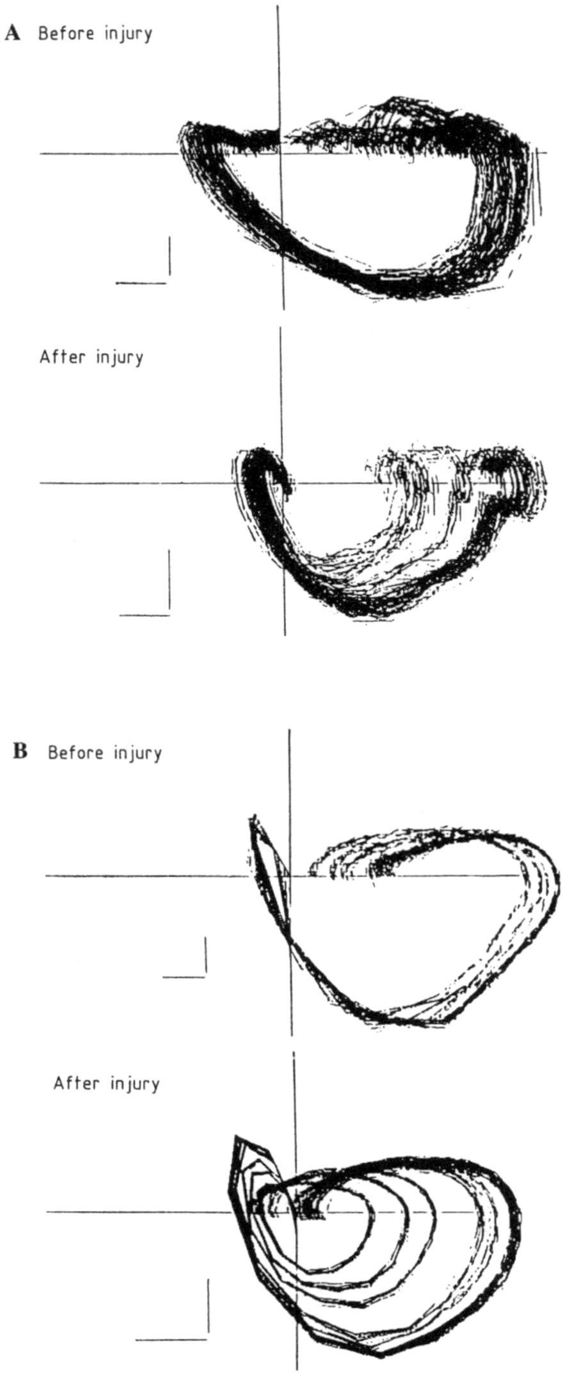
A
Before injury
After injury
B
Before injury
After injury

when stimulus intensity was reduced towards threshold. Third, Musha et al. [5] found that the temporal relationships of action potentials propagating along a squid axon differ from those evoked by Gaussian impulse trains. These observations support our CAP results showing that impulse conduction of the dorsal column normally fluctuate to some degree.

The dorsal root CAP had smaller fluctuations compared with the dorsal column CAP. This finding suggests that impulse conductivity of primary afferent fibers changed after entering the dorsal column (CNS) from the dorsal root (PNS). Two possible explanations of the difference in impulse conductivity between the CNS and PNS should be considered. First, once the primary afferent fiber enters in the CNS, it gives multiple branches to gray matter and ascending and descending branches. The conduction of impulses in the primary afferent fibers may be modulated at branchpoints within the spinal cord. The longitudinal current driving saltatory conduction may be variably shunted into these branches. Second, the axonal diameter may change before and after branchpoints, possibly contributing to variability of conduction velocity, amplitude, and refractory period. Third, differences in microenvironment between the CNS and PNS may cause the changes of impulse conduction. For example, it has been reported that stimulation of the afferent fibers produces extracellular potassium accumulation in the spinal cord [7]. Extracellular potassium changes may modulate spinal cord transmission, especially in injured spinal cord.

After injury, the dorsal column and dorsal root CAP recorded distal to the lesion site showed larger fluctuations than before injury. CAP recorded below the lesion did not change significantly after injury. These results indicate that the increased fluctuation of the dorsal column and dorsal root CAP was caused by injury. Since transmission of information is mediated in part by frequency modulated spike trains, such abnormal behaviors of injured fibers may result in transmission of inappropriate messages to the higher centers.

Fig. 5A,B. Dorsal column and dorsal root CAP in the phase space diagram before and after injury. In the phase space, the potential amplitude V_t (*abscissa*) was displayed against its time derivative of each point, dV/dt (*ordinate*). Note that the plus direction of the abscissa indicates minus value of amplitude, and the plus direction of the ordinate indicates plus value of time derivative. Therefore, the trajectory in the *lower half of the graph* corresponds to the portion from initial positive peak to negative peak. Before injury, the dorsal column and dorsal root CAP show a relatively narrow bundle of the trajectories. However, after injury, they formed a wide bundle of the trajectories forming several sub-bundles. The *horizontal bar* indicates 0.1 mV (**A**) and 0.5 mV (**B**), and the *vertical bar* indicates 1 μV/μsec (**A**) and 5 μV/μsec (**B**)

Characteristics of the Dorsal Column and Dorsal Root CAP at Different Stimulus Frequencies

Before spinal cord injury, the dorsal column CAP amplitude decreased only slightly with increased interpulse frequency (Fig. 4A). The dorsal root CAP amplitude did not change with interpulse frequencies up to 333 Hz before injury (Fig. 4B). However, after injury, CAP amplitudes at high interpulse frequencies were significantly smaller than those at lower stimulus frequencies in both dorsal column and root. CAP recorded below the lesion did not change after injury, indicating that the conduction failure at high interpulse frequencies was caused by injury. These results suggest the presence of three types of axons, i.e., "abnormal axons" that can not be activated at any stimulus frequency, "normal axons" that can be activated at a high stimulus frequency, and "subnormal axons" that can be activated only at a low stimulus frequency.

After spinal cord injury, the dorsal column CAP amplitudes activated at 20–125 Hz exceeded those at 12.5 Hz (Fig. 4A). We also found such a paradoxical augmentation after train stimulations at the same stimulus frequencies [6]. Such paradoxical augmentations were not observed before spinal cord injury. Recruitment of conducting dorsal column axons may explain this phenomenon. Specifically, we propose that depolarization changes the resting potential of abnormal dorsal column axons close to threshold, allowing transmission at moderate stimulus frequencies (20–100 Hz). Recently, we found that repetitive sciatic nerve stimulation produced significant spinal cord extracellular K^+ rises in myelin deficient mutant rat, but small or no changes in the normal rat [8]. We suggest that injured axons not only release more K^+ during action potentials but are more sensitive to changes in extracellular K^+ activity than normal axons.

Dorsal root responses showed a different pattern of the frequency response after injury (Fig. 4B). CAP amplitude progressively decreased at interpulse frequencies greater than 100 Hz. However, the paradoxical augmentation between 20–125 Hz did not occur in injured dorsal root. The reason for this difference is not clear. Injured dorsal root axons may not produce enough extracellular K^+ rises to depolarize adjacent axons. Alternatively, they may be less sensitive to extracellular K^+ rises than the injured dorsal column axons.

Diagnostic Value of Measurement of SEP Fluctuation

Double pulse stimulation places greater physiological demands on axons. It thus should be a more sensitive approach to detecting abnormal axonal behavior than averaging responses evoked by a constant low frequency stimulation. Our results suggest that the double pulse method can detect the presence of mildly injured or injured subpopulations of axons. It has been reported that high frequency stimulus trains are useful for detecting axonal dysfunction [3]. Although high frequency stimulus trains are sensitive to impaired axonal conduction, they are not well suited for continuous monitoring of SEP for long periods. Furthermore, such approaches do not allow the responses of the axons to be assessed simultaneously over a wide stimulus frequency range. Finally, continuous high frequency trains introduce an additional variable of fatigue that would not be as prominent with double pulses delivered at a constant low frequency.

The phase space diagram demonstrates CAP fluctuation more clearly than simple amplitude versus time displays (oscillographic waveform; Fig. 5). The phase space has been employed for the analysis of dynamic systems in the field of physics [1]. Assuming that action potentials represent the trajectory of dynamic responses, the phase space diagram allows direct visualization of response fluctuations that are otherwise not evident. The combination of the phase space diagram and continuous double pulse stimulation provides new powerful physiological diagnostic methods for detecting axonal dysfunction.

References

1. Aihara K, Matsumoto G (1986) Chaotic oscillations and bifurcations in squid giant axons. In: Holden AV (eds) Chaos. Princeton Univ Press, New Jersey, pp 257–269
2. Derksen HE, Verveen AA (1966) Fluctuations of resting neural membrane potential. Science 151:1388–1389
3. Edwards MSB, Bolger CA, Jewett DL (1983) Diagnosis of acute nerve compression in the cat with high frequency nerve trains evoked responses. Neurosurgery 12:1–6
4. Lass Y, Abeles M (1975) Transmission of information by the axon: 1. Noise and memory in the myelinated nerve fiber of the frog. Biol Cybern 19:61–67
5. musha tT Kosugi Y, Matsumoto G, Suzuki M (1981) Modulation of the time relation of action potential impulses propagating along an axon. IEEE 28:616–623
6. Sakatani K, Ohta T, Shimo-oku M (1987) Conductivity of dorsal column fibers during experimental spinal cord compression and after decompression at various stimulus frequencies. Central Nervous System Trauma: 4:161–179
7. Syková E (1987) Modulation of spinal cord transmission by changes in extracellular K^+ activity and extracellular volume. Can J Physiol Pharmacol 65:1058–1066
8. Young W, Rosenburth J, Wojack JC, Sakatani K, Kim H (1989) Extracellular potassium activity and axonal conduction in spinal cord of the myelin-deficient mutant rat. Exp Neurol 106:41–51

Physiological and Pharmacological Aspects of the Modulation of Nociceptive Transmission at the Dorsal Horn Level of the Spinal Cord

J.-M. BESSON[1]

As proposed by Melzack and Wall in the "gate control" theory, the transmission of nociceptive messages at the spinal relays is subject to both segmental and supraspinal controls.

Segmental Controls

Segmental controls are characteristically inhibitory effects produced by large-diameter fibers on responses of spinal neurons to nociceptive stimulation [see 2, 9]. The inhibitions can be produced on both nociceptive specific and nonspecific (multireceptive) neurons and can also be obtained by dorsal column stimulation. In this latter situation, the resultant inhibition is produced by activation of collaterals of primary afferent fibers in the posterior columns passing into the spinal gray matter, since the effects are still present after a total spinal section. Nevertheless, in intact animals the additional involvement of a supraspinal loop cannot be excluded, especially since Gerhart et al. [6] have demonstrated inhibition of primate spinothalamic neurons produced by stimulation of the ventrobasal nucleus of the thalamus. These experimental findings relative to the inhibition induced by large afferent fiber activation can partly explain the therapeutic effectiveness of the electrical stimulation (weak intensity, high frequency) of peripheral nerves or dorsal columns. Interestingly, the natural activation of large afferent fibers (mechanical vibratory stimulation) induces a reduction in chronic pain and an increase in the pain threshold in normal subjects. There is, however, a major divergence between the clinical and experimental findings, namely, that patients are relieved for several hours after several minutes of stimulation of large afferent fibers, whereas the inhibitions observed on dorsal horn neurons only last for seconds or minutes. This differences is difficult to explain, but

[1] Unité de Recherches de Physiopharmacologie du Système Nerveux de l'INSERM (U. 161), 2 rue d'Alésia, 75014 Paris, France

K. Shimoji et al. (Eds.)
Spinal Cord Monitoring and Electrodiagnosis

may result from fundamental differences in the responsiveness of neuronal mechanisms (whether spinal or supraspinal) implicated in nociception in normal animals compared with those in chronic pain patients.

Both pre- and postsynaptic events have been proposed to explain the inhibitory effects induced by stimulation of large afferent fibers, and cells in the substantia gelatinosa (SG: lamina 2 of the dorsal horn) were considered responsible for the modulation of the transmission. Indeed, since most neurons in SG do not project toward supraspinal structures, an interneuronal system implicated in the transfer and modulation of nociceptive messages seems likely. Unfortunately, despite ultrastructural analysis of marked physiologically identified SG neurons, the *precise* involvement of these neurons again remains hypothetical. This is mainly due to the morphological, electrophysiological and neurochemical heterogeneity of this area. Whatever the case, the SG has been implicated in segmental controls of the transmission of nociceptive messages. At least a part of this control results from presynaptic mechanisms presumed to be acting via axoaxonic synapses where primary afferent terminals are postsynaptic.

From the pharmacological point of view, several substances have been suggested, notably gamma-aminobutyric acid (GABA) and the endogenous opioids, as playing a role in the mechanisms of presynaptic control.

Much pharmacological evidence favors a role for GABA in presynaptic inhibition. Immunohistochemical studies have demonstrated GABAergic cells in the dorsal horn often forming axoaxonic synapses with afferent fibers, and electrophysiological studies have shown GABA A and GABA B receptors on A and C afferent fibers. Interestingly, there is a significant decrease in GABA B binding in the dorsal horn of rats treated neonatally with capsaicin. Whatever the mechanisms (pre- or postsynaptic) GABA has been shown to be involved in the inhibitory effects induced by dorsal column stimulation on dorsal horn nociceptive neurons [4].

The direct depressive effects of morphine and enkephalins on the transmission of nociceptive messages at the spinal level are clearly established, since their systemic (in spinal animals for morphine) or iontophoretic administration selectively reduces the responses of dorsal horn neurons to nociceptive inputs [see 3]. The presence of opioid receptors on terminals of thin primary afferent fibers projecting into the superficial laminae of the dorsal horn, suggests a role for endogenous opioids in presynaptic controls on the transmission of nociceptic messages. However, immunocytochemical and anatomical evidence has not supported the initial proposition of presynaptic effects of opioids on fine primary afferent fibers since axoaxonic contacts have not been clearly demonstrated. In contrast, the postsynaptic action of opioids is supported by immunocytochemical and electrophysiological studies. Whatever its mechanism(s) of action, the functional implication of the spinal opioid system is well supported by the powerful and long-lasting analgesic effects seen in animals after intrathecal administration of morphine. Clinical application of this finding was immediate, and clinicians use this route of administration for the relief of intractable pain. Clinical effects clearly demonstrate the activation of spinal opioid receptors inhibits the transmission of nociceptive messages. It seems essential to identify precisely the respective roles of the different opioid receptors, which could differ according to different types of pain. Indeed, in animals, different opioid agonists applied intrathecally produce analgesic effects that depend on the type of cutaneous nociceptive stimulus (mechanical, thermal, and chemical), visceral or muscular, used to gauge the effects [see 10].

A major problem relating to the functional significance of the intrinsic opioid systems of the dorsal horn is to understand how these systems are triggered. In humans, the analgesia induced by high frequency and low intensity transcutaneous stimulation is segmentably distributed and generally not naloxone reversible. In contrast the analgesia induced by low frequency-high intensity peripheral stimuli, also activating small afferent fibers, seems to be associated with a release of endogenous opioids. This is in good agreement with several neurochemical studies in animals in which nociceptive stimulation has been shown to increase levels of met-enkephalinlike material in the cerebrospinal fluid. These results indicate that the spinal cord opioid systems could be triggered in a complex manner (activation of segmental mechanisms and involvement of an extrasegmental loop) by nociceptive inputs.

Supraspinal Modulation of Spinal Transmission of Noxious Messages

Descending pathways from the higher centers of the central nervous system can modify sensory or motor functions at the spinal level. Study of these influences on the activity of dorsal horn neurons driven by noxious inputs started at the end of the 1960s and presently has clearly established the effectiveness of descending inhibitory systems in the modulation of nociceptive messages. The mechanisms involved in these phenomena, mainly originating in the cortex and the brain stem, are still being studied extensively in an attempt to establish a neurophysiological basis for the behavioral and clinical findings of stimulation-produced antinociception (SPA) and as an explanation of some of the mechanisms underlying the analgesic effects of morphine.

This paper is essentially devoted to brain stem, whose role in the modulation of pain is strongly supported by both behavioral and electrophysiological studies [see 1, 2, 5, 9].

After the original report of Reynolds showing that stimulation of the periaqueductal gray matter (PAG) produced analgesia in the rat, we performed an extensive mapping study in the cat which includes 300 sites of stimulation spread over the brain stem between the mesencephalon and the medulla (Fig. 1). It appears that two main areas play a major role in SPA: the ventral PAG including the dorsal raphe nucleus and more posteriorly the

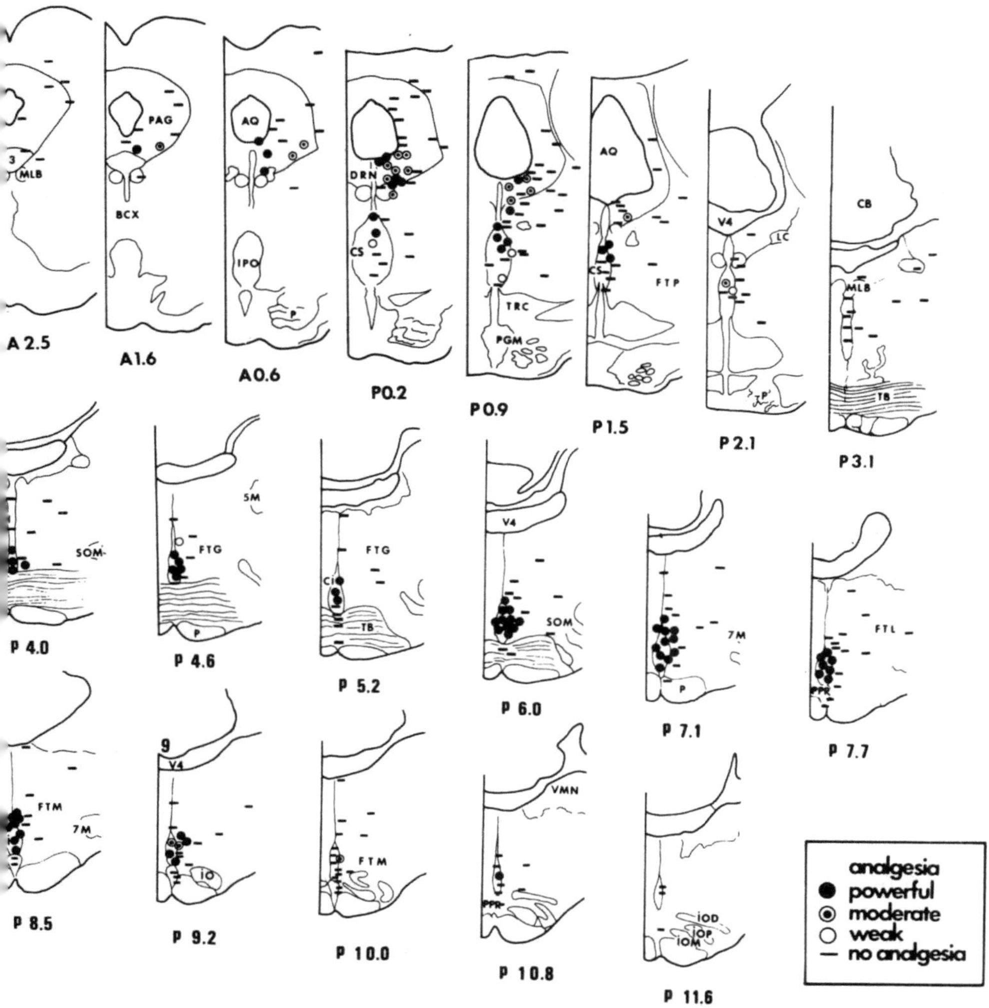

Fig. 1. Localization of stimulation-produced analgesia (SPA) efficacious sites throughout the mesencephalon and the medulla in the awake, freely moving cat. The most powerful analgesic effects were triggered by stimulation of nuclei raphe dorsalis (*DRN*), centralis superior (*CS*) and centralis inferior (*CI*, nucleus raphe magnus). *5M*, motor trigeminal nucleus; *FTL*, lateral tegmental field; *7M*, facial nucleus medial division; *FTM*, magnocellular tegmental field; *IO*, inferior olive; *VMN*, vestibular medial nucleus; *PPR*, postpyramidal raphe; *IOD*, dorsal nucleus of the inferior olive; *IOP*, principal nucleus of the inferior olive; *IOM*, medial nucleus of the inferior olive; *MLB*, medial longitudinal bundle; *PAG*, periaqueductal grey; *BCX*, decussation of the brachium conjunctivum; *AQ*, aqueduct; *IPO*, posterior interpenduncular nucleus; *TRC*, tegmental reticular nucleus; *PGM*, pontine grey matter; *DRN*, dorsal raphe nucleus; *FTP*, paralemniscal tegmental field; *V4*, 4th ventricle; *LC*, locus coerulus; *CB*, cerebellum; *TB*, trapezoid body; *SOM*, medial nucleus of the superior olive. From [8]

nucleus raphe magnus (NRM) from which the most prominent effects were obtained. PAG stimulation is used for the treatment of severe chronic pain in humans. However, there are many discrepancies between the various neurosurgical groups working in this field, and recent observations in animals could partly explain some of these contradictory results. Indeed, our recent investigations in the rat raise the problem of the possible involvement of stress-induced analgesia in most of the analgesic effects elicited by PAG stimulation and illustrate the relatively limited extent of "pure analgesic areas", and, consequently, one could presume that these zones are difficult to locate in humans.

We initially reported strong depressive effects induced by PAG stimulation on the responses of spinal dorsal horn neurons to noxious cutaneous stimulation and suggested that analgesia could be partly mediated by the activation of descending inhibitory systems. This is in good agreement with the observation of Basbaum and coworkers, showing that lesions of the dorsolateral funiculus (DLF) block the analgesia induced by PAG stimulation [see 1].

All subsequent studies confirmed that both PAG and NRM stimulation inhibits to a marked extent the responses of dorsal horn neurons to nociceptive stimulation. Several groups have shown that the antinociceptive action of PAG is at least in part exerted through the NRM which sends, via the DLF, direct projections onto the spinal cord. However, an important discrepancy between electrophysiological and behavioral studies is that in contrast to restricted zones supporting analgesia, inhibitions of dorsal horn nociceptive neurons can be elicited from large areas of the brain stem, including those where aversive effects in awake animals are produced.

The pharmacology of descending systems is extraordinarily complex and involves 5-HT, Norepinephrine, endogenous opioids, and certainly various other putative substances, as revealed by recent immunocytochemical techniques. Very often discrepancies can be found in the literature. These may be due, at least in part, to methodological problems such as route of drug administration and the variety of tests used in evaluating the behavioral effects. In addition, massive and frequently unspecific action of the electrical or chemical stimulation may contrast with the graded and differential activation observed under physiological conditions.

From the behavioral point of view, the localization of effective stimulation sites for SPA in or close to nuclei where 5-HT containing neurons had been demonstrated are unquestionably in favor of the participation of this monoamine. Tolerance to PAG stimulation was initially reported in the rat, and we found in the cat that the analgesia induced by electrical stimulation of the feline NRM progressively disappeared when repeated prolonged stimulation was applied. This phenomenon seems to be sustained by serotoninergic mechanism since the effectiveness of central stimulation was clearly restored after administration of the serotonin precursor, 5-hydroxytryptophan (5-HTP). The time course of the effects of 5-HTP on SPA was comparable to the time course of 5-HT synthesis after administration of the precursor. The recovery of the efficacy of central stimulation after 5-HTP indicated that tolerance was not due to tissue damage caused by intense periods of stimulation. It is also of considerable interest that tolerance to PAG stimulation in humans is reduced by the oral administration of L-tryptophan.

In addition, numerous behavioral, electrophysiological, anatomical, and neurochemical data strongly support the involvement of 5-HT systems in the modulation of the spinal transmission of nociceptive messages. In brief:

- From the anatomical point of view, direct projection from the NRM onto the spinal cord have been shown. The serotoninergic nature of a great number of these raphe spinal fibers have been shown, and direct spinal 5-HT terminals onto nociceptive dorsal horn neurons at the origin of the spinothalamic tract have been recently demonstrated.
- From the electrophysiological point of view it has been shown that PAG or NRM stimulation induced inhibitory effects at the spinal level could be due, in part, to the activation of serotoninergic raphe spinal fibers.
- From the neurochemical point of view, several studies suggest the hypothesis of a release of 5-HT during SPA induced by NRM stimulation: (1) In the anesthetized rat, stimulation of NRM under conditions similar to those producing analgesia provoked a significant increase (+50%) in 5-HT synthesis in the spinal cord on the basis of an increased rate of 5-HTP accumulation after the blockade of central 5-HTP decarboxylase by benserazide; (2) in vivo studies using electrochemical detection of 5-HT have shown that NRM stimulation in the anesthetized rat produces a release of extracellular 5-HT in the dorsal horn; (3) more recently a direct demonstration of an increase in 5-HT and its metabolism in the CSF has been described after stimulation of the NRM area. Such a demonstration of a release of 5-HT during NRM stimulation is of importance since iontophoresis of 5-HT in the vicinity of dorsal horn neurons strongly depresses their responses to noxious stimuli; the analgesic effects induced by electrical stimulation of the NRM are reduced by intrathecal 5-HT antagonists; and the intrathecal administration of 5-HT can produce analgesia. These effects of serotonin were potentiated by blockade of monoaminoxidase or reuptake inhibition, while they were antagonized by receptor blockers such as methysergide and cyproheptadine.

A large body of experimental evidence has been provided during the past decade to document the modulation of nociceptive processes by noradrenergic neurons located in the brain stem. The origins of the brain stem-descending NE fibers have been located; these fibers arise mostly from the locus coeruleus, the subcoeruleus, and the A5 group and project to the level of the dorsal horn. Stimulation of these sites produces antinociception and inhibits the activity of spinal cord dorsal horn neurons activated by noxious cutaneous stimulation. These approaches agree with various studies using intrathecal administration and iontophoresis of norepinephrine, showing a spinal depression of nociceptive messages. The analgesic effects induced by intrathecal administration of norepinephrine are mediated by activation of alpha-2-adrenoceptors.

However, the mechanisms of activation of these descending noradrenergic systems during PAG and NRM stimulation remain unclear. Periaqueductal gray matter or NRM must directly or indirectly activate the locus coeruleus or other noradrenergic cells projecting to the spinal cord.

The involvement of endogenous opioid substances in the brain stem control of the transmission of nociceptive messages is supported by several lines of evidence:

- Naloxone, a specific opiate antagonist, reverses to a partial degree the analgesic effects of NRM and PAG electrical stimulation.
- The brain stem regions from which analgesia can be evoked by electrical stimulation are also involved in the analgesic action of opiates.
- Biochemical and immunohistochemical techniques have clearly shown that opioid receptors and/or neurons containing various opioids exist at the PAG, NRM, and dorsal horn levels.

In humans, an increase in the level of endorphinlike material in the ventricular CSF has been reported by some authors. Recently this increase has been linked to the ventricular injection of contrast medium and not to the analgesic effect. There is thus an obvious need for further animal studies of this problem. In addition, there are contradictions in the literature with regard to the relative importance of endogenous opioids in the behavioral and electrophysiological effects induced from PAG and NRM. These discrepancies and those relating to serotoninergic and noradrenergic mechanisms indicate a neurochemical heterogeneity of these areas, as suggested by immunocytochemical studies. Studies are needed on the interactions, at both spinal and supraspinal levels, among monoamines, opioids,and other neuronal types.

To explain the functional significance of descending systems, the hypothesis most often put forward postulates the existence of an endogenous analgesic system: the PAG and NRM and their spinal projections make up a negative feedback loop activated by nociceptive stimulation, producing an inhibition of the spinal transmission of nociceptive messages [see 1, 2, 5]. An argument cited in favor of this hypothesis is the activation of raphe magnus neurons by nociceptive stimulation. Results reported by Le Bars and coworkers [7] describing diffuse noxious inhibitory controls (DNIC) could be an important step forward in understanding, at least in part, the activation of this system by natural stimulation.

In addition to their activation by various forms of natural nociceptive stimuli, it has been suggested that the activity of these descending brain stem inhibitory controls is increased by morphine. Several arguments have been advanced to support this hypothesis: (1) Analgesia can be evoked by microinjection of morphine within the PAG and NRM; (2) analgesic effects of systemic morphine are reduced after DLF, NRM, or PAG lesions; (3) serotoninergic mechanisms are involved in morphine analgesia; and (4) the excitatory effects of morphine after various routes of administration on NRM neurons, some of them projecting to the spinal cord, agree with such an hypothesis. However, in these latter studies, the effect of morphine is not unequivocal, since the activity of a not negligible proportion of these cells is depressed by morphine. Again from the electrophysiological point of view, there is much controversy regarding the effect of morphine. Some authors reported a depressive effect on the activity of nociceptive dorsal horn neurons after its microinjection in the PAG or NRM; however, there is accumulating evidence from Le Bars et al. [7] that morphine does not increase tonic descending inhibitory controls, but rather decreases them. Since the time course of these effects is related to behavioral analgesia, these findings are discussed with reference to the functional role of DNIC in pain processes.

In conclusion, despite the many gaps in our knowledge of the physiology and pharmacology of the various controls systems, their existence has had important consequences in the study of pain. Thus, the various techniques used to interrupt or destroy the pathways or relays involved in pain have gradually been abandoned. These neurosurgical approaches are progressively being replaced by other techniques (neurostimulation, intrathecal or intracerebral administration of drugs, etc.) which aim at reinforcing the activity of inhibitory controls with the advantage of not producing irreversible lesions.

References

1. Basbaum AI, Fields HL (1984) Endogenous pain control systems: brainstem spinal pathways and endorphin circuitry. Ann Rev Neurosci 7:309–338
2. Besson JM, Chaouch A (1987) Peripheral and spinal mechanisms of nociception. Physiol Rev 67:67–186
3. Besson JM, Lazorthes Y (1985) Substances opioides médullaires et analgésie (spinal opioïds and the relief of pain). INSERM, Paris, colloque, vol 127
4. Duggan AW, Foong FW (1985) Bicuculline and spinal inhibition produced by dorsal column stimulation in the cat. Pain 22:249–259
5. Fields HL, Besson JM (1988) Pain modulation. Progress in Brain Res, vol 77. Elsevier, Amsterdam New York Oxford
6. Gerhart KD, Wilcow TK, Chung JM, Willis WD (1984) Inhibition of primate spinothalamic tract neuons by stimulation in ventral posterior lateral (VPLc) thalamic nucleus: possible mechanisms. J Neurophysiol 49:450–466
7. Le Bars D, Dickenson AH, Besson JM, Villanueva L (1980) Aspects of sensory processing through convergent neurons. In: Yaksh TL (eds) Spinal afferent processing. Plenum Press, New York, pp 467–504
8. Oliveras JL, Besson JM (1988) Stimulation-produced analgesia in animals: behavioral investigations. In: Fields HL, Besson JM (eds) Progress in Brain Research, vol 77. Elsevier Amsterdam New York Oxford, pp 141–157
9. Willis WD (1985) The pain system. The neural basis of nociceptive transmission in the mamalian nervous system. Karger, Basel
10. Yaksh TL, Noveihed R (1985) The physiology and pharmacology of spinal opiates. Annu Rev Pharmacol Toxicol 25:433–462

B. Human Studies

Etomidate Augmentation of Scalp Recorded Somatosensory Waves: Time Course, Reproducibility, and Dose Effect

R.W. McPherson[1] and R.C. Levitt

Abstract

Bolus etomidate (ETOM) transiently increases the amplitude of scalp recorded somatosensory evoked potentials (SSEP). The reproducibility of this augmentation and its dose-response relationship are unknown. In unpremedicated patients, we studied the effect on the SSEP of repetitive administration of single doses of ETOM (0.1 mg/kg, iv bolus) in six patients and increasing doses of ETOM in six additional patients. Anesthesia was induced with fentanyl (15–20 μg/kg, iv) plus thiopental (1–2 mg/kg, iv) and maintained with 0.4%–0.8% isoflurane in oxygen, and the surgical incision was infiltrated with bupivacaine (0.5% without epinephrine). ETOM administration was delayed for 30 min following anesthesia induction. In group 1, 0.1 mg/kg ETOM was administered iv as a bolus three times at 30 min intervals. In group 2, 0.05, 0.1, or 0.2 mg/kg was administered at 30-min intervals in random order in each patient. SSEP were measured immediately before and once each minute for 5 min after ETOM administration following nondominant median nerve stimulation. In group 3 ($n = 5$) the time of wave augmentation was determined using continuous SSEP stimulation following 0.2 mg/kg ETOM.

In group 1, administration of 0.1 mg/kg ETOM (three trials) increased the latency of N20 and P23 by 1.0–1.4 ms and 1.3–2.6 ms, respectively ($P <$ 0.05). P15N20 amplitude was increased by about 50% ($P < 0.05$). and N20P23 amplitude increased to 174% of control ($P < 0.05$). The amplitude increase was similar during the three ETOM administrations for both P15N20 and N20P23. Latency remained elevated by about 1.5 ms and amplitude remained elevated (P15N20 = 138%; N20P23 = 150%) 5 min following injection. Mean arterial blood pressure (MABP) was unchanged by 0.1 mg/kg etomidate. In group 2, 0.05 mg/kg ETOM altered neither amplitude nor latency. However, 0.1 mg/kg and 0.2 mg/kg increased N20P23 amplitude to

[1] Department of Anesthesiology and Critical Care Medicine, The Johns Hopkins Medical Institutions, 600 North Wolfe Street, Baltimore, MD 21205, USA

K. Shimoji et al. (Eds.)
Spinal Cord Monitoring and Electrodiagnosis

161 ± 33 and 230 ± 10% of control ($P < 0.05$), respectively. N20 and P23 latency were increased by ETOM 0.1 mg/kg by about 1.0 ms, while 0.2 mg/kg increased N20 latency by 1.0 ms and P23 latency by 1.5 ms. In group 3, the time for SSEP augmentation to reach 50% of maximum was about 2 min. Bolus administration of ETOM (0.1 mg/kg) reproducibly increased SSEP amplitude, and a larger dose (0.2 mg/kg) further increased amplitude augmentation. Thus, intermittent injection of ETOM can be used to augment small SSEP waves with reproducible increases in wave amplitude.

Introduction

Somatosensory evoked potentials (SSEP) are monitored intraoperatively in an attempt to decrease the risk of neurological injury during spinal, intracranial, and cerebral vascular surgery. Volatile anesthetic gases decrease SSEP amplitude [5, 8], while barbiturates and synthetic narcotics (fentanyl) cause relatively small changes in both SSEP amplitude and latency [1, 6]. Anesthetic depression of SSEP waveforms may increase the time necessary to produce characteristic SSEP waves because of lowered signal to noise ratio.

Etomidate (ETOM) is potentially useful in neuroanesthesia because of decreased cerebral blood flow, cerebral oxygen consumption, and intracranial pressure [7]. Both bolus injection of ETOM and a short infusion transiently increase SSEP amplitude by 200%–600% [3, 6] and thus may reverse SSEP wave depression produced by spinal cord compression or depression by volatile anesthetic gases.

We assessed the reproducibility and dose–response relationship of intravenous ETOM and scalp recorded SSEP wave augmentation during fentanyl-isoflurane anesthesia. We also determined the time of onset of augmentation using a method of continuous SSEP monitoring.

Methods

Three groups of unpremedicated patients undergoing lumbar surgery were studied. In group 1 ($n = 6$), ETOM (0.1 mg/kg, iv bolus) was assessed three times at 30-min intervals, in group 2 ($n = 6$), ETOM (0.05, 0.1, and 0.2 mg/kg, iv bolus) was given in random order at 30-min intervals, and in group 3 ($n = 5$) 0.2 mg/kg ETOM was given during stable anesthesia. Anesthesia was induced with thiopental (1–2 mg/kg, iv) plus fentanyl (15–20 μg/kg, iv), followed by pancuronium (0.1 mg/kg, iv), and anesthesia was maintained with isoflurane 0.4%–0.8% in oxygen. MABP, heart rate, end tidal carbon dioxide tension, and temperature were measured. Bupivacaine (0.5% without epinephrine) was used to infiltrate the surgical site. Initial ETOM injection was postponed at least 30 min following the start of isoflurane administration. In all patients, the inspired isoflurane concentration was maintained constant throughout the study period. Fentanyl (50–100 μg, iv) was occasionally necessary to augment anesthesia. End tidal carbon dioxide tension was maintained constant throughout the study in each patient at about 35 mmHg.

SSEP were recorded using a Nicolet Med 80 or Nicolet Compact 4 (Nicolet Biomedical, Madison, WI, USA). The nondominant median nerve was stimulated via sterile 23 gauge needles placed percutaneously in a location at the wrist which produced a digital twitch. Responses were recorded over the contralateral scalp (2 cm posterior to C3 and C4) with reference to the forehead (F_{PZ}) according to the international 10–20 system of electrode placement. Stimuli (128; 5.8/s) were delivered at an intensity (250 V; 1.9–4.9 mA) just sufficient to cause a digital twitch. The same stimulus intensity was used throughout the study in each patient. The responses were amplified, filtered (5–1500 Hz), and stored on magnetic disk for later analysis. The waves were evaluated for P15, N20, and P23 latencies and the peak-to-trough amplitude from P15 to N20 and the amplitude from N20 to P23.

SSEP waves were assessed prior to anesthesia induction, immediately after induction and 1 min before and each minute after ETOM injection for 5 min. The times are designated relative to bolus injection of ETOM into a rapidly flowing intravenous line in the extremity contralateral to the stimulated extremity.

In group 3 ($n = 5$), we used continuous accumulation of SSEP (stimulus rate = 5.9/s) to determine the time course over which ETOM augments the scalp recorded SSEP waves. To accomplish this goal, we used an in house algorithm which uses selective filtering to allow continuous generation of the scalp recorded wave. Basically this program allows continuous generation and rapid change in the average wave form as the individual responses to stimulation change. Stimulation (5.9 s; 19 mA) was provided by a Nicolet Compact 4.

Results

MABP was unchanged by any ETOM dose (0.05, 0.1, or 0.2 mg/kg). The amounts of thiopental and fentanyl administered were similar in the three groups. Inspired isoflurane concentration was similar in the three groups.

In group 1 when ETOM (0.1 mg/kg) was administered three times, P15 latency did not increase, whereas N20 and P23 both increased transiently about 1 s with each ETOM administration.

Anesthesia induction in group 1 did not alter P15N20 amplitude, but isoflurane decreased amplitude by about 40% ($P < 0.05$) (Fig. 1). The increase in P15N20 amplitude with the three administrations of ETOM (0.1 mg/kg) varied between 36% and 76%. Isoflurane decreased N20P23 amplitude by

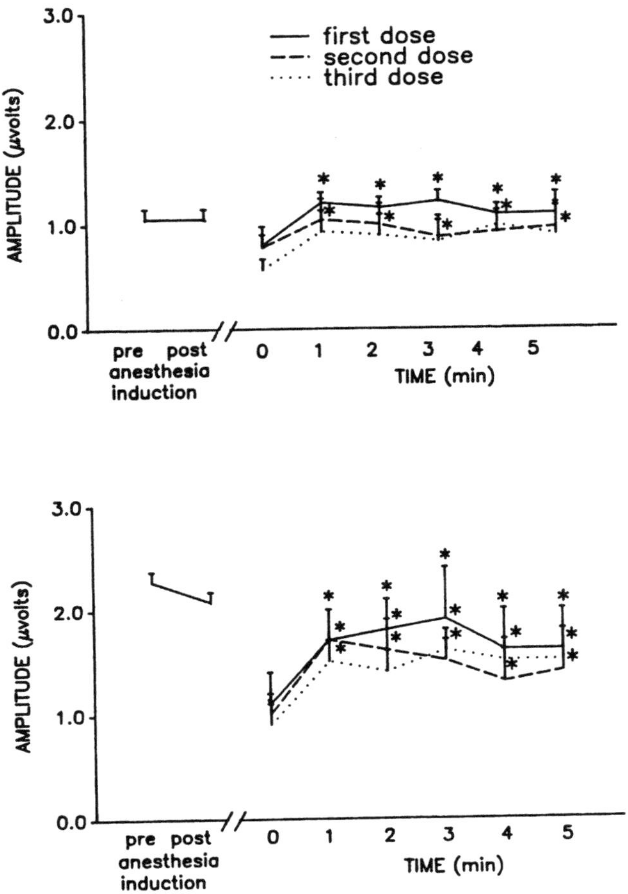

Fig. 1. The effect of three doses of ETOM (0.1 mg/kg, iv) administration 30 min apart on P15N20 amplitude (*upper panel*) and N20P23 amplitude (*low panel*) is shown. * $P < 0.05$ compared to 0 time. Mean ± SEM. *Time 0* represents data obtained immediately prior to ETOM administration

about 50% ($P < 0.05$). The increase in N20P23 amplitude by ETOM was about 74% (range 70%–78%; $P < 0.05$) and was similar during the three ETOM administrations.

Figure 2 shows changes in latency of P15, N20, P23 following 0.05, 0.1 and 0.2 mg/kg (i.v. bolus). P15 latency was not increased with any dose of ETOM. N20 latency was increased ($P < 0.05$) consistently with 0.1 and 0.2 mg/kg ETOM for the time of the study. The increase in latency of P23 was more pronounced with 0.2 mg/kg.

In group 2, P15N20 amplitude was not increased by 0.05 mg/kg ETOM, but increased significantly ($P < 0.05$) with both 0.1 mg/kg and 0.2 mg/kg

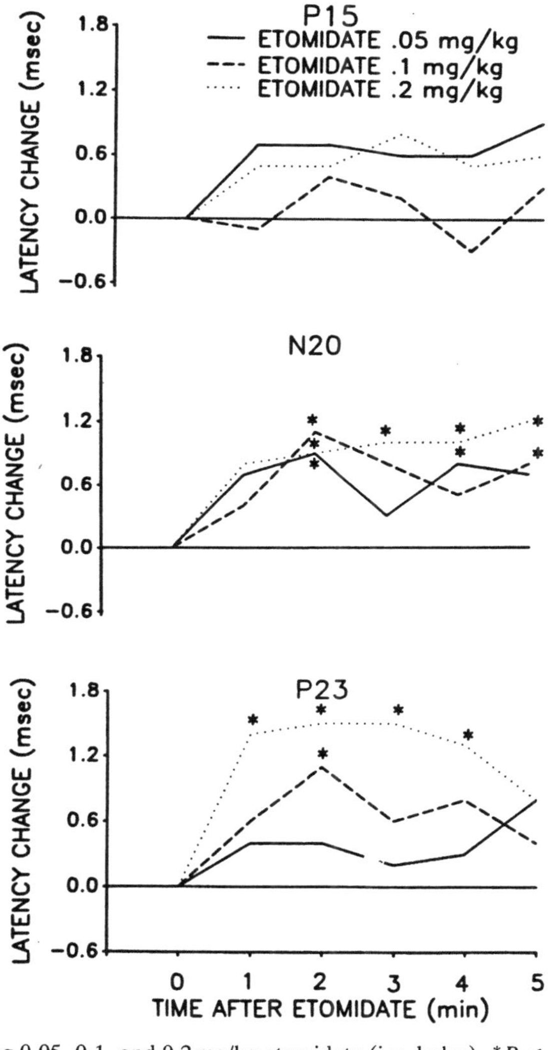

Fig. 2. Changes in latency following 0.05, 0.1, and 0.2 mg/kg etomidate (i.v. bolus). $*P < 0.05$ compared with control

(Fig. 3). N20P23 amplitude progressively increased as 0.05, 0.1, and 0.2 mg/kg were administered. N20P23 amplitude increased to 129% of control with 0.05 mg/kg, 161% of control with 0.1 mg/kg, and 230% of control with 0.2 mg/kg ETOM. The increase in amplitude persisted throughout the study (5 min) with both 0.1 and 0.2 mg/kg ETOM.

In group 3, the time to maximum increase in amplitude was about 2 min. Figure 4 shows the rapid change in SSEP, using continuous generation of SSEP.

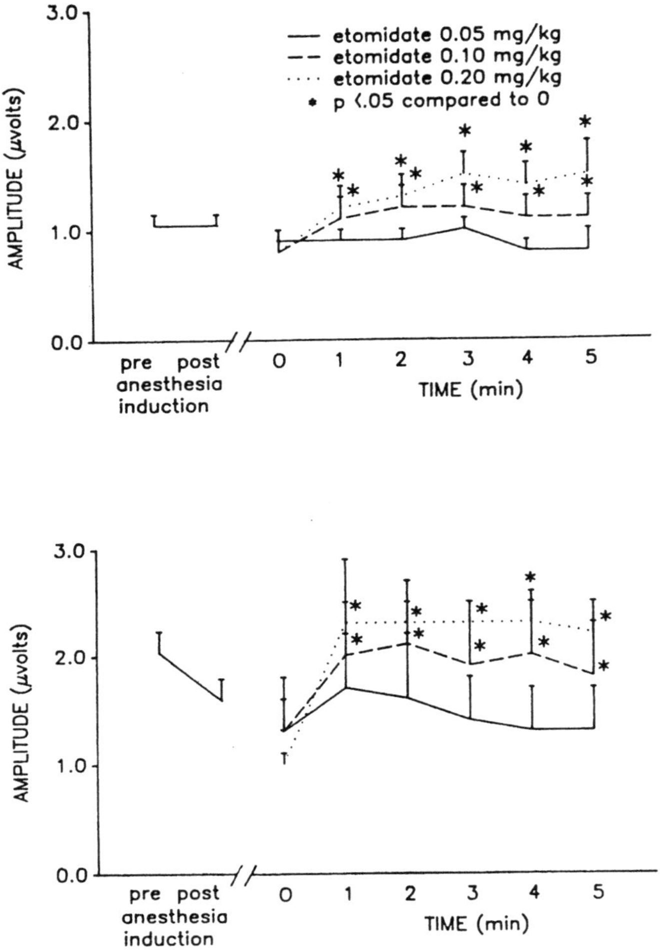

Fig. 3. The effect of 0.05, 0.1 and 0.2 mg/kg ETOM administered at 30-min intervals on P15N20 (*upper panel*) and N20P23 amplitude (*lower panel*) is shown. $^*P < 0.05$ compared to 0 time. Mean ± SEM

Discussion

Our finding that ETOM (0.1 mg/kg) reproducibly augmented SSEP in a dose related fashion is important for its potential application in modifying anesthetic techniques and to enhance intraoperative SSEP monitoring.

Volatile anesthetic gases [5, 8] and nitrous oxide [5] reduce the amplitude of scalp recorded waves and potentially make SSEP monitoring more difficult under those circumstances and ETOM may be used to reverse this depression. In our study, ETOM almost completely reversed SSEP amplitude

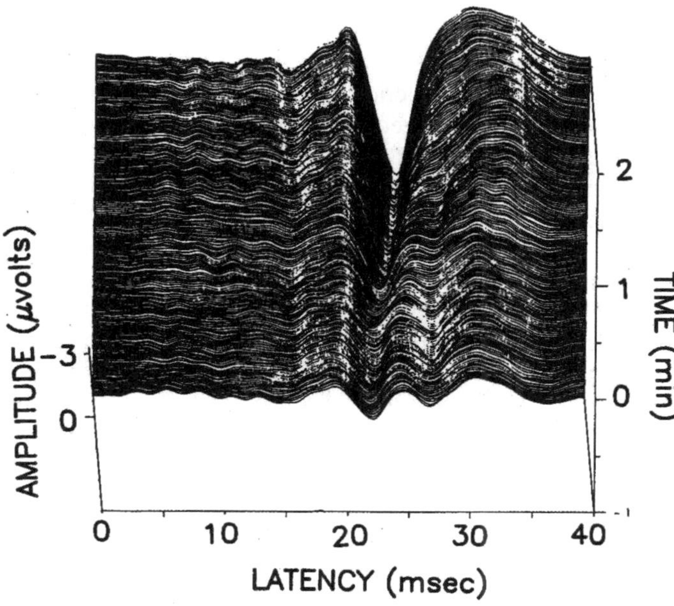

Fig. 4. The scalp recorded SSEP waves during continuous stimulation is shown in a pseudo three-dimensional plot. Each *trace* represents the updated scalp record with the addition of a single new response. *Time 0* represents the injection of ETOM (0.2 mg/kg) as a bolus into a peripheral intravenous line

depression caused by isoflurane. ETOM can easily be substituted for short acting barbiturate in patients at risk of neurological injury due to its beneficial effects on the brain and minimal cardiovascular depression. ETOM (0.2 mg/kg) decreases intracranial pressure and depresses cerebral metabolism [7]. ETOM may be administered by bolus injection or by infusion to augment anesthesia. An initial rapid ETOM infusion followed by a maintenance infusion at a lower rate maintains stable anesthesia [2], but ETOM infusion prolongs awakening compared to inhalational agents [2].

Our results corroborate those of Koht et al. [4] who used a short (25 min) ETOM infusion which augmented scalp recorded amplitude and increased latency. The relatively small increase in amplitude found in the present study compared with previous reports [3, 6] may be related to the basal anesthetic (fentanyl-barbiturate-isoflurane). We excluded N_2O to maximize waveform amplitude and thereby avoided any possible interaction of nitrous oxide and ETOM. Thus our results may be of importance only when N_2O is eliminated from the anesthetic technique.

Our results can be divided into two areas which are important for clinical monitoring. First, a dose of ETOM with minimal cardiovascular effects (0.1 mg/kg) reproducibly augments SSEP amplitude, suggesting that intermittent doses of ETOM can be used to increase the size of abnormally small waves.

Second, augmentation is enhanced by increasing the bolus dose, suggesting that the effect is greater with a higher serum level. Thus, our data suggest that intermittent bolus injection is useful in reproducibly augmenting amplitude while avoiding prolonged awakening associated with ETOM infusion.

During preparation of this manuscript, the efficacy of ETOM in allowing SSEP monitoring, which was prevented by usual anesthetic management, was reported. Sloan et al. [9] presented a patient in whom ETOM was useful (bolus plus infusion) in facilitating SSEP monitoring and allowing demonstration of reversible neurological impairment.

In conclusion, intermittent bolus injection of ETOM can be used to augment SSEP waves in an anesthetic technique which excludes N_2O. Our data suggest that intermittent doses of ETOM can be used to enhance SSEP amplitude, thus avoiding prolonged awakening associated with a continuous ETOM infusion. ETOM doses which are effective in increasing wave amplitude do not decrease MABP.

References

1. Drummond JC, Todd MM, Sang H (1985) The effect of high dose sodium thiopental on brain stem auditory and median nerve somatosensory evoked responses in humans. Anesthesiology 63:249–254
2. Fragen RJ, Avram MJ, Henthorn TK, Caldwell NJ (1983) A pharmacokinetically designed etomidate infusion regimen for hypnosis. Anesth Analg 62:654–660
3. Kochs E, Treede RD, Schulte-am-Esch J (1986) Increase of somatosensory evoked potentials during anesthesia induction with etomidate. Anaesthesist 35:359–364
4. Koht A, Schultz W, Schmidt G, Schramm J, Watanabe E (1988) Effects of etomidate, midazolam and thiopental on median nerve somatosensory evoked potentials and the additive effects of fentanyl and nitrous oxide. Anesth Analg 67:435–441
5. McPherson RW, Mahla M, Johnson R, Traystman RJ (1985) Effects of enflurane, isoflurane, and nitrous oxide on somatosensory evoked potentials during fentanyl anesthesia. Anesthesiology 62:626–633
6. McPherson RW, Sell B, Traystman RJ (1986) Effects of thiopental, fentanyl, and etomidate on upper extremity somatosensory evoked potentials in humans. Anesthesiology 65:584–589
7. Moss E, Powell D, Gibson RM, McDowall DG (1979) Effect of etomidate on intracranial pressure and cerebral perfusion pressure. Br J Anaesth 51:347–352
8. Samra SK, Vanderzant CW, Domer PA, Sackellares JC (1987) Differential effects of isoflurane on human median nerve somatosensory evoked potentials. Anesthesiology 66:29–35
9. Sloan TB, Bonai AK, Toleikis JR, Koht A (1988) Improvement of intraoperative somatosensory evoked potentials by etomidate. Anesth Analg 67:582–585

Four Components and Their Origins of the Ascending Spinal Potential Evoked by Stimulation at the Conus Medullaris Level

S. Hirabayashi[1], L. Kenkow, and T. Kurokawa

Introduction

The epidural spinospinal evoked potential is well known to have two major components conducting in the lateral and the dorsal columns, respectively, when weak stimulation is given on the dorsum of the dura mater. But, when the stimulation is given at lower thoracic levels, the ascending spinal potential has four rather than two components, a fact which has not been reported before. In animal experiments, four similar components are observed in cats with stimulation of the lower segments of the spinal cord. The purpose of this study is to describe such a phenomenon and to offer suggestions as to the origins of those components.

Methods

In patients, the ascending evoked spinal potential was measured during surgery after placing both stimulating and leading electrodes percutaneously in the posterior aspect of the thoracic or lumbar epidural space. Stimulation by rectangular pulses of 0.2 ms in duration with a frequency of 3 HZ was given by a bipolar electrode. The potential evoked was also picked up by a bipolar electrode. The potential was studied, changing the level and strength, and compared with the descending spinal potential.

In cats, bipolar electrodes were placed on the dorsum of the dura mater exposed by laminectomy at the thoracic or lumbar levels. After confirming the presence of four components similar to those in humans, the susceptivity of those components to partial resection of the spinal cord or fila radicularia was studied. Then, cross-sectional areas of the spinal cord and fila were

[1] Department of Orthopedic Surgery, Tokyo University Hospital, 7-3-1 Hongoh, Bunkyo-ku, Tokyo, 113 Japan

K. Shimoji et al. (Eds.)
Spinal Cord Monitoring and Electrodiagnosis

measured with a planimeter on photoprints of H.E. stained sections to visualize geometrical features of the neural tissue which was stimulated.

Results

Ascending Spinal Potential in Patients

The ascending spinal potential in patients only had more than two components when the stimulating electrodes were placed between the 10th thoracic (T10) and the 1st lumbar (L1) vertebral level. When the stimulating electrodes were moved more cranial to the T9 level, the two components reported by Kurokawa [2] and Imai [1] were observed, and when the stimula-

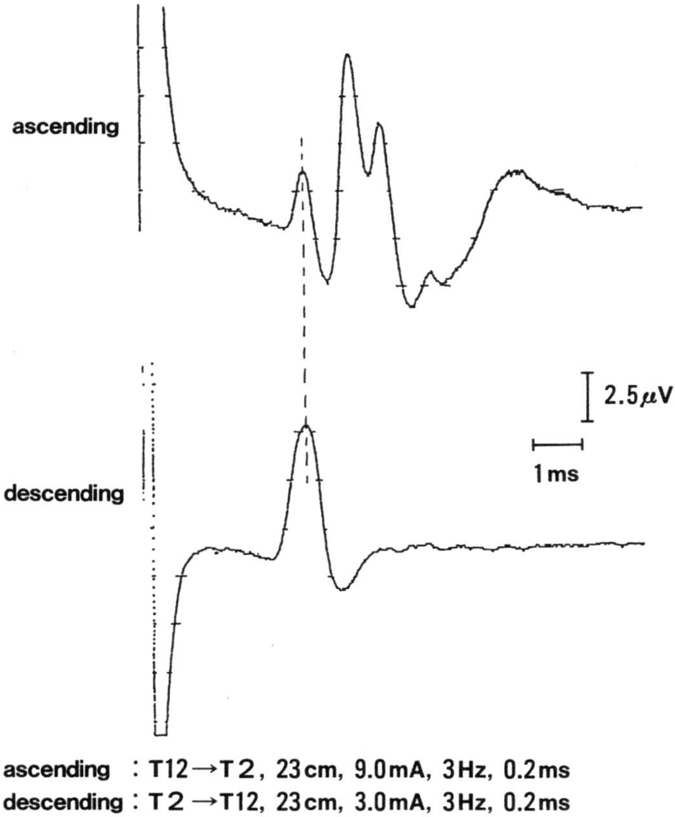

Fig. 1. In a 14-year-old girl with idiopathic scoliosis the latency of the first component of the ascending spinal potential is the same as that of the descending one

tions were below the L2 level, two negative peaks of a different pattern were measured.

In a patient in whom the conduction distance was long, four components were observed as the stimulation became stronger. In one in whom the conduction distance was short, three rather than four components were observed as the stimulation became stronger. The latency of the first component of the ascending spinal potential was the same as that of the descending one (Fig. 1).

Ascending Spinal Potential in Cats

In cats, also, the ascending spinal potential had three or four components, but only when the stimulating electrodes were between the L2 and L5 vertebral levels. When the stimulating electrodes were cranial to the L1 level, the two components reported in literature [1–3] were observed, and when the stimulation was caudal to the L6 level, two negative peaks of different feature were measured.

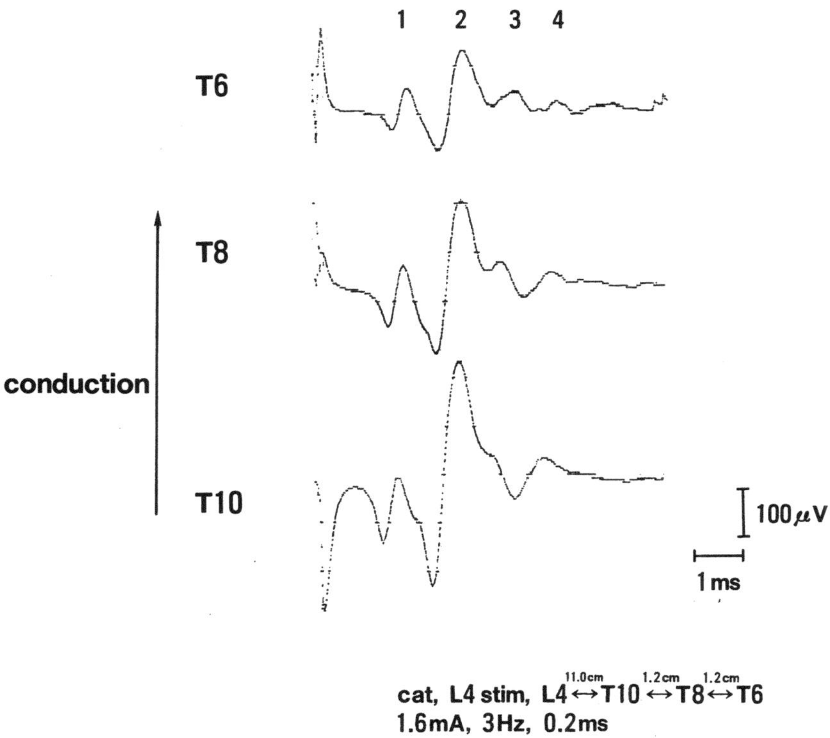

Fig. 2. Having a longer distance, the second component divides into two peaks, resulting in four components as a whole

When the conduction distance was short, three rather than four components were observed. With a longer conduction distance, the second component divided into two peaks, resulting in four components as a whole. Consequently, the ascending spinal potential evoked by stimulation at the lower lumbar vertebral level had four components, which could be numbered in the order of their latencies (Fig. 2).

When the potential was evoked by stimulation on the dorsum of the dura mater, the thresholds of the first and the third components were higher than those of the second and the fourth ones. When the fila radicularia were resected, the second and the fourth components disappeared.

In contrast, the first and the third components disappeared when the spinal cord rather than the fila was resected. When the posterior columns of the spinal cord were resected, the third and the fourth components disappeared. The first and the second components disappeared as the lateral columns were resected.

Geometrical Features of the Lower Segments of the Spinal Cord in Cats

At the L1 level, there were a few fila radicularia around the spinal cord. The fila increased and the spinal cord decreased in their volume at the lower levels. Below the L4 level, the volume of the fila was larger than that of the spinal cord.

Discussion

There were similarities in humans and cats in terms of the pattern of the potential and the level of the stimulation to evoke those components. When the stimulation was given at levels where the volume of the spinal cord was much more than that of the fila, the two components known in the literature [1–3] were observed both in humans and in cats. Only when the stimulation was near the conus medullaris level, could three or four components, according to the conduction distance, be measured (Fig. 3).

The results of animal experiments suggested a hypothesis in terms of the origins and the conducting pathways of four components as follows. The first component was evoked by direct stimulation of the spinal cord and conducted along the lateral column, the second by stimulation of the fila and conducted along the lateral column, the third by direct stimulation of the spinal cord and conducted along the posterior column, and the fourth by stimulation of the fila and conducted along the posterior column.

In measuring the human ascending epidural spinal potential, the influence of the fila stimulated is considered little at the thoracic level, where

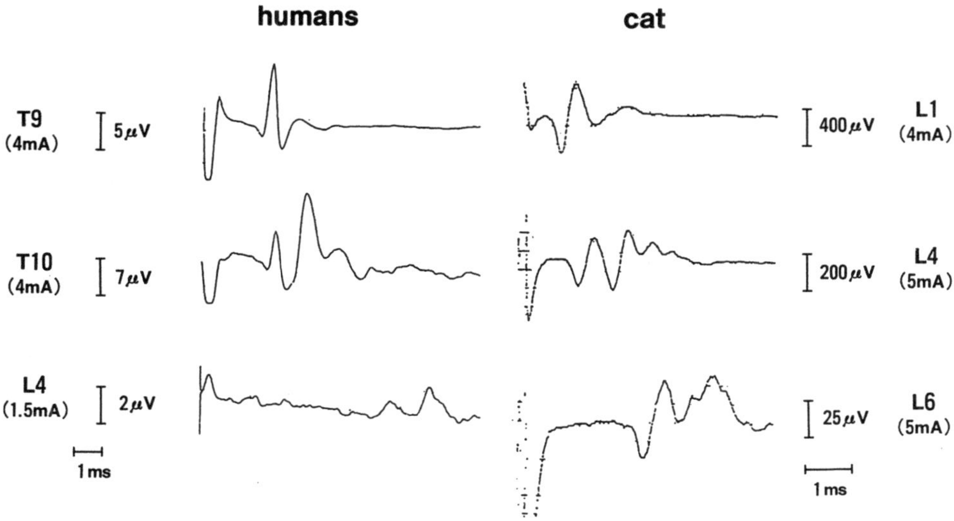

Fig. 3. There are simlarities in humans and cat in terms of the pattern of the potential and the level of the stimulation to evoke those components. Only when the stimulation is near the conus medullaris level (T10 in humans, L4 in cat), can three or four components, according to the conduction distance, be measured

the volume of the fila is smaller than that of the spinal cord. But its influence cannot be neglected at the conus medullaris level where many fila surround the spinal cord.

Taking those similarities into account, the first component of the human ascending potential was likely to be evoked by direct stimulation of the spinal cord, because the latency of this component was the same as that of the descending potential which was evoked by direct stimulation of the spinal cord. Whether this hypothesis pertains to humans still remains to be studied since there is no direct visual evidence for stimulation to the spinal cord or fila radicularia.

Conclusions

1. The ascending epidural spinal potential has four components when epidural stimulation is given to the dorsal aspect of the conus medullaris.
2. From the results of animal experiments, two components are suggested to be evoked by direct stimulation of the spinal cord and the others to be evoked by stimulation of the fila radicularia.
3. In measuring the human ascending epidural spinal potential, we must pay attention to the potential evoked by stimulation of the fila radicularia when the stimulation is given near the level of conus medullaris.

References

1. Imai T (1976) Spinal cord action potentials evoked by epidural stimulation of cord – basal waveforms and their meanings –. Nippon Seikeigeka Gakkai Zasshi (J Jpn Orthop Assoc) 50:1037–1056
2. Kurokawa T (1972) Evoked spinal cord action potentials by epidural stimulation of cord. Jpn J Electroencephalogr Electromyogr 1:64–66
3. Tsuyama N et al. (1978) Clinical application of spinal cord action potential measurement. International Orthop (SICOT) 2:39–46

Somatotopical Evoked Responses from the Spinal Cord and Cerebral Cortex to Finger Stimulation

H. Takahashi[1], I. Suzuki, and B. Ishijima

Introduction

With somatosensory evoked potentials (SEP) or evoked spinal cord potentials (ESCP) to the stimulation of each finger, it becomes possible to examine the sensory functions of clinically important dermatomes of C5/6–C8/T1 segments which can not be well evaluated by SEPs to peripheral nerve stimulation. The purpose of the present study is to obtain fundamental knowledge about somatotopical responses of the cervical cord or the cerebral cortex to finger stimulation.

Materials and Methods

ESCP to finger stimulation were recorded in nine patients who had undergone wide laminectomies for cervical lesions. Four or six channel electrodes were placed on the dorsal surface of the cervical cords with the reference being placed in the paravertebral muscles.

As to the nomenclature of ESCP to finger stimulation, the names of P1 (the first positive wave), N1 (the first negative wave following P1), and P2 (the slow positive wave after N1) were used following the description of ESCP to ulnar nerve stimulation by Shimoji et al. [9, 10].

SEP to finger stimulation were recorded from the cerebral cortex intraoperatively in 11 patients, and postoperatively in 4 who had skull defects over the frontoparietal region. These four individuals all had exhibited organic intracranial lesions with increased intracranial pressure and had undergone decompressive craniectomies. Recordings were made when the patients' condition had stabilized and no signs of sensory disturbances were observed. SEP from the area of the skull defects showed 4–5 times larger amplitudes

[1] Department of Neurosurgery, Tokyo Metropolitan Neurological Hospital, 2-6-1 Musashidai, Fuchu-shi, Tokyo, 181 Japan

K. Shimoji et al. (Eds.)
Spinal Cord Monitoring and Electrodiagnosis

than those recorded from the scalp in normal control subjects, and they demonstrated almost the same waveforms as cortical SEP if the sensory pathways or cortex remained intact.

Scalp SEP to finger stimulation were recorded in 14 normal subjects, and in another 11, SEP to median nerve stimulation were recorded as controls. In the present study, all SEP components to finger stimulation were referred to using the SEP nomenclature with respect to median nerve stimulation although components of SEPs to finger stimulation had longer peak latencies. SEP from the parietal area were named N19, P22, and P23 [11]. Frontal SEP consisted of P20 and N25 [11].

Each finger was stimulated, one by one, using ring electrodes at a stimulus current 300%–400% of the sensory threshold with a pulse width of 0.2 ms. The median nerve was stimulated at the wrist. Stimulus current was 120%–160% greater than the motor threshold with a pulse width of 0.2 ms.

Recording of SEP and ESCP were made using a Nihonkoden Neuropack, Neuropack 8 or Atak 450, and the frequency band was set between 15 to 3 kHz.

Results

Cervical ESCP to Finger Stimulation

The waveform of ESCP to finger stimulation was essentially the same as that of ESCP to median nerve stimulation. But, the amplitude of the former was smaller than the latter, and the peak latencies of the former were 2–3 ms longer than those of the latter.

The ESCP to finger stimulation recorded from the ventral epidural space showed waves whose polarity reversed to N1 and P2 components from the dorsal surface, as in the case of ESCP to median nerve stimulation.

In six of nine patients the regions of C5–C7/8 segment were exposed and multichannel recordings were possible.

Figure 1 demonstrates examples of ESCP to finger stimulation in a 45-year-old man with intractable arm pain due to peripheral nerve injury who underwent DREZtomy (of dorsal root entry zone lesions) for pain relief. Six electrodes were placed on C6 to T1 segments and altogether 30 ESCP to finger stimulation were recorded. The ampltidues of these N1 components changed according to the sites of recording electrodes and to the sites of stimulation. The maximum N1 to finger I stimulation appeared at the C6 segment and the maximum N1 to finger II, III, IV, and V stimulation appeared at the C7, C7, C7/8, and C8 segment, respectively. Figure 2 shows distribution patterns of normalized amplitude of N1. Since the amplitude of ESCP varied from each finger, normalized values of N1 are given in Fig 2. For example, the amplitude of the largest N1 of six ESCPs to finger I stimula-

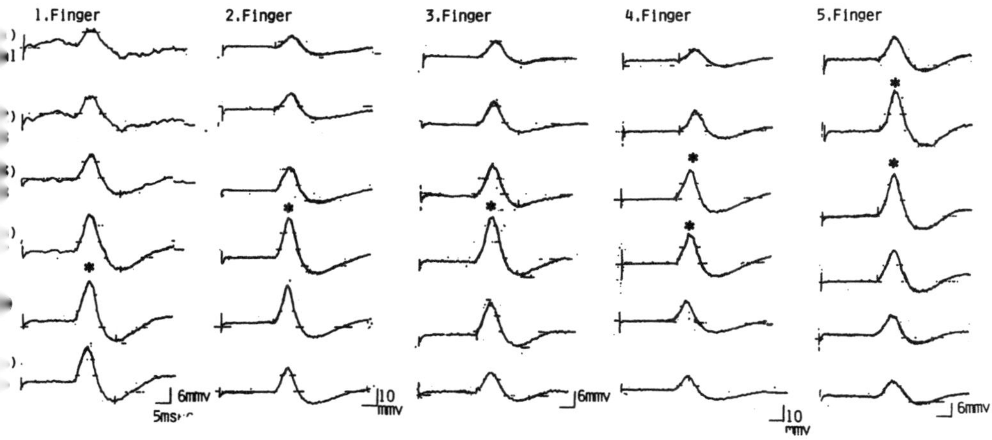

Fig. 1. ESCP to each finger stimulation recorded with six channel electrodes placed on the dorsal surface of the cervical cord in a 45-year-old man with intractable pain. *, maximum N1

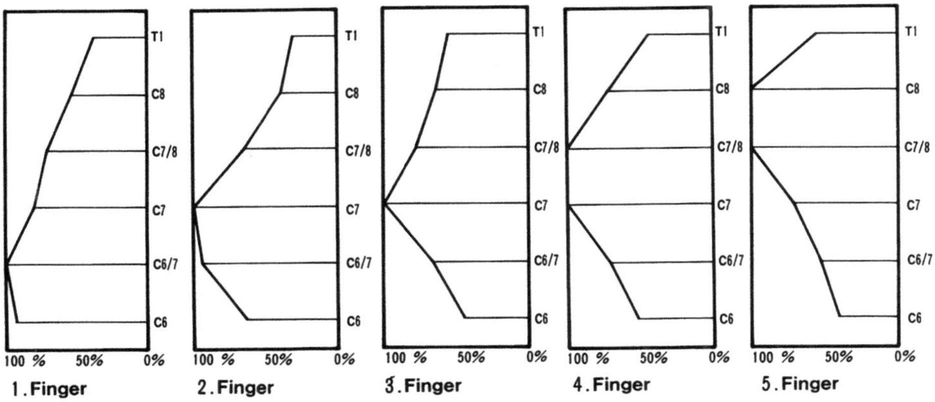

Fig. 2. The distributions of percent amplitude of N1 of the same data as shown in Fig. 1

tion was taken as 100, and the remaining five ESCPs for N1 were calculated as a percentage to the largest response (% amplitude). Figure 2 demonstrates somatotopical N1 responses on the dorsal surface of the cervical cord to finger stimulation, although some response overlaps exist.

The other five patients showed also essentially the same somatotopical responses on the dorsal surface of cervical cord to each finger stimulation.

Scalp SEP in Control Patients

SEP following stimulation of each finger showed patterns analogous to those produced by median nerve stimulation. However, the average peak latencies of the early components were 2–3 ms longer than the latter components. Eight needle electrodes were placed on the scalp over the sensory cortex every 10 mm from 35 mm to 105 mm lateral to the midline, and SEP were recorded in response to each finger stimulation. However, these SEP did not show clear somatotopical responses.

SEP to Finger Stimulation in a Patient with Skull Defect

SEP recorded over the area with a skull defect showed amplitudes three to five times larger than those of scalp SEP in normal patients and their forms were similar to those of SEP from the cortex, as already reported [11].

The following case is representative. The patient was a 23-year-old man with glioma in the left parietal area posterior to the postcentral gyrus. Decompressive craniectomy and biopsy were performed, followed by chemoradiotherapy. SEP to finger stimulation were recorded when the patient's state was stable and no overt neurological disturbances were noted.

The sensory cortex was first determined by median nerve SEP and then 15 needle electrodes were placed linearly every 5 mm from 35 to 105 mm lateral to the midline over the sensory cortex. SEP to finger stimulation were recorded for each finger. All SEP records are shown in Fig. 3. Analysis time was 50 ms. Calibration of SEP to the fifth finger stimulation indicated 2.5 μV, and calibrations to the other four digits indicated 5 μV.

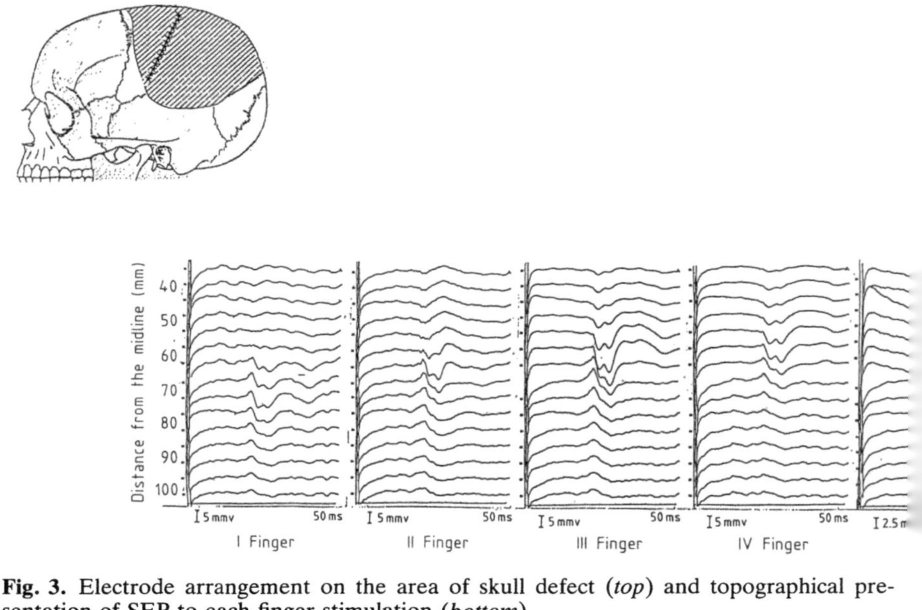

Fig. 3. Electrode arrangement on the area of skull defect (*top*) and topographical presentation of SEP to each finger stimulation (*bottom*)

Figure 3 shows that the location of the largest response moves from lateral to medial when the stimulation site moves from the thumb to the fifth finger, although some response overlap was observed. In this case, SEP to the third finger stimulation were largest and most similar to those of median nerve SEP.

The distribution patterns of percent amplitude for N19, P22, and P23 demonstrated somatotopical responses of each component, as shown in N1 of ESCP (Figs. 4–6). As for N19, the largest response to thumb stimulation appeared 75 mm lateral to the midline, while that to the 5th finger appeared 60 mm lateral to the midline (Fig. 4). Thus, although N19 showed a slight somatotopical response, a pattern was not clearly evident.

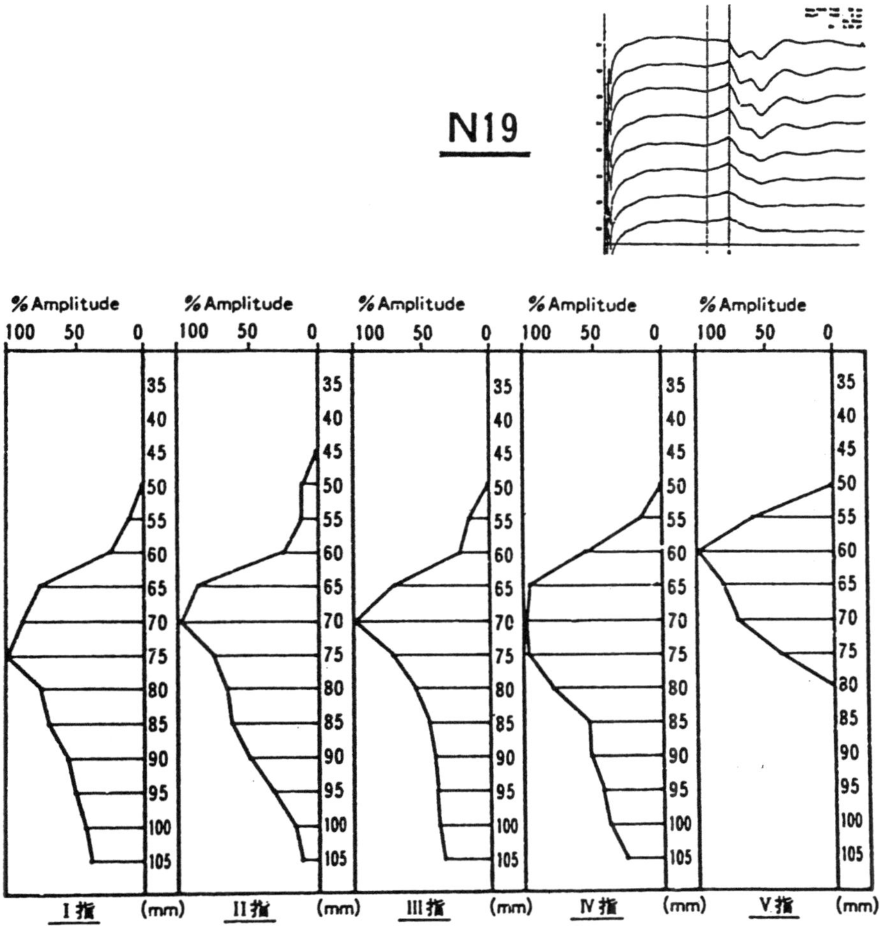

Fig. 4. The distributions of percent amplitude of N19 of the sensory cortex

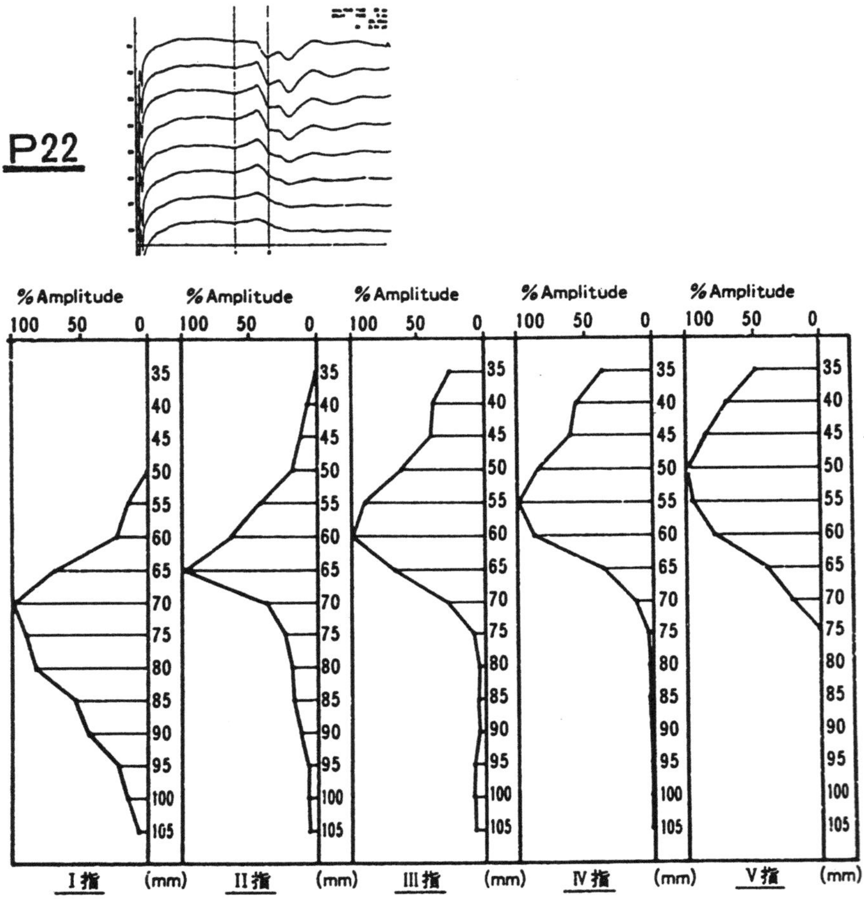

Fig. 5. The distributions of percent amplitude of P22 on the sensory cortex

On the other hand, as shown in Fig. 5, the P22 component showed a clearer somatotopical responses than N19. In addition, P22 responses were largest 5–10 mm medial to those of N19 responses, probably indicating different locations for the dipoles for these two components. P23 also showed better somatotopy than N19, but the largest responses appeared about 5 mm lateral to those of P22 (Fig. 6).

In a cadaver brain, the length of the postcentral gyrus from the midline to the sylvian fissure was 95 mm. Assuming this length to be correct, maximum SEP to finger stimulation appeared chiefly over the lateral three quarters of the postcentral gyrus. SEP recorded from the cortex intraoperatively also showed essentially the same results, and P22 always demonstrated better somatotopical responses than N19 [3, 6, 14].

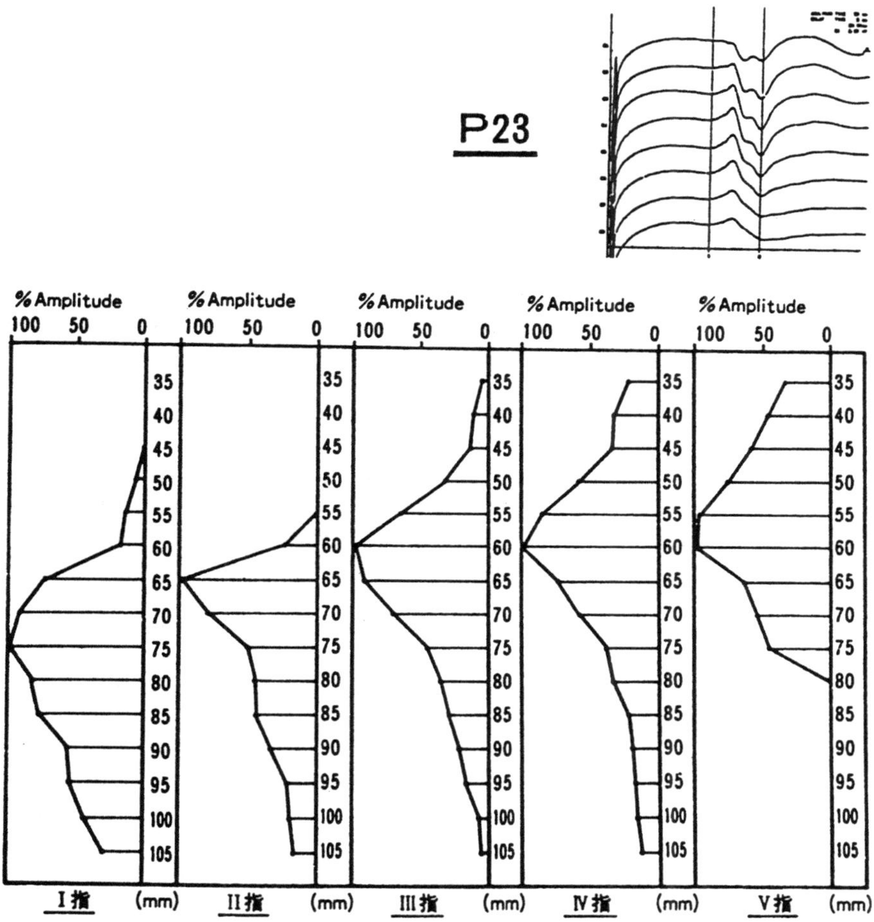

Fig. 6. The distributions of percent amplitude of P23 on the sensory cortex

P20 in a Case with Frontal Bone Defect

The distribution pattern of P20 was examined in a patient with a ruptured aneurysm in the right middle cerebral artery. Craniotomy and clipping were performed during the acute stage. Severe vasospasms followed by increased intracranial pressure made frontal decompression necessary. Frontal SEP were recorded to fingers I, III, and V stimulation after the patient recovered and showed no neurological signs. The largest response to finger I stimulation was located most laterally and that of finger V stimulation most medially and responses to finger III stimulation were located between these two.

Discussion

According to Desmedt et al. [4, 5], median nerve stimulation might activate not only afferent fibers from the skin, joints, and muscles but also motor fibers antidromically. However, finger stimulation activates pure sensory fibers. The wave form of ESCP and SEP to finger stimulation are analogous to those of median nerve ESCP and SEP, but the peak latencies of each components of ESCP and SEP to finger stimulation were about 3 ms longer. The delay might reflect the impulse conduction time from the finger to the wrist [4, 5]. The smaller amplitudes of finger ESCP and SEP might also be due to a smaller number of excited nerve fibers than those of median nerve SEP.

N1 of ESCP to finger stimulation showed somatotopical responses on the dorsal surface of the cervical cord. N1 is thought to be generated by horizontal dipoles made by synaptic potentials of dorsal horns [9, 10]. And this indicated changes of locations of these dipoles in accord to alterations of stimulated cervical segments or fingers. But, the real location of these dipoles can be determined only when both dorsal and ventral recordings can be obtained simultaneously.

SEP to finger stimulation showed clear somatotopical responses in a case with a skull defect, but revealed that the cortical locations of the maximal amplitude of N19, P22, and P23 differed. These results indicate that their respective dipoles may be located in different areas and have different vectors [1, 2, 3, 7, 12].

It has already been vertified by magnetoencephalography that N19/P20 are generated by the dipole located in the area 3b deep in the fissure of Rolando [12]. Thus, N19 is distributed widely over the parietal cortex and a clear somatotopical response could not be obtained because of its horizontal dipole. On the other hand, P22 showed the clearest somatotopical responses, which is in agreement with the theory that this component is produced by the vertical dipole existing in the postcentral cortex [1, 3, 7, 14]. Median nerve SEP also showed that P22 appeared in the restricted area of the post-central gyrus medial to the area where N19 is distributed [11, 13]. These results all indicated that P22 showed clearer somatotopy than N19, as first observed by Woolsey et al. [14] and more recently by Duff et al. [6] and Deiber et al. [3].

Summary and Conclusion

ESCP and cortical SEP to finger stimulation were recorded intraoperatively or in patients with skull defects. N1 of ESCP and P22 of SEP showed better somatotopical responses to finger stimulation than the other components.

The maximum N1 responses to finger I stimulation were obtained over C6 segment and the maximum P22 to finger I stimulation appeared on the postcentral gyrus about 70 mm lateral to the midline.

In contrast, as to finger V stimulation, the maximum N1 appeared on C7 to C8 segment and the maximum P22 on the postcentral gyrus 50 mm lateral to the midline.

The maximum responses to stimulation of other fingers were located somatotopically between those to finger I and V stimulation on the cervical cord as well as on the cerebral cortex.

The somatotopical responses coincided well with those of homunculus made by Penfield et al. [8]. These findings were not always reproducible when recorded over the scalp or the cervical skin, but can be a substantial basis to clinical applications of these evoked potentials.

References

1. Allison T (1982) Scalp and cortical recordings of initial somatosensory cortex activity to median nerve stimulation in man. Ann NY Acad Sci 338:671–677
2. Broughton RJ (1967) Somatosensory evoked potentials in man: cortical and scalp recordings. Thesis, Mc Gill Univ, Montreal, Quebec Canada
3. Deiber MP, Giard MH, Mauguiere F (1986) Separate generators with distinct orientations for N20 and P22 somatosensory evoked potentials to finger stimulation? Electroencephalogr Clin Neurophysiol 65:321–334
4. Desmedt JE, Bourguet M (1985) Color imaging of parietal and frontal somatosensory potential fields evoked by stimulation of median or posterior tibial nerve in man. Electroencephalogr Clin Neurophysiol 62:1–17
5. Desmedt JE, Nguyen TH, Bourguet M (1987) Bit-mapped color imaging of human evoked potentials with reference to the N20, P22, P27 and N30 somatosensory reponses. Electroencephalogr Clin Neurophysiol 68:1–19
6. Duff TA (1980) Topography of scalp recorded potentials evoked by stimulation of the digits. Electroencephalogr Clin Neurophysiol 49:452–460
7. Lueders H, Lesser RP, Hahn JH et al. (1983) Cortical somatosensory evoked potentials in response to hand stimulation. J Neurosurg 58:885–894
8. Penfield W, Jasper H (1954) Epilepsy and the functional anatomy of the human brain. Little, Brown & Co, Boston
9. Shimoji K, Higashi H, Kano T (1971) Epidural recording of spinal electrogram in man. Electroencephalogr Clin Neurophysiol 30:235–239
10. Shimoji K, Matsuki M, Shimizu H (1977) Wave-form characteristics and spatial distribution of evoked spinal electrogram in man. J Neurosurg 46:304–313
11. Takahashi H, Yasue M, Suzuki I, Ishijima B (1988) Cortical and subcortical somatosensory evoked potentials to median nerve stimulation in man. Brain Nerve 40: 275–283 (in Japanese)
12. Wood CC, Cohen P, Cuffin BX et al. (1985) Electrical sources in human somatosensory cortex: identification by combined magnetic and potential recordings. Science 227:1051–1053
13. Wood CC, Spencer DD, Allison T et al. (1988) Localization of human sensorimotor cortex during surgery by cortical surface recordings of somatosensory evoked potentials. J Neurosurg 68:99–111
14. Woolsey CN, Erickson TC, Gilson WZ (1950) Study of the postcentral gyrus of man by evoked potential technique. Trans Neurol Ass 57:50–52

Spinal Evoked Potentials and Abnormal Sensory Modalities in Spinal Cord Injuries and Peripheral Neuropathies

K. SEKO[1], T. FUKUDA, K. SIRASU, T. HISA, H. SIOTSU, and T. FUJIWARA

Summary

Cz-Fpz SEP components [P1(P36), N1(N48), P2(P60), N2(N75)] following stimulation of the posterior tibial nerve at the ankle were statistically analyzed with respect to the grade of sensory disturbances (thermal and vibratory sensation) in 27 patients with diabetic neuropathy, incomplete spinal cord injury and hemiplegia. Thermal sensation was weakly associated with P1 and P2, whereas vibratory sensation was weakly associated with P1.

These findings suggest that all sensory modalities are associated with these SEP components to some extent.

Introduction

Spinal evoked potentials (SEPs) consist of several components with different latencies. The developmental process of these components is not fully understood. In the case of SEPs following stimulation of the posterior tibial nerve at the ankle, P1(P36), N1(N48), P2(P60), and N2(N75) were recorded. According to the study by Ziganow [7] on 25 patients with incomplete spinal cord injuries, joint position sensation conveyed by faster conducting, more myelinated fibers (via posterior column of spinal cord) might be related to P1/N1 amplitude, whereas pin-prick sensation (via spinothalamic pathway) conveyed by slower conducting, less myelinated fibers might be related to P2/N2 amplitude. However, this theory has not been verified. Therefore, SEPs (Cz-Fpz) were recorded in 27 patients with diabetic neuropathy, spinal cord injury, and hemiplegia following stimulation of the posterior tibial nerve at the ankle, and each component of SEP was analyzed statistically in relation to the grade of the patient's sensory disturbance.

[1] Department of Neurology and Internal Medicine, Sakuma Hospital, 673 Azakobama, Amino-cho, Takeno-gun, Kyoto, 629-31 Japan

K. Shimoji et al. (Eds.)
Spinal Cord Monitoring and Electrodiagnosis

In a patient who underwent gangliectomy, Hosobuchi [7] found a significant loss of unmyelinated fibers in the ventral roots taken for histological examination several years after the gangliectomy had been performed. In the last 15 years extensive work in animals by Coggeshall has shown morphological and functional evidence of nonmotor ventral root fibers [1].

One possible explanation is that these fibers are pain fibers whose cell bodies arise in the dorsal root ganglion and whose axons pass up the ventral root instead of the dorsal. This is supported by the fact that careful pulling of exposed ventral roots in awake patients produces pain [6]. This article is our attempt to confirm the existence of afferent fibers within the ventral roots of humans.

Methodology

In a group of 25 patients with cerebral palsy undergoing selective posterior rhizotomy for relief of spasticity, we stimulated the ventral roots of the surgically exposed cauda equina and recorded spinal evoked potentials with epidural electrodes placed at T2 (Fig. 1). The spinal potentials were recorded with quadripolar catheter electrodes Neuromed Ere 4 (Neuromed, Fort Lauderdale, FL, USA), sufficiently semirigid to allow their placement in the epidural space without a stylet. Each of the four recording surfaces was 3 mm long, separated by 6-mm spaces. The distance between the most distal recording surface and the most proximal surface was 30 mm. The midline position of the electrode was confirmed by showing equally sized responses to left and right tibial nerve stimulation at the ankle (Fig. 2). In a few cases we placed an additional epidural electrode above the conus. After surgical separation of the lumboscaral roots into the dorsal and ventral components, each root was separately stimulated with stainless steel hook electrodes, consisting of two outer cathodes and one inner anode (Fig. 3). Square wave stimuli of 0.1 ms duration and 1–4 mA intensity were delivered at a 13.3 Hz rate. To obtain the evoked potential, 100–200 responses activated by ventral or dorsal root stimulation were averaged, using a Sentinel-4 evoked potential instrument (Axon System, Inc., Deer Park, NY, USA), with the following parameters: 20- or 40-ms analysis time, 20 Hz low frequency filter and 1700 Hz high frequency filter, and 256 sampling points per channel.

During root stimulation special care was taken to prevent current spread from the stimulated ventral root to the dorsal root, or vice versa. Morphological and functional criteria were used to distinguish dorsal from ventral roots within the dural sack. These were: ventral position of the ventral roots at the intervertebral foramina and the dorsal position of dorsal roots, and the lusher vasa nervorum of the ventral roots as compared with the dorsal roots. Another criterion for distinguishing dorsal from ventral roots was their response to electrical stimulation [5]. Ventral roots have a lower threshold to electrical stimulation, at least in children with cerebral palsy.

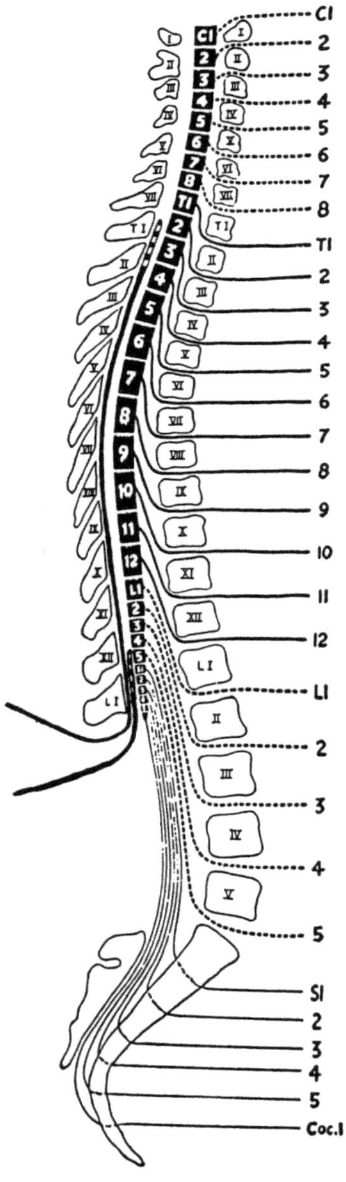

Fig. 1. The surgically exposed cauda equina with recording electrodes over the epidural space of the thoracic and sacral spinal cord

To reduce possible artifact from paravertebral muscle activity, Pavulon was administered to induce complete muscle relaxation. To confirm the absence of any muscle activity, we monitored the EMG from the paravertebral muscle with needle electrodes inserted into the muscle at T2. We

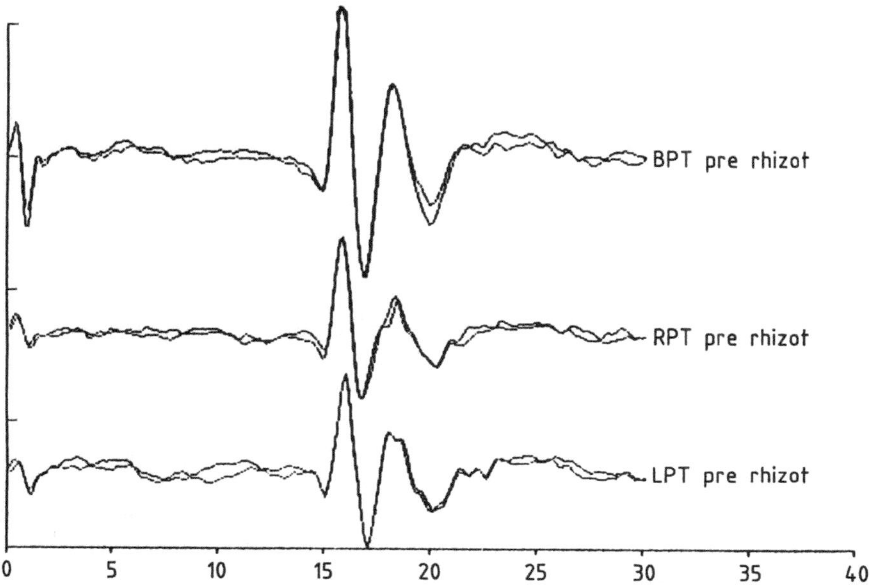

Fig. 2. Spinal somatosensory evoked potentials recorded epidurally over T2 level of spinal cord showing responses to the left, right and bilateral tibial nerve stimulation at the ankle. *Divisions* on the *ordinate* are 5 μV

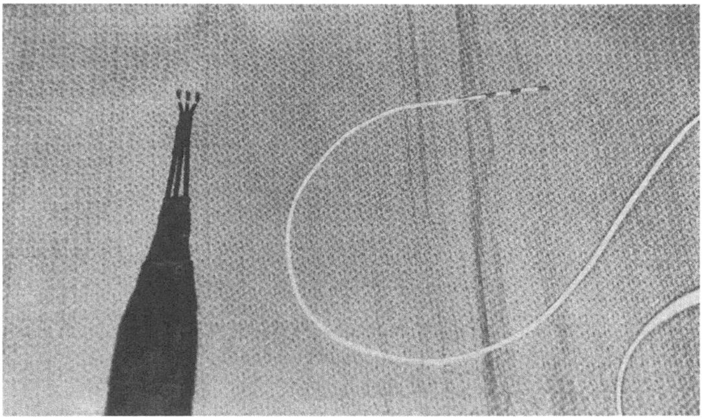

Fig. 3. Three-hook stainless steel electrode for root stimulation (*left*) and epidural electrode with four recording surfaces (*right*)

point out that this part of the methodology is critical because in spastic patients a huge activation of the paravertebral muscle can be seen after stimulation of the either the dorsal or ventral roots. This muscle activity can be picked up by the epidural electrodes, yielding false-positive results.

Results

Twenty-five patients with lower extremity spasticity underwent surgery and were evaluated for ventral root spinal evoked potentials (VRSP). In nine children with cerebral palsy and one child with tethered cord syndrome, we encountered spinal potentials evoked by ventral root stimulation. In two of these patients, the dorsal roots had been completely cut, and ventral root stimulation resulted in large spinal evoked potentials recorded over the T2 spinal level. Spinal potentials evoked by ventral root stimulation resembled responses evoked by dorsal root stimulation in terms of response latency, waveform, and number of peaks. However, ventral root spinal evoked potentials (VRSP) were to smaller amplitude than dorsal root spinal evoked potentials (DRSP) for comparable stimulus intensities. We did not detect dorsal root activity associated with ventral root stimulation after placing recording electrodes over the dorsal root. In contrast, dorsal root stimulation generally produced large and reproducible ventral root reflex activity.

In one patient, ventral root stimulation resulted in a well-defined VRSP over the conus as well as over the T2 level of the spinal cord. After the corresponding dorsal root was cut, only the electrical stimulation of the proximal stump of this dorsal root generated a spinal evoked response (Fig. 4). Stimulation of the distal stump did not. In the other 15 patients, we did not

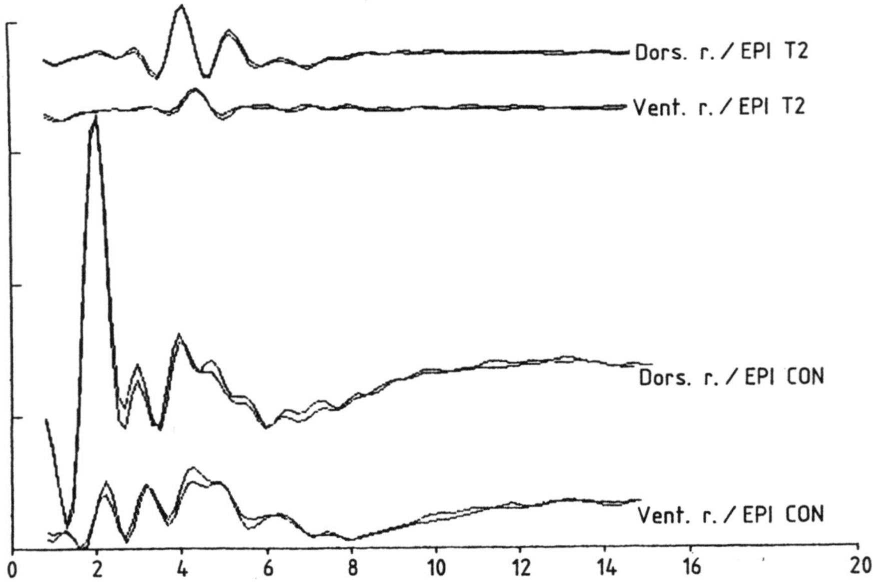

Fig. 4. Ventral and dorsal root spinal evoked responses recorded from the epidural space of the conus (*two lower traces*) and over the T2 level of the spine (*two upper traces*). Responses from the corresponding dorsal root were obtained after stimulation of the proximal stump of cut root. *Divisions* on the *ordinate* are 10 μV

observe VRSP. Technical limitations in the first five patients studied preclude a firm conclusion concerning their absence. We are, however, reasonably certain that at least 10 of the 25 patients studied did not have VRSP. We have not yet had an opportunity to examine nonspastic patients for the occurrence of VRSP.

Discussion

The mechanism of the VRSP is unclear at the present. A few possibilities should be considered. First, VRSP may arise from antidromic activation of motor axons resulting in Renshaw cell interneuronal activity. This type of activity, perhaps through abnormal connections present in spastic patients, may then cause VRSPs. Second, the VRSP may be due to the activation of ventral root afferents, as reported in animal studies since the time of Sherrington. In general, ventral root afferents, as described in many mammalian species, are small unmyelinated fibers and relatively few in number as compared to dorsal root afferents. The robust nature and amplitude of the spinal potentials argue against the possibility that these small unmyelinated fibers are the source of the VRSP activity. However, we have not ruled out the possibility that the ventral roots of patients with cerebral palsy have larger myelinated ventral root afferents. The third possibility, as offered by Coggeshall [1], is that ventral root afferents are fibers whose cell bodies are in the dorsal root ganglion and whose axons only loop into the ventral root. This seems less probable because the VRSP persists after the dorsal root has been severed. Also stimulation of the distal stump of the severed dorsal root does not generate a spinal evoked response. In two patients in whom stimulation of the ventral roots generated VRSP after the corresponding dorsal roots were cut, there still is the theoretical possibility that the looping fibers in the ventral roots enter into the spinal cord via the adjacent noncut dorsal root. If this were the case, one would expect longer latencies after the stimulation of ventral roots as compared with stimulation of the corresponding dorsal roots due to the additional length of the fibers. This was not case in our measurements.

Finally, this phenomenon has to be examined in nonspastic patients to confirm whether this phenomenon is the biological rule, or only a feature of spasticity.

References

1. Coggeshall RE (1980) Law of separation of function of the spinal roots. Physiology Rev 60:716–755

2. Coggeshall RE, Coulter JD, Willis WD (1974) Unmyelinated axons in the ventral roots of the cat lumbosacral enlargement. J Comp Neurol 153:39–58
3. Coggeshall RE, Applebaum ML, Fazen M, Stubbs, TB, Sykes MT (1981) Unmyelinated axons in human ventral roots, a possible explanation for the failure of dorsal rhizotomy to relieve pain. Brain 98:157–166.
4. Cranefield PE (1974) The way in and the way out. Futura Mount Kisco, New York
5. Fasanò VA, Barolat-Romana G, Zeme S, Sguazzi A (1979) Electrophysiological assessment of spinal circuits in spasticity by direct dorsal root stimulation. Neurosurgery 4: 146–151
6. Frykholm R, Hyde J, Norlen G, Skoglund CR (1953): On pain sensations produced by stimulation of ventral roots in man. Acta Physiol Scand [Suppl] 106:455–459
7. Hosobuchi Y (1980) The majority of unmyelinated afferent axons in human ventral roots probably conduct pain. Pain 8:167–180
8. Sherrington CS (1894) On the anatomical constitution of nerves of skeletal muscles: With remarks on recurrent fibers in fibers of the ventral spinal nerve root. J Physiol (Lond) 17:211–258

Spinal Evoked Potentials and Spine-Scalp Propagation Velocities: Origins, Conduction Characteristics, and Clinical Observations

R.Q. Cracco[1], J.B. Cracco, and L. Pelosi[2]

In the ensuing paragraphs research concerning spinal evoked potentials and spine-scalp propagation velocities which has been carried out in our laboratories is reviewed.

Spinal Evoked Potentials

Evoked potentials to stimulation of nerves in the lower extremities which arise in the cauda equina and in spinal cord afferent pathways can be recorded from surface electrodes placed over the lumbar, thoracic, and cervical spines [4]. These potentials progressively increase in latency rostrally (Figs. 1, 2). In bipolar leads attached to the skin over the lumbar spine, the response to common peroneal nerve stimulation at the knee consists of an initially positive triphasic potential which reflects potentials arising in the roots of the cauda equina. In lower thoracic leads which overlie caudal spinal cord segments, the response is greater in amplitude and duration and more complex in configuration than in more rostral or more caudal leads. In adults this response is usually initially positive and triphasic, the negative component of which has several peaks or inflections (Fig. 1). In infants and young children it usually consists of a large positive-negative diphasic potential followed by a broad negative and then a positive potential (Fig. 2) [1]. Investigations of similar potentials recorded in cats and monkeys demonstrate that the initial diphasic potential arises in the intramedullary continuations of dorsal root fibers, and the subsequent potentials reflect synaptic and post synaptic activity which is concerned with local reflex mechanisms rather than with the propagation of the response to more rostral levels [5, 8, 11].

Over rostral cord segments the response in children and adults consists of very small (often < 2 μV in rostral spinal leads), initially positive triphasic potentials, with poorly defined positive phases which progressively

[1] State University of New York Health Science Center at Brooklyn, Department of Neurology, 450 Clarkson Avenue, Brooklyn, NY 11203, USA
[2] *Present Address*: Department of Clinical Neurophysiology, II. School of Medicine, Naples, Italy

K. Shimoji et al. (Eds.)
Spinal Cord Monitoring and Electrodiagnosis

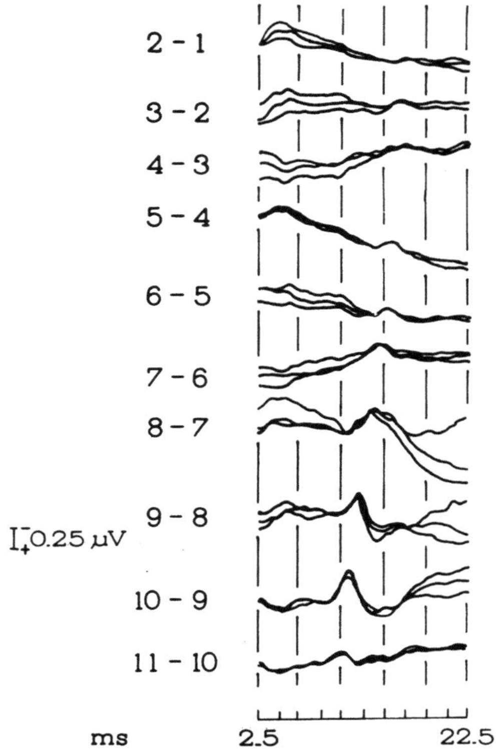

Fig. 1. Bipolar recordings of the spinal response to left peroneal nerve stimulation. Inter-electrode distance is 4.5 cm. Electrode 11 is placed over the second lumbar spine and electrode 1 over the 3rd cervical spine. There is a delay of 2.5 ms between the shock and the sweep onset. Three recordings are superimposed in each trace. The response progressively increases in latency at more rostral recording locations. In the lead in which the caudal electrode is placed over the 9th thoracic spine (4th trace from *bottom*), a complex response with three negative components is recorded. Onset latency differences are much greater between this and the 5th trace from *bottom* than between more caudal or more rostral leads [from 4]

decrease in amplitude rostrally (Figs. 1, 2) [1, 2, 4]. This amplitude decrement reflects temporal dispersion and the greater distance between the skin recording electrode and the spinal cord at rostral thoracic and cervical sites. Investigations in cats and monkeys suggest that these potentials arise in multiple, rapidly conducting afferent pathways, including the dorsal and dorsolateral columns which lie primarily ipsilateral to the stimulated peripheral nerve [5, 8, 11].

The onset of the negative potential at each recording location serves as the latency indicator, since this reflects the approximate time the fastest fibers contributing to a response arrive under the recording electrode in reference leads or the recording electrode nearest the approaching volley

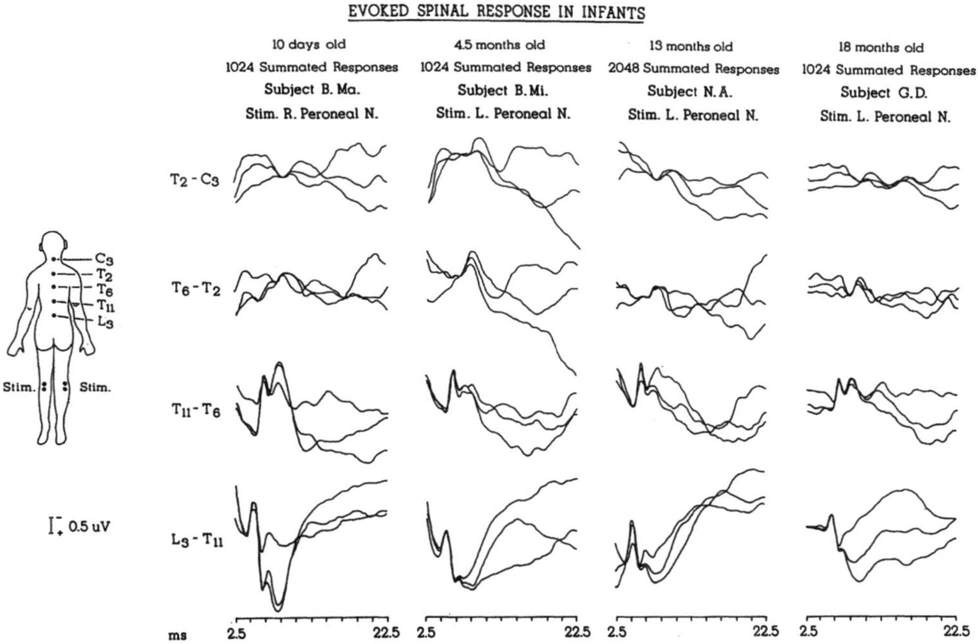

Fig. 2. Evoked spinal response in infants

in bipolar leads. The conduction velocity up the spine of the response evoked by peroneal nerve stimulation is nonlinear; it is slower over caudal spinal cord segments than over the cauda equina or rostral spinal cord (Fig. 3) [1, 2, 4]. Similar observations have been made with stimulation of the sciatic nerve and nerve to the medial head of the gastrocnemius muscle in both surface and dural recordings performed in cats and monkeys [5, 8, 11]. This slowing over caudal cord reflects branching of dorsal root fibers and synaptic activity. The increase in speed of conduction over rostral cord segments suggests that large diameter second order afferent fibers contribute to this response, since the speed of conduction in first order fibers which ascend the length of the cord would not be expected to increase in rostral cord segments.

In normal adults the mean conduction velocity of the response to peroneal nerve stimulation from lumbar to cervical recording locations is about 70 m/s [4]. Segmental conduction velocities are about 65 m/s over peripheral nerve and cauda equina (point of stimulation in popliteal fossa to L3 spine), 50 m/s over caudal spinal cord (T12 to T6 spines) and 85 m/s over rostral spinal cord (T6 to C7 spines; Fig. 3) [2, 9]. In the newborn human infant, these segmental conduction velocities are about half those obtained in the adult. They progressively increase with age. Peripheral conduction velocities are largely within the adult range by 3 years, whereas velocities over caudal and rostral spinal cord do not reach adult values until the 5th year (Fig. 3) [2]. This sug-

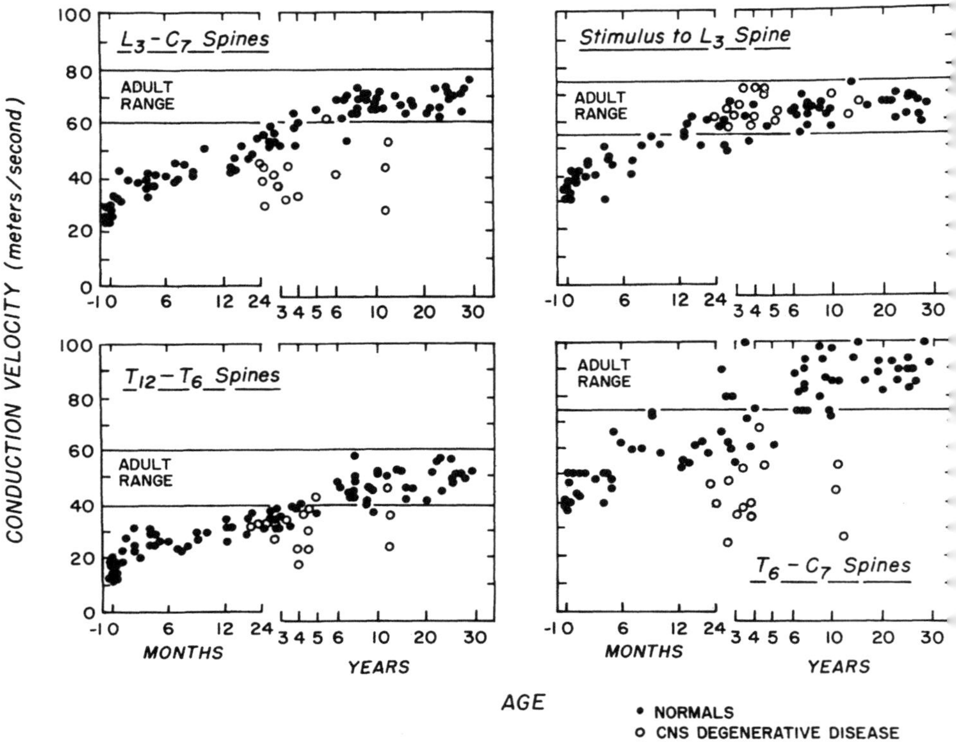

Fig. 3. Distribution of the overall spinal conduction velocities (L3–C7 spines) and the segmental conduction velocities over peroneal nerve and cauda equina (stimulus to L3 spine), caudal spinal cord (T12–T6 spines), and rostral spinal cord (T6–C7 spines) in normal infants, children, and adults (*closed circles*) and patients with CNS degenerative disease (*open circles*). In normal subjects conduction velocities are slower over caudal cord segments than over peripheral nerve-cauda equina or rostral spinal cord; peripheral conduction velocities reach adult values by 3 years, whereas velocities over spinal cord do not reach adult values until 5 years. In patients peripheral conduction velocity is normal, but spinal cord conduction velocity is slowed and this effect is *greatest* over rostral cord segments

gests that maturation of human rapidly conducting spinal afferent pathways proceeds at a slower rate than maturation of rapidly conducting peripheral sensory fibers.

We have recently compared spinal responses to stimulation of the posterior tibial nerve at the ankle with those previously obtained with common peroneal nerve stimulation at the knee [9]. With stimulation of both nerves the speed of conduction over peripheral nerve and spinal cord is nonlinear, being slower over caudal cord. The common peroneal nerve spinal evoked response is conducted significantly faster than the tibial nerve response over peripheral nerve and cauda equina and rostral spinal cord, but these values are similar over caudal cord segments [9]. Hence, the degree of nonlinearity in speed of conduction up the spine is much greater with common peroneal nerve stimulation. This indicates that the spinal response to common peroneal nerve stimulation is mediated by more rapidly conducting pathways. This is not unexpected since rapidly conducting peripheral afferent fibers which supply muscle spindles and tendon organs would be expected to be more numerous at the knee than at the ankle. Additionally, it has been shown in primates that the ratio of afferent fibers arising from deep vs cutaneous receptors is greater in proximal hind limb than in distal hind limb [13, 14]. These afferents from deep receptors include the large group I fibers which synapse in caudal cord nuclei such as Clarke's column which give rise to large rapidly conducting second order fibers. Second order cutaneous fibers such as those which ascend in Morin's tract could also contribute to the increase in conduction velocity over rostral spinal cord segments.

In work just recently completed [10], spinal evoked potentials to stimulation of the tibial nerve at the knee and ankle and the sural nerve at the ankle were recorded from needle electrodes inserted into the ligamentum flavum over multiple spinal levels during surgery for scoliosis in neurologically normal subjects. Responses were similar, but much greater in amplitude and more complex in configuration than potentials recorded from surface leads. As in surface leads, the speed of conduction up the spine was nonlinear whether the mixed tibial nerve was stimulated at the ankle or the knee or whether the cutaneous nerve was stimulated [10]. Nonlinearity in speed of conduction up the spine has not previously been reported with stimulation of pure cutaneous nerves in humans. This finding suggests that second order afferent fibers such as those which ascend in Morin's tract contribute significantly to spinal potentials evoked by stimulating cutaneous nerves as well as mixed muscle and cutaneous lower extremity nerves when recorded in this way.

Clinical Studies

Clinical studies of spinal evoked potentials to peroneal nerve stimulation carried out on patients using surface leads are summarized in the ensuing paragraphs. Most studies were performed on infants and young children since these responses are much greater in amplitude and better defined in the young pediatric age group. In adult patients with clinically evident complete spinal cord lesions, spinal evoked potentials recorded from surface electrodes caudal to the lesion were similar to those obtained in normal subjects. No response was recorded in leads rostral to the lesion [4]. In a study of a group of infants and children with myelodysplasia from whom spinal evoked potentials and cerebral evoked potentials to peroneal nerve stimulation were recorded, there was correlation between the degree of evoked potential abnormality and the clinical status of the patients [6]. In a few of these patients it was possible to diagnose caudal displacement of the spinal cord (tethered cord); in these patients the large complex spinal potentials which are normally recorded over the lower thoracic spine were recorded over lumbar spinous processes. In several children with myelomeningocele a positive potential was recorded in leads immediately rostral to the lesion. This potential progressively decreased in amplitude but did not change in latency rostrally [6]. This is a nonpropagated volume conducted potential which is consistent with physiological transection of the spinal cord. Similar positive potentials have been recorded rostral to spinal cord transections in cats and monkeys [5, 8, 11].

Spine and scalp recorded SEP to peroneal nerve stimulation were studied in a group of children with degenerative disease of the central nervous system (Fig. 3). The conduction velocity over the peroneal nerve was normal. Responses were recorded over the spinal cord in most patients, but conduction velocities over the cord were slowed. Velocities over the spinal cord were slower over rostral cord segments than they were over the caudal cord (Fig. 3). This may suggest that a "dying back" process, similar to that known to occur in some peripheral neuropathies, is operative in the spinal afferent pathways of these patients [3].

Peroneal nerve and spinal conduction velocities were also found to be slowed in some neurologically asymptomatic juvenile diabetics [7]. Peripheral nerve or spinal conduction velocities alone were slowed in some patients, while in others both peripheral and spinal conduction velocities were slowed. As in children with CNS degenerative disease, spinal conduction velocity over rostral spinal cord segments was chiefly affected.

Spine-Scalp Propagation Velocities

The major technical problem in recording spinal evoked potentials is the small size of these potentials, particularly in rostral spinal leads in adults where these potentials reach only a small fraction of a microvolt. This problem can be minimized by recording both spine and scalp potentials to stimulation of lower extremity nerves and dividing the distance between spine and scalp recording sites by the latency differences of potentials recorded at these locations. In this way, spine to scalp propagation velocities can be determined [9, 12]. The conduction velocity of the response over stimulated peripheral nerve fibers from the point of stimulation to the lumbar spine recording electrode can also be determined. Hence, this method takes into account the length of the somatosensory pathway and permits differentiation between peripheral and central disorders of afferent conduction.

With peroneal nerve stimulation at the knee, the propagation velocities of responses from the L3, T12, T6 and C7 spinous processes to the scalp P27 potential are approximately 40, 37, 35 and 24 m/s [9, 12]. With posterior tibial stimulation at the ankle (using scalp P37) these values are about 34, 33, 29, and 22 m/s, respectively [9]. The standard deviations of these velocities are of the order of 1–2 m/s with peroneal nerve stimulation and 2–3 m/s with tibial nerve stimulation. These small standard deviations suggest that this method is sensitive and reliable. Using this method, abnormality may be defined on the basis of (1) the absence of potentials always recorded in normal subjects (L3 spine and peak of P27 or P37 or onset of N35 or N45) and (2) L3, T12, and T6 to scalp propagation velocities which are 3 S.D. below the mean obtained from normal control subjects.

This method has the advantage that even when responses cannot be recorded over the rostral spinal cord because of their small signal size, responses recorded over more caudal spinal regions may be used as latency indicators and conduction characteristics from these locations to the scalp-recorded SEP may be reliably determined. One theoretical disadvantage to this method is that the generator sources of the spinal evoked potentials and the spinal afferent pathways which transmit the scalp response are not likely to be identical. A more important problem is that slowed spine to scalp propagation velocities or absent scalp potentials may result from pathology occurring at many sites including cerebral, brainstem or spinal levels. This makes the precise localization of a lesion difficult using this method. Nevertheless, this technique provides important information concerning the integrity of spinal afferent pathways in many patients with spinal cord disease.

Clinical Data

Forty patients with disease of the spinal cord or peripheral nervous system were studied [12]. Focal incomplete spinal cord compressive lesions generally resulted in slowing of spine to scalp propagation velocities (Fig. 4). Severe diffuse or multifocal pathology usually resulted in absence of scalp responses (Fig. 4). Although there was no consistent correlation of the evoked potential findings with the sensory exam, there was as a correlation between the SEP findings and the clinical prognosis [12]. This data suggests that this technique will prove useful in the evaluation of some patients with spinal cord pathology.

Summary

Studies of spinal evoked potentials carried out in our laboratories on human infants, children and adults are reviewed. The effects of maturation on these potentials and clinical studies in the pediatric age group are summarized. Recent studies comparing the conduction characteristics of spinal evoked potentials to peroneal, posterior tibial, and sural nerve stimulation recorded from surface and depth leads are related, and the neural substrates which underlie these potentials are emphasized. Finally, investigations concerning

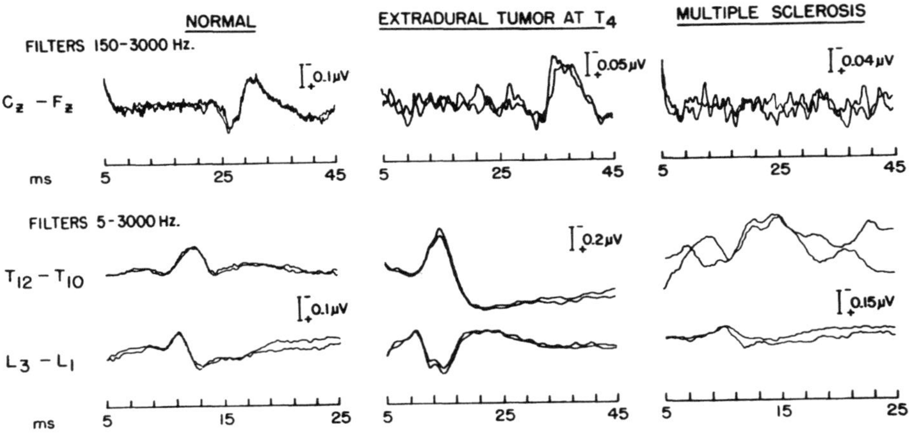

Fig. 4. Comparison of spine and scalp recordings to peroneal nerve stimulation in a normal subject, a patient with extradural tumor at T4 and a patient with clinically definite multiple sclerosis. Two recordings are superimposed in each trace. Scalp potential is delayed in patient with the extradural tumor and spine-scalp propagation velocity was slow. Scalp potential is absent in the patient with multiple sclerosis

spine scalp propagation velocities and the clinical utility of this method are discussed.

References

1. Cracco JB, Cracco RQ, Graziani LJ (1975) The spinal evoked response in infants and children. Neurology (Minneap) 25:31–36
2. Cracco JB, Cracco RQ, Stolove R (1979) Spinal evoked potential in man: A maturational study. Electroencephalogr Clin Neurophysicol 46:58–64
3. Cracco JB, Bosch VV, Cracco RQ (1980) Cerebral and spinal somatosensory evoked potentials in children with CNS degenerative disease. Electroencephalogr Clin Neurophysiol 49:437–445
4. Cracco RQ (1973) Spinal evoked response: Peripheral nerve stimulation in man. Electroencephalogr Clin Neurophysiol 35:379–386
5. Cracco RQ, Evans B (1978) Spinal evoked potential in the cat. Effects of asphyxia, strychnine, cord section and compression. Electroencephalogr Clin Neurophysiol 44:187–201
6. Cracco JB, Cracco RQ (1982) Spinal somatosensory evoked potentials: Maturational and clinical studies. Ann NY Acad Sci 388:526–537
7. Cracco JB, Castells S, Mark E (1984) Spinal somatosensory evoked potentials in juvenile diabetes. Ann Neurol 15:55–58
8. Feldman MH, Cracco RQ, Farmer P, Mount F (1980) Spinal evoked potential in the monkey. Ann Neurol 7:238–244
9. Pelosi L, Cracco JB, Cracco RQ (1987) Conduction characteristics of somatosensory evoked potentials to peroneal, tibial and sural nerve stimulation in man. Electroencephalogr Clin Neurophysiol 68:287–294
10. Pelosi L, Caruso G, Cracco RQ, Cracco JB, Balbi P (1990) Intraoperative recordings

of spinal somatosensory evoked potentials to tibial and sural nerve stimulation. Muscle Nerve (In Press)
11. Sarnowski RJ, Cracco RQ, Vogel HB and Mount F (1975) Spinal evoked response in the cat. J Neurosurg 43:239–336
12. Schiff JA, Cracco RQ, Rossini PM, Cracco JB (1984) Spine and scalp somatosensory evoked potentials in normal subjects and patients with spinal cord disease: Evaluation of afferent transmission. Electroencephalogr Clin Neurophysiol 59:374–387
13. Werner G, Whitsel BL (1967) The topology of dermatomal projection in the medial lemniscal system. J Physiol (Lond) 192:123–144
14. Whitsel BL, Petrucelli LM, Sapiro G (1969) Modality representation in the lumbar and cervical fasciculus gracilis of squirrel monkeys. Brain Res 15:67–78

Evoked Potential Patterns in Chronic Spinal Cord Lesions*

M. SARZYŃSKA[1], I. HAFTEK, and J. HAFTEK

Introduction

Evoked potential recordings are widely used for evaluation of the central nervous system status. They seem to be particularly useful in evaluating spinal cord function in acute and chronic spinal cord lesions (Wise 1984, Rowed 1982, Greenberg and Ducker 1982). Moreover, SEP have been proved to be an early detector of disturbances of spinal cord function during surgeries (Schramm 1985).

The main point of interest in our study was to find SEP patterns characteristic for the site of injury and to develop a method for a very exact electrophysiological localization of a spinal cord lesions. This would be especially valuable in multilevel myelopathies.

In our previous study we found that the most characteristic finding for acute spinal cord injury is the so-called injury potential – a deep positive waveform recorded at the lesion site. An example of such an evoked potential recorded recently is shown in Fig. 1. This is a case of a 59-year-old man with fracture dislocation at D8–9 level. Partial, but very profound injury to the cord with complete motor paralysis was noted, but some superficial and deep sensations were preserved below the level of the lesion. Tibial nerve stimulation was applied, The recordings below the lesion were normal for this level (Fig. 1A). At the lesion site a big injury potential was present (Fig. 1B), above the lesion a conducted waveform carried by preserved fibers was seen (Fig. 1C). This study was performed to investigate the SEP response patterns from epidural space to peripheral nerve stimulation in chronic spinal cord lesions.

* This study was partly sponsored by Vivian L. Smith Foundation, Houston, USA
[1] Electrophysiological Lab. Metropolitan Rehabilitation Center, ul. Wierzejewkiego 12, 05–511 Konstancin, Poland

K. Shimoji et al. (Eds.)
Spinal Cord Monitoring and Electrodiagnosis

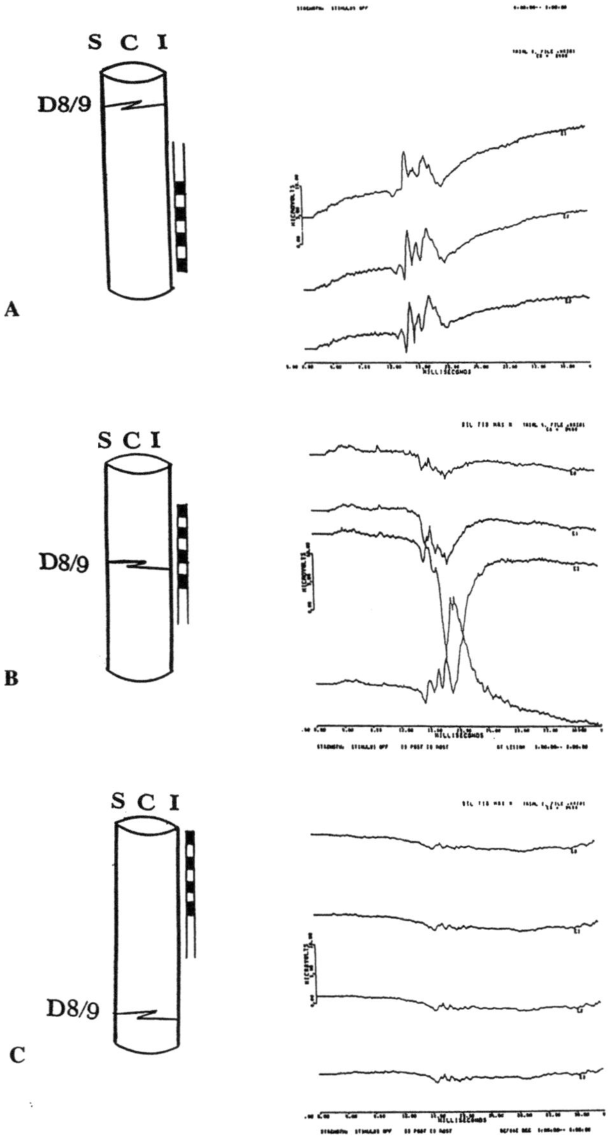

Fig. 1 A–C. Epidural recordings done in acute incomplete spinal cord lesion at D8–9 level. Tibial nerve stimulation. **A** Below the lesion, normal polyphasic response is seen. **B** At the lesion site, the injury potential is visible. **C** Above the lesion, a polyphasic waveform of low amplitude is recorded. The schematic drawings of the spinal cord lesion and position of electrode are adapted after John Halter

Clinical Materials and Method

Twenty-two patients with a variety of chronic spinal cord lesions (cervical spondylotic myelopathies, spinal cord ischemic changes, spinal cord compressions due to congenital malformations of spine, spinal cord tumors, spinal cord posttraumatic compressions) were investigated. A neuromed J1980 electrode with four contacts spaced 1 cm was positioned in the epidural space or directly on the spinal cord along the cord, midline, anteriorly or posteriorly during surgery. The reference electrode was a crocodile clip inserted into a muscle close to the recording electrode. The recordings were done usually above, below, and at the lesion site. Tibial nerve stimulation at the popliteal fossa or median nerve stimulation at the wrist with 0.2-ms square pulses was applied. The frequency was up to 7 Hz. Biological Travaller Two and Mystro 5+ were used as recording equipment.

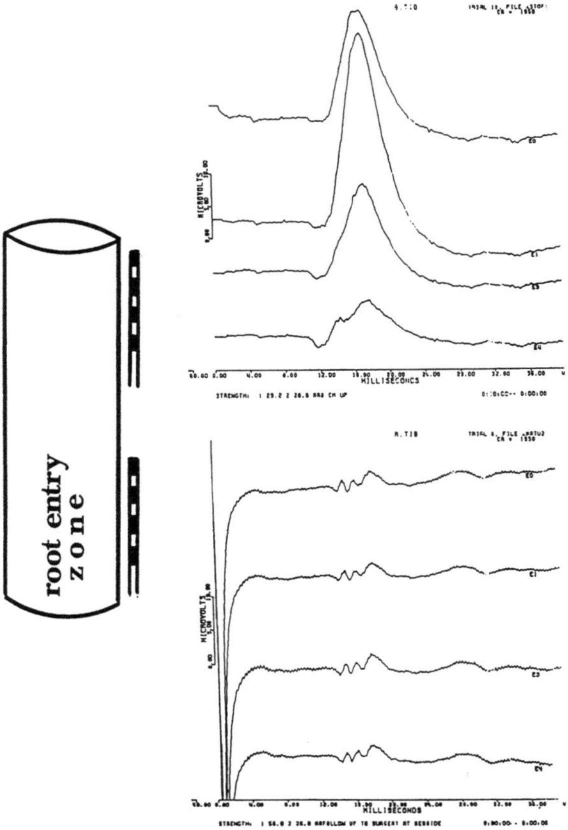

A
B

Fig. 2A,B. Recordings usually obtained in epidural space of healthy subjects. Tibial nerve stimulation. **A** At the root entry zone, a big negative potential is seen. **B** Above the root entry zone, a polyphasic conducted waveform is recorded

Results

Recordings performed in our laboratory allowed us to find out what the typical SEP recorded in a healthy subject were. They are different depending on the level (Fig. 2). At the level of the root entry zone a big negative potential is observed (Fig. 2A). Most probably this is a response of interneurons for excitation coming from the periphery. With tibial nerve stimulation the amplitude is maximum at the L1 level and then decreases rostrally, taking a polyphasic pattern of D11–12 level (Fig. 2B). It keeps a polyphasic character up to cervical level, although the polyphasic nature is not always recordable.

A typical injury potential was also noted in a patient after diagnostic puncture of the spinal cord (Fig. 3). The patient was a 16-year-old male with a suspected spinal cord tumor. He had developed progressive motor deficits, especially in the right upper limb, without any disturbances in sensation. Myelo-CT-scan suggested an intraspinal lesion. After opening the dura the recording was done at the C6 level with tibial nerve stimulation applied. The waveforms recorded were fairly normal (Fig. 3A). A spinal cord puncture using a very fine needle was performed to check whether there was cyst inside the cord. Neither cyst nor tumor was found. Thereafter the EP record-

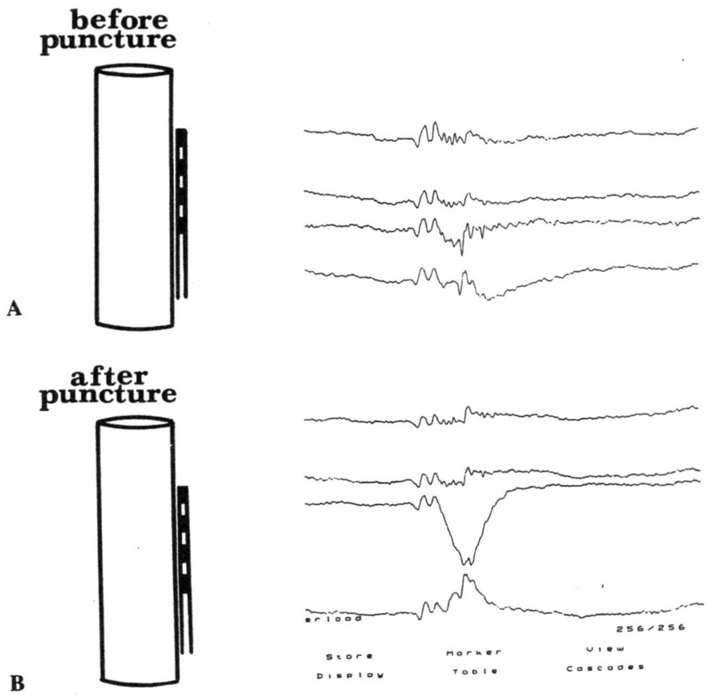

Fig. 3A,B. Recordings obtained directly from the cord surface from C6 level before (**A**) and after (**B**) spinal cord puncture

ings were repeated at the same level. A positive injury potential was recorded at only one electrode contact (Fig. 3B). This shows that the presence of injury potential might be detected with an accuracy of about 1 cm.

In chronic spinal cord lesions some characteristic SEP patterns have been observed. First, an injury potential was found to exist in chronic lesions of the spinal cord. In Fig. 4 the recordings performed in a 60-year-old woman with an unclear diagnosis are shown. Arteriovenous malformation was suspected. Paraplegia from D8–9 developed. There was complete motor paralysis accompanied by severe spasticity and pain. Some traces of touch and deep sensation were still present. The dura was opened, and no malformation was found but the signs of ischemia were visible. Recordings from this region using tibial nerve stimulation were performed. A sudden change of conducted waveform into an injury potential was observed (Fig. 4A). Above the lesion only a far-field positive potential was noted (Fig. 4B). No conducted wave-

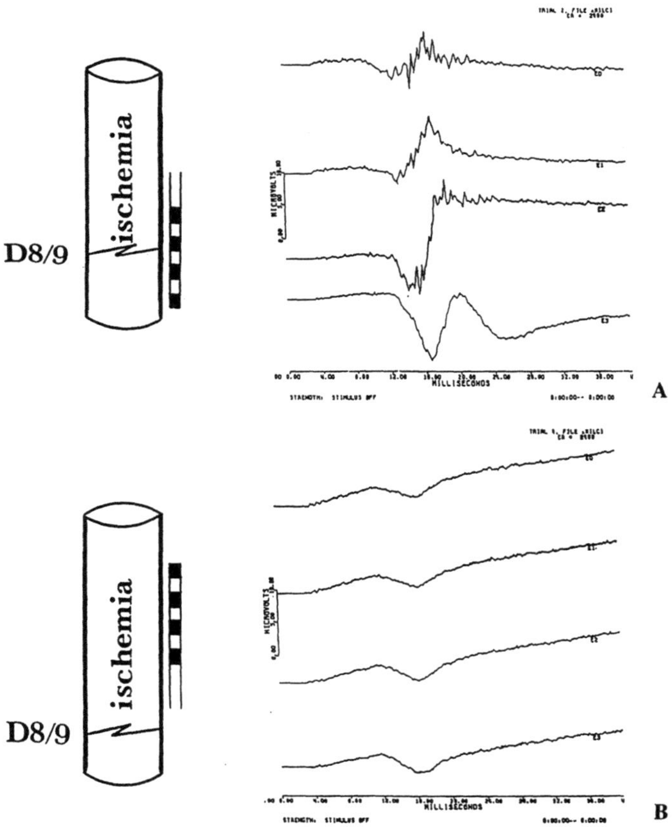

Fig. 4A,B. Epidural SEP observed in spinal cord ischemia at D8/D9 level. Tibial nerve stimulation. **A** At the lower edge of ischemic area a sudden change into an injury potential is visible. **B** Well within the ischemic area, only a far-field injury potential was recordable

form was recordable at this level. Another example was a 60-year-old man with cervical spondylotic myelopathy. Profound tetraparesis, but only slight sensory deficit were noted. C5 corporectomy was done from an anterior approach. Epidural recordings with tibial nerve stimulation were done (Fig. 5). The traces below the lesion were normal (Fig. 5A) while above the lesion (Fig. 5B) only a far-field injury potential was seen.

Secondly in chronic cases the disappearance of conducted waveforms or one of the components at the root entry zone was seen. An interesting result was obtained in a 61-year-old woman with an intraspinal tumor (astrocytoma) in the lower cervical cord. Deep tetraparesis with profound disturbances of all kinds of sensations were noted. After opening the dura the recording electrode was positioned directly on the spinal cord, and tibial nerve stimulation was applied. A polyphasic waveform was recorded at the lesion (lowest

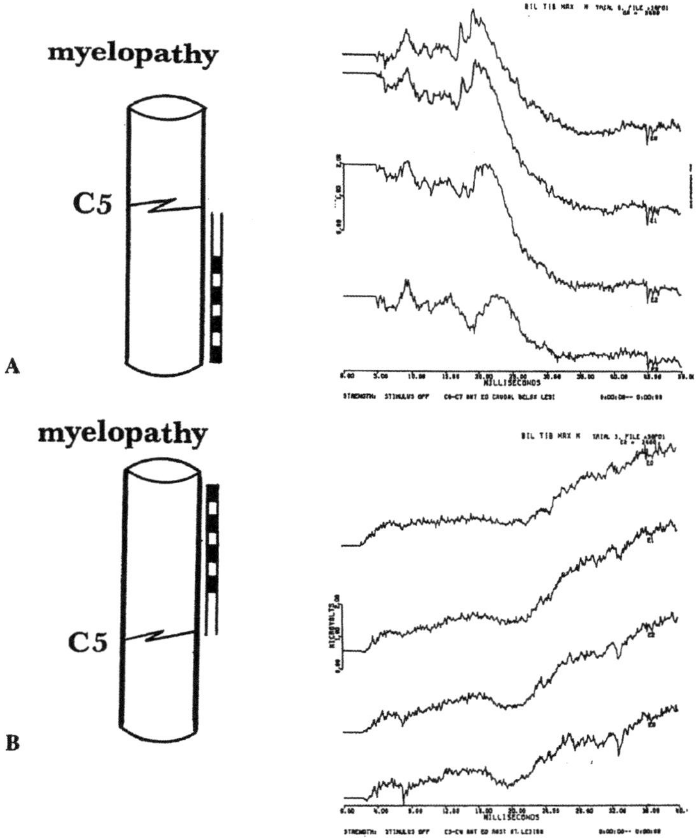

Fig. 5A,B. Epidural SEP at C5 level in cervical spondylotic myelopathy. Tibial nerve stimulation. **A** Below the site of compression, fairly normal waveforms are seen. **B** Above the compression place, only a far – field injury potential is observed

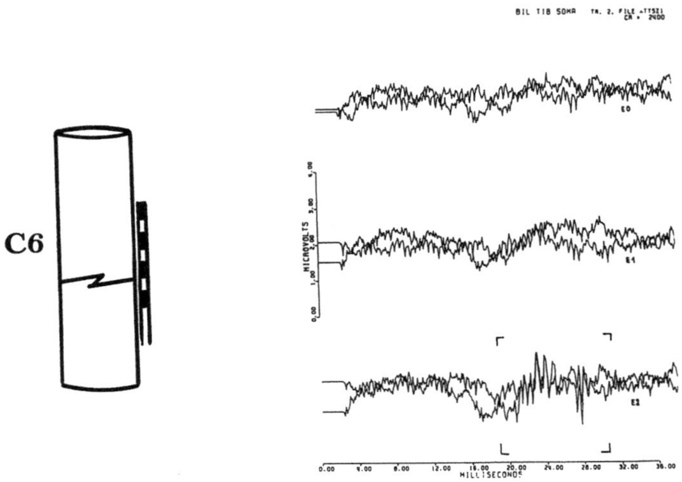

Fig. 6. SEP at C6 level directly from the spinal cord surface in a patient with intraspinal tumor, epidural recordings. Tibial nerve stimulation. In the *last trace* the low amplitude polyphasic waveform is seen. This pretty normal waveform disappears (*two upper traces*) with no sign of injury potential visible

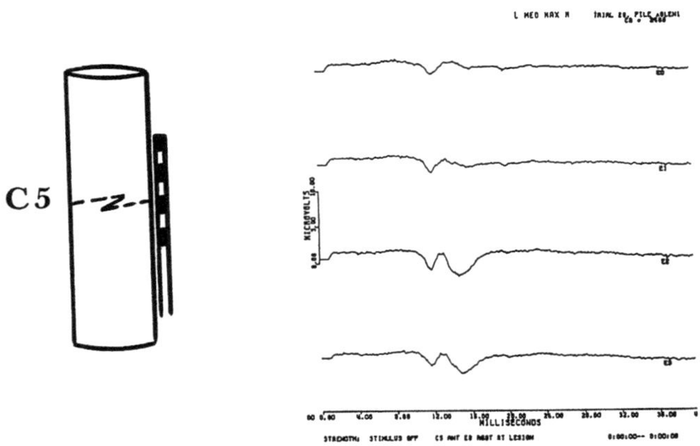

Fig. 7. SEP recording done by median nerve stimulation at C5 level in chronic spinal cord posttraumatic compression, at the site of compression. Rostral decrease of negative component is observed

trace in Fig. 6) and it just disappeared 1 cm higher with no sign of injury potential (two higher traces in Fig. 6). Another example (Fig. 7) was a 31-year-old man with chronic traumatic instability and subluxation of C5–6. No neurological deficit except decreased sensation in the left hand was noted.

During surgical decompression done from an anterior approach, the signs of compression of the dura at C5 level were seen. Epidural recordings at that point were performed with median nerve stimulation. Steady rostral diminishing of the main negative component with a tendency to disappear was noted.

Our third observation was that the injury potential and the disappearance of a component in root entry zone might occur together. Figure 8 shows the recordings obtained in an 11-year-old boy with a congenital malformation of the spine at L1/D12 level spinal cord and cauda equina chronic compression, flaccid paraparesis, and pain problems as well as neurogenic bladder developed. Various sensations were preserved. Decompressive laminectomy and fusion of the spine were performed. Epidural recordings from the exposed region with tibial nerve stimulation were done. At the level of L1 (Fig. 8A) a steady diminishing and disappearance of the main negative component was observed, whereas at D12 level a positive, most probably far-field, injury potential with superimposed small negativity was recorded (Fig. 8B).

Discussion

Our study provided evidence that an injury potential was recorded in the epidural space in chronic spinal cord lesions, as it has been observed in acute cases. This deep positivity most probably manifests the presence of destroyed nerve fibers in spinal cord fibers in spinal columns, and its size may mirror the size of the injury. The change in epidural waveform shape is very well defined, so it gives an opportunity to very exactly localize the lesion along the spinal cord, with about a 1 cm accuracy. Such a situation is illustrated in Fig. 4. This might be a very useful diagnostic indication in multi-level spinal cord lesions, especially myelopathies.

The shape of the injury potential in acute and chronic cases is very similar. This suggests that the nature of the positive waveform in the spinal cord is the same in both types of lesions. In chronic cases, however, we have noted a disappearance of the conducted waveform or one of the waveform components which was not recorded in acute cases. The mechanism responsible for these phenomena is not clear. Most probably it is connected with ischemia of the spinal cord. In most of the investigated patients the main cause of injury was a slowly developing ischemia and microcirculation disturbances. In acute cases the lesion consists mainly of disrupted fibers.

There are still some doubts about the mechanism of generation of the injury potential itself. It is suggested (Schramm et al. 1983b) that the disturbance in the transmembrane flow of ions and blockade of propagation of action potentials along the fibers at the lesion site before it reaches the recording electrode, are responsible for generation of the positive potential. It is also

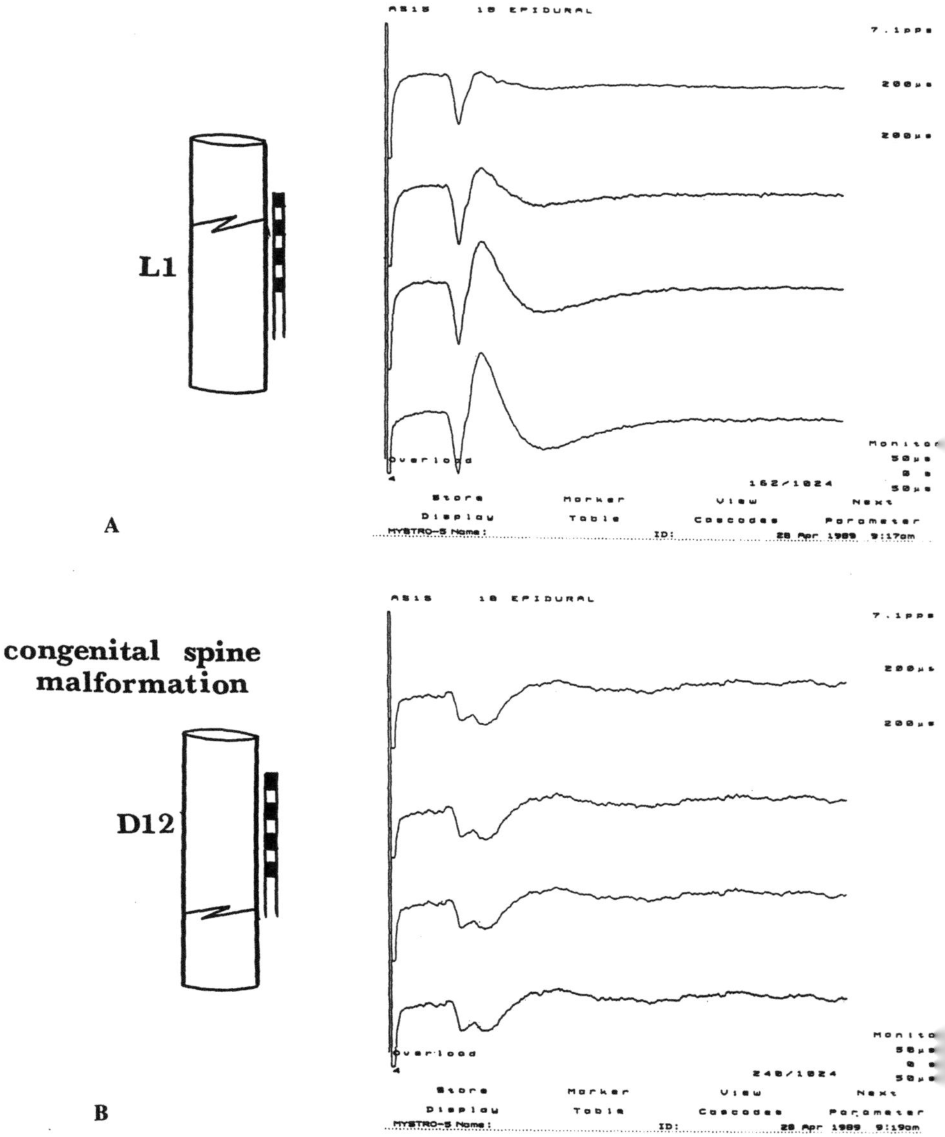

Fig. 8A,B. Epidural SEP in a case of spinal cord compression due to congenital malformation at L1/D12 level. Tibial nerve stimulation. **A** A decrease and eventually disappearance of the main negative component is observed. **B** A few centimeters higher an injury potential with a slight negativity superimposed was recorded

well known that a static positive membrane potential develops in destroyed axons (Goodman et al. 1985). Interesting studies by Stegeman et al. (1987) using a computerized simulation technique have shown that the slowing down of propagation of action potentials along the nerve fibers may cause development of positive stationary potentials. The bigger the velocity difference, the higher the potential amplitude.

Whatever the mechanism of the potential is, we certainly do observe the positive waveform near the lesion site, and it may be an indicator of spinal cord pathology. It also may localize very accurately the site of the lesion. The recordings done above the injured part of spinal cord may provide more information in monitoring the preservation of spinal cord pathways.

In some chronic spinal cord lesions the injury potential is not recordable at the most "suspicious" regions of the cord, and we do not know why. The disappearance of the main negative component at the root entry zone seems to indicate a spinal cord lesion at this level. An example of this is presented in Fig. 8A. In the same patient at the D12 level only a far-field injury potential is seen (Fig. 8B), and this result could confirm the existence of a spinal cord lesion at the L1 level. At the same time the question arises as to why the injury potential was not recorded at the level of L1.

At a cervical level, by median nerve stimulation, the situation is even more difficult to interpret since the changes of SEP patterns observed are not very well manifested. In the case of posttraumatic spinal cord compression shown in Fig. 7, the negative component decreases rostrally at the C5 level, and it might be regarded as normal. In other patients, however, we have still observed quite a big negativity at the C2 level (Fig. 9). Therefore we

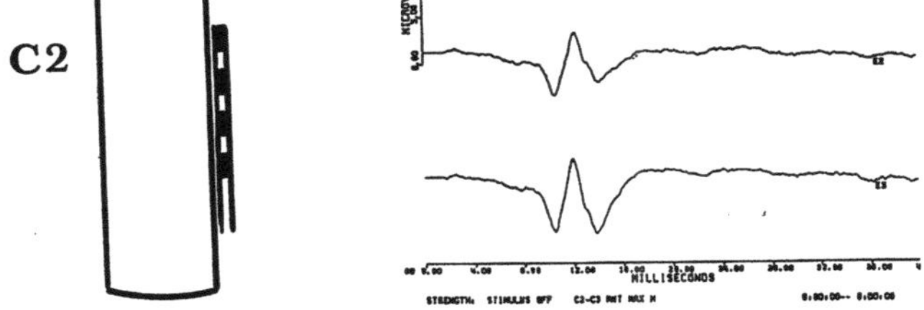

Fig. 9. Epidural SEP recorded by median nerve stimulation at C2 level in a patient with a healthy spinal cord. The main negative component is still present

would rather regard such a decrease in amplitude of recorded waveform as a pathological finding.

Our clinical material is too limited to draw any definite conclusions, but our study has provided some features of epidural SEP recorded in chronic spinal cord lesions.

Conclusions

1. An injury potential is observed in humans in chronic spinal cord lesions and is similar in morphology to that found in acute lesions.
2. The injury potential may localize the lesion with an accuracy of about 1 cm along the cord.
3. The recordings of epidural SEP may be useful in clinical practice to localize the most severe pathological changes in diffuse spinal cord lesions.
4. The disappearance of the conducted waveform and/or the abnormal changes in SEP patterns above the lesion site may serve as a source of information about the preserved spinal cord fibers and so they can be useful in spinal cord monitoring.

References

1. Deecke L, Tator CH (1973) Neurophysiological assessment of afferent and efferent conduction in the injured spinal cord of monkeys. J Neurosurg 39:65–74
2. Goodman R, Wachs K, Keller S, Black P (1985) Spontaneous spinal cord "injury potential" in the rat. Neurosurgery 17:757–759
3. Greenberg RP, Ducker TB (1982) Evoked potentials in clinical neurosciences. J Neurosurg 56:1–18
4. Rowed DW (1982) Value of somatosensory evoked potentials for prognosis in partial cord injuries. In: Tator CH (ed) Early management of acute spinal cord injury. Raven Press, New York
5. Schramm J (1985) Spinal cord monitoring: current status and new developments. Central Nervous System Trauma 3:207–227
6. Schramm J, Hashizume K, Fukushima T, Takahashi H (1979) Experimental spinal cord injury produced by slow, graded compression. J Neurosurg 50:48–57
7. Schramm J, Shigeno T, Brock M (1983a) Clinical signs and evoked response alterations associated with chronic experimental cord compression. J Neurosurg 58:734–741
8. Schramm J, Krause R, Shigeno T (1983b) Experimental investigation on the spinal cord evoked injury potential. J Neurosurg 59:485–492
9. Shinomiya KI, Furuya K, Yamamura I, Kaneda A, Yokoyama M (1985) Clinical study of cervical spondylotic myelopathy using evoked spinal cord potentials. In: Schramm J, Jones SJ (eds) Spinal cord monitoring Springer, Berlin Heidelberg New York Tokyo, pp 290–301
10. Stegeman DF, Van Oosterom A, Colon EJ (1987) Far field components induced by a propagating generator: computational evidence. Electroencephalogr Clin Neurophysiol 67:176–187

11. Tamaki T, Noguchi T, Takano H, Tsuji H, Nakagawa T, Imai K, Inoue S (1984) Spinal cord monitoring as a clinical utilization of the spinal evoked potential. Clin Orthop 184:58–64
12. Young W (1982) Correlation of somatosensory evoked potentials and neurological findings in spinal cord injury. In: Tator CH (ed) Early management of acute spinal cord injury. Raven Press, New York

Observation of the Circulation of Cauda Equina in Spinal Canal Stenosis Following Electrical Stimulation of Peripheral Nerves

A. DEZAWA[1], S. MORIYA, H. KITAHARA, K. TAKAHASHI, and M. MURAKAMI

Introduction

The pathogenesis of intermittent claudication in spinal canal stenosis has attracted increasing attention as a blood flow problem in the spinal cord and cauda equina. It has been noted that the increased density and dilatation of blood vessels in the lumbar spinal cord or the ipsilateral limb can be induced after electrical stimulation of a limb.

Tamaki found that spinal canal stenosis patients often could walk further after peripheral nerves were stimulated. We applied a newly developed "ultrathin fiberoptic myeloscope system" to observe the change on spinal cord and cauda equina surface.

The purpose of this study is to observe directly the change in circulation in cauda equina with a fiberoptic ultrathin myeloscope in spinal canal stenosis following the electrical stimulation of different nerves such as tibialis posterior, peroneal nerve, or femoral nerve in the lower extremity.

Materials and Methods

Our myelofiberscope with a microscope-telescope focusing system was used successfully in visualizing the microblood vessels of spinal cord and cauda equina in 24 adults (14 men and 10 women, aging from 33 to 65. Simultaneously, segmental spinal evoked potentials were recorded at the level of conus medullaris (Fig. 1).

This newly developed ultrathin fiberoptic myeloscope (myelofiberscope) system is fine enough to pass through an 18-gauge Tuohy needle. The fiberscope is very flexible and has an insertion tube with an outer diameter of

[1] Department of Orthopedic Surgery, Chiba University, School of Medicine Chiba, 280 Japan

K. Shimoji et al. (Eds.)
Spinal Cord Monitoring and Electrodiagnosis

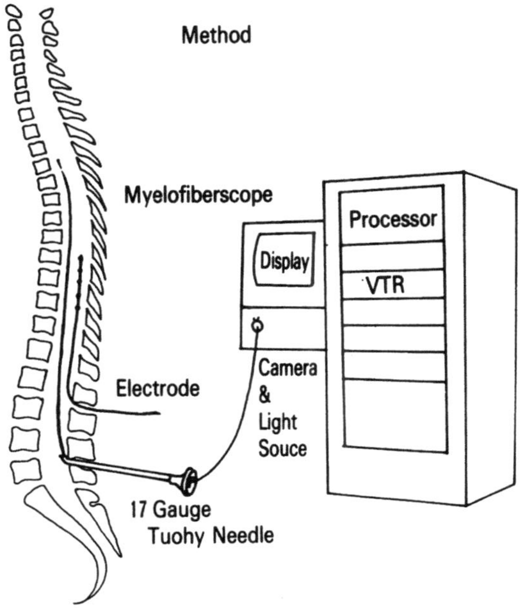

Fig. 1. Myelofiberscope procedure and the recording system

0.75 mm and a working length of 120 cm. It is designed to go forward cranially to observe from cauda equina to cisterna magna.

This instrument has 3000 quartz fiber (5.5 μ) optic image guides and 35 light guides (50 μ). The maximum flexion circle is 10 mm. The angle of view in the air is 40°–70° and the depth of field 3–30 mm (Fig. 2).

The lumbar subarachnoid space was identified by the hanging drop method with an 18-gauge Tuohy needle under local anesthesia in the lumbar interspace L4–5 using the midline approach in lateral position (Fig. 3). Myelofiberscope puncture is carried out in about the same manner as an ordinary lumbar puncture. It is often difficult to perform in the patient with cauda equina involvement, but in most individuals the myelofiberscope can be easily moved cranially. The tip of our myelofiberscope was developed so as to be visible through a fluoroscope (Fig. 4). Almost all patients complained of mild headache. There were close similarities between this headache and that experienced after a lumbar puncture.

Simultaneously, segmental spinal evoked potentials elicited by peripheral nerve stimulation were recorded with five electrodes placed over the conus medullaris. The stimulus was delivered to the femoral, peroneal, and tibialis posterior nerves.

The responses were evoked by peripheral stimulation 1.5 times the motor threshold using rectangular electrical pulses with a width of 0.2 ms at a frequency of 2 per second from a constant current stimulator. The resulting responses were clear and well defined with little residual noise, after averaging 256–512 responses with Medelec MS92B.

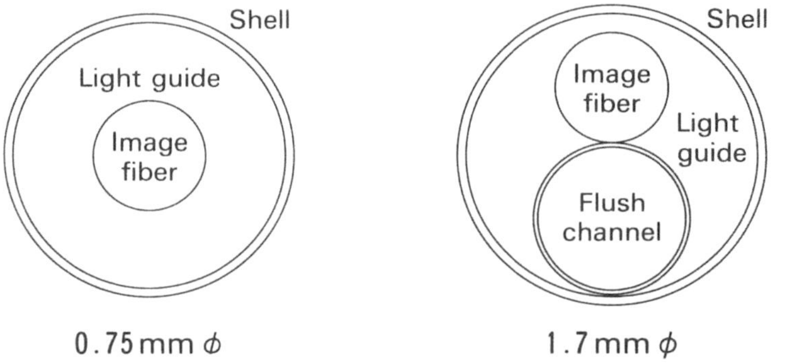

Fig. 2. Cross section of myelofiberscope

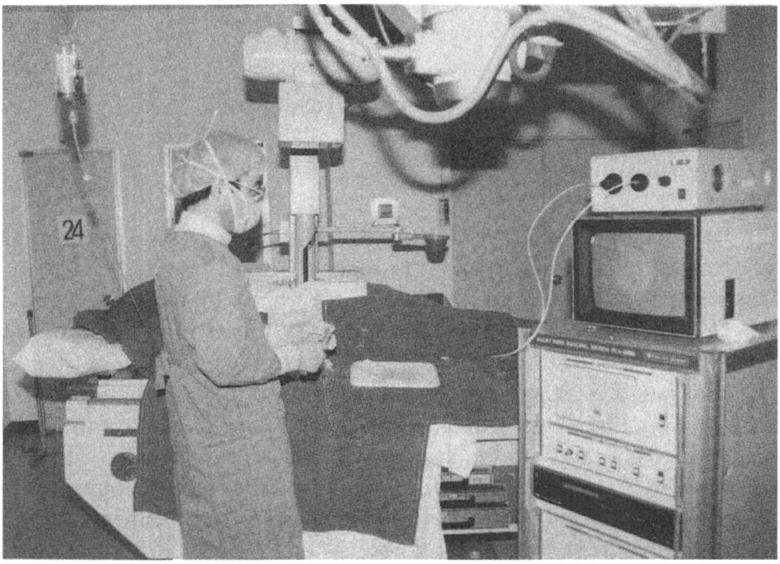

Fig. 3. Practical use of the equipment

Results

Case 1

A 71-year-old man presented with a 7-year history of paresthesia ascending from feet to hips and difficulty in standing or walking after a short distance. Symptoms were relieved by electrical stimulation of the peripheral nerve at his left knee for 3 min. Many spinal nerve roots with accompanying blood vessels can be seen with the myelofiber-scope (Fig. 5).

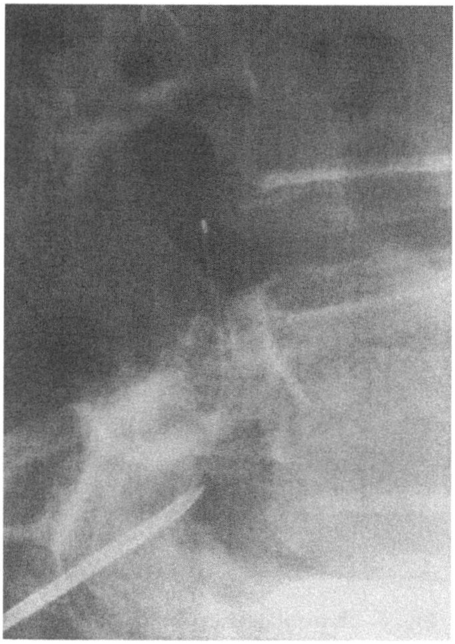

Fig. 4. X-ray of the tip of a myelofiberscope taken through a fluoroscope

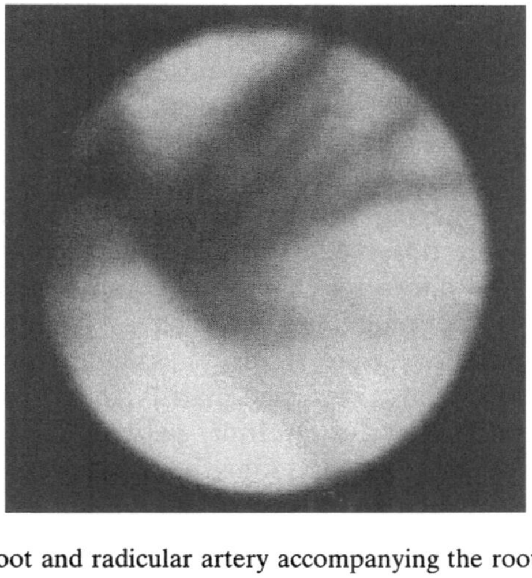

Fig. 5. Human lumbosacral spinal nerve root and radicular artery accompanying the root from caudal to cranial

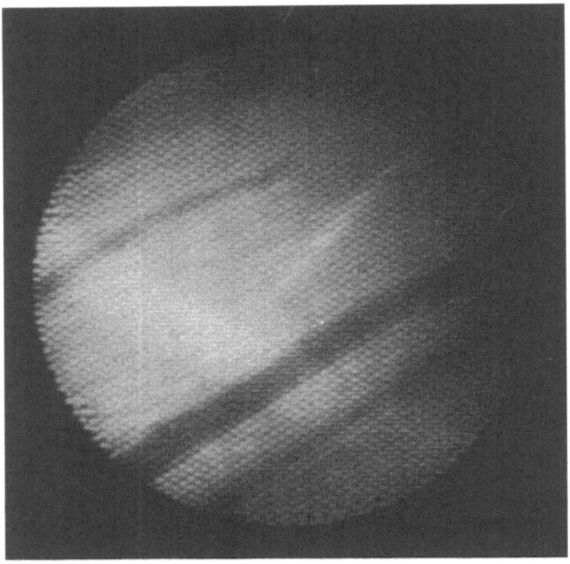

Fig. 6. Radicular artery coursed proximally from the distal end of the cauda equina

Case 2

A 64-year-old man had experienced pain in the left calf after walking for 800 m. On entering the subarachnoid space, the spinal cord and its roots appeared vivid yellowish-white in colour, with their arteries and veins easily distinguished, and we observed the cauda equina with thin radicular arteries coursed proximally from the distal side (Fig. 6). The movement of the cauda equina was in accord with the pulsation of the patient's heart. The direction and speed of blood flow could not be found during coughing and sneezing.

Changes in circulation in 12 spinal canal stenosis patients of 24 adults were observed. There was an increase in one patient, no change in three and uncertain changes in eight. Segmental spinal evoked potentials elicited by peripheral nerves such as femoral nerve, peroneal nerve, and tibialis posterior were recorded simultaneously or sequentially with five electrodes over the conus medullaris. An increase in the waveform of the segmental SEP after 1 s electrical peripheral nerve stimulation for 5 min was noticed in three patients, the other cases were unchanged (two patients) or uncertain (7 patients).

Discussion

No definite mechanism has ever been proved showing a corelationship between an increase in circulation on the cauda equina and an increase in amplitude of s-SpEP following electrical stimulation of peripheral nerves in SCS patients. Observations on the microvasculature of human lumbosacral spinal nerve roots in spinal canal stenosis support the concept that the appearance of signs and symptoms of intermittent claudication after electrical simulation of peripheral nerves can probably be explained by an increase of blood flow in radicular arteries as well as an increase in the degree of compression on nerve roots. It is thus suspected that the relative ischemia of conducting cauda equina roots may be caused by mechanical impairment of conduction through these roots.

References

1. Blau JN, Logue V (1961) Intermittent claudication of the cauda equina. An unusual syndrome resulting from central protrusion of a lumbar intervertebral disc. Lancet 1: 1081–1086
2. Burmann MS (1931) Myeloscopy for the direct visualization of spinal cord. J Bone Joint Surg 13:695
3. Ooi Y et al. (1973) Myeloscopy possibility of observing lumbar intrathecal space by use an endoscope. Endoscopy 5:91
4. Pool JL (1938) Direct visualization of dorsal nerve roots of cauda equina by means of a myeloscope. Arch Neurol Psychiatry 39:1308

1. Electrical Stimulations

Short Latency Somatosensory Evoked Potentials to Stimulation of Different Lower Extremity Nerves: Nature of Generator Sources as Inferred from Scalp Topography

L. Pelosi[1], J.B. Cracco, R.Q. Cracco[2], and N.F. Hassan

Scalp recorded short latency somatosensory evoked potentials (SSEP) to stimulation of the posterior tibial (PTN), peroneal (PN) and sural (SN) nerves at the ankle and common peronal nerve (CPN) at the knee have been described [1, 3, 7–9]. However, SSEP to stimulation of these different lower extremity nerves have not been obtained and compared in the same individual, so that similarities and differences in the waveform characteristics and scalp topography of these SSEP have not been systematically investigated. Therefore, we obtained SSEP to stimulation of the PTN, PN, and SN at the ankle and the CPN at the knee in eight normal subjects. The configuration, amplitude, scalp topography, and intersubject variability of these potentials were compared. The study was carried out to obtain information concerning the generator sources of these potentials.

Materials and Methods

Eight normal adults (five men and three women, mean age 24.5 years) were studied. Recordings were performed with the subjects prone and relaxed, but awake. Stimulating electrodes were located 3 cm apart (cathode proximal) over the left CPN at the knee and the left PTN, PN and, SN at the ankle. The nerves were stimulated independently (duration 0.2 ms) at rates of 1.7–4.7/s. Stimulus intensity varied between 2 and 2.5 times sensory threshold which produced a twitch of muscles innervated by each of the mixed nerves and paresthesia on the foot area supplied by the SN.

Recordings were obtained from 15 cephalic electrodes attached widely over the scalp and to the ears. All cephalic electrodes were referred to a noncephalic reference electrode placed on the right shoulder. Two eight channel montages were employed. A common lead (C_z-shoulder) was employed with each eight channel montage to compare the variability between

[1] *Present Address*: Department of Clinical Neurophysiology, II. School of Medicine, Naples, Italy
[2] Department of Neurology, State University of New York Health Science, Center at Brooklyn, 450 Clarkson Ave., Brooklyn, NY11203, USA

K. Shimoji et al. (Eds.)
Spinal Cord Monitoring and Electrodiagnosis

recordings. Two trials of 1000–2000 averages were superimposed. A frequency band width of 5–1500 Hz (−3 dB) and an analysis time of 100 ms were used.

The peak latencies of the first prominent positive and following negative potential (P27, N35 for CPN and P37, N45 for PTN, PN and SN stimulation) were identified at each scalp recording site. Peak to peak amplitudes (P27 to N35 and P37 to N45) and amplitudes of P27 and P37 from onset to peak were measured. Mean peak latencies of P27 and P37 were determined in vertex-shoulder recordings.

Results and Discussion

The configuration and distribution of SSEPs to stimulation of the three nerves at the ankle were similar across subjects, but CPN-SSEP were lower in amplitude, and their configuration and scalp distribution showed much greater intersubject variability (Figs. 1, 2). This suggests that complex mechanisms which variably interact with one another are reflected in scalp SSEP to CPN stimulation at the knee, and these are discussed in the ensuing paragraphs.

With stimulation of nerves at both the ankle and the knee, the initial prominent positive and negative components were maximal in amplitude at the vertex (Figs. 1, 3). These potentials were greater in amplitude at recording sites ipsilateral to the stimulated leg in all eight subjects with PTN, PN, and SN stimulation (Fig. 1) and in five subjects with CPN stimulation. This paradoxical localization has been observed previously and has been explained on the basis of dipole sources located on the mesial surface of the contralateral hemisphere, oriented towards the ipsilateral hemisphere [7–9].

In six subjects with PTN stimulation and in three with PN and SN stimulation, P37 was recorded with opposite polarity over central parietal scalp regions on the two sides (Figs. 1, 3). This has been reported for both CPN-P27 and PTN-P37 [3, 9]. It has been proposed that the cortical sources of P27 and P37 produce horizontal dipoles which are oriented tangential to the scalp surface and project with the positive end ipsilateral and the negative end contralateral to the stimulated side [3,9]. However, time-locked out of phase activity over the two hemispheres was not observed in one subject with PTN, in four with PN and SN and in two with CPN stimulation. In these the contralateral negative potential was shorter in latency than P27 and P37, as has been observed by others [1, 9]. Therefore, this contralateral negative component more likely reflects a separate generator located in the hemisphere contralateral to the stimulated side [1, 9].

The configuration and distribution of SSEP to stimulation of the three nerves at the ankle were similar across subjects (Fig. 1). In contrast, the configuration and distribution of CPN SSEP showed great intersubject vari-

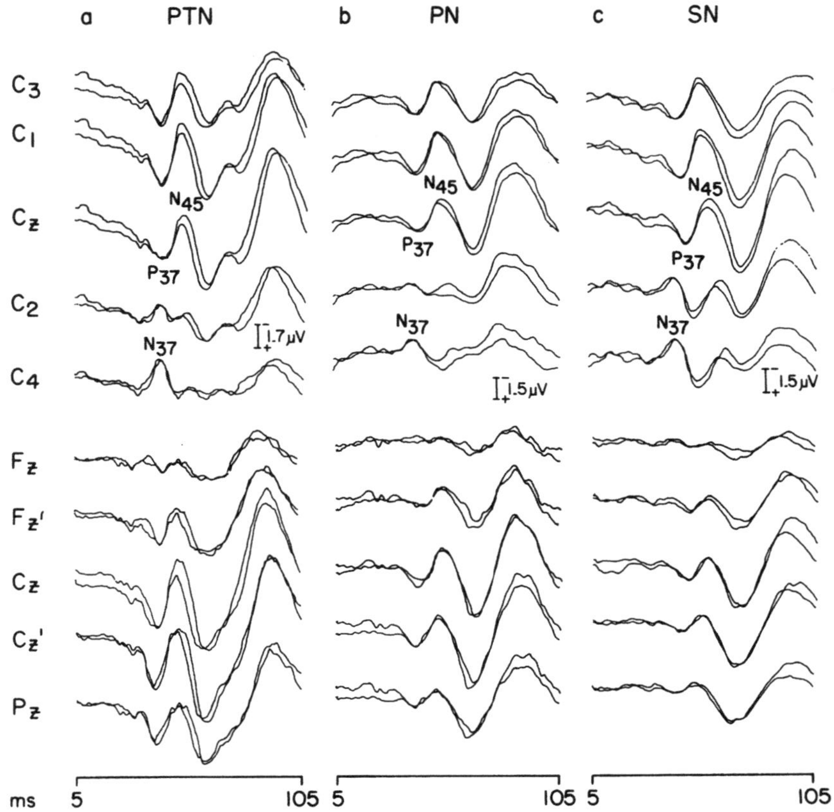

Fig. 1a–c. Distribution in the coronal (*top traces*) and midsagittal (*bottom traces*) planes of SSEPs to stimulation of the left (**a**) posterior tibial nerve (*PTN*), (**b**) peroneal nerve (*PN*), and (**c**) sural nerve (*SN*) in the same subject. F_z' and C_z' were placed midway between *F* and *C* and *C* and *P*, respectively. Note similarity in configuration of components [From 4]

ability (Fig. 2). The amplitude of CPN-SSEP was much less than that of SSEPs to stimulation of all three nerves at the ankle, including the PN. This is surprising since the CPN contains all the sensory axons present in the PN as well as additional contributions from its superficial, deep, and articular branches. Since the cortical projection area of sensory afferents from the calf is along the curvature of the interhemispheric fissure, its dipole sources likely project more tangential to the surface than a generator limited to the foot projection area which is deep in the interhemispheric fissure [6, 11]. Therefore, the interaction of dipole generators from the foot and the calf projection areas with different orientations could result in partial cancellation of scalp activity evoked by CPN stimulation and in lower amplitude CPN-SSEPs. This would not be expected to occur with PN, SN, or PTN stimulation because only the foot cortical projection area is activated.

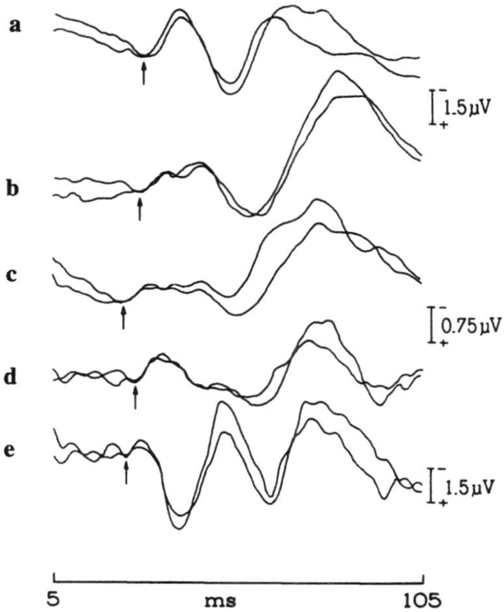

Fig. 2a–e. C_z noncephalic reference (shoulder) recordings of SSEP to left common peroneal nerve stimulation in five subjects. P27 is indicated by *arrows*. Note the variability in configuration, amplitude, and duration of components. [From 4]

No significant differences between P37 peak latencies were found for the three nerves stimulated at the ankle. We have already shown that the conduction velocities over peripheral nerve and propagation velocities from cauda equina to cortex of SSEP to PTN, PN, and SN are similar, but these values are significantly slower than those obtained with CPN stimulation [5]. This is consistent with a greater group I muscle afferent contribution to the CPN. Therefore interaction with partial cancellation of evoked activity arising in generator sources activated independently by cutaneous and muscle afferents might also contribute to the lower amplitude and greater variability of CPN-SSEP. Additionally, a greater convergence of input from muscle and cutaneous afferents at cortical and subcortical levels with CPN stimulation could result in a greater degree of reciprocal inhibition [2]. Conversely, stimulation of nerves at the ankle probably result in more synchronous activation of cortical generators and therefore yield more stable and larger amplitude SSEP.

The recent findings by Wang et al. [10] that the initial cortical component (P26) evoked by stimulation of the femoral nerve at the inguinal ligament is lower in amplitude than P37 evoked by posterior tibial nerve stimulation and is greatest in amplitude near the midline or at scalp locations contralateral to the side of stimulation is consistent with our conclusions. The femoral nerve is a mixed nerve which contains a large proportion of muscle afferents,

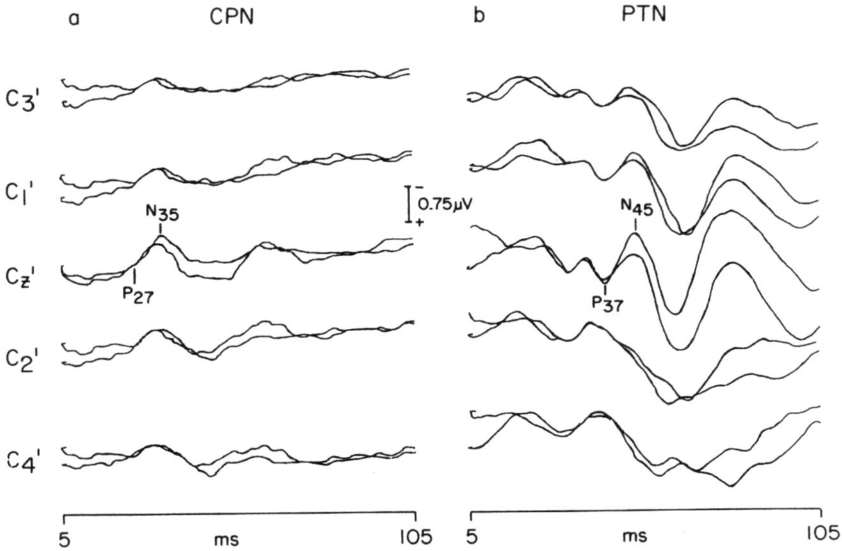

Fig. 3a,b. Coronal distribution of SSEP to stimulation of the left (**a**) common peroneal nerve (*CPN*) and (**b**) posterior tibial nerve (*PTN*) in the same subject. Electrodes were placed 3 cm behind standard placements. *CPN-P27* is barely detectable and *N35* shows similar configuration and amplitude at sites ipsi- and contralateral to stimulation. In contrast, *PTN-P37* and *N45* are maximal at C_z' and C_1' and a negative potential (N37) is seen at contralateral sites [From 4]

and its cortical representation area is lateral to that of CPN, PTN, PN, and SN.

The greater complexity of the CPN-SSEP results in a lower amplitude response which is more variable in its waveform and topography than SSEP obtained with stimulation of nerves at the ankle. For these reasons, CPN-SSEP are less desirable for routine clinical use. The fact that PTN stimulation consistently yielded stable responses which were greater in amplitude than those obtained with PN and SN stimulation indicates that this nerve is the most suitable for routine clinical testing.

Summary

SSEP to stimulation of the common peroneal nerve (CPN) at the knee and posterior tibial (PTN), peroneal (PN) and sural (SN) nerves at the ankle were recorded from the scalp and compared. CPN-SSEP were lower in amplitude and their configuration and scalp distribution showed much greater inter-subject variability than SSEP obtained with stimulation of nerves at the ankle. This suggests that complex mechanisms which variably interact with one another are reflected in scalp SSEP to CPN stimulation at the knee. The larger amplitude and minimal intersubject variability in morphology and topography of PTN-SSEP indicate that this nerve is the most suitable for routine clinical use.

References

1. Beric A, Prevec TS (1981) The early negative potential evoked by stimulation of the tibial nerve in man. J Neurol Sci 50:299–306
2. Burke D, Gandevia SC, McKeon B, Skuse NF (1982) Interactions between cutaneous and muscle afferent projections to cerebral cortex in man. Electroencephalogr Clin Neurophysiol 53:349–360
3. Cruse R, Klem G, Lesser RP, Lueders H (1982) Paradoxical lateralization of cortical potentials evoked by stimulation of posterior tibial nerve. Arch Neurol (Chic) 39:222–225
4. Pelosi L, Cracco JB, Cracco RQ, Hassan NF (1988) Comparison of scalp distribution of short latency somatosensory evoked potentials (SSEPs) to stimulation of different nerves in the lower extremity. Electoencephalogr Clin Neurophysiol 71:422–428
5. Pelosi L, Cracco JB, Cracco RQ (1987) Conduction characteristics of somatosensory evoked potentials to peroneal, tibial and sural nerve stimulation in man. Electroencephalogr Clin Neurophysiol 68:287–294
6. Penfield W, Rasmussen T (1950) A clinical study of localization of function. In: The Cerebral Cortex of Man. Macmillan, New York
7. Rossini PM, Cracco RQ, Cracco JB, House WJ (1981) Short-latency somatosensory evoked potentials to peroneal nerve stimulation: scalp topography and the effect of different frequency filters. Electroencephalogr Clin Neurophysiol 52:540–552
8. Tsumoto T, Hirose N, Nonaka S, Takahashi M (1972) Analysis of somatosensory evoked potentials to lateral popliteal nerve stimulation in man. Electroencephalogr Clin Neurophysiol 33:379–388
9. Vas GA, Cracco JB, Cracco RQ (1981) Scalp-recorded short-latency cortical and subcortical somatosensory evoked potentials to peroneal nerve stimulation. Electroencephalogr Clin Neurophysiol 52:1–8
10. Wang G, Cohen LG, Hallett M (1989) Scalp topography of somatosensory evoked potentials following electrical stimulation of femoral nerve. Electroencephalogr Clin Neurophysiol 74:112–123
11. Werner G, Whitsel BL (1968) Topology of the body representation in somatosensory area I of primates. J Neurophysiol 31:856–869

Reflexes Evoked in Various Human Muscles During Voluntary Activity

T. TANI[1], K. KIDA, H. YAMAMOTO, and J. KIMURA

Although the H reflex is theoretically recordable from all muscles, in practice it is consistently obtained only by recording from the triceps surae while stimulating the tibial nerve at the knee. It is known that voluntary muscle contraction makes it possible to evoke the H reflex from those muscles in which the reflex is normally unelicitable at rest, because the excitability of the corresponding motoneurones is increased by descending volitional impulses. In contrast to the H reflex, the T reflex, which is the monosynaptic spinal reflex elicited by a mechanical tap to the tendon and is equivalent in many respects to the H reflex, has not been sufficiently investigated in voluntarily contracting human muscles. In the present study, with a tendon tap during effort, the authors could consistently obtain the reflex responses from the tibialis anterior muscle (TA muscle), the abductor pollicis brevis muscle (APB muscle), and the paraspinal muscles. In addition to mechanical elicitation, electrical nerve stimulation was also employed to induce the responses in the TA and the APB muscles, allowing the further assessment of the nature of the responses by comparing the mechanically elicited reflex responses with the electrically elicited ones.

Methods

Subjects

The study of each muscle was performed in a different group of subjects during separate sessions. We studied 50 healthy subjects and 18 patients with isolated lumbosacral radiculopathies for an investigation of the TA muscle, 13 healthy subjects for the APB muscle, and 14 healthy subjects for the paraspinal muscles.

[1] Department of Orthopedic Surgery, Kochi Medical School, Nanko-ku, Kochi, 781-51 Japan

K. Shimoji et al. (Eds.)
Spinal Cord Monitoring and Electrodiagnosis

Stimulation

A reflex hammer with an electrode at the tip was used for the study of the TA muscle and the paraspinal muscles. A trigger pulse was obtained when a circuit was completed at the moment of contact of the hammer tip with a piece of copper foil placed on the tapping site. For the study of the APB muscle another hammer with a microswitch for triggering was used, where there was a delay of a millisecond or so between first contact and the closing of the microswitch. The tap was delivered manually with blows of average strength to the tendon of the TA muscle, to the proximal phalanx of the thumb, or to the stick which was pressed to an interspinous space being held by the examiner. An electrical stimulus of 1 ms duration and of submaximal intensity was also applied percutaneously to the common peroneal nerve at the knee or to the median nerve at the palm, wrist, or elbow with the cathode placed proximally to the anode.

Recording

The responses recorded through surface electrodes with a bandwidth of 10 Hz–10 KHz were averaged for a period of 100 ms. Both mechanical and electrical elicitation were made during the same session, without changing the placements of the recording electrodes.

Tibialis Anterior Muscle

The subject was positioned supine. A pair of electrodes was placed over the end-plate zone of the muscle and distal tibia. The subject was instructed to contract the muscle steadily at about 10% of maximum force throughout the stimulatory period using a machine (Musukulator GT-50) which allowed the muscle tension generated every moment to be transformed into a digital readout. Usually, 128 responses were averaged.

Abductor Pollicis Brevis Muscle

The subject was seated comfortably, and the left hand was always tested. The active and the reference electrodes were placed over the end-plate zone of the muscle and at the base of the proximal phalanx, respectively. Although accurate standardization of the muscle tension generated was not attempted, it was kept at some 20% of maximum force by the application of external resistance, being roughly monitored by means of visual or auditory EMG feedback. Usually 256 responses were averaged.

Paraspinal Muscles

The subject was positioned prone. Eight active electrodes were placed over the right paraspinal muscles, 3.5 cm from the midline, at four thoracic and lumbar interspinous levels each (T5–6 – T8–9 and L1–2 – L4–5). Eight reference electrodes were placed 9 cm further laterally at the same levels as the active electrodes. A four-channel averager allowed simultaneous recording through four out of eight pairs of electrodes with taps to an interspinous space. Each test set was an average of 64 summated responses during moderate active extension of the trunk.

Results

Tibialis Anterior Muscle

Both the H and T wave were present in all normal subjects tested. The mean onset latencies of the H wave on the right and left sides in 50 subjects were 28.3 ± 1.8 ms and 28.4 ± 1.9 ms, respectively, between which there was no significant statistical difference. The corresponding values of the T wave in 29 subjects were 33.5 ± 2.4 ms and 33.7 ± 2.5 ms, which were not significantly different, either. Both the H and T latencies in subjects of all ages were highly correlated with height ($r = 0.68$ and 0.78). They also had a high correlation with age ($r = 0.53$ and 0.52) in 28 subjects with the height between 165 and 175 cm. An analysis of the simultaneous regression of the latencies on height and age revealed the formulas: H latency (ms) = 0.186 height (cm) + 0.074 age (years; $r = 0.783$; $P < 0.01$); T latency = 0.267 height + 0.094 age ($r = 0.819$; $P < 0.01$). It was also revealed based on the study of normal subjects that a side-to-side latency difference of the H and T wave greater than 1.5 ms and 2.3 ms, respectively, and/or low amplitude less than 26% and 39% of the contralateral side, respectively, would be abnormal. Eighteen patients with isolated compressive lumbosacral radiculopathies and with no obvious wasting of the TA muscle were tested: two L4, twelve L5, and four S1 radiculopathies. According to these criteria, none of the patients with unilateral L4 or S1 radiculopathy showed altered latency or amplitude of the response on the affected side compared to the normal side, whereas 8 out of 12 patients (67%) with unilateral L5 radiculopathy showed abnormalities (Fig. 1).

Abductor Pollicis Brevis Muscle (Fig. 2)

With mechanical elicitation, the initial event labeled SP (with a mean onset and peak at 27.7 ± 1.6 ms and 32.5 ± 1.7 ms, respectively) which was considered to correspond to the tendon jerk was recorded in all subjects.

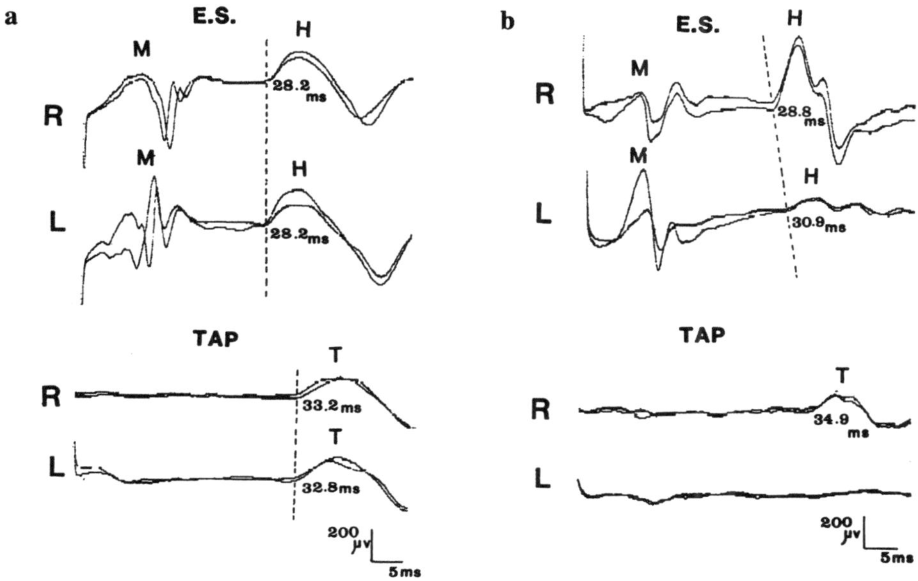

Fig. 1a. A 40-year-old man with an absent left knee-jerk and with no obvious TA muscle wasting. Myelographic and operative data confirmed the presence of a left L4 radiculopathy. Both the *H* and the *T* reflex elicited from the TA muscle during voluntary activity showed no alterations on the affected side compared with the normal side in both latency and amplitude. **b** A 74-year-old woman with symmetric knee and ankle jerks and with no obvious TA muscle wasting. Myelographic and operative data confirmed the presence of a left L5 radiculopathy. The *T* reflex elicited from the TA muscle during voluntary activity was absent on the affected side, whereas the *H* reflex was still present but delayed. *R*, right; *L*, left; *E.S.*, electrical stimulation; *M*, M response

Subsequently, there were responses labeled A(T) (with a mean peak at 46.1 ± 2.2 ms) and B(T) (with a mean peak at 56.7 ± 2.8 ms). The A(T) response was present in only 50% of the subjects, whereas the B(T) response was obvious in 92% of the subjects. With electrical shocks of submaximal intensity applied to the median nerve at the palm, the shortest latency response recorded is a direct motor response (M wave). Subsequently there were responses labeled SP (with a mean onset and peak at 27.2 ± 1.8 ms and 31.2 ± 1.8 ms, respectively), A(E) (with a mean peak at 44.2 ± 3.0 ms), and B(E) (with a mean peak at 55.9 ± 2.6 ms) as seen with mechanical stimulation. Whereas the SP wave with a latency similar to the F wave and the B(E) wave were more constantly observed (100% and 83%, respectively), the A(E) response was detectable in fewer subjects (67%), as was the case with mechanical elicitation. Assessing the nature of the A and the B responses, a measurement of peak intervals between SP and A, and SP and B was made in both mechanically and electrically evoked responses. Consequently, mean intervals between SP and A elicited by a tap and an electrical stimulus in 7 subjects were 13.3 ± 0.9 ms and 13.7 ± 1.5 ms, respectively. There

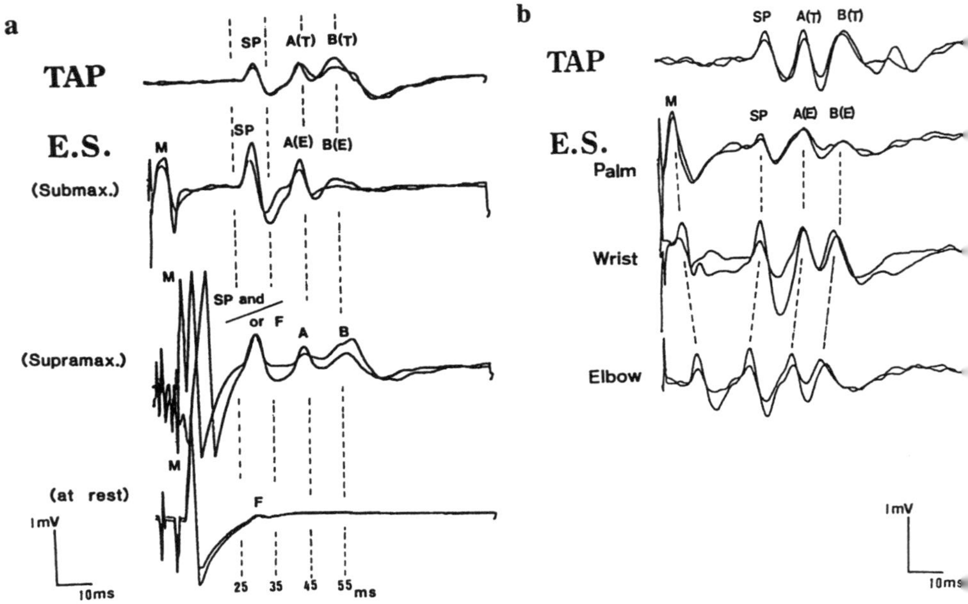

Fig. 2a. A comparison between mechanically and electrically elicited averaged responses from the APB muscle during voluntary contraction in a 33-year-old man. The median nerve was stimulated at the palm using shocks of a submaximal intensity (*second trace from top*) and a supramaximal intensity (*third trace from top*) during voluntary contraction of the muscle. The bottom tracing shows an averaged *F* wave evoked by shocks of a supramaximal intensity at rest. **b** Averaged responses recorded from the APB muscle with median nerve stimulation at three points during voluntary contraction in a 35-year-old man. The latencies of the *A(E)* and the *B(E)* waves as well as the *SP* wave shortened progressively as the stimulus site was moved proximally from the palm to the wrist and finally to the elbow.

was no significant statistical difference between them. The corresponding values between SP and B in 11 subjects were 25.0 ± 1.8 ms and 25.4 ± 1.6 ms, which were not significantly different either. These might suggest that the B(T) and B(E) as well as A(T) and A(E) share similar pathways. Furthermore, the latencies of the A(E) and B(E) waves as well as the SP wave shortened progressively as the stimulus site was moved proximally from the palm to the wrist and finally to the elbow (Fig. 2b). Using the latency difference and the distance between stimulating electrodes at the palm and elbow sites, the afferent conduction velocity of each response was calculated as follows: SP: 63.9 ± 6.7 m/s ($n = 10$), A(E): 65.8 ± 10.2 m/s ($n = 5$), and B(E): 56.4 ± 9.1 m/s ($n = 6$).

Paraspinal Muscles

The reflex responses were recorded from all eight pairs of electrodes with taps to one of the interspinous spaces (Fig. 3a). Usually, at least two distinctive

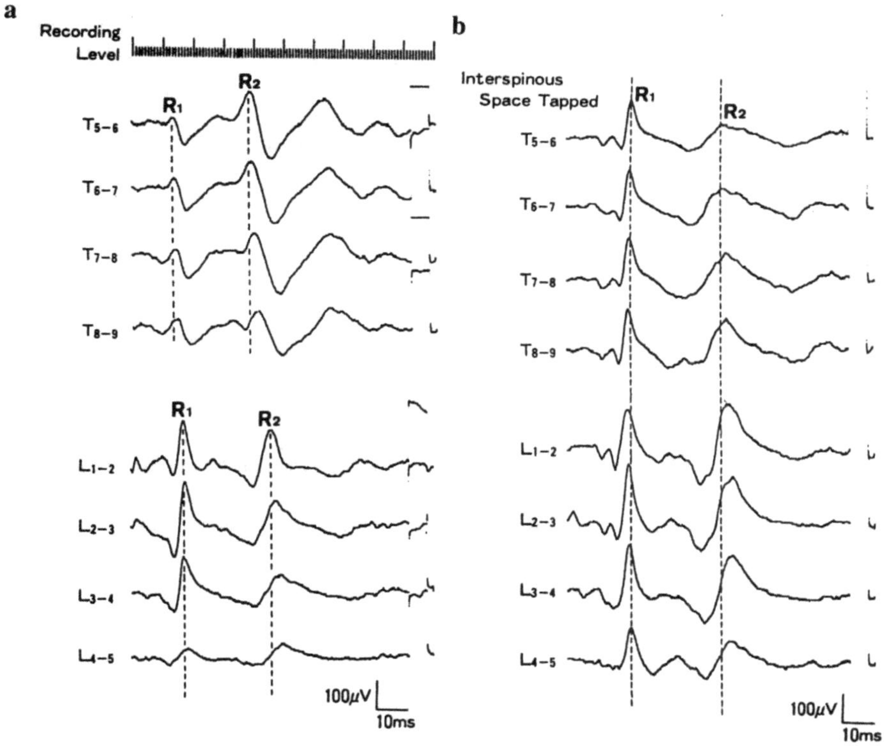

Fig. 3a. Averaged reflex responses elicited from paraspinal muscles with taps to the L1–2 interspinous space during voluntary activity in a 28-year-old man. Both the *R1* and the *R2* waves were recordable at various segmental levels concurrently with taps to one of the interspinous spaces. **b** Averaged reflex responses elicited from paraspinal muscles at the L4–5 level with taps to the various spinal levels during voluntary activity in a 27-year-old man. The latencies of the *R2* waves tended to be longer with more caudal stimulation, whereas that was not the case with the *R1* wave

types of responses labeled R1 and R2, respectively, were registered. Both of them always showed shorter latencies with the thoracic recordings than the lumbar recordings, regardless of the spinal levels tapped, which is compatible with the difference of spinal reflex arc length between the thoracic and the lumbar segments. However, the latencies of the responses evoked at the same segmental level varied, depending on toward which interspinous space we tapped (Fig. 3b). The R2 response evoked in a lumbar segment tended to have a shorter latency with more rostral stimulation, whereas that was not the case with the R1 response. The latency of the R1 response tended to be a little longer or even unaltered when the stimulus site was moved away from the recording site. The mean peak latencies of the R1 and R2 responses elicited from the paraspinal muscles at the same levels as the taps are shown in Table 1 and Fig. 4. The R1 waves showed longer latencies with more caudal recordings at the thoracic levels, which is also compatible with the longer reflex arc in the lower thoracic segment. However, that was not the case with lumbar recordings, in which the R1 latencies showed no alterations among the three recordings from the bottom, which could be explained by more considerable overlap of innervation at the lumbar levels. On the other hand, the R2 waves always showed longer latencies with more caudal recordings even in the three lumbar recordings from the bottom.

Discussion

The present study started as a search for the reflex elicited from human paraspinal muscle with taps to the interspinous spaces. The reflex was originally reported in 1979 by Trontelj et al. [9] who described the asymmetric amplitude of the response in scoliosis patients as well as the method they developed. Dimitrijevic et al. [2] in 1980 revealed early and late reflex responses of lumbar paraspinal muscle to tapping, referring to the favorable effect of tonic postural activity of the muscle while standing. However, sufficient attention was paid neither to the importance of voluntary muscle contraction to evoke the response consistently, nor to the conduction-time difference between the reflexes evoked at various segmental levels in normal subjects.

It is known from the work of Hagbarth [3] that H reflexes can be potentiated and become elicitable from hand muscles and the tibialis anterior muscle during voluntary contraction. The potentiation is considered to be brought about due to the background facilitation of motoneurons imposed by the descending motor pathways [10]. In contrast to the H reflex, only a few papers have dealt with the T reflex during voluntary activity. In the present study, with tendon taps during effort as well as with an electrical nerve stimulation, we were able to consistently obtain the reflex responses from various human muscles in which these reflexes are normally unelicitable at rest. Since a background of voluntary muscle contraction was necessary

Table 1. Mean peak latencies of R1 and R2 and mean peak intervals between them at various segmental levels

Level of recording and tap	Peak latency (ms) R1	R2	R2–R1
T5–6	12 ± 1 (*n* = 13)	37 ± 3 (*n* = 10)	25 ± 3 (*n* = 9)
T6–7	12 ± 1 (*n* = 11)	38 ± 2 (*n* = 13)	25 ± 2 (*n* = 11)
T7–8	13 ± 1 (*n* = 13)	40 ± 3 (*n* = 11)	26 ± 3 (*n* = 10)
T8–9	14 ± 1 (*n* = 14)	41 ± 3 (*n* = 12)	27 ± 3 (*n* = 12)
L1–2	18 ± 1 (*n* = 14)	49 ± 4 (*n* = 12)	31 ± 3 (*n* = 11)
L2–3	19 ± 1 (*n* = 13)	51 ± 4 (*n* = 9)	32 ± 3 (*n* = 9)
L3–4	19 ± 1 (*n* = 13)	54 ± 2 (*n* = 11)	34 ± 2 (*n* = 11)
L4–5	19 ± 2 (*n* = 13)	55 ± 1 (*n* = 9)	36 ± 3 (*n* = 9)

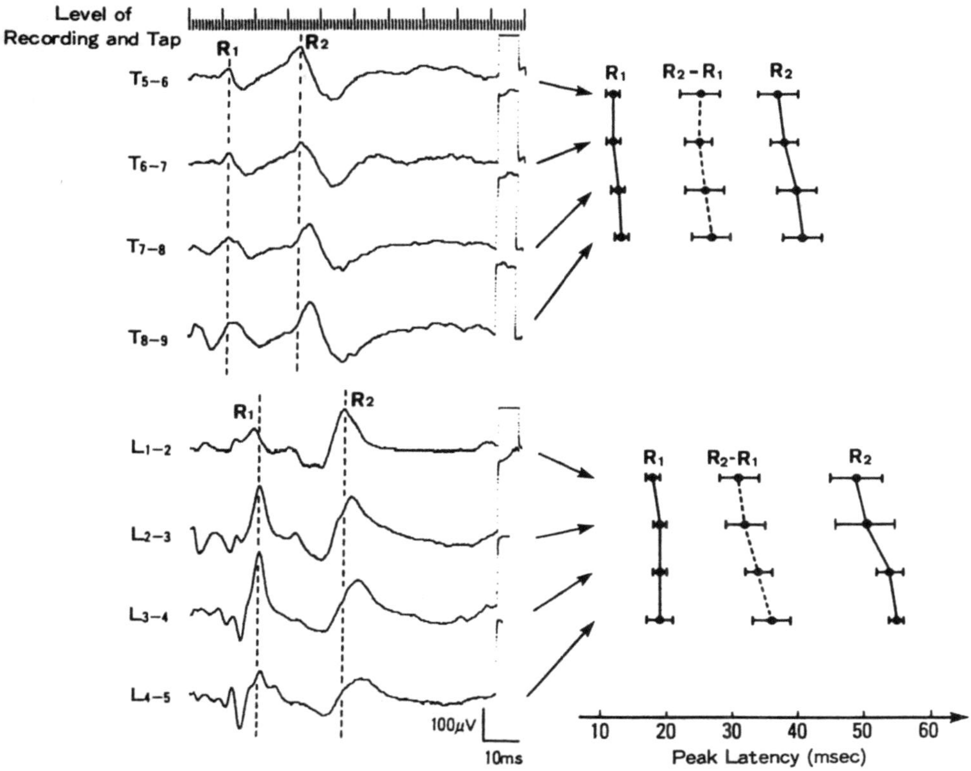

Fig. 4. Averaged reflex responses elicited from paraspinal muscles at the same levels as the taps during voluntary activity in a 31-year-old man (*left*). Mean peak latencies of the *R1* and the *R2* waves, and mean peak intervals between them at various segmental levels in 14 subjects tested (*right*)

to elicit the reflexes, consecutive responses were required to be averaged to differentiate time-locked responses from random voluntary motor unit activities which tended to disturb the analysis of the reflexes. Despite voluntary activity during elicitation, the latencies of these reflexes from the TA muscle in 50 normal subjects were highly correlated with height and age, when standardization of the muscle tension generated was attempted (about 10% of maximum force in the present study). The formulas were therefore produced by an analysis of the simultaneous regression of these latencies on height and age, as was demonstrated in the H reflex from the triceps surae by Braddom and Johnson [1]. It was also concluded based on the study of 50 normal subjects that a side-to-side latency difference of the H and T wave from the TA muscle greater than 1.5 ms and 2.3 ms, respectively, and/or low amplitude less than 26% and 39% of the contralateral side, respectively, would be abnormal, which was confirmed to be of clinical value in diagnosis of unilateral L5 radiculopathy. The T reflex tended to be absent on the affected side when the H reflex was still present but delayed. Considering that L5 is the most common nerve root involved, and an asymmetry of reflexes on physical examination is one of the most reliable indicators of lumbosacral nerve root compression, the monosynaptic spinal reflex evoked in voluntarily contracting TA muscle presented in this study could be a useful diagnostic means, because there is no commonly accepted reflex from L5 innervated muscles available.

In the study of the APB muscle, when the voluntary contraction was kept at some 20% of maximum force, there were long-latency EMG responses labeled A and B following the initial event labeled SP with both mechanical taps and electrical median nerve stimulation (Fig. 2). Although these designations followed the paper of Marsden [6] in which the responses of the contracting long thumb flexor to stretch of the top joint of the thumb were described, most of the other authors have ignored the earlier A component, possibly because it is often burried in an inhibited part of the response following the SP wave. Whereas the SP responses are widely agreed to be of monosynaptic spinal reflex origin, the origin of subsequent A and B responses is more open to conjecture: (1) volitional or polysynaptic reflex inputs to motoneurons, (2) supraspinal transmission, (3) resonance hypothesis, (4) cutaneous afferents, (5) slowly conducting muscle afferents. In the present study we have documented that the respective delays of A and B after the monosynaptic response with mechanical elicitation are equal to those with electrical elicitation, which implies that mechanically elicited long-latency responses share similar pathways with electrically elicited ones. The same kind of idea was put forward recently [7], however, they used the first dorsal interosseous muscle for mechanical elicitation and the thenar muscle and the median nerve for electrical elicitation. This implication could be an argument against some of the theories for long-latency responses, above all resonance hypothesis [4], that is, long-latency responses may be the consequence of mechanical oscillation of stretched muscle, sensed by the primary muscle spindle endings. If the muscular oscillations are also caused by electrical nerve stimulation, they must be initiated by electrically induced twitches, that is, M response and/or H response. If the resonance-induced afferent spindle volleys caused by M response were responsible for long-latency responses, the responses evoked by more proximal stimulation would appear with longer latencies. However, the result of more proximal stimulation was that both the A and B waves as well as the SP wave were evoked at shorter latencies (Fig. 2b). If the resonance-induced afferent spindle volleys caused by the H response were responsible for long-latency responses, the responses evoked by stimulation at any site should have latencies more than twice as long as the H wave. Both the A and B waves in the present study, however, had latencies less than twice as long as the H wave in every subject. It therefore seems unlikely that the etiology of the long-latency responses in the present study is explicable in terms of the resonance hypothesis.

On the basis of the studies of limb reflexes so far presented, it would be reasonable to consider that reflex responses are also elicitable from the paraspinal muscles during voluntary contraction with mechanical taps. In fact two distinctive types of responses labeled R1 and R2 following the paper of Dimitrijevic et al. [2] were registered with mechanical taps to the interspinous spaces during antigravity upper back extension in the prone position. When the tap was delivered to the same spinal levels as the recording, the reflex time which was represented by the peak latency of the major negative deflection varied in such a way that the latencies tended to be longer with the more caudal recording in both the R1 and R2 responses, which is compatible with the longer spinal reflex arc in the lower segment. The tendency was, however, more pronounced in the R2 than in the R_1 response, because the discrepancy between the latencies of the two became progressively greater as the segmental level tested was moved more caudally from T5–6 to L4–5 level (Fig. 4). These results together with the reflex time itself of each response suggest that the R1 response must be an oligosynaptic spinal reflex, whereas the R2 response might be mediated by more slowly conducting afferents and/or a reflex more rostral in origin, although we were unable to entirely exclude voluntary potentials in the midst of the silent period [8]. Both the R1 and the R2 responses were recordable at various segmental levels concurrently with taps to one of the interspinous spaces, even at thoracic segments with taps to a lumbar level and vice versa, because of the increased excitability of the entire group of paraspinal muscle motoneurones, in addition to the spread of the mechanical event (Fig. 3a). When the tapping level was moved away from the recording level, the latency of the R1 response showed no alteration or only a small prolongation, probably due to the time required for the mechanical event to propagate. In contrast, the latency of the R2 response evoked at a lumbar segment tended to be shorter as the level of tapping was moved from a lumbar to a thoracic level in spite of the increasing distance between the tapping and recording site (Fig. 3b). This might be a favorable finding for the hypothesis of a cutaneous reflex with supraspinal transmission [5] as the etiology of the R2 response, but further elucidation will be needed.

References

1. Braddom RI, Johnson EW (1974) Standardization of H reflex and diagnostic use in S_1 radiculopathy. Arch Phys Med Rehabil 55:161–166
2. Dimitrijevic MR, Gregoric MR, Sherwood AM, Spencer WA (1980) Reflex responses of paraspinal muscles to tapping. J Neurol Neurosurg Psychiatry 35:1112–118
3. Hagbarth KE (1962) Post-tetanic potentiation of myotatic reflexes in man. J Neurol Neurosurg Psychiatry 25:1–10
4. Hagbarth KE, Hägglund JV, Wallin EU, Young PR (1981) Grouped spindle and electromyographic responses to abrupt wrist extension movements in man. J Physiol 312:81–96
5. Jenner JR, Stephens JA (1982) Cutaneous reflex responses and their central nervous pathways studied in man. J Neurol 333:405–419

6. Marsden CD, Merton PA, Morton HB, Adam JER Hallett M (1978) Automatic and voluntary responses to muscle stretch in man. Neurophysiol 4:167–177
7. Noth J, Podoll K, Friedemann HH (1985) Long-loop reflexes in small hand muscles studied in normal subjects and in patients with Huntington's disease. Brain 108:65–80
8. Shahani BT, Young RR (1973) Studies of the normal human silent period. New Developments in Electromyography and Clinical Neurophysiology 3:589–602
9. Trontelj JV, Pecak F, Dimitrijevic MR (1979) Segmental neurophysiological mechanisms in scoliosis. J Bone Joint Surg [B]:310–313
10. Upton ARM, McComas AJ, Sica REP (1971) Potentiation of 'late' responses evoked in muscles during effort. J Neurol Neurosurg Psychiatry 34:699–711

Peripheral Electrophysiological Correlates to Magnetic Transcranial Stimulation

J.R. Toleikis[1] and T.B. Sloan

Introduction

It is generally agreed that the early diagnosis and management of spinal cord injuries plays a major role in the long range prognosis of patients with these injuries. Although the clinical exam remains the best means for evaluating the extent of an injury, for many patients a functional assessment based on this exam may be difficult, if not impossible to make. For this reason, physicians have relied on the objective findings that can be obtained from evoked potential testing. Somatosensory evoked potentials (SSEP) were introduced as a means for objectively evaluating the sensory pathways in spinal cord injured patients several years ago [6]. Although it is generally believed that SSEP are primarily mediated by the dorsal columns and therefore provide information only about a select portion of the posterior cord, the use of these signals for the evaluation of spinal cord injured patients has remained widespread because of the close correlation that appears to exist between the extent of an injury and the degree of normalcy of the SSEP signals. However, the retention of motor function is the primary concern for these patients and as useful as they have been, SSEP are a reflection only of sensory pathway function. Until relatively recently, a technique for the noninvasive assessment of motor pathway function had not been developed.

In 1980, it was demonstrated that it was possible to stimulate the motor cortex by means of electrical pulses applied through scalp electrodes placed strategically over the precentral gyrus [4]. However, because of the large amount of current that was shunted through the scalp and calvarium, it was necessary to apply high levels of current in order to stimulate the brain. In awake patients, these current levels were found to be painful. This presented a significant problem in terms of using the technique for the evaluation of awake patients. More recently, it was demonstrated that it was possible to stimulate the brain painlessly and to elicit peripheral motor responses using a

[1] Department of Anesthesia, Northwestern University Medical School, 303 East Superior, Chicago, IL 60611, USA

K. Shimoji et al. (Eds.)
Spinal Cord Monitoring and Electrodiagnosis

rapidly changing magnetic field [1]. In terms of evaluating spinal cord injured patients, the technique provides a means for diffusely stimulating the brain in order to make a functional assessment as to whether spinal motor tracts are intact, transected, or compromised. However, in order to do so effectively requires a knowledge of (1) the optimal stimulus sites and intensities for eliciting peripheral responses from a particular extremity and (2) the expected response parameters. Two studies have already addressed these issues with limited success [3, 5]. The purpose of this study was to acquire additional information in these areas so that the technique could be confidently used to test patients with severely compromised spinal cord function.

Methods

Fifty paid volunteers were tested as part of a study approved by the Northwestern University Human Subjects Committee, the Northwestern Memorial Hospital Research Committee, and the Federal Drug Administration. Volunteers with a past medical history of any neurological disorder or who had undergone a cranial surgical procedure were excluded from the study. Informed consent was obtained in all cases.

A montage of 35 points centered around Cz were marked on each volunteers' scalp (Fig. 1). Marks were placed at seven points along the midline. The points were spaced 2 cm apart and extended from 6 cm anterior to 6 cm posterior of Cz. Four additional rows of seven points each were also

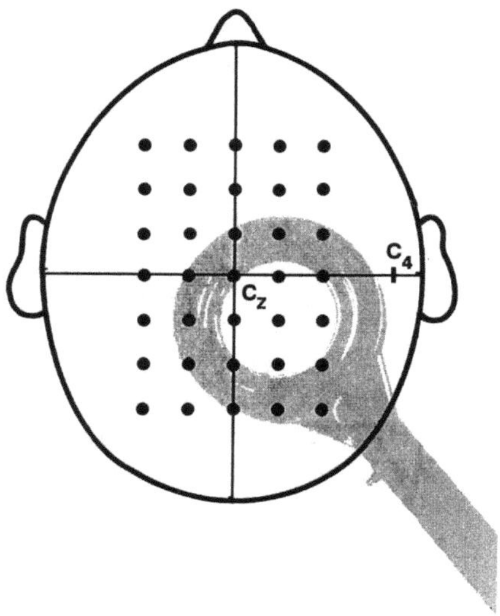

Fig. 1. Montage of 35 stimulation sites. The sites are located 2 cm apart. The magneto-electric stimulation coil is placed tangential to the surface of the scalp and is centered over each location where four stimuli are presented

marked, two rows placed 2 cm apart on each side of the midline. Magnetic stimulation was produced using a Cadwell MES-10 magnetoelectric stimulator (Cadwell Laboratories Inc., Kennewick, WA, USA). When the stimulus was applied at locations further anterior or lateral than those in the montage, it was found to be uncomfortable. The coil was oriented so that the direction of current flow in the stimulation coil was counterclockwise for the first 36 volunteers (Group I) and clockwise for the remaining 14 (Group II). A circular clear plastic disk marked with a cross hair was placed in the center of the stimulation coil as an aid in the accurate placement of the center of the coil over each marked location. Four stimuli were presented at each location with the coil placed tangential to the scalp.

Nine millimeter metallic disk EMG surface electrodes were placed bilaterally on the upper and lower extremities. Compound muscle action potentials (CMAP) were recorded with minimal voluntary contraction from the abductor digiti minimi muscles in the hands and the abductor hallucis muscles in the feet. The CMAP were recorded on a Pathfinder I (Nicolet Instrument Company, Madison, WI, USA) with a bandpass of 100–3000 Hz and a timebase of 60–80 ms.

On the basis of the results of preliminary work which suggested that lower extremity CMAP were difficult to obtain, the first six volunteers were tested with a constant stimulus intensity of 90% of maximum (at 100% of maximum, the magnetic flux density at the center of the coil is 2.3 Tesla). Although useful data were obtained using this protocol, some of the volunteers complained that although the stimulus was not painful, it was uncomfortable, particularly at the more frontal stimulation locations. An analysis of the initial data from these first six volunteers indicated that the stimulus location 4 cm anterior to Cz was consistently associated with well-defined lower extremity CMAP. Therefore, testing for each of the remaining volunteers began at the location 4 cm anterior to Cz with the stimulus intensity set at 50% of maximum. Stimuli were presented and the CMAP from each extremity were recorded. The intensity of the stimuli was increased in 10% increments to a level at which well-defined, repeatable lower extremity CMAP were obtained. The intensity of the stimuli was then kept constant at that level for the rest of the study. For each of the 35 marked stimulation sites, four stimuli were presented, and the peripheral responses from each of the four extremities were recorded. These were stored in the Pathfinder I for future analysis.

Each volunteer's data consisted of the amplitude of each extremity's maximal and average (the average amplitude of the four responses) CMAP which resulted from stimuli presented at each marked scalp location. The absolute values of the CMAP amplitudes for a given extremity varied considerably from volunteer to volunteer. In order to treat each set of data equally, it was therefore necessary to normalize the data of each volunteer for each extremity. By taking into account only the normalized and not the absolute CMAP amplitudes from stimulation at a particular site, the

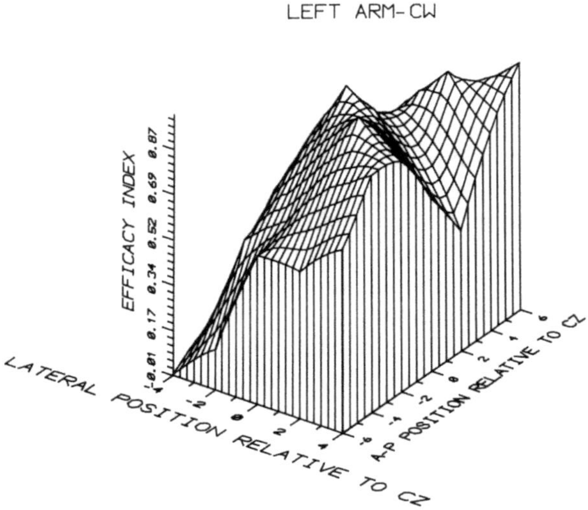

Fig. 2. Surface map of normalized left arm CMAP amplitudes when current flow in the stimulation coil is clockwise. The normalized amplitude drops sharply at 4 cm to the right of and 2 cm anterior to Cz. It is maximal at locations approximately 4 cm from this location. Amplitudes drop off rapidly beyond this distance

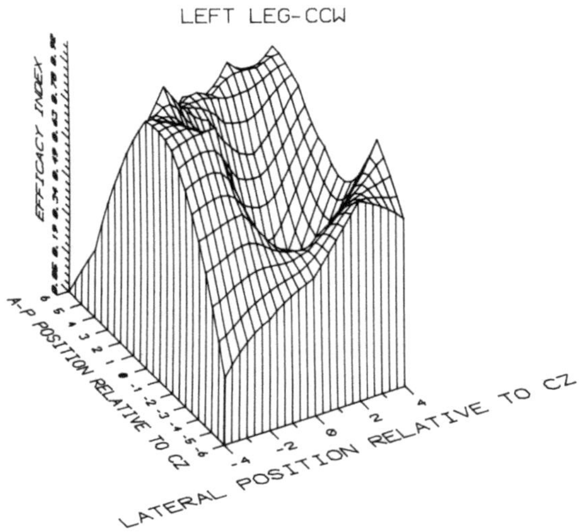

Fig. 3. Surface map of normalized left leg CMAP amplitudes when current flow in the stimulation coil is counterclockwise. The response amplitude is minimal at 1 cm to the right of and 1.5 cm posterior to Cz.; the approximate anatomical location for the left leg motor area. The maximal amplitudes occur from stimulation of some of the sites about 4 cm from this location

normalized data represent the relative efficacy of the 35 stimulation sites for eliciting responses from an extremity. In order to establish population trends, the normalized data from each volunteer in each group were then summed and renormalized.

A software package (Surfer, Golden Software, Golden, CO, USA) was then used to plot topographical and three-dimensional surface maps (Figs. 2 and 3) of the normalized data. These maps describe the relative sizes of the CMAP elicited from a given extremity when the center of the stimulation coil is located over a particular location on the scalp.

Results

The 50 volunteers consisted of 24 men and 26 women with a mean age of 26.1 (range: 13–45 years). The mean height was 67.26 inches (range: 62–75 inches) and the mean weight was 147.84 pounds (range: 95–230 pounds).

Bilateral upper and lower extremity CMAP were obtained from all the volunteers tested. Generally, it was very easy to obtain CMAP from the upper extremities at stimulus intensities of 50% of maximum and above. The lower extremities required at least 60% of maximum stimuation intensity to elicit a CMAP, and occasionally 90% of maximum was required. The average intensity for the 50 volunteers was 78% of maximum (78.1 for group I and 77.6 for group II).

For each direction of current flow, there was one stimulus location which was optimal for eliciting a CMAP from a specific extremity. The maximal and average amplitudes and latency values of the CMAP resulting from stimulation at each of these locations were tabulated for each volunteer. The range of amplitude values between volunteers was very wide with the mean values for the upper extremities always larger and in most cases about twice that for the lower extremities. This was true regardless of whether we were comparing maximal or averaged responses (Table 1) and was true regardless of the

Table 1. Mean amplitudes of averaged and maximum data

Averaged data	LUE	RUE	LLE	RLE
Group I	1337 (605)	1602 (748)	698 (837)	553 (498)
Group II	1024 (511)	1025 (689)	521 (405)	602 (377)
Maximum data	**LUE**	**RUE**	**LLE**	**RLE**
Group I	1972 (858)	2290 (864)	1091 (972)	962 (795)
Group II	1588 (687)	1606 (937)	987 (678)	1094 (756)

Values are measured in microvolts. Values in *brackets* are standard deviations. *LUE*, left upper extremity; *RUE*, right upper extremity; *LLE*, left lower extremity; *RLE*, right lower extremity

Table 2. Onset latencies

	LUE	RUE	LLE	RLE
Group I	19.53 (1.48)	19.80 (1.39)	38.80 (3.28)	38.68 (3.39)
Group II	20.43 (0.99)	20.74 (0.98)	39.43 (2.48)	39.15 (2.04)
ALL	19.78 (1.42)	20.07 (1.36)	38.98 (3.08)	38.82 (3.06)

Values are measured in milliseconds. Values in *brackets* are standard deviations. Abbreviations as in Table 1

current direction in the coil. Latency values demonstrated much less variability (Table 2).

Both topographical and three-dimensional surface plots suggest that the degree of stimulation is dependent upon the relative position of the stimulator coil and the anatomical loci of the cortical motor areas for each extremity. Those stimulation sites which appeared to be approximately one coil radius from the location of a given extremity's cortical motor region produced maximal CMAP from that extremity. On the other hand, when the stimulator coil was approximately centered over an extremity's cortical motor region, minimal CMAP were obtained from that extremity. Figure 2 is the surface map for left hand responses when clockwise current flow in the coil is used. It indicates that the ability to elicit CMAP from the left hand drops sharply at the location (4, 2) which is 4 cm to the right of and 2 cm anterior to Cz. It also indicates that the sites for eliciting maximal CMAP from the left hand are situated in an arc about 4 cm from this location. This was the case regardless of the direction of current flow in the coil. The same results were found for the right abductor digiti muscle responses.

When lower extremity CMAP were plotted in a similar fashion (Fig. 3), the point of minimal stimulation occurred when the center of the coil was located approximately over (1, −1.5); 1 cm to the right of and 1.5 cm posterior to Cz. This location closely approximates where the cortical motor region for the left leg would be. Some stimulation sites about one coil radius from this location produced maximal CMAP, while others produced only moderate stimulation. None of the volunteers were tested with both clockwise and counterclockwise current flow in the coil. However, the maps suggest that the direction of current flow makes a difference only for eliciting lower extremity CMAP.

Discussion

The results of this study suggest that in a population of normal individuals, peripheral activation of a given extremity appears to result from direct stimulation of the cortical motor region for that extremity. Repeatable

CMAP can be predictably obtained from all four extremities, and their amplitudes appear to depend on the placement of the stimulation coil in relation to an extremity's cortical motor region. The shape of the maps suggests that it is the edge of the coil which provides the maximal stimulation, whereas the center provides minimal excitation. Therefore, when the cortical motor region for a specific extremity is encircled by the coil, minimal cortical stimulation will result, and, hence, minimal amplitude CMAP will be elicited from the respective extremity.

However, not all sites that approximate one coil radial distance from the cortical motor region elicit equal responses, suggesting that the amplitude of the CMAP is also dependent upon the interaction of the induced current flow in the cortex and the cortical cytoarchitecture. Differences in direction of current flow do not appear to affect the locations for minimal CMAP from any of the extremities, but do cause changes in the locations for maximal responses from the legs. This may again suggest that the amplitude of CMAP is dependent upon the direction of induced current flow in the brain and the orientation of the cytoarchitecture. The mapping technique highlights the minimal regions for leg stimulation. These minimi correspond to the general anatomical locations for each leg's motor region. These do not change regardless of the current direction that is used. However, the minimal areas for hand responses do not appear to correlate anatomically with the anatomical locations of the cortical motor regions for the hands. Based on the International 10–20 system and the fact that the average A1–A2 distance for our population of volunteers is about 35.5 cm, this would place C3 or C4 about 20% of this distance off the midline, or about 7.1 cm. The inside radius of the coil is 2.5 cm. Therefore, when the coil is located 4 cm lateral to the midline, the central portion of the coil will extend to a distance of about 6.5 cm. This distance probably encompasses a portion of the motor region for the hand, particularly the motor region for the abductor digiti minimi, which lies closer to the midline than the rest of the muscles of the hand. In addition, the coil has an outside radius of 4.75 cm, and the area around its edge is thought to provide the maximal stimulus. If the center of the coil is placed at those locations on the map which elicit maximal CMAP from the hands, then the location of the point on the edge of the coil which is common for all the stimulation sites suggests that the anatomical location which is responsible for producing the maximal responses is probably located lateral to the minimum location (4, 2) on the map.

References

1. Barker AT, Jalinous R, Freeston IL (1985) Non-invasive magnetic stimulation of human motor cortex. Lancet 1:1106–1107
2. Barker AT, Freeston IL, Jalinous R, Jarratt JA (1987) Magnetic stimulation of the human brain and peripheral nervous system: An introduction and the results of an initial clinical evaluation. Neurosurgery 20:100–109

3. Hess CW, Mills KR, Murray NMF (1987) Methodological considerations for magnetic brain stimulation. In: Barber C, Blum T (eds) Evoked potentials III. The third international evoked potentials symposium. Butterworths, Boston, pp 456–461
4. Merton PA, Morton HB (1980) Stimulation of the cerebral cortex in the intact human subject. Nature 285:227
5. Mills KR, Murray NMF, Hess CW (1987) Magnetic and electrical transcranial brain stimulation: Physiological mechanisms and clinical applications. Neurosurgery 20:164–168
6. Perot PL (1973) The clinical use of somatosensory evoked potentials in SCI. Clin Neurosurg 20:367–381

Motor Evoked Potentials Following Transcranial Magnetic Stimulation in Cervical Myelopathy

Y. Nishijima[1], M. Okada, T. Izumi, N. Toda, and Y. Yamazaki

Introduction

Transcranial magnetic stimulation is now considered a useful electrophysiological technique for evaluating central motor tracts. We investigated the latency of the motor evoked potential (MEP) following magnetic stimulation in normal subjects and in patients with cervical myelopathy. It was suggested from our results that the latency of the MEP following scalp or spinal stimulation increased in patients with cervical myelopathy. The purpose of this report is to show the results of measurement of the magnetic MEP latency after scalp, spinal, Erb's point, and elbow stimulation in healthy subjects and in patients with cervical myelopathy.

Subjects and Methods

Subjects

The volunteer control group consisted of 33 persons who gave informed consent to the procedure. There were 26 males and 7 females. The mean age was 33.5 years and the mean height was 167.9 cm. None had a history of neurological disease. The patient group consisted of 26 patients with cervical myelopathy; 15 had cervical spondylotic myelopathy, 5 had traumatic myelopathy, 5 had ossification of the posterior longitudinal ligament, and 1 had cervical cord tumor. There were 18 males and 8 females, the mean age was 51.3 years old and the mean height was 158.5 cm. The mean age of the patient group was significantly higher and the mean height was less than in the control group. Motor function of upper extremity in the patients was evaluated by the

[1] Department of Orthopedic Surgery, Kanazawa Medical University, Uchinada-machi, Ishikawa-ken, 920–02 Japan

K. Shimoji et al. (Eds.)
Spinal Cord Monitoring and Electrodiagnosis

scoring system of the Japan Orthopedic Association (JOA) to evaluate the relationship between MEP latency and the severity of the myelopathy.

Methods

MEP following magnetic stimulation were recorded from the surface electrodes placed over both abductor pollicis brevis muscles (APB). The interelectrode distance was 2.5 cm. Responses were amplified (the bandwidth of the amplifier was 10 Hz–10 kHz) using an electromyograph machine (Evomatic 8000). MEP following a single stimulation was recorded, and an averaging system was not used.

The scalp, the 5th cervical spinous process, Erb's point, and the median nerve at the elbow were stimulated with the magnetic stimulator (Dantec Magnetic Stimulator, Model 9016E006). The stimulating coil had a 2.7-cm inner diameter. It was placed over the midline of the vertex or 5 cm to either side for scalp stimulation. The cervical spine was stimulated at the posterior midline or 1 cm to either side at the level of the 5th cervical spinous process. The scalp and the cervical spine were stimulated with the maximal output of the stimulator. Erb's point and the median nerve were stimulated with 80% of the maximal output. The responses following scalp stimulation were recorded for both slightly contracted and for relaxed muscles in the healthy controls. In the patient group, the response was recorded only with the muscles relaxed.

Somatosensory evoked potentials (SSEP) elicited by median or tibial nerve stimulation were recorded in the patient group with the standard method.

Results

Volunteers

Typical MEP evoked in the APB by stimulation of the elbow, Erb's point, the 5th cervical spinous process, and the scalp are seen in Fig. 1. Responses were evoked bilaterally by midline stimulation and contralaterally by stimulation 5 cm from the midline when the scalp was stimulated. In cervical spine stimulation, a bilateral MEP was not always evoked by midline stimulation.

For the scalp stimulation, the latency of the MEP became shorter and the amplitude became greater when the output of the magnetic stimulator was increased. The latency of the MEP recorded from voluntarily contracted muscles was shorter than that from relaxed muscles. Facilitation by voluntary contraction was clearly seen in the MEP produced by scalp stimulation. The

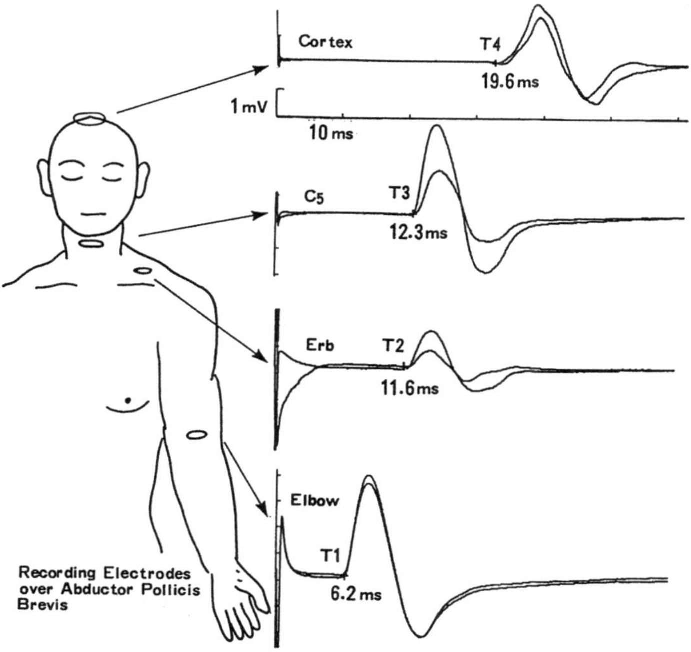

;. **1.** A representative recording of the motor evoked potential (MEP) produced by ıgnetic stimulation in a 23-year-old healthy woman

ıble **1.** Motor evoked potential latency in volunteer controls

	Left (ms)	Right (ms)
bow – APB	7.1 ± 0.9 (*n* = 33)	7.1 ± 0.8 (*n* = 33)
b's point – APB	12.1 ± 1.0 (*n* = 33)	12.4 ± 0.9 (*n* = 33)
; – APB	13.0 ± 0.2 (*n* = 18)	13.1 ± 1.0 (*n* = 21)
ırtex – APB	21.0 ± 1.6 (*n* = 33)	21.4 ± 1.6 (*n* = 23)

ᴾB, abductor pollicis brevis

ent periods from the time of stimulus to the onset of the MEP are shown in ıble 1 as the motor conduction time between each stimulus point and the PB. The conduction time from the scalp to the APB includes both central ıd peripheral motor conduction times. There was no difference of latency tween the right and left sides at any point.

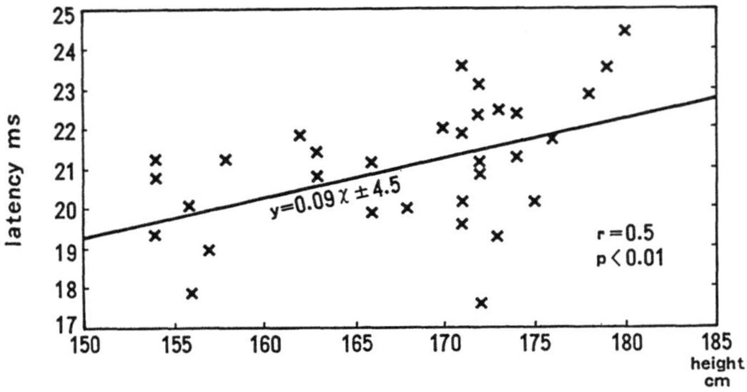

Fig. 2. The correlation between the MEP latencies and the heights of the volunteer controls. A good correlation was observed ($r = 0.5$)

Cervical Myelopathy

MEP was not obtained in four cases on the left side and in three cases on the right side by scalp stimulation in the patients with cervical myelopathy. In addition, with cervical spine stimulation, no MEP was seen in 10 cases on the left side and in 13 cases on the right side. When an MEP was obtained, the waveform showed a tendency to have a multiphasic form. The conduction times (± SD) between the cortex and the APB was 24.3 ± 2.7 ms for the left side and 24.4 ± 2.7 ms for the right side. These values were significantly different from those obtained in the control group ($p < 0.01$). The conduction time between the 5th cervical spine and the APB is also signficantly different ($p < 0.01$) from that in the controls. The latency of the MEP between Erb's point or the elbow and the APB was not significantly different from that in the volunteer controls. The correlation between motor function score in the upper extremities, as evaluated with JOA scoring system, and the MEP latency was examined. Of six patients with scores 0 or 1, three had no detectable MEP, and the values of the other three patients were outside the mean ± 2 SD of the controls. Of 20 patients with scores of 2, 3, or 4, 11 had latencies over the mean ±2 SD of the controls and the other nine had latencies within that range. Regarding SEP recordings, the latency of N19 by median nerve stimulation and P38 by posterior tibial nerve stimulation were not significantly different from that in control data (Table 3).

Case Presentations

Case 1
A 59-year-old man suffered from cervical myelopathy due to C5–6 and C6–7 cervical spondylosis. Motor function of the upper extremity was rated as 2 by the JOA scoring system. C6 radiculopathy and myelopathy in the upper and

lower extremities of both sides were revealed by neurological examination. In the MEP recordings, as shown in Fig. 3, the latency after scalp or 5th cervical spinous process stimulation was increased to 26.0 ms or 18.4 ms, respectively.

Case 2
A 64-year-old woman suffered from incomplete spastic quadriplegia due to ossification of the posterior longitudinal ligament from C3 to C6. The motor function of the upper extremities was 1 point. In the MEP recordings as shown in Fig. 4, the latency after scalp stimulation was prolonged to 33.7 ms. The latency of the MEP after elbow, Erb's point, or cervical spine stimulation was not prolonged.

Table 2. Motor evoked potential latency with cervical myelopathy

	Left (ms)	Right (ms)
Elbow – APB	7.7 ± 1.0 (NS) ($n = 26$)	7.9 ± 1.5 (NS) ($n = 26$)
Erb's point – APB	12.5 ± 1.9 (NS) ($n = 26$)	13.0 ± 1.2 (NS) ($n = 26$)
C5 – APB	14.4 ± 1.8* ($n = 16$)	14.8 ± 1.9* ($n = 13$)
Cortex – APB	24.3 ± 2.7** ($n = 22$)	24.4 ± 2.7 ($n = 23$)

APB, abductor pollicis brevis: *NS*, not significant; *$p < 0.01$; **$p < 0.001$

Table 3. Somatosensory evoked potential latency in patients with cervical myelopathy

	Latency (ms)	Control (ms)
N19	19.2 ± 1.3 (NS) ($n = 25$)	(18.5 ± 0.9) ($n = 28$)
P38	39.5 ± 3.5 (NS) ($n = 24$)	(37.1 ± 2.2) ($n = 28$)

NS, not significant

Discussion

The conduction time from the scalp to the APB in our controls coincided well with the previous report of Barker [2]. Facilitation in voluntarily contracted muscle and increasing latency with higher output stimulation was also recognized. A good correlation between MEP latency to scalp stimulation and

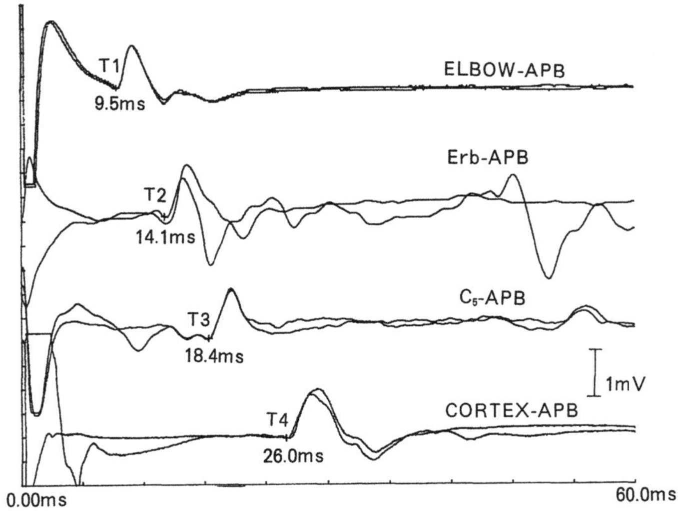

Fig. 3. MEP recorded in a 59-year-old man with cervical spondylotic myelopathy (Case 1). The latency following scalp or the 5th cervical spinous process stimulation was increased to 26.0 ms and 18.4 ms, respectively. *APB*, abductor pollicis brevis

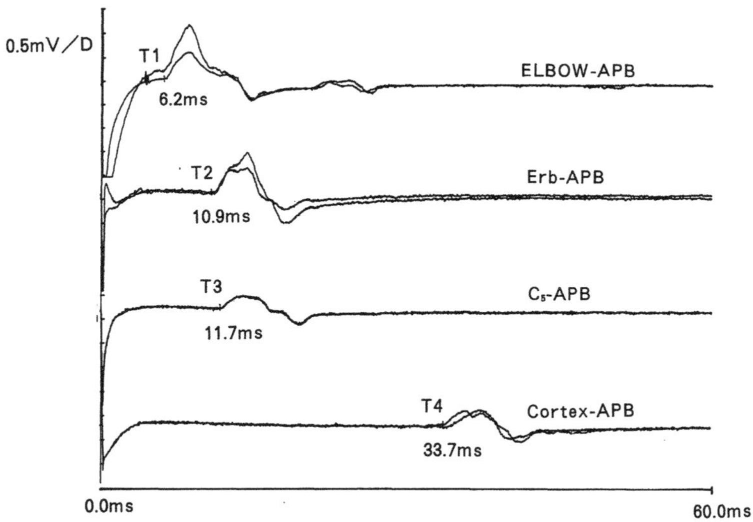

Fig. 4. MEP recorded in a 64-year-old woman with incomplete spastic quadriplegia due to ossification of the posterior longitudinal ligament from C3 to C6 (case 2). The latency following scalp stimulation was prolonged to 33.7 ms. The latency of the MEP after elbow, Erb's point, or cervical spine stimulation was not prolonged. *APB*, abductor pollicis brevis

the height of the controls indicated that the latency demonstrates the conduction time of central motor tracts with considerable precision. Some authors [1, 7, 8] have reported central conduction time or conduction velocity from the time difference in the latency following cortical stimulation and cervical spinal stimulation. However, we show only the latency from the stimulus to the recording site since it was difficult in some cases to obtain an MEP following 5th cervical spine stimulation.

An abnormality in the MEP following scalp or cervical spinal stimulation can indicate either dysfunction of central motor tracts or the spinal nerve root if the MEP for the peripheral nerve stimulation recorded at the same time shows no change.

The clinical usefulness of the MEP obtained following cortical stimulation has been recognized in MS [6], motor neuron disease [4], hemiplegia [3], and cervical spondylosis [1, 7, 8]. Prolonged MEP latency following cortical stimulation was reported in cervical spondylosis by several authors [1, 7, 8]. Abbruzzese [1] reported increased MEP latency in all cases of myelopathy and in 40% of radiculopathy cases in his patients with cervical spondylosis. However, previous reports were based on electric cortical or spinal stimulation techniques which cause some discomfort due to the electric shock. Pain-free stimulation has become possible with the recent development of the magnetic stimulator, and hence cortical stimulation has become acceptable for routine use. The clinical application of the magnetic MEP was investigated in MS and an increase in the latent period was reported [2]. The clinical value of the magnetic MEP is undetermined in other neurological disorders. Our result may be the first series reported on the use of the magnetic MEP in cervical myelopathy.

Essentially similar results were obtained for our magnetic and the electric MEP of other authors [1, 8] The magnetic MEP in severely myelopathic patients who were 0 or 1 point on the JOA scoring system could not be elicited or its latency was more prolonged than in the patient with a JOA score of 2, 3, or 4 points. From these results, one might say that it is possible to judge from the MEP not only the presence or absence of nerve conductivity, but also degree of dysfunction in the central motor pathways in patients with cervical myelopathy.

Previously, the SEP was considered to have little correlation with motor function in myelopathy [9]. On the other hand, the function of the posterior column of the cord is well diagnosed with SEP [5]. In our study, the SEP latency for both median and tibial nerve stimulation was not statistically prolonged. These results do not entirely suggest MEP superiority over the electrical method of diagnosis in cervical myelopathy. These data seemed to derive from the greater susceptibility of the motor tracts to be damaged in cervical myelopathy than the sensory tracts in cervical myelopathy.

It will become possible to make a more accurate diagnosis in cervical myelopathy by using both the MEP by magnetic cortical or spinal stimulation and the SEP.

Summary

MEP (motor evoked potentials) following magnetic stimulation were studied in 33 volunteer controls and in 26 patients with cervical myelopathy.

The conduction time from the scalp to the APB (abductor pollicis brevis) was 21.0 ± 1.6 ms and that from the 5th cervical spinous process to the APB 13.0 ± 1.2 ms on the left side in the controls. These values are coincide well with previous reports. Facilitation in voluntarily contracted muscle was recognized for scalp stimulation. A good correlation was found between the MEP latency to scalp stimulation and the height of the controls. In patients with cervical myelopathy, the latency of the MEP following scalp or cervical stimulation was prolonged. The MEP was either not elicited at all or the latency was further prolonged in severe myelopathy.

References

1. Abbruzzese G, Dall'agata D, Morena M, Simonetti S, Spadavecchia L, Severi P, Candrioli G, Favale E (1988) Electric stimulation of motor tracts in cervical spondylosis. J Neurol Neurosurg Psychiatry 51:796–802
2. Barker AT, Freeston IL, Jalinous R, Jarrat JA (1987) Magnetic stimulation of human brain and peripheral nervous system: An introduction and results of an initial clinical evaluation. Neurosurgery 20:100–109
3. Berardelli A, Inghilleri M, Manfredi M, Zamponi A, Cecconi V, Dolce G (1987) Cortical and cervical stimulation after hemispheric infarction. J Neurol Neurosurg Psychiatry 50:861–865
4. Berardelli A, Inghilleri M, Formisano R, Accornero N, Manfredi M (1987) Stimulation of motor tracts in motor neuron disease. J Neurol Neurosurg Psychiatry 50:732–737
5. Cusick JC (1988) Monitoring of cervical spondylotic myelopathy. Spine 13:877–880
6. Mills KR, Murray NMF (1985) Corticospinal tract conduction time in multiple sclerosis. Ann Neurol 18:601–6055
7. Rossini PM, Caramia MD, Zalora F (1987) Mechanisms of nervous propagation along central motor pathways: Noninvasive evaluation in healthy subjects and in patient with neurological disease. Neurosurgery 20:183–191
8. Thompson MB, Dick JPR, Asselman P, Griffin GB, Day BL, Rothwell JC, Sheehy MP, Marsden CD (1987) Examination of motor function in lesions of the spinal cord by stimulation of the motor cortex. Ann Neurol 21:389–396
9. Yiannikas MB, Shahani BT, Young RR (1986) Short-latency somatosensory-evoked potentials from radial, median, ulnar, and peroneal nerve stimulation in the assessment of cervical spondylosis. Arch Neurol 43:1264–1271

Magnetic Transcutaneous Stimulation of the Motor Pathway in Spinal Cord Disorders

T. Iizuka[1], H. Azuma, H. Tanaka, H. Sugioka, and T. Kondo

Summary

Motor evoked potential (MEP) from small hand and foot muscles was measured by the transcranial magnetic stimulation according to Barker's method. Central motor conduction time (CMCT) was calculated by stimulating over the scalp and then over the cervical and lumbar areas. In healthy subjects, the CMCT (head to neck) was 8.3 ± 1.2 (SD) ms (n = 50) in relaxed muscles and was 6.0 ± 1.4 (SD) ms (n = 29) in partially voluntarily activated muscles, and the CMCT (head to lumbar) was 16.1 ± 3.0 (SD) ms (n = 50) in relaxed muscles and was 14.5 ± 1.2 (SD) ms (n = 29) in partially activated muscles. F wave study was done in 14 patients and (F + M − 1)/2 was calculated according to Kimura's formula. Clinical study showed that MEP and both CMCT with and without facilitation, and CMCT(F) were significantly prolonged in 18 patients with chronic compression cervical myelopathy and were abnormal in 5 patients with spinal cord injury (n = 23). The CMCT and clinical motor functions were well correlated. The pre- and postoperative data (n = 22) were compared and showed the reasonable results of this new method for measuring the motor function. Side effects were not encountered.

Introduction

The functional involvement of sensory pathways in spinal cord disorders can be assessed by use of somatosensory evoked potentials or spinal evoked potentials. An equivalent investigation for the motor pathways would be desirable. The conventional methods of EMG, M wave, and F wave studies have limitations when it comes to the assessment of central motor pathways.

[1] Department of Orthopedic Surgery, Saitama Medical School, 38 Moroyama, Iruma, Saitama, 350-04 Japan

K. Shimoji et al. (Eds.)
Spinal Cord Monitoring and Electrodiagnosis

In 1985, Barker et al. successfully recorded motor evoked potentials (MEP) by transcranial stimulation with a noninvasive magnetic stimulator [1]. This study evaluated the reliability and usefulness of the magnetic MEP for assessment of central motor pathways in spinal cord disorders.

Materials and Methods

We measured MEP from 120 cases, including 43 patients with cervical myelopathy. For this present study, we selected patients who had undergone surgery and investigated 18 (11 males and seven females; mean age, 62 years; mean height, 156 cm) with cervical myelopathy and five patients (four males and one female, aged 14–63 years; mean age, 41.0; mean height, 167 cm) with cervical cord injury. Fifty healthy subjects (25 males and 25 females; mean age, 36 years; mean height 160 cm) were used for the control study. The diagnosis was confirmed in all cases by plain radiographs of the cervical spine, computed tomography, myelography, and magnetic resonance imaging.

Decompression and stabilization surgery was performed in patients with cervical myelopathy and in all but one of the laminoplasty (Kurokawa) operations, and four anterior decompression and fusion procedures were performed [7]. For assessment of cervical myelopathy, the evaluation method of the Japanese Orthopaedic Association (JOA score) was used. The motor function of the upper and lower extremities was graded from 0 to 4: 4, normal; 3, slightly disturbed; 2, moderately disturbed; 1, severely disturbed; and 0, cannot eat with spoon or cannot walk even with support. Frankel's criteria were used for the assessment of spinal cord injury.

Electrophysiological Methods

Transcutaneous magnetic stimuli were delivered over the scalp and cervical (or lumbar) regions using a magnetic stimulator (Novametrix Model 200 MAGSTIM). MEP were recorded by surface electrodes placed over both the abductor digiti minimi (ADM) and abductor hallucis (AH) muscles at the same time. The band-pass filter of the amplifier was set to 20–2000 Hz (DISA 1500 or Nicolet Viking). MEP were recorded while the subjects were relaxed and also while they voluntarily contracted the target muscle (facilitations).

The onset-latency, duration, and peak-to-peak amplitude of the MEP were measured. Central motor conduction times (CMCT) to the spinal cord segments innervating arm or leg muscles were calculated by subtracting the latency after cervical (or lumbar) stimulation from the latency after cortical stimulation. F wave studies were performed in 14 patients for measurement of the latent period from anterior horn cells of the spinal cord to the hand or foot muscles, calculated according to Kimura's formula (F + M − 1)/2. Then the CMCT(F) was calculated in the same manner as mentioned above. The criteria for an abnormal response were: no response, or the occurrence of MEP latency, and CMCT values exceeding the upper confidence limit (mean + 2.5 SD) of the control population. The results were processed statistically using Student's *t* test.

Results

Conduction Time from Relaxed Muscles (Table 1)

The mean latencies of the cortical MEP, cervical MEP (F + M − 1)/2, CMCT, and CMCT(F) were determined in 18 patients with cervical myelopathy and in 50 normal subjects. The mean latencies of the cortical MEP, CMCT, and CMCT(F) obtained from both hand and foot muscles were significantly increased compared with normal subjects. Figure 1 shows significant delay in MEP in a patients with cervical myelopathy.

Conduction Time from Voluntarily Contracted Muscles (Table 2)

The mean latencies of the cortical MEP, CMCT, and CMCT(F) obtained from both hand and foot muscles were significantly increased compared with normal subjects. The mean latency of the cortical MEP was shortened a few milliseconds by facilitation in both normal subjects and myelopathy patients.

Table 1. Latency without facilitation and CMCT (mean ± SD)

Target muscle	Normal subjects	Cervical myelopathy	Statistical significance
ADM	$n = 50$	$n = 18$	
Head	20.9 ± 1.6	28.7 ± 4.9	**
Cervical	12.5 ± 1.6	12.8 ± 1.0	
(F + M − 1)/2	12.8 ± 1.0	13.9 ± 1.0	
CMCT	8.3 ± 1.2	16.0 ± 4.8	**
CMCT (F)	8.2 ± 1.3	14.4 ± 5.1	**
AH			
Head	36.9 ± 2.5	48.1 ± 12.4	*
Lumber	20.9 ± 1.6	21.9 ± 2.0	
(F + M − 1)/2	22.9 ± 1.3	24.9 ± 1.9	
CMCT	16.1 ± 3.0	25.6 ± 10.6	*
CMCT (F)	14.4 ± 2.0	23.8 ± 11.1	*

ADM, abductor digiti minimi: *AH* abductor hallucis; *CMCT*, central motor conduction time
$^{*}p < 0.05$; $^{**}p < 0.01$

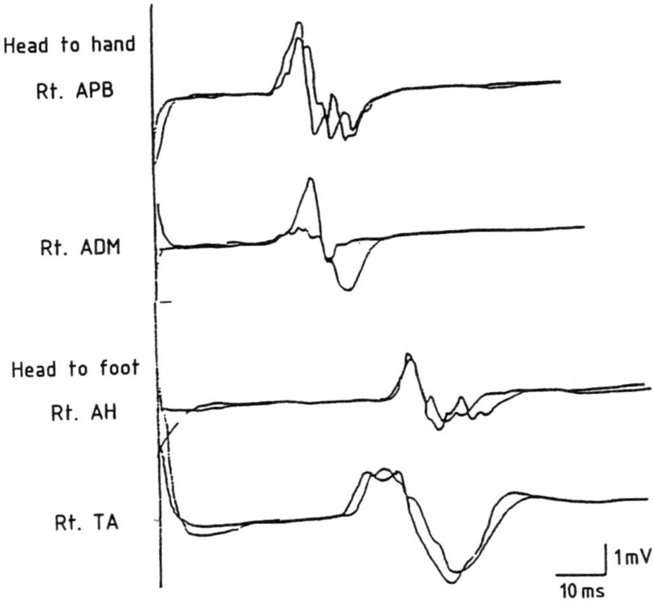

Fig. 1. A 71-year-old woman with cervical myelopathy. MEP from hand and foot muscle without facilitation were recorded. Significant delay of latency was observed. *Rt*, right; *APB*, abductor pollicis brevis; *ADM*, abductor digiti minimi; *AH*, abductor hallucis; *TA*, tibialis anterior

Table 2. Latency with facilitation and CMCT (mean ± SD)

Target muscle	Normal subjects	Cervical myelopathy	Statistical significance
ADM	$n = 50$	$n = 18$	
Head	19.1 ± 1.3	27.0 ± 5.0	**
Cervical	12.5 ± 1.6	12.8 ± 1.0	
(F + M − 1)/2	12.8 ± 1.0	13.9 ± 1.0	
CMCT	6.0 ± 1.4	14.4 ± 5.1	**
CMCT (F)	5.5 ± 1.2	13.6 ± 4.8	**
AH			
Head	35.5 ± 1.4	43.6 ± 5.2	**
Lumber	20.9 ± 1.9	21.9 ± 2.0	
(F + M − 1)/2	22.9 ± 1.3	24.9 ± 1.9	
CMCT	14.5 ± 1.2	22.3 ± 4.5	**
CMCT (F)	11.9 ± 1.7	19.8 ± 3.9	**

ADM, abductor digiti minimi; *AH*, abductor hallucis, *CMCT*, central motor conduction time $^{**}p < 0.01$

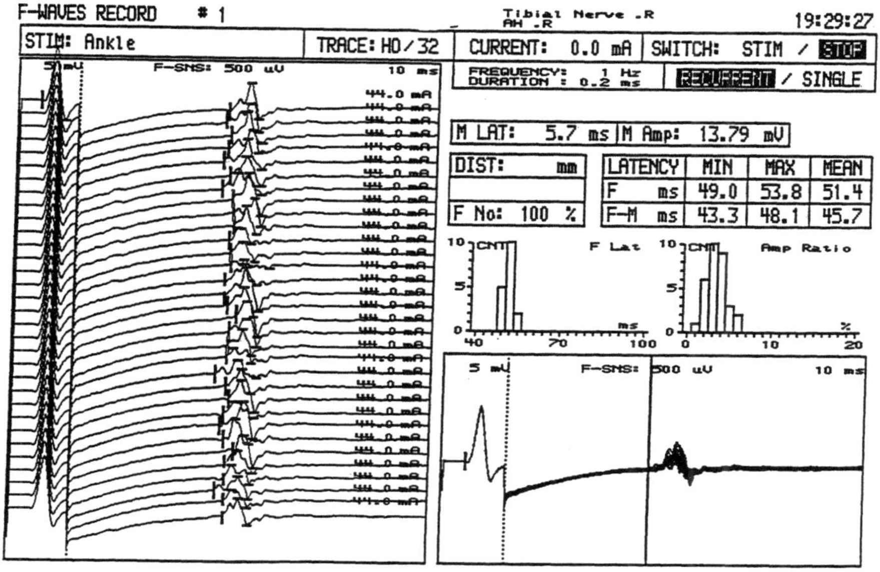

Fig. 2. F wave study: 32 consecutive F waves were averaged in a patient with cervical myelopathy. (F + M − 1)/2 was calculated: 28.1 ms

F Wave Study (Fig. 2, Tables 1, 2)

Figure 2 shows the results from a patients with cervical myelopathy. As the mean latency of the F wave was 51.4 ms, (F + M − 1)/2 was 28.1 ms. The mean value of (F + M − 1)/2 was about 1 ms longer than the cervical MEP and 2–3 ms longer than the lumbar MEP.

Motor Function and CMCT in Cervical Myelopathy (Table 3)

The mean CMCT in patients with severely disturbed motor function (JOA score: 0 or 1) was increased significantly over that in patients with slightly disturbed motor function (JOA score: 2 or 3) for both upper and lower extremities.

Preoperative and Postoperative JOA Score and CMCT (Table 4)

Figure 3 shows the preoperative and postoperative MEP from a 60-year-old woman patient with cervical myelopathy. Postoperatively, shortening of latency was observed, by 3 ms for head-to-head latency and by 5.8 ms for head-to-foot latency. The mean JOA score improved by about 1 point, postoperatively, and the mean CMCT was shortened 2 to 4 ms.

Table 3. Motor function and CMCT (ms)

	CMCT($\bar{s}$) Mean ± SD	(*n*)	CMCT($\bar{c}$) Mean ± SD	(*n*)
JOA score of upper extremities				
0, 1	19.9 ± 0.9	(4)	17.4 ± 13.8	(5)
2, 3	13.5 ± 4.8	(8)	13.6 ± 4.6	(11)
Normal	8.3 ± 4.8	(50)	6.2 ± 1.4	(41)
JOA score of lower extremities				
0, 1	37.3 ± 12.1	(3)	28.5 ± 8.7	(6)
2, 3	20.0 ± 2.6	(6)	20.8 ± 3.0	(8)
Normal	16.1 ± 3.0	(45)	14.9 ± 2.0	(38)

$\bar{c}$ and $\bar{s}$, with and without slight voluntary contraction of the target muscles, *CMCT*, central motor conduction time; *JOA*, Japanese Orthopadeic Association

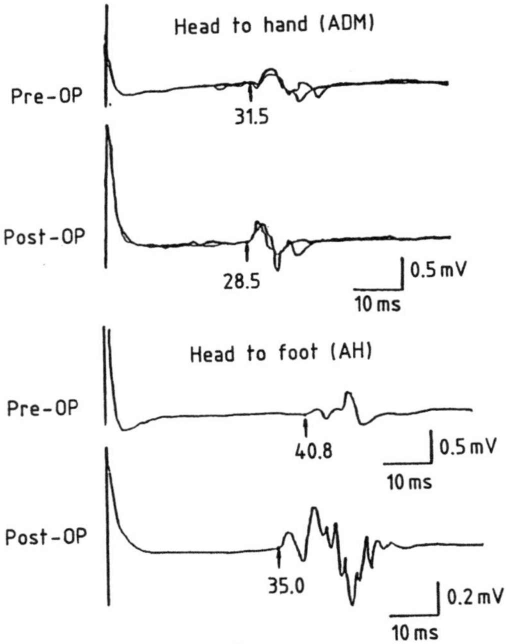

Fig. 3. Pre- and postoperative MEP from hand and foot muscles in a cervical myelopathy patient with ossification of the posterior longitudinal ligament. Postoperatively, shortening of the latencies was observed. *ADM*, abductor digiti minimi; *AH*, abductor hallucis

Table 4. Pre- and postoperative JOA score and CMCT

	JOA score pre	post	CMCT($\bar{s}$) pre	post	CMCT($\bar{c}$) pre	post
Upper Extremities						
mean	1.8	2.7	15.9	12.4	14.8	12.0 (ms)
SD	0.7	0.7	5.1	3.8	4.8	3.6
(n)	(18)	(18)	(11)	(11)	(16)	(16)
Lower Extremities						
mean	1.4	2.2	22.5	20.6	24.1	20.4
SD	0.8	1.0	5.4	4.4	7.3	4.7
(n)	(18)	(18)	(7)	(7)	(14)	(14)

$\bar{c}$ and $\bar{s}$, with and without slight voluntary contraction of the target muscles; *JOA*, Japanese Orthopaedic Association; *CMCT*, central motor conduction time

Results in Cervical Cord Injury (Table 5)

Abnormalities of the MEP were observed in all 5 patients, and the degree of abnormality was well correlated with the severity of the neurological deficit.

Discussion

Since Barker introduced a revolutionary noninvasive method of magnetic transcranial stimulation of the central motor pathways in 1985, many clinical investigations have been performed using this procedure. Daube (1988) described the recent status of the magnetic MEP in clinical application and pointed out that the usefulness for multiple sclerosis was confirmed by several authors, but that it had not yet been investigated well for other spinal cord disorders [2, 3].

There have been only a few reports describing changes in the magnetic MEP in cervical myelopathy [4, 5, 9]. Jarratt showed abnormality of MEP from hand muscle in one out of a few patients with cervical spondylosis. Mills also reported changes in the MEP in one patient with cervical myelopathy. Significant delays in the mean MEP and CMCT obtained from hand and foot muscles in 10 patients with cervical myelopathy were reported by us.

We found a significant delay or absence of response in the MEP determined in 18 patients with cervical myelopathy and 5 patients with cervical cord injury compared with 50 normal controls. The changes were present regardless of the site of recording (hand or foot), the condition of the target muscle (with or without facilitation), or the method of calculation of the CMCT (cervical and lumbar stimulation or F wave study). The mean CMCT and clinical motor function correlated well with each other. Preoperative and

Table 5. Results of the patient with cervical cord injury

Patients	Level	Type	Treatment	Frankel	MEP	Period after injury (months)
1. YA	C5	Transverse	Laminoplasty	B → C	None/delay (+++)	2
2. MK	C4	Transverse	Laminoplasty	B → C	None/delay (+++)	3
3. HT	C5	Transverse	Anterior fusion	C → D	Delay (++)	4
4. MT	C5	Central	Anterior fusion	D → E	Delay (±)	24
5. SY	C6	Central	Skeletal traction	C → E	Delay (±)	1

postoperative studies also indicated the reliability of this method for assessing the motor pathways in cervical spinal cord disorders.

Regarding safety, no serious complications of magnetic transcranial stimulation have been reported. No side effects were noted in this study of 73 patients (including our experiences as subjects). Although this method seems to be very safe further prospective studies should be performed for the definitive investigation of its safely [8, 10].

Conclusions

1. MEP and CMCT were significantly prolonged in 18 cervical myelopathy patients compared to 50 healthy subjects.
2. CMCT and the degrees of motor disturbance were well correlated.
3. Postoperatively, mean CMCT were shortened a few milliseconds from pre-operative latency.
4. MEP were absent or delayed in all the cervical cord injury patients and well correlated to the motor function.
5. Assessment of the central motor system in cervical myelopathy and cervical cord injury may be possible by the magnetic MEP.

References

1. Barker AT, Freestone IL, Jalinous R, Merton PA, Morton HB (1985) Magnetic stimulation of the human brain. J Physiol 369:3
2. Daube JR (1988) Selected application of magnetic stimulation AAEE/AEEGS Joint Symposium, pp 31–34
3. Hess CW, Mills KR, Murray NMF (1987) Responses in small hand muscles from magnetic stimulation of the human brain. J Physiol 388:397–419
4. Iizuka T, Tanaka H, Sugioka H, Kondo T (1989) Assessment of the motor pathway in cervical myelopathy by transcutaneous magnetic stimulation. Sekitsui Sekizui Janaru (Spine & Spinal Coral) 2:611–616
5. Jarrat JA (1986) Magnetic stimulation for motor conduction. AAEE Symposium, pp 27–33
6. Kimura J (1983) Electrodiagnosis in diseases of nerve and muscle: Principle and practice Davis, Philadelphia
7. Kurokawa T, Tsuyama N, Tanaka H, Kobayashi M, Machida H, Nakamura K, Iizuka T, Hoshino Y (1982) Enlargement of spinal canal by sagittal splitting of spinous process. In: Onomura T, Hayashi K (eds) Progress in cervical spine surgery Bessatsu Seikei Geka (Orthopedic Surgery), Vol. 2. Nankodo, Tokyo, pp 234–240
8. Jou I, Matsuda H, Nakata N et al (1989) Experimental study on safety of transcranial stimulation J Electrodiagnosis of the spinal cord 11:13–16
9. Mills KR, Murray NMF, Hess CW (1987) Magnetic and electrical transcranial brain stimulation; Physiological mechanism and clinical applications. Neurosurgery 20: 164–168
10. Tsubokawa T, Katayama Y, Yamamoto T et al. (1987) Assessment of spinal cord injury with multimodality evoked spinal cord potentials. Part 2: Correlation with neurological outcome in clinical spinal cord injury. Neuro-orthopedics 3:82–89

Clinical Application of Motor Evoked Potentials in Disorders of the Spine

J. Dvořák[1], J. Herdmann[2], and R. Theiler[1]

Summary

Transcranial and spinal magnetic stimulation was performed on 62 healthy adults to obtain normal values of central motor conduction times to cervical and lumbar segments innervating limb muscles. The mean central motor latency (CML) was 5.1 ms when recording from the biceps brachii and 5.2 ms when recording from the abductor pollicis brevis or abductor digiti minimi muscles. The mean CML to lower limb muscles were 13.0 ms to the quadriceps femoris, 13.3 ms to the anterior tibial, and 13.4 ms to the extensor digitorum brevis muscles. Conduction times higher than "mean latency + 2 × standard deviation" may indicate a pathological conduction slowing.

Some 136 patients with cervical spine disorders were examined. Of all patients 76% showed a pathological delay of their motor evoked potentials (MEP) to upper limb muscles. Of 52 patients with lumbar spine disorders 63% showed pathologically delayed MEP to lower limb muscles.

Magnetic stimulation of the motor system is a painless neurophysiological technique which enables us to examine the central motor pathways in awake subjects with respect to disorders of the spine. This method offers important additional information in conditions such as compression of the spinal cord or of the motor roots.

Introduction

Since the beginning of neurophysiology it has been a great challenge to stimulate neuromuscular tissue, and especially the brain. Since 1980 it has been

[1] Department of Neurology, Spine Unit, W. Schulthess Hospital, Neumünsterallee 3, 8008 Zürich, Switzerland

[2] Department of Neurosurgery, Heinrich-Heine-Universität, Moorenstr. 5, 4000 Düsseldorf 1, FRG

K. Shimoji et al. (Eds.)
Spinal Cord Monitoring and Electrodiagnosis

possible to stimulate the cortex transcranially by means of the high-voltage electrical technique [7] and since 1985 by means of the magnetic stimulator [2]. The outstanding advantage of magnetic stimulation compared to high-voltage electrical stimulation is the painlessness. Thus, transcranial brain stimulation has become widely applicable to awake patients, some of whom are severely disabled and suffer from intense pain. The aim of this study was to assess normal values and to simplify and standardize stimulation and recording methods in order to examine patients with various disorders affecting the spinal cord and the motor roots. Patients with cervical spine disorders suffered from advanced cervical myelopathy, on the one hand, and soft tissue disease of the cervical spine without neurological deficit, on the other hand. Patients with lumbar spine disorders showed spinal stenosis or single nerve root compression syndromes. MEP were used as a diagnostic tool in order to classify and localize the lesion and in order to determine the indication for a surgical procedure.

Methods

Physical, physiological, and methodological principles of transcranial magnetic stimulation have been described in detail by Barker et al. [1]. A DANTEC Magnetic Stimulator or a Novametrix Magstim ME 200 was used. For cortical stimulation the center of the coil was placed over the vertex in order to record potentials from upper limb muscles and 4–6 cm ventral of the vertex in order to record potentials from lower limb muscles, respectively. In order to stimulate the motor roots the coil was centered in the midline over the 4th to 7th cervical and over the 3rd lumbar to 3rd sacral vertebra, respectively. Potentials were picked up by surface electrodes placed over the muscle belly and over joint or bone (reference). MEP were recorded from the biceps brachii, abductor pollicis brevis, abductor digiti minimi, quadriceps femoris (rectus femoris), tibialis anterior, and extensor digitorum brevis muscles. Dantec Counterpoint or Schwarzer Picker Basis 8000 E.M.G. machines were used to record the potentials for on- or off-line analysis. Further data evaluation and statistical analysis was performed on Apple Macintosh systems using Microsoft Excel and Abacus Concepts Stat View software.

For transcranial stimulation a threshold intensity was determined with the muscle at rest. Stimulation was then performed with an intensity of "1.5 × threshold" with the muscle slightly contracted in order to facilitate the response [5]. For root stimulation the intensity was adjusted in order to obtain a potential with a sharp negative deflection [3].

Furthermore, a conventional neurography of the median, ulnar, and peroneal nerves including F wave recordings was performed. This allowed us to calculate the conduction time of the lower motor neuron from anterior horn cell to muscle. The formula used was (M latency + F latency − 1)/2. Due to the turn-round time of the antidromic excitation in the motor neuron,

1 ms has to be subtracted from the sum of M and F wave latencies. If this peripheral latency (PL) was subtracted from the cortical latency (CL) after transcranial magnetic stimulation, we obtained a central motor latency (CML-F) from cortex to anterior horn cell. If the central motor conduction time was calculated as the difference between CL and root latency (RL) after magnetic stimulation over the spine, this "central" motor latency (CML-M) included the conduction over the motor root to its exit from the intervertebral foramen [3]. This is especially important in lumbar spine diagnostics where motor roots, i.e. cauda fibers, are 10–20 cm long.

Some 62 healthy adults, 136 patients with cervical spine disorders, and 52 patients with lumbar spine disorders were examined. Normals and patients showed an equal distribution with respect to age and size. Normals were aged 18–75 years and 150–191 cm tall. Patients were aged 16–80 years and 148–190 cm tall. All subjects were grouped according to age (young: $\leqslant$ 44 years; old: $\geqslant$ 45 years) and to size (small: $\leqslant$ 174 cm; tall: $\geqslant$ 175 cm). The results of the normal subjects did not show any significant differences with respect to sex or age groups. Except for the CL and RL of the biceps brachii and the RL of the anterior tibial muscle, all CL, RL, and PL showed a significant size-group related difference. With respect to the CML, only the CML-M to the anterior tibial muscle showed a significant size-group related difference (see Table 1).

Results

Normal Values

Table 1 gives the CML-M and CML-F values obtained from 62 healthy subjects. Besides mean value and standard deviation (SD) we have calculated "mean + 2 × SD". We propose that latencies exceeding this value may be regarded as pathologically increased latencies.

Patients with Cervical Spine Disorder

Some 136 patients with cervical spine disorders were examined. According to case history and clinical and radiological findings they were diagnosed as having rheumatoid arthritis (n = 51), degenerative disease (n = 56), or cervical spine trauma (n = 29). Radiological and clinical findings ranged from normal (especially in patients with cervical spine trauma) to extreme compression of the cervical cord with severe tetraparesis.

Most of the patients with rheumatoid arthritis suffered severe joint pains and were disabled due to their primary disease. Therefore it was not possible to perform an extensive examination of many muscles on both sides. In these

Table 1. Central motor latencies

	CML-M Mean ± SD (ms)	Mean + 2 × SD (ms)	CML-F Mean ± SD (ms)	Mean + 2 × SD (ms)
BB	5.1 ± 1.0	7.1		
APB	5.2 ± 0.6	6.4	4.3 ± 0.8	5.9
ADM	5.2 ± 0.9	7.0	4.0 ± 0.8	5.6
QF	13.0 ± 1.4	15.8		
TA small	12.8 ± 1.4	15.6		
tall	14.0 ± 1.3	16.6		
EDB	13.4 ± 1.7	16.8	11.3 ± 1.7	14.7

In order to obtain normal values 62 healthy subjects (24 female, 38 male) were examined. Central motor latencies were calculated by means of the fractionated magnetic stimulation only (CML-M) and by means of the additional F wave recordings (CML-F). See text. Besides mean ± standard deviation the proposed upper limit of the normal range (mean + 2 × SD) is given. Only for the CML-M to the anterior tibial muscle was there a statistically significant difference between small (≤ 174 cm) and tall (≥ 175 cm) subjects ($p < 0.01$). *BB*, biceps brachii; *APB*, abductor pollicis brevis; *ADM*, abductor digiti minimi, *QF*, quadriceps femoris; *TA*, tibialis anterior; *EDB*, extensor digitorum brevis muscle

patients it was most important that the examination itself was more or less painless. For this reason the high-voltage electrical stimulation technique was never used. Of the patients with rheumatoid arthritis involving the cervical spine 57% showed significantly increased central motor conduction times. The highest CML-M and CML-F values to the abductor pollicis brevis and to the abductor digiti minimi muscles were 16.0 and 14.9 ms (moderately affected patient with spinal stenosis between C3 and C5) and 20.0 and 17.5 ms (patient with spastic tetraparesis with spinal stenosis due to inflammatory tumor within the atlanto-axial ligaments), respectively (Fig. 1).

The central motor conduction times of 56 patients summarized as having a "degenerative" cervical spine disorder were significantly increased in 75% of the patients. Eighty-six percent of the patients with a neurological deficit and 57% of those without clinical signs showed prolonged conduction times. The highest CML-M and CML-F values to the abductor pollicis brevis and to the abductor digiti minimi were 10.0 and 8.6 ms (patient with median discal herniation C5/C6 with enhanced lower limb tendon jerks) and 15.8 and 14.7 ms, respectively (Fig. 2).

Thirteen of 29 patients with cervical spine trauma did not show any significant radiological findings. Only 7 of the 29 patients showed minor neurological signs like asymmetric reflexes or like a circumscribed sensory deficit. The remaining 22 patients complained of more or less diffuse pain in the neck, in the shoulders, in the arms, or drawing to the head. However, 6 of the 7 patients with a neurological deficit (86%) and 14 of the 22 neurologically unaffected patients (64%) showed increased central motor conduction times. The highest CML-M and CML-F values to the abductor pollicis brevis and to the abductor digiti minimi was 9.1 and 7.5 ms (patient without neurol-

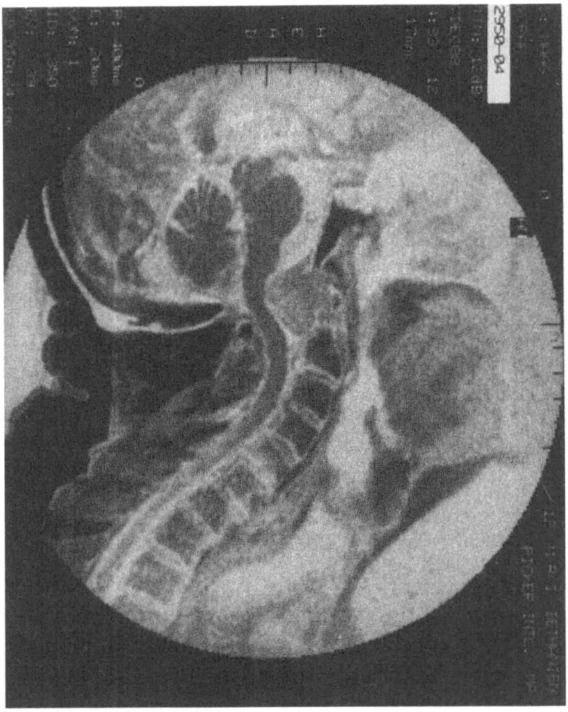

A

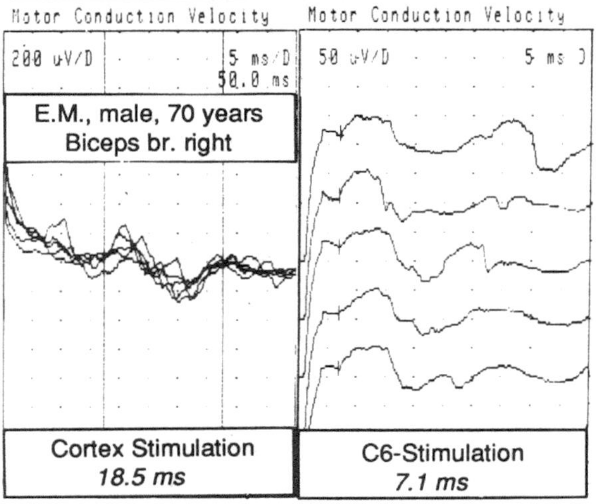

B

Fig. 1. A MRI scan of a 70-year-old patient who suffered a long-standing rheumatoi arthritis with marked joint deformity. It shows severe inflammatory involvement of th C1/C2-segment. He presented with a tetraspastic syndrome and had been confined to wheelchair for half a year. **B** Original recordings from the patient's right biceps brach muscle. The affection of the motor fibers showed increased CML to all muscles examinec

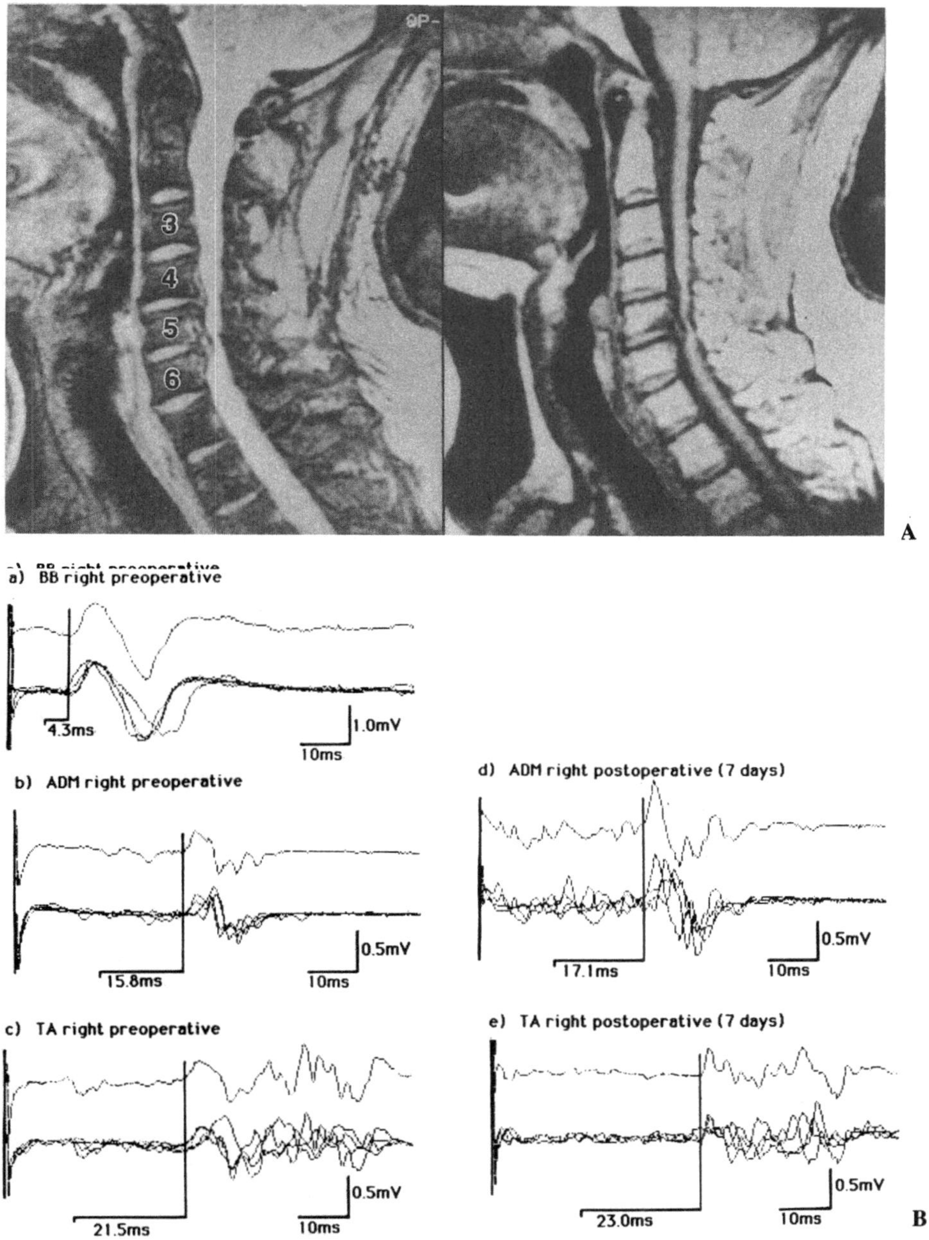

Fig. 2. A MRI scan of a 41-year-old patient with cervical myelopathy due to severe degenerative changes of the cervical spine, most pronounced at the C5–6 level. He presented with parapasticity, mild motor dysfunction of both hands, and pain drawing from the neck to the right hand. **B** Original recordings from the right biceps brachii (*BB*), abductor digiti minimi (*ADM*) and anterior tibial (*TA*) muscles. The potentials to the biceps were completely normal on both sides. CML to the abductor digiti minimi and to the anterior tibial muscles were markedly increased (the individual values are indicated below the *bars*). The potentials were small and split up. Seven days after an anterior fusion of the C5–6 segment there was not yet any significant change, although the patient was free of pain and his gait was already improving

ogical deficit, C4/C5 instability) and 9.3 and 8.2 ms (patient with enhanced tendon jerks, no radiological findings), respectively.

Patients with Lumbar Spine Disorder

Fifty-two patients with disorders affecting the lumbar spine were examined. According to the clinical and radiological findings they were divided into two groups: lumbar stenosis ($n = 35$) and nerve root compression syndromes ($n = 17$).

Of all patients with lumbar stenosis 60% presented a significant delay of MEP to one or more muscles of the lower limbs. Eight-six percent of the patients with a significant neurological deficit and 54% of those without neurological signs showed motor conduction times exceeding "mean normal value + 2 × SD". The highest CML-M values to the anterior tibial and to the extensor digitorum brevis muscles were 20.6 and 30.0 ms, respectively (Fig. 3).

Of the patients with nerve root compression syndromes 71% showed pathological MEP latencies. The highest CML-M values to the anterior tibial and to the extensor digitorum brevis muscles were 18.1 and 20.0 ms, respectively (Fig. 4).

In 13 of 15 patients in whom it was possible to record an F wave besides MEP of significantly increased CML-M, the calculated CML-F was normal. This clearly localized the conduction slowing within the motor roots (see Fig. 4). In the two patients in whom both CML-M and CML-F were increased, the difference between both was distinctly higher than in normals.

Discussion

Overall 67% of the patients with cervical spine disorders, 72% of those with, and 60% of those without neurological deficits showed significantly increased central motor conduction times. This raised the question whether the limits set for the normal range were too low [4]. Thus the limit was set to "mean value plus 3 SD" instead of 2 and the data was recalculated. However, the figures changed only very little. Fifty-one percent of all patients, 56% of those with, and 45% of those without neurological signs revealed central motor conduction times exceeding the mean normal value by more than 3 SD. This means that the increase of central motor conduction time of the patients examined was generally highly significant. These results are in accordance with those of Masur et al. [6] who found pathologically increased central motor conduction times in 10 of 12 patients with cervical spinal stenosis and neurological deficit.

The results of the patients with cervical spine trauma are surprising as the lack of objective findings in other examinations often contrasting with the

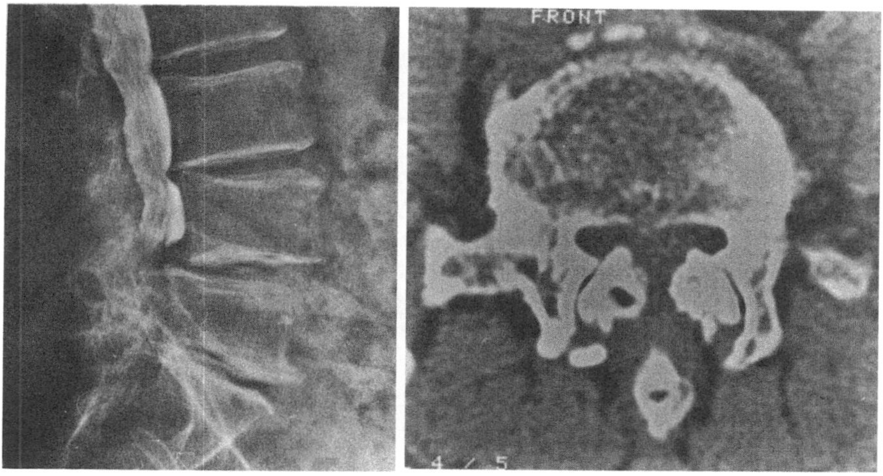

A

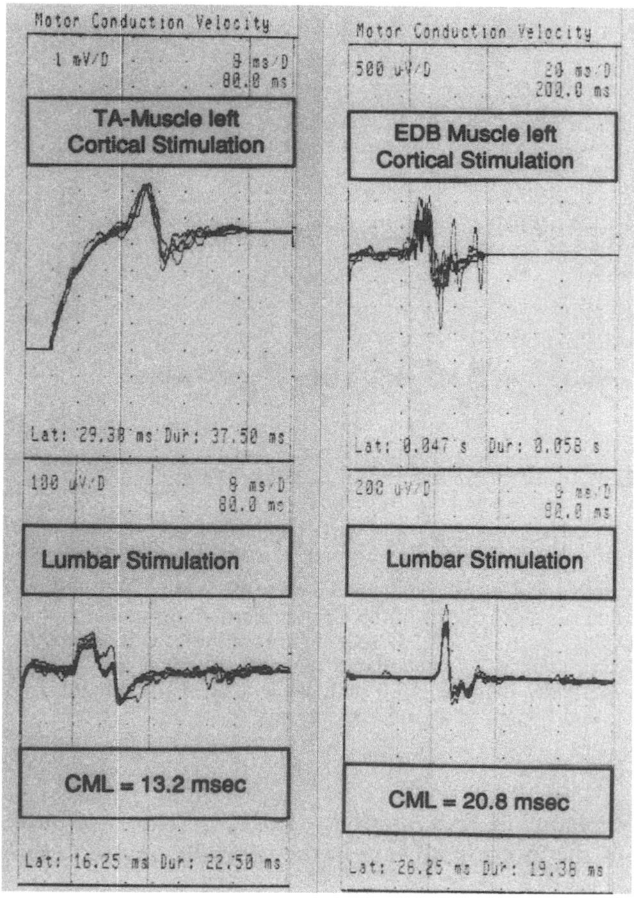

B

Fig. 3. A Myelogram and CT scan of a 75-year-old patient with lumbar stenosis: there is a complete stop and stenosis at the L4/L5 level. **B** Original recordings from the left anterior tibial (*TA*) and extensor digitorum brevis (*EDB*) muscles. Only the central motor latency (*CML*) to the EDB was delayed

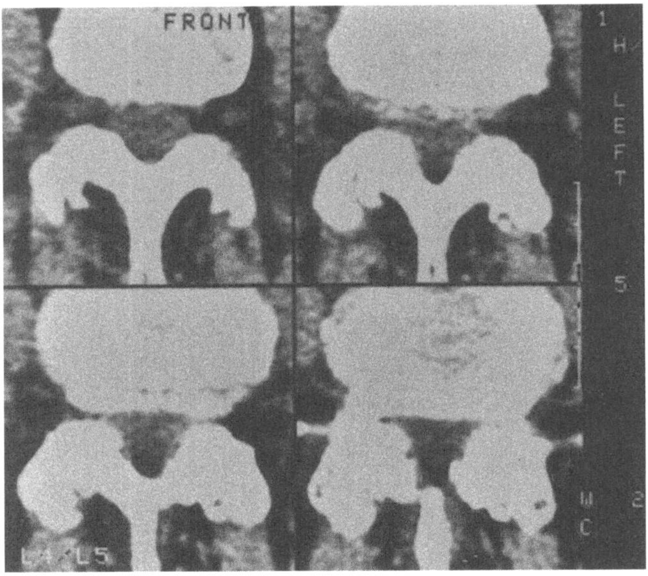

A

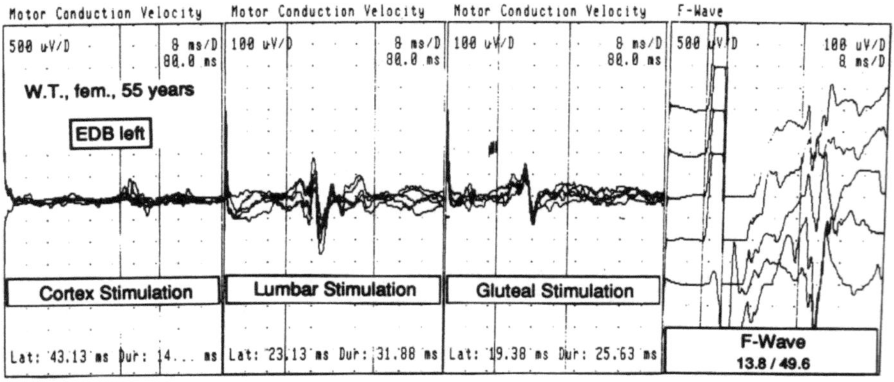

B

Fig. 4. A CT scan of a 55-year-old woman who presented with an L5 sensory-motor deficit on the left side due to osteoarthritis of the intervertebral joints and lateral stenosis L4–5. **B** Original recordings from the left extensor digitorum brevis muscle (*EDB*). The central motor latency calculated using fractionated magnetic (cortical and lumbar) stimulation (CML-M) was markedly delayed (20.0 ms), while CML-F (calculated by means of M latency, 13.8 ms, and F latency, 49.6 ms was normal (11.9 ms). See text

numerous complaints of these patients. MEP, however, often indicated an involvement of the motor pathways in the clinically unaffected patients with soft tissue disease.

In patients with lumbar spine disorders the affection of the motor pathways refers to a lesion of the nerve roots, i.e., the axons of the lower motor neuron. Here it is especially important to perform a neurography and F wave recordings in order to calculate the PL, i.e., the conduction time from anterior horn cell to muscle. If this is subtracted from the magnetically obtained CL, the resulting CML-F does not include conduction time over the motor roots, i.e., cauda fibers, and should thus be normal in disorders only affecting the motor pathways within the lower part of the lumbar spine. Since the RL, obtained by means of magnetic stimulation over the spine, measures the conduction time of the motor fibers from their exit from the intervertebral foramen to the muscle [3, 8], the CML-M, i.e., the difference between CL and RL, includes the conduction time over the intraspinal part of the lower motor neuron. CML-M and CML-F may be used to localize the site the conduction slowing. Often, however, F waves cannot be recorded due to the motor root lesion.

A grouping of the patients according to the degree of clinical affection or to the degree of radiological findings was not performed in this study. However, central motor conduction times seemed to be the more increased, the more the patients were disabled. The correlation with radiological findings seemed to be less tight, especially in the less severely affected patients.

Assuming that the results given here are confirmed by future studies, this would mean that the sensitivity of MEP is higher than the sensitivity of conventional EMG. Certainly MEP are less painful as long as the magnetic stimulator is used. Furthermore, in comparison to electromyography the latencies measured from MEP to a certain extent allow a quantification of the lesion. Besides being easy to reproduce MEP allow the course of the disease to be followed.

Conclusions

These results suggest that MEP are a sensitive means of detecting involvement of the motor pathways in cervical as well as lumbar spine disorders. With respect to the easy handling and to the painlessness of magnetic stimulation, MEP are very suitable as a screening technique preceding neuroradiological methods. MEP may indicate the degree of affection of the motor fibers, and especially in soft tissue disease they may help to clarify the various complaints of the patients. MEP help to localize the lesion and allow the course of the disease to be followed. Furthermore, they may help planning of surgical procedures, including the intraoperative monitoring of the motor pathways.

References

1. Barker AT, Freeston IL, Jalinous R, Jarratt JA (1987) Magnetic stimulation of the human brain and peripheral nervous system: an introduction and the results of an initial clinical evaluation. Neurosurgery 20:100–109

2. Barker AT, Freeston IL, Jalinous R, Merton PA, Morton HB (1985) Magnetic stimulation of the human brain. J Physiol 369:3P
3. Britton TC, Meyer B-U, Herdmann J, Benecke R (1990) Clinical use of the magnetic stimulator in the investigation of peripheral conduction time. Muscle Nerve 13:396–406
4. Dvořák J, Herdmann J, Theiler R (1990) Magnetic transcranial brain stimulation: painless evaluation of central motor pathways. Normal values and clinical application in spinal cord diagnostics upper extremities. Spine 15:155–160
5. Hess CW, Mills KR, Murray NMF (1986) Magnetic stimulation of the human brain: facilitation of motor responses by voluntary contraction of ipsilateral and contralateral muscles with additional observations on an amputee. Neurosci Lett 71:235–240
6. Masur H, Elger CE, Redner K, Fahrendorf G, Ludolph AC (1988) Function of the spinal cord in patients with cervical spinal stenosis – an electrophysiological study. EEG EMG 19:263–266
7. Merton PA, Morton HB (1980) Stimulation of the cerebral cortex in the intact human subject. Nature 285:227–228
8. Ugawa Y, Rothwell JC, Day BL, Thompson PD, Marsden CD (1989) Magnetic stimulation over the spinal enlargements. J Neurol Neurosurg Psychiatry 52:1025–1032

Motor Evoked Potentials: Issues in Safety and Operative Monitoring

W.J. LEVY[1], J, ORO, and M. TRAAD

Introduction

The monitoring of the motor system is an important goal for spinal cord evaluation. Ideally we need a safe procedure, easily used, reliable, and able to indicate problems with the upper or lower motor neuron pathways. Until a few years ago little was available in such a direction. However, the advent of both spinal cord stimulation and transcranial motor pathway stimulation has offered the opportunity to develop such testing. A particularly useful type of transcranial stimulation in the awake patient is the magnetic stimulator, first described in 1985 by Dr. Anthony Barker. This stimulator was initially limited by the large surface area of brain which it stimulated, and by the fact that it was just powerful enough to stimulate awake patients. Therefore it was often not reliable on an anesthetized patient with a higher stimulus threshold. The technology of these devices is evolving rapidly, and more powerful stimulators are being tested by Novametrix and Cadwell Labs, as well as some other groups. The one which has been initially evaluated for the operating room is the Cadwell MES-10. This can stimulate patients in the operating room under some conditions.

Separately, an evaluation of the safety of this method continues. Encouraging for the safety of the test is that a large number of patients and volunteer investigators have been stimulated. Many of the latter have had a large number of stimuli, into the tens of thousands. Some have had biochemical, EEG, and neuropsychological evaluation, and have not shown evidence of side effects. No definite complications have resulted to date, and the incidence is low. However, caution in use is important, particularly for the possibility that the test can directly or indirectly lead to a seizure in a seizure prone patient.

Additionally an important new use of the magnetic stimulator is being developed that is not intraoperative, but rather as an operative planning guide.

[1] University of Pittsburgh, Pittsburgh, Pa, and University of Missouri, Columbia, Mo

K. Shimoji et al. (Eds.)
Spinal Cord Monitoring and Electrodiagnosis

This builds on the traditional neurosurgical use of direct cortical stimulation, done via direct brain stimulation. Such uses have been essential in seizure surgery and surgery near eloquent cortical areas. Such mapping has been restricted to surgery or unpleasant transcranial electrical stimulation. However, the magnetic stimulator with a figure-of-eight coil can stimulate transcranially with approximately a 1-cm accuracy. There is little or not noxious sensation. The resulting maps fit the known distribution of the cortical motor homunculus, and they are dramatically changed in some injury conditions. This suggests the possibility that these methods can help in surgical planning as well as disease investigation.

Concomitantly with this development of the MEP, our group has also worked on the development of a test of the lower motor neuron and peripheral nerves which does not rely on stimulation, but makes use of spontaneous EMG. This is the same type of test which has come into common use for the monitoring of the seventh nerve, but here it is applied to the spinal cord This offers the advantages of instantaneous feedback, both auditory and visual. It monitors the axons actually being injured, and can evaluate a large number of different spinal roots by proper selection of the muscles tested. Its value is both ease of use the relevance to root and nerve manipulation.

Methods

For transcranial stimulation in safety testing a Novametrix stimulator was used with a 13-cm diameter coil. Volunteers were healthy subjects with no past history of neurological disorders, including seizures. They received a standard EEG prior to and 1 h after the stimulation paradigm. In addition, their blood pressure, pulse, serum cortisol, serum prolactin, and neuropsychometric status were monitored. For the latter counting backward by serial sevens, naming days of the week, Boston naming, story recall, and visual perception were tested before the after the stimulation, as well as some tasks during the stimulation.

For the stimulation four stimuli were administered at each of four sites. These were right temporal, right frontal, left frontal, and left temporal. Appropriate neuropsychological tests of the brain location being stimulated were applied at each of the sites. Following the stimulation the subjects were followed, and at approximately 1 year had follow-up EEGs and inquiries about their general status and career progress.

For operative monitoring the Cadwell MES-10 has been used due to the increased stimulation strength it has over the original Novametrix model. This was positioned over the motor cortex, using a pointed or "tear-drop" shaped coil. The edge of the coil was positioned so as to lie over the motor cortex. Anesthesia in our cases was with nitrous oxide and narcotics, which was kept as light as possible in order to prevent depression of motor cortex responsiveness. Muscle relaxants were given by continuous infusion, either atricurium bromide or pavulon. A Cadwell Quantum signal averaging computer was the recording device.

For spontaneous EMG, the recording device is a Cadwell Quantum or 7400 four-channel signal averaging computer, which has EMG capability on each channel, with a separate audio feedback system. The filters are 100 Hz low cut and 10000 Hz high cut, with a gain of 100 per division. The recording electrodes are EEG needles, sterile, which are placed with the active in one muscle supplied by the root, and the reference in another part of that muscle, or another muscle supplied by that root which is physically separate and likely to improve the sampling procedure of the monitoring. Usually we monitored the appropriate muscles for the root bilaterally.

Records were visually inspected at the time of surgery and selected responses printed. The current machines do not allow a continuous record of several EMG channels to be stored, but this would be desirable.

Anesthesia was inhalational or intravenous, with inhalation using nitrous oxide and low concentrations of forane. In order to enable the monitoring of spontaneous EMG a less than complete muscle block must be used. The exact level for good detection of responses has not been determined, but even a response only to a train of four peripheral stimuli is usually regarded as an 80% block. Some level of block which allows detection of EMG, but not movement, or a less than acceptable degree of muscle relaxation would be optimal. The cases without paralysis during the monitoring were either done with an initial muscle relaxant induction of succinylcholine or of pavulon, which was no longer effective by the time the monitoring section of the case was done. The case with a low level of block were given a continuous infusion of atricurium bromide, which was adjusted so as not to provide a high grade block to a train of four impulses from a peripheral nerve stimulus, and therefore to allow a low level EMG activity to be observed.

For transcranial mapping in awake patients, volunteers, either normals or quadriplegics, were given a detailed informed consent procedure. On all subjects Cz was identified and marked with an eye shadow pencil. Then markings were made at one CM intervals on the midline from 8 cm anterior, to 4 cm posterior to Cz. Markings were then made laterally to the right side, at 2-cm intervals that were on a line between Cz and the external auditory meatus. For stimulation, the anterior aspect of the coil, at the intersection, of the figure of eight was placed at the measurement points, with the coil handle pointed anteriorly and the coil placed tangential to the scalp in a balanced manner between the two halves. The stimulator was a Cadwell MES-10, with a tape wrapped butterfly coil. The stimulator was used at 100% stimulation intensity and at a cycle time which did not produce coil warming.

For recording a Cadwell Spectrum was used, with EMG recordings done by paired surface electrodes over the deltoid, biceps, triceps, APB, and ADM muscles. All data were recorded onto a laser printer output, and duplicated for record keeping purposes. All subjects were relaxed, without voluntary contraction. The EMG latency and amplitude were measured and plotted onto a matrix with the stimulation site for each point. The entire area of the scalp which produces responses was mapped, in order to define the area involved as well as possible. The exceptions to this were some lateral points over the superficial temporal muscle, where for comfort reasons the stimulation was not extended further laterally.

Results

We tested eight patients in the safety protocol and had to evidence of changes present after the test which were induced by the stimulation in the tests [9]. The only apparent change was during the stimulation, where some evidence of short term interference with the counting of serial sevens and the naming of days of the week was present. These were significant at the < 0.01 and 0.03 levels respectively. They suggest that a very short term effect is present, which does not persist to the end of the test. The fact that the effects were not present in each test situation may further suggest that the timing for the effect is a key element.

We have followed these subjects for up to 2 years, with EEG repeated at approximately 1 year posttest. No abnormalities in the EEG were seen, and no neurological illnesses (including seizures) or performance impairments in their careers have been reported. This is a small set of subjects, but nonetheless they have not had any effects demonstrating a deleterious side effect to the testing situation.

Latencies were consistent with those reported for these muscles previously. These values are consistent with D wave activation of the muscles, but could represent I wave activation.

Case 1 was a 22-year-old male who was injured at C4 2 years previously, with a complete quadriplegia leaving some biceps function that had improved. The biceps strength was 4/5 and the sensory level was at C6. He had had extensive physical therapy treatment. He had also had biofeedback training, which primarily involved the biceps muscle. He had a present but significantly diminished sensation to touch in the upper arms bilaterally, and a lack of position sense in the digits His cortical map was made, and with the biceps, triceps, APB and ADM muscles monitored we were able to produce responses only in the biceps These responses come from a large area of cortex, covering 82 cm^2.

Case 2 was a 47-year-old male who was paralyzed 22 months previously at C1–2 by a wave during swimming. He was at the time completely quadriplegic, but subsequently recovered to antigravity function (3/5) strength, in both the upper and lower extremities. He had been through extensive physical therapy and biofeedback training, the latter for the deltoid muscle. He had voluntary control of the deltoid at a 4/5 level. In the case of the other extremities he was able to use his left hand well enough to control a joystick on a power wheelchair, but not well enought to raise an eating utensil usefully, for example. His cortical stimulation showed easily elicited responses from the deltoid (Fig. 1) but not from the biceps, triceps, ADB, or ADM muscles The response came from an area of 72 cm^2. There are shorter latency responses in the center and longer latency responses at some points of stimulation which are more peripheral to the sites eliciting deltoid movement in the volunteers.

In the normal control subjects responses to EMG stimulation were obtained from each of the monitored muscles, although the triceps was less easily responsive. Consistently the deltoid, biceps and triceps were located medially to the area of the scalp producing responses in the APB and ADM muscles. Figure 2 shows a typical deltoid and Fig. 3 a typical APB representation of the responding scalp area. The area was 12 cm, compared with 4 cm^2 for the biceps and 2 cm^2 for the deltoid in this same subject. This is 67% to 83% larger for the hand muscles than the proximal muscles. This figure would be conservative due to the limitations in this study of measuring the lateral extent of the hand area.

Regarding the spontaneous EMG of the spinal roots, 23 patients were monitored. There were 8 herniated discs and 15 procedures for spondylosis or canal stenosis. There were 6 cervical procedures and 17 lumbar procedures.

Responses were observed to have an initial quiet EMG period with only noise activity on the screen, or an occasional discharging motor unit (Fig. 4). We did not observe activity above this level prior to surgery in the area of the root, and also did not usually observe EMG with the laminectomy itself in these cases. In three cases there was a discharge during dissection around the facets and ligamentous attachments to the spine. When the root was being manipulated, EMG activity often occurred (Fig. 5). This originated from single units in some instances and from multiple units firing separately in

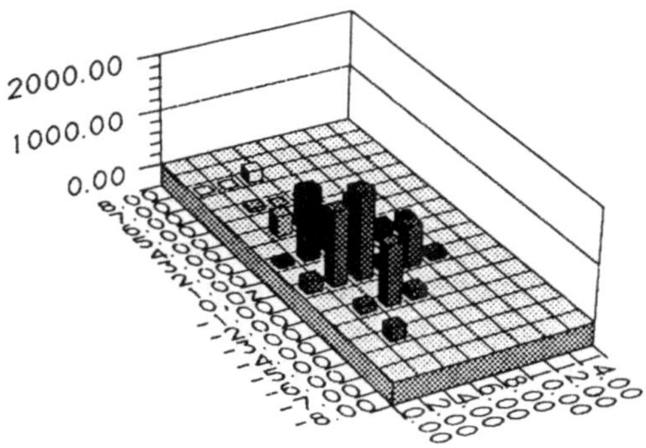

Fig. 1. TW deltoids EMG amplitude P1. A cortical magnetic stimulator map of the deltoid in a patient 2 years after spinal cord injury at C1–2, whose lowest muscle that responds with significant strength is the deltoid. Muscles in the upper and lower limbs have weak movement but are barely of antigravity strength or significantly less than antigravity strength. A very large area of the cortex is devoted to the deltoid

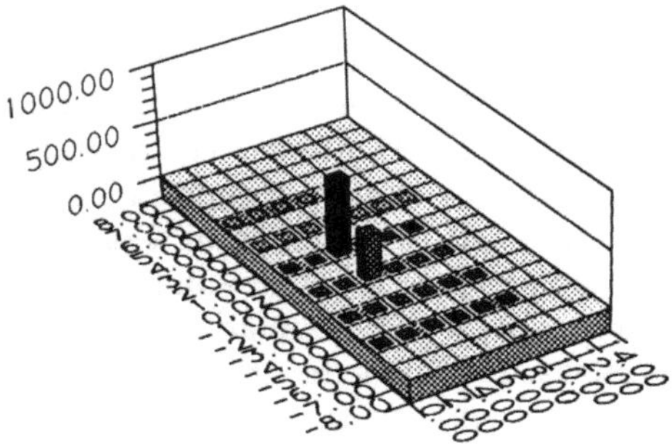

Fig. 2. BA deltoids EMG amplitude P1. A cortical map of the right half of the cortex showing the sites which produced deltoid contraction in a normal volunteer. Those sites which did not respond are shown as squares without height, while those which did have a "skyscraper" bar with amplitude

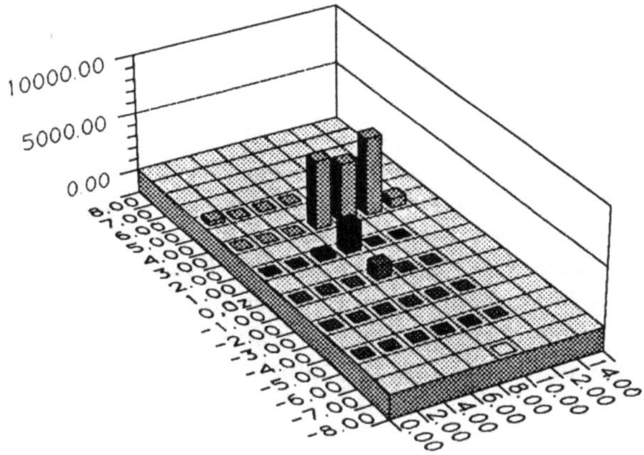

Fig. 3. BA APB EMG amplitude P1. A cortical magnetic stimulation map of the responses of a normal subject's abductor pollicis Brevis (thumb). Notice that the representation is more lateral and has more area that the biceps. This fits with the normal pattern that distal muscles have larger representations

others. Some cases had bursts of discharges from motor units which appeared to be related in their firing pattern. More extensive irritation was signaled by prolonged discharges of motor units which lasted several seconds. We did not observe very long discharges, e.g., tens of seconds of minutes, in any of our cases to date.

The surgical activity which often produced the discharge was direct manipulation of the root, or more commonly manipulation of a structure to which the root was attached or adjacent. These included adhesions, cotton patties used for exposure and hemostasis, and work on the opposite root. In two instances gentle pressure on the root resulted in firing of the motor units prior

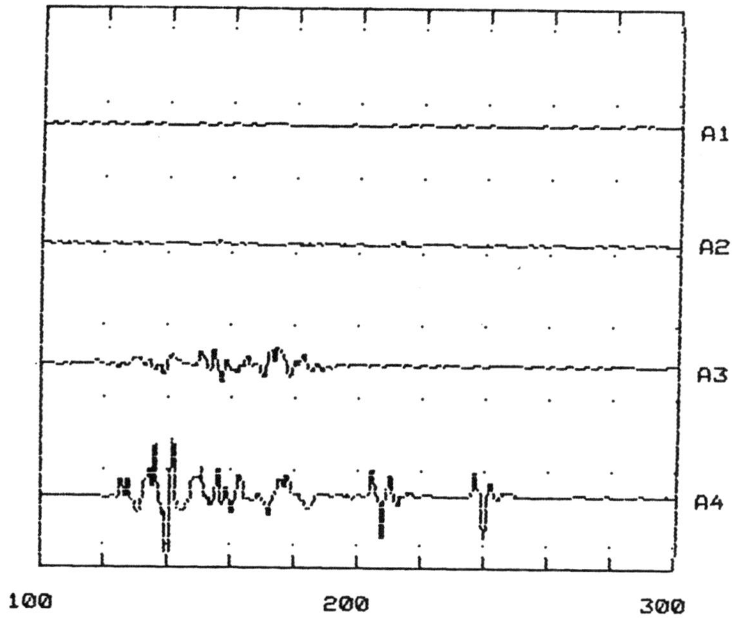

Fig. 4. Some repetitive firing of several motor units is seen on the *bottom trace*. This is suggestive of potential irritation of the nerve, but may be a baseline activity

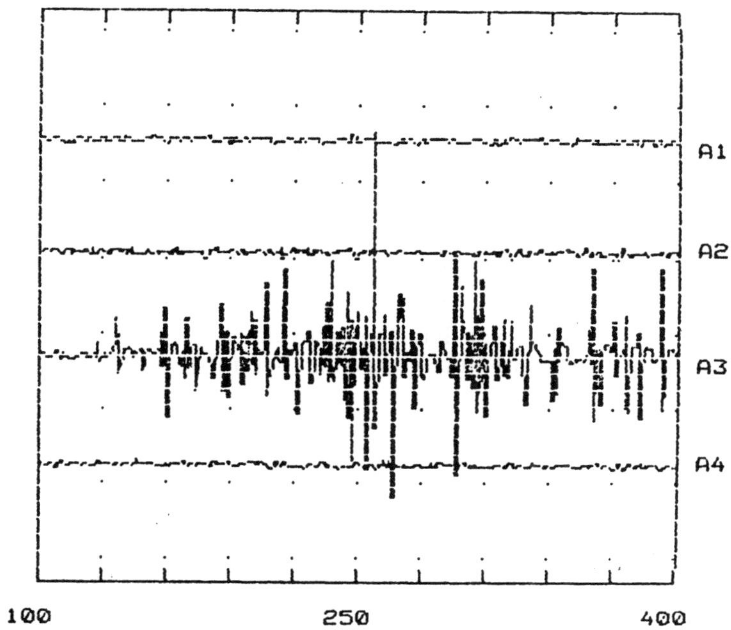

Fig. 5. Continued discharges of a number of motor units are seen in the third channel, the right gastrocnemius as the root is being manipulated. This discharge, especially when temporarily linked to the manipulation of a root, is a potential warning

o decompression of the nerve by removal of a disk, but not after its removal. Ve also observed contralateral root firing a few cases. Our overall impression if the relation of surgical manipulation to EMG firing was that the responses iccurred often with the onset of the disturbance of the root, and did not ılways continue. For instance, if the root was compressed by the nerve root etractor it would fire initially and at removal of the retractor, but not during he retraction. Furthermore, traction on the root seemed to produce more eaction than compression did, and this suggests that longitudinal shearing orces are involved.

At times, unquestionable muscle movement occurred from root mani-ıulation in a limb or paraspinal muscle which was not detected by the EMG. This demonstrates that volume conducted EMG signals from distant muscles vill not necessarily be detected. That in turn underscores the need for appro-ıriate sampling of the muscles representing that root.

)iscussion

;afety of magnetic stimulation is a crucial step to clarify in the development ıf this new field. There is a basis of experience from studies on direct elec-rical stimulation the brain to set damage conditions. These are based on :urrent density and charge per half wave. Magnetic stimulation appears to ıe well below these current limits [7]. Electrical stimulation is also below he limits, but care in the choice of electrode and amount of current applied s important.

Published clinical studies in Europe and the US have so far not revealed ıdverse effects from the stimulation. The largest published US series is Bridges vho found no adverse effects to the magnetic stimulation in 40 patients. This is encouraging and continued follow-up of these patients has so far also evealed no further problems.

Our own experience in cats and rats whith transcranial electrical stimula-ion of a higher level than the equivalent in a human was without any evi-lence of deficit in EEG. Evans blue dye for blood brain barrier breakdown, ır microscopic evaluation. The only circumstance in which we have been ıble to produce damage was that of a hypertension in cats caused by higher ;timulation rates. This was clearly both rate and anesthetic dependent, mak-ng it likely that considerable attention needs to be paid to choice of anesthetic.

There is an important role in the risk-benefit ratio in evaluating the safety ıf a transcranial stimulation test. If the information gained in the monitoring ıf the nervous system can provide a reduction in risk, then any risk of the test ess than that gain is not a large issue. In high risk cases paralysis may be a ;everal percent — can help in risk, even probable. The early warning of the MEP reducing the danger. While the extent of the reduction is not specified ⁄et, it surely is greater then the risk of the MNEP. The clinical and animal experience with transcranial stimulation suggests that a less than 1% incidence of problems is present. This is encouraging for the application of the test in risk reducing situations, such as the operating room. However in screening applications where the questions of disease and treatment are less clear, the tradeoffs are hard to define and experience must be accumulated even more carefully to make decisions about use of the test. It remains possible that seizure prone individuals could have a seizure as a result of magnetic stimulation, although no cases have been reported to date. Photic stimulation alone can provoke seizures. Overall continued safety evaluation and cautious use of transcranial stimulation are needed, but the safety is encouraging.

Regarding cortical mapping, its use to localize the important functional areas of the brain is a well established technique with exposed cortex at surgery. It has played an important enough role in some types of cases to be indicated, either under local anesthesia or by an operation to create implanted grid arrays. Extension of this method to a noninvasive test would provide useful information more widely and with less risk and inconvenience to the patient.

In evaluating this method here, we found that to our surprise the most proximal muscle affected by the injury which had useful strength dominated the representation. This contrasted with the expected pattern in three normals.

The ability to map the motor homunculus transcranially is a new development and much basic work on methodology and reproducibility needs to be carried out. However, the changes seen in the quadripegic patients here in distinction to the normals are indicative of responses to trauma which are probably important in recovery from injury, but which also be affected by pharmacologic and rehabilitative therapy.

This supports previous work that substantial cortical plasticity is present and adapts to the injury, perhaps playing an important role in the recovery process. An important question is whether the absence of sensory input is a determining factor in this rearrangement. Another question is how the physical therapy a and biofeedback directed at the remaining functioning muscles affects this plasticity process. It may be that in a patient with remaining movement of a small degree in important hand muscles and of greater degree in more proximal muscles the plasticity rearrangement can be affected more favorably by special attention to the more distal but less functional muscles.

We are reporting a method for localizing the motor cortex areas of specific muscles transcranially. This method is a valuable tool for assessing cortical localization questions.

Regarding the spontaneous EMG test, the optimal monitoring system for roots and peripheral nerves should reflect irritation that may increase risk at an early stage, give prompt feedback, should monitor the axons at jeopardy, should be roughly quantificable to predict the level of risk, and should be easy to use. In this regard the SEP, either from peripheral nerve or dermatomal stimulation, has not yet established an accepted role in this matter. It is not specific to the root when a peripheral nerve is used, and it may evaluate different axons than the ones at risk. With the addition of a focal stimulating electrode the root can be localized in difficult anatomical situations where identification is not straightforward.

Motor evoked potentials from transcranial stimulation could eventually play an important role, but they are still evolving and are difficult to use. As an intermediate test the spontaneous EMG has some advantages. It reflects events occurring at the injury site and in the axons which are stressed by events. It provides feedback in milliseconds, and it is easy to use. In this latter regard the feedback is both visual and auditory, which offers advantages in the attention consuming operating room environment. The muscles which respond provide localization clues about the root irritated and the site of irritation. This method is similar to that for facial nerve monitoring, which is becoming a standard, if it is not so in many centers already.

There are limitations which this system shares with facial nerve monitoring. One is that it is not easily quantifiable. It is difficult to know on the one hand when a discharge signals approaching irreversible damage. On the other hand, a sectioned axon might discharge once or several times, and then fall silent, but the silence is an ominous event. Prolonged discharges, and multiple unit discharges, represent stress to some of the axons that should not be applied to others, and indicate that the surgeon should evaluate the events. Another issue is that of adequate sampling in the distribution of the root being monitored. It is possible to use two separated muscles predominantly supplied by the root. These can either be monitored on separate channels or yoked together in a single channel by having the active electrode in one and the reference in another. However a systematic study of muscles optimal for a given root should improve the detection of events.

We have found this system to be of value in these cases, and think that it merits development and evaluation. The favorable experience with this method in the facial nerve is particularly encouraging. Our present efforts are to gain experience in a large series of patients, and to determine optimal monitoring sites for the principal roots which provide the best sample of activity in terms of muscle mass, function and accessibility.

References

1. Agnew, McCreery (1987) Safety in transcranial stimulation. Neurosurgery 20(1): 100–109
2. Barker AT, Jalinous R, Freeston IL (1985) Non-invasive magnetic stimulation of human motor cortex. Lancet i:1106–1007
3. Bridge SL, Delaney RC (1989) Transcranial magnetic stimulation: An Assessment of Cognitive and other cerebral effects. Neurology 39:417–419, 1989
4. Brown Rh, Nash CL, Lorig RA, Schatzinger LA (1977) Spinal cord monitoring during operative treatment of the spine. Clin Orthop 126:100–105

5. Edmunds HL, Paloheimo MPJ, Backman M, Johnson JR, Holt RT, Schields CB (1989) Transcranial magnetic motor evoked potentials (tcMMEP) for functional monitoring of motor pathways during scholiosis surgery. Spine 14:683–686
6. Levy WJ (1987) SEP vs MEP debate: the monitoring of motor pathways. Clin Neurosurg 34:239–260
7. Levy WJ, McCaffrey M, York DH, Tanzer F (1984) Motor evoked potentials from transcranial stimulation of the motor cortex in humans. Neurosurgery 15:214–227
8. Levy WJ, McCaffrey M, York DH, Tanzer F (1984) Motor evoked potentials from transcranial stimulation of the motor cortex in cats. Neurosurgery 15:287–302
9. Levy WJ, Oro J, Tucker D (1989) Safety studies of electrical and magnetic stimulation for the production of motor evoked potentials. In: Magnetic stimulation in clinical neurophysiology. Butterworth, London
10. Moller AR (1987) Electrophysiological monitoring of cranial nerves in operations in the skull base. In: Sekar LN, Schramm V (eds) Tumors of the cranial base: diagnosis and treatment. Futura, Mt. Kisco
11. Moller AR, Jannetta PJ (1984) Preservation of facial nerve function during removal of acoustic neuromas. Use of monopolar constantp-voltage stimulation adn EMG. J Neurosurg 61:757–760
12. Moller AR, Jannetta PJ (1985) Monitoring of facial nerve function during removal of acoustic tumor. Am J Otol [November Suppl]:27–29
13. Prass R, Luders H (1986) Acoustic (loudspeaker) facial electromyographic monitoring Part 1. Neurosurgery 19:392–400
14. Raudzens PA (1982) Intraoperative monitoring of evoked potentials. In: Bodis-Wollner I (ed) Evoked potentials. New York Academy of Sciences, New York, pp 308–325

II. Clinical Applications

A. Monitoring

Prognostication of Surgical Outcome in Cervical Spondylotic Myelopathy Using Evoked Spinal Cord Potentials

K. Shinomiya[1], Y. Kurosa, M. Fuchioka, H. Sato, M. Yokoyama, H. Komori, and K. Furuya

Introduction

The diagnosis of cervical spondylotic or ossification of the posterior longitudinal ligament (OPLL) myelopathy has made rapid progress as a result of using several new medical imagings such as CT and MRI. It is still difficult, however, to establish a correct diagnosis of the location and extent of intraspinal cord lesions and to predict postoperative results.

We have been using evoked spinal cord potentials (ESCP) in order to make a correct level diagnosis since 1982 because we believed that these potentials revealed spinal cord function directly. Decompressive surgery for cervical myelopathy has been performed on the basis of the results of the electrophysiological method and the new medical imagings, and the postoperative results have been encouraging. It is apparent, however, that not all surgical cases yield satisfactory results, in spite of a sufficient extent of decompression as indicated by the several diagnostic methods mentioned above. The purpose of this study was to estimate the severity and extent of the spinal cord disorder and to identify the most important factor in prognosticating postoperative results using ESCP.

Materials and Methods

One hundred and twenty-one cases of cervical myelopathy with long tract signs subjected to surgical treatment from 1983 to 1988 were the subjects of this report. None of the patients had any history of cervical spinal cord injury. Of this group, 101 patients were treated, either using an anterior one-level fusion procedure (39) or anterior decompression over two levels (62), 18 using a posterior procedure consisting of laminectomy or laminoplasty, and 2 using a combined procedure.

[1] Department of Orthopedic Surgery, Tokyo Medical and Dental University, 1-5-45 Yushima, Bunkyo-ku, Tokyo, 113 Japan

K. Shimoji et al. (Eds.)
Spinal Cord Monitoring and Electrodiagnosis

The day before surgery, evoked spinal cord potentials were recorded via a cervical epidural electrode with five poles spaced 15 mm apart. Stimulation was applied to the thoracic cord (conductive ESCP), the median nerves (segmental ESCP), and finger surfaces (segmental D-ESCP). The duration of stimulation was 0.2 ms and intensity ranged from more than twice threshold to supramaximum if patients could tolerate it. Bandwidth was from 20 to

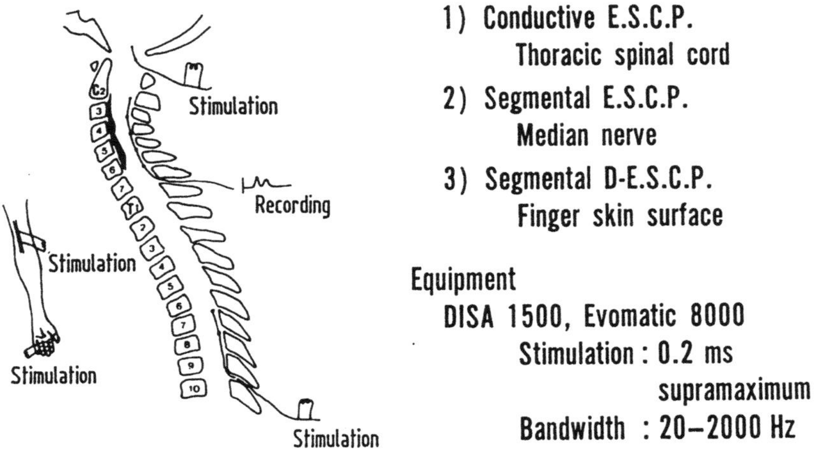

Fig. 1. Evoked spinal cord potentials (*ESCP*) system

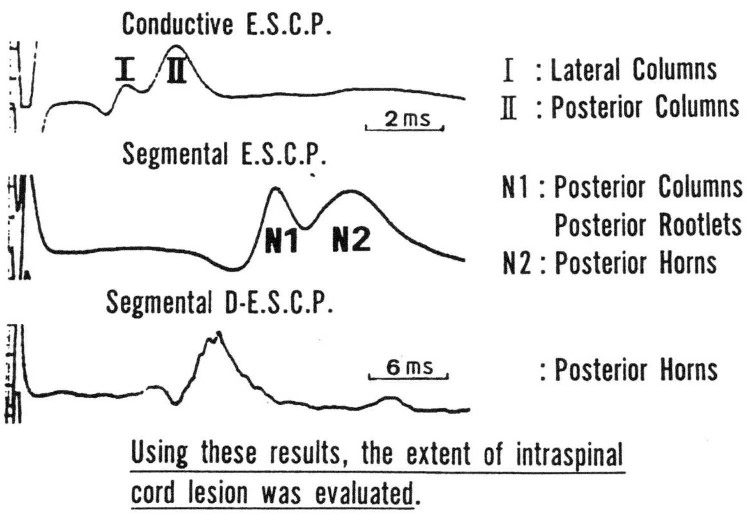

Fig. 2. Pattern of evoked spinal cord potentials (*ESCP*)

2000 Hz. About 30–50 stimuli for the conductive and segmental ESCP and 100–500 stimuli for the segmental D-ESCP were applied and averaged, using DISA 1500 or Evomatic 8000 (Dantec, Denmark). Five monopolar recordings were obtained simultaneously from the five poles of the recording electrode. A reference electrode was placed in the subcutaneous tissue of the neck. The extent of the spinal cord disorder or lesion responsible was diagnosed on the basis of these recordings because the recording electrode covered from C3 to C7 in most cases (Fig. 1).

We have reported that the conductive ESCP and the N1 of the segmental ESCP originate in the white matter and that the N2 of the segmental ESCP and most of the segmental D-ESCP originate in the gray matter [2, 4]. On the basis of these results, we evaluated the extent of the patients' intraspinal cord lesions (Fig. 2).

In Japan, cervical spondylotic myelopathy is assessed using an evaluation score, the JOA score, devised by the Japan Orthopedic Association (maximum score is 17). This score was used to evaluate the patients' function before and after surgery (Table 1). The recovery rate was also calculated by the JOA system.

Table 1. Japan Orthopaedic Association (JOA) score

1) Motor function of upper extremity
   - 0: Patient cannot eat with chopsticks or spoon
   - 1: Patient can eat with a spoon, but not with chopsticks
   - 2: Patient can eat with chopsticks, though not easy
   - 3: Patient eats with chopsticks, but in an awkward manner
   - 4: Normal

2) Motor function of lower extremity
   - 0: Patient cannot walk
   - 1: Patient needs a support to walk on a flat area
   - 2: Patient needs no support on a flat area, but needs it when going up and down the stairs
   - 3: Patient can walk without a support, but in an awkward manner
   - 4: Normal

3) Sensation
   - A: Upper extremity
     - 0: Distinct sensory disturbance
     - 1: Slight sensory disturbance
     - 2: Normal
   - B: Lower extremity same evaluation as for upper extremity
   - C: Trunk same evaluation as for upper extremity

4) Bladder function
   - 0: Retention of urine
   - 1: Severe urinary disturbance
   - 2: Mild urinary disturbance
   - 3: Normal

Results

Difference Between Anterior and Posterior Procedures in Terms of Postoperative Results

First of all, we attempted to determine whether postoperative results are affected by the surgical procedure or not. The postoperative results of both anterior and posterior procedures were as follows:

Anterior surgery:
- Preoperative JOA score : 8.7 ± 3.3
- Postoperative JOA score: 14.5 ± 2.5
- Recovery rate : 70%

Posterior surgery:
- Preoperative JOA score : 8.5 ± 3.4
- Postoperative JOA score: 13.0 ± 2.5
- Recovery rate : 53%

According to these results, anterior surgery seems to be slightly superior to posterior surgery, but there were only some statistical differences between the two procedures ($P < 0.05$).

Difference Between Surgery on One Level (One-Level Surgery Group) and over Two Levels (Multilevel Surgery Group)

Among the anterior surgery patients, 39 underwent anterior fusion at one level, and the other 62 had anterior decompression over two levels. The multilevel surgery group consisted of this anterior decompression group (62), a posterior surgery group of 18 patients, and a combined group of 2 patients, for a total of 82 patients. The one-level surgery group contained 39 anterior fusion cases. The postoperative results of both the one-level and the multilevel surgery groups were as follows:

One-level surgery group:
- Preoperative JOA score : 9.1 ± 3.5
- Postoperative JOA score: 14.7 ± 2.5
- Recovery rate : 71%

Multilevel surgery group:
- Preoperative JOA score : 8.4 ± 3.3
- Postoperative JOA score: 14.0 ± 2.7
- Recovery rate : 65%

In general, it is believed that multilevel-lesion patients tend to have more serious symptoms and recover less than one-level lesion cases. There is no significant statistical difference between the one-level and multilevel surgery groups. We believe this is because not all patients in our multilevel surgery group had multilevel lesions. The multilevel surgery group consisted of both posterior procedure patients, whose responsible lesion was on one level, and anterior procedure cases, in which precautionary anterior fusion was performed on a disc adjacent to the responsible lesion.

Diagnosis of the Extent of the Lesion Using Evoked Spinal Cord Potentials

In order to make a diagnosis of the extent of the lesion conductive ESCP, segmental ESCP, and segmental D-ESCP have all been used. Patients in whom all the ESCP indicated marked changes at the same level resulted in the diagnosis of a one-level lesion. The others showed inconsistencies in wave changes among these three ESCP or between the right and left segmental ESCP, so diagnosis of a multilevel lesion was made. All 39 patients in the one-level fusion group were counted as the one-level lesion group. In the multilevel surgery group, only 29 individuals were included in the multilevel lesion group, which the other 53 were added to the one-level lesion group. Ultimately, it was concluded that the one-level lesion group consisted of 92 patients (76%), while the multi-level lesion group included 29 (24%). The differences between the two groups in postoperative outcome were as follows:

One-level lesion group (92 patients):
- Preoperative JOA score : 9.3 ± 3.2
- Postoperative JOA score: 15.0 ± 1.9
- Recovery rate : 74%

Multilevel lesion group (29 patients):
- Preoperative JOA score : 6.4 ± 3.0
- Postoperative JOA score: 11.5 ± 2.7
- Recovery rate : 48%

The one-level lesion group had a statistically higher postoperative JOA score and a recovery rate superior to the multilevel-lesion group ($P < 0.01$). These results appear to be of extreme importance for spinal surgeons since they can correctly predict outcome before surgery.

Case Presentation

Case 1: One-Level Lesion

A 65-year-old woman (Fig. 3) had a preoperative JOA score of 0, 2, 1, 1, 1, 1 = 6. Myelography revealed indentation at C3–4, C4–5, C5–6, and C6–7.

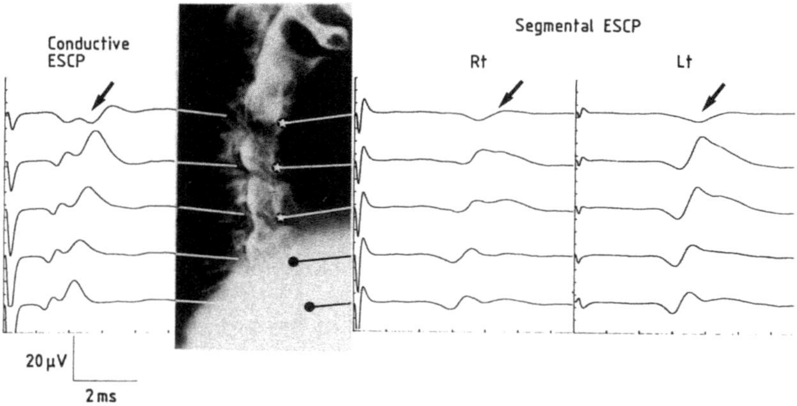

Fig. 3. Case 1. Conductive and segmental evoked spinal cord potentials (*ESCP*) in a 65-year-old woman with cervical spondylotic myelopathy. *RT*, right; *LT*, left

Conductive and segmental ESCP showed a marked decrease in amplitude at C3–4 and normal wave patterns at C4–5, C5–6, and C6–7. According to these results, we diagnosed the lesion responsible as being at only C3–4. However, precautionary anterior decompression from C3 to C7 was performed because there was marked degeneration at the adjacent disks and malalignment of the cervical spine. Her postoperative JOA score was 4, 4, 1, 2, 2, 3 = 16, and recovery rate was satisfactory (91%).

Case 2: Multilevel-Lesion

A 69-year-old woman (Fig. 4) had a preoperative JOA score of 1, 0, 0, 0, 2, 1 = 4. Myelography showed an indentation at C4–5, C5–6, and C6–7. The primary lesion appeared to be at C4–5 on the basis of findings of conductive and segmental ESCP. However, the wave pattern of the right segmental ESCP at C5 and C6 apparently changed compared with the left segmental ESCP. In order to confirm responsibility of the C5–6 and C6–7 discs, segmental D-ESCP were also recorded, and marked wave changes were observed at C5 and C6. According to these results, she was diagnosed as having a multilevel lesion. After anterior decompression from C4 to C7, she recovered gradually, but her final JOA score was 3, 2, 1, 2, 2, 2 = 12, and recovery rate was only 61%, which could not be considered as a satisfactory results.

Discussion

There have been several reports related to prediction of the outcome of cervical spondylotic myelopathy. Our long-term follow-up study of cervical

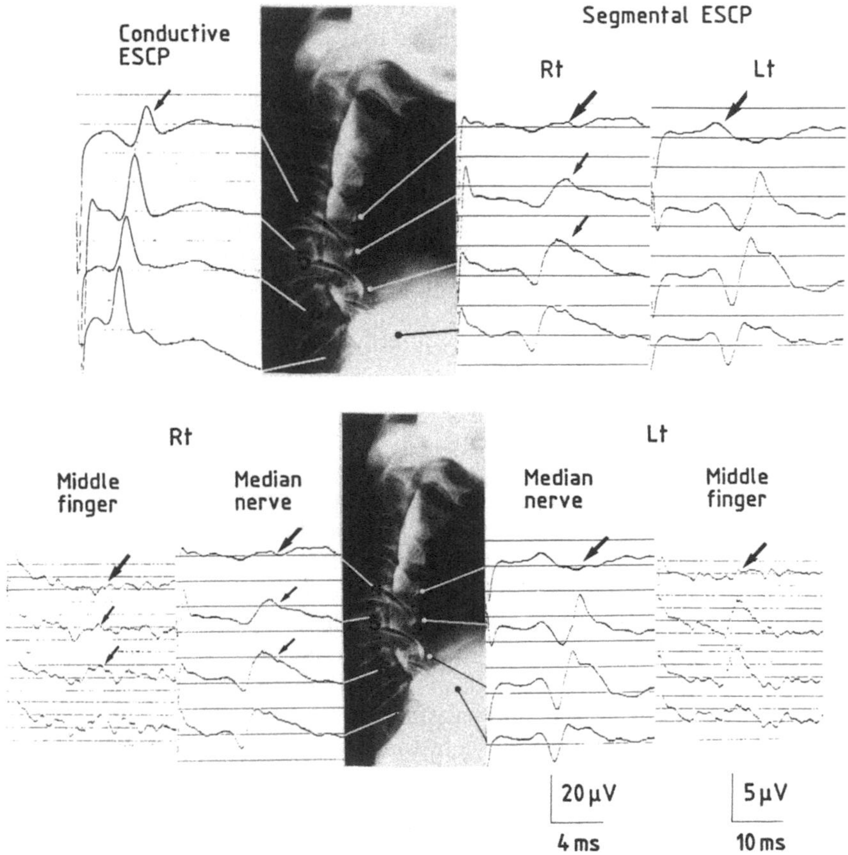

Fig. 4. Case 2. Conductive and segmental evoked spinal cord potentials (*ESCP*) in a 69-year-old woman with cervical spondylotic myelopathy. *RT*, right; *LT*, left

spondylotic myelpathy concluded that clinical factors, such as the duration of the period from onset to surgery and a history of spinal cord injury, were closely related to outcome after surgery [3]. It is difficult, however, to prognosticate postoperative results using these clinical factors since there were, in fact, wide variations in patient recovery.

We have been carrying out examinations of evoked spinal cord potentials in order to determine the extent and severity of spinal cord disorders because we believe that this electrophysiological examination directly and more correctly reflects the spinal cord function than other examinations, including clinical neurologic examination and patients' histories. At first, we applied this examination to making a level diagnosis. We began to notice, however, that severe wave changes, such as decreases in conduction velocity and positive wave changes, had an influence upon outcome after surgery as we continued to perform these examinations. Moreover, it became clear that these severe wave changes are intimately related to the extent of the spinal disorder, as shown in this study.

Our method appears to have significant diagnostic value with respect to the extent of the spinal cord disorder. There were a few cases, however, in which the responsible lesion could not be determined using this method because the evoked spinal cord potentials disappeared at a lower cervical level, and evaluation above the region could not be achieved. In these few cases, especially the continuous OPLL cases, we performed the following procedure as well: another stimulating electrode was placed in the cisterna magna, and a descending conductive ESCP was recorded in order to determine the upper limit of the spinal cord lesion [5].

We encountered unsatisfactory multioperated patients with severe cervical canal stenosis. In these cases, postoperative results after initial anterior fusion surgery on one level were satisfactory, but recurrent symptom worsening due to lesions at adjacent discs did not respond satisfactorily to repeated surgery. This means that repeated compressions of the spinal cord at multiple levels tend to result in extensive lesions and severe damage. Ono et al. reported five cases of severe cervical spondylotic myelopathy and multiple levels of spinal cord compression in all five patients at autopsy [1]. This pathological report is in accordance with our own electrophysiological findings.

The following conclusions are indicated on the basis of this electrophysiological diagnostic study: First, most of the patients with cervical myelopathy (76%) had localized lesions and satisfactory postoperative results. Second, fewer patients (24%) had extensive lesions and unsatisfactory results regardless of the surgical procedure. The discrepancy in postoperative recovery between the two groups was striking. As shown in case 1, it was difficult to discriminate between these groups on the basis of other medical imagings.

This method is invasive and risky if performed by an inexperienced physician and without patients being fully conscious. We have not experienced any significant complications, however, and extremely important findings can be obtained using our method. These findings have meant rapid progress in the accurate diagnosis of cervical myelopathy. We are, nevertheless, currently developing a new, less invasive method for level diagnosis.

Summary

1. Evoked spinal cord potentials have been used to diagnose the extent of spinal cord disorders in cervical myelopathy.
2. Of cervical myelopathy patients 76% were found to have localized lesions and a satisfactory outcome after surgery.
3. Of the patients 24% had extensive lesions and unsatisfactory results regardless of excellent surgical procedures.

References

1. Ono K, Ota H, Tada K, Yamamoto T (1977) Cervical myelopathy secondary to multiple spondylotic protrusions. Spine 2:109–125
2. Shinomiya K, Furuya K, Yamaura I et al. (1984) Spinal cord monitoring of spinal cord function using evoked spinal cord potentials. In: Honma S, Tamaki T (eds) Fundamentals and clinical application of spinal cord. Saikon Publ, Tokyo, pp 161–173
3. Shinomiya K, Sato R, Sato M et al. (1984) Long-term results following anterior cervical fusion for cervical spondylotic myelopathy. Orthopedic Surgery 35:509–517
4. Shinomiya K, Furuya K, Sato H et al. (1988) Diagnosis of cervical myelopathy using segmental evoked spinal cord potentials obtained by stimulating finger surface. In: Ducker TB, Brown RH (eds) Neurophysiology and standards of spinal cord monitoring. Spinger, Berlin Heidelberg New York Tokyo, pp 62–70
5. Shinomiya K, Furuya K, Sato R et al. (1988) Electrophysiologic diagnosis of cervical OPLL myelopathy using evoked spinal cord potentials. Spine 13:1225–1233

Intraoperative Spinal Cord Monitoring Using Evoked Spinal Cord Potentials on Spinal Cord Tumors

Y. Fukui[1], K. Satomi, T. Okuma, K. Kenmotsu, M. Kamata, T. Ikai, and K. Hirabayashi

Summary

Intraoperative spinal cord monitorings using evoked spinal cord potentials (ESCP) have been performed in 19 patients with spinal cord tumors, including five intramedullary tumors and two arteriovenous malformations (AVM) since 1984. There were 10 men and 9 women between the ages of 22 and 73 years, with an average of 49 years. ESCP resulting from median or tibial and peroneal nerve and/or spinal cord stimulation were used for monitoring according to the spinal level of tumors.

Fifteen patients were monitored by conductive ESCP which traveled over the site of the tumor in the spinal cord. There were no remarkable changes in ESCP while monitoring during surgery in 13 patients; however, in one patient, ESCP increased in latency after removal of the tumor, and in another patient with AVM, ESCP disappeared at the time of clipping of the feeder artery. Although conductive ESCP could not be recorded in four patients, spinal cord was monitored by MN-ESCP in one, which resulted in failure.

Spinal cord monitoring using ESCP is necessary for the surgery of spinal cord tumors, especially in intramedullarly tumors and AVM.

Introduction

Recently, the value and importance of electrophysiological investigations in the surgery of the spinal cord have been well emphasized. Spinal cord tumor is one of the diseases which can only be cured by surgical removal by neurosurgeons and orthopaedic surgeons. However, it is a great problem if paresis is exaggerated after the removal of the spinal cord tumor. In order to prevent postoperative neurological problems, intraoperative spinal cord

[1] Department of Orthopedics, Keio University School of Medicine, 35 Shinano-machi, Shinjuku-ku, Tokyo, 160 Japan

K. Shimoji et al. (Eds.)
Spinal Cord Monitoring and Electrodiagnosis

monitorings using evoked spinal cord potentials (ESCP) have been performed in 19 patients with spinal cord tumors since 1984.

The purpose of these monitorings is to evaluate the possibility of the ESCP as an indicator of neurological prognosis during surgery of spinal cord tumors. Five different potentials were recorded. Spinal cord-ascending evoked spinal cord potentials (SC-AESCP) were obtained by stimulating the spinal cord at the thoracolumbar region and recording rostrally in the cervical region. Spinal cord-descending evoked spinal cord potentials (SC-DESCP) were obtained by stimulating the spinal cord at the cervical region and recording caudally in the thoracolumbar region. Median nerve evoked spinal cord potentials (MN-ESCP) were obtained by unilateral stimulation of the median nerve. Spinal cord-evoked peroneal (tibial) nerve potentials [SC-EPN(TN)P] were obtained by stimulating the spinal cord at the lower thoracic region and recording in the peroneal (tibial) nerve.

Materials and Methods

There were 10 men and 9 women between the ages of 22 and 73 years, with an average age of 49 included in the study. Pathological diagnoses of tumors were as follows: neurinoma in 11 patients, ependymoma in 3, astrocytoma in 1, lipoma in 1, hemangioma in 1, and arteriovenous malformation (AVM) in 2. According to site, the tumors in the spinal canal were intradural extramedullary tumor (EMT) in six patients, intramedullary tumor (IMT) in five, intra-and extradural tumor (so called dumbbell tumor, IET) in three, extradural tumor (EDT) in one, cauda equina tumor (CET) in two, and AVM in two (Table 1).

ESCP resulting from median or tibial and peroneal nerve and/or spinal cord stimulation were used for monitoring the spinal cord according to the spinal level of tumors. The stimulus was a square-wave constant current or voltage impulse 0.2 ms in duration. The intensity was about five times motor threshold. ESCP were recorded from monopolar electrodes introduced into the dorsal epidural space or the yellow ligament (Fig. 1).

ESCP were recorded every five min from the beginning of the laminectomy until closing of the dura, and any alteration of ESCP in amplitude or latency was carefully observed. Clinical symptoms were examined neurologically after surgery.

Results

Fifteen patients were monitored by conductive ESCP which traveled over the site of the tumor in the spinal cord. There were no remarkable changes in ESCP monitored during surgery in thirteen patients; however, in one, ESCP

Table 1. Materials and results

Patient	Age	Sex	Pathological diagnosis	Site	Level	ESCP	Stimulation	Recording	Result
1	66	F	Neurinoma	EMT	C3–5	SC-DESCP	C2–3	T12	Within normal limit (WNL)
2	49	M	Neurinoma	EMT	C6–T1	SC-DESCP	T10	L4–5	WNL
3	47	F	Neurinoma	EMT	C7–T2	SC-AESCP	T8–9	C4–5	WNL
4	49	F	Neurinoma	EMT	C3–4	SC-AESCP	L1–5	C1–2	WNL
5	58	M	Neurinoma	EMT	T10–11	SC-AESCP	L4–5	T9	WNL
6	43	F	Neurinoma	EMT	T12–L2	SC-DESCP	C5	L4–5	ESCP not recorded
7	65	M	Ependymoma	IMT	T8–10	SC-DESCP	C4–5	T11–12	WNL
8	31	M	Lipoma	IMT	C6–T3	SC-DESCP	C5–6	T12	WNL
9	52	M	Ependymoma	IMT	C6–T5	MN-ESCP	MN	C5–6	Monitoring false negative
10	42	M	Ependymoma	IMT	C4–6	SC-AESCP	T9	C2–3	Latency extended at the excision
11	38	M	Astrocytoma	IMT	T10–12	SC-DESCP	T9	L1	ESCP not recorded
12	45	F	Neurinoma	IET	C4–5	SC-AESCP	L2	C2–3	WNL
13	41	F	Neurinoma	IET	C5–6	SC-AESCP	L2	C2–3	WNL
14	73	M	Neurinoma	IET	C4–5	SC-AESCP	T11	C1–2	WNL
15	56	F	Hemangioma	EDT	T5	SC-DESCP	T11	L1–2	WNL
16	44	M	Neurinoma	CET	L2	SC-ETNP	T12–1	TN	Conductive ESCP not recorded
17	65	F	Neurinoma	CET	T12–L2	SC-DESCP	T9	L4–5	WNL
18	22	F	AVM	AVM	C6	SC-AESCP	T9	C2–3	WNL
19	43	M	AVM	AVM	T1–L2	SC-DESCP	T4	T11	ESCP disappeared at the clipping

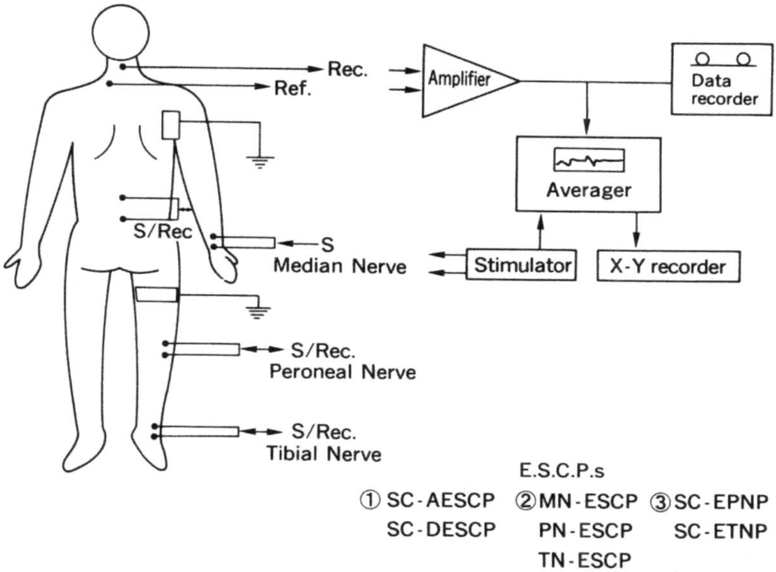

Fig. 1. The recording system. Stimulation is applied either to the median, peroneal, and tibial nerve or to the spinal cord. Recording is acquired from electrodes introduced into interlaminar ligaments or from bipolar electrodes introduced into the dorsal epidural space. Reference (*Ref*) needle electrode is put in the paraspinal muscle. *S*, stimulation; *Rec*, recording

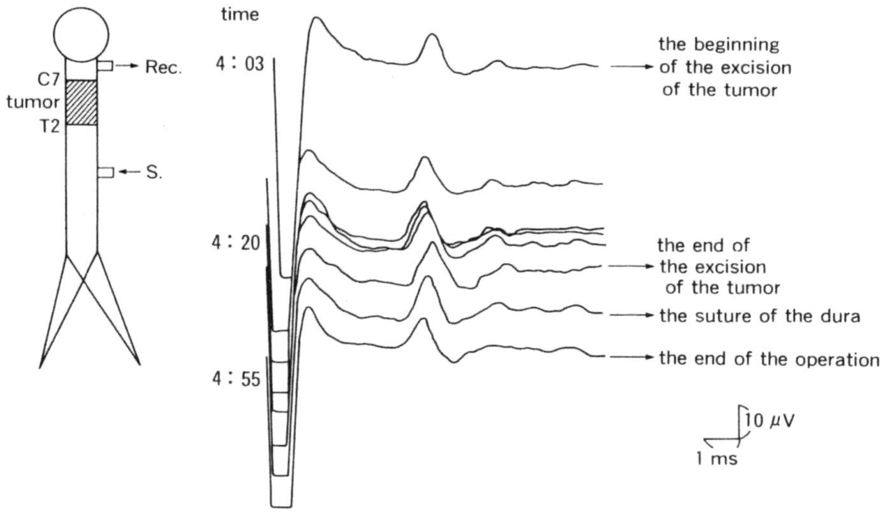

Fig. 2. Spinal cord monitoring using SC-AESCP recorded from a patient with intradural extramedullary tumor (Case 3). ESCPs consisted of a large negative potential and subsequent small potentials, and they had no remarkable change in amplitude and in latency during the surgery. *S*, stimulation; *Rec*, recording

increased in latency after the removal of the tumor and in another patient with AVM, ESCP disappeared at the time of clipping of the feeder artery. Although conductive ESCP could not be recorded in four patients, spinal cord was monitored by MN-ESCP in one, which resulted in failure. Results in detail according to the site of tumors are shown below.

Intradural Extramedullary Tumor

In one patient, ESCP were not recorded, but in the other five, the ESCP were almost normal during surgery, and neurological improvements after surgery were remarkable in these patients.

Case 3. The patient was a 47-year-old woman with intradural extramedullary tumor between C7 and T2. Spinal cord was monitored using AESCP, which had no remarkable change in amplitude and in latency during surgery (Fig. 2).

Intramedullary Tumor

In four patients, the ESCP were almost within normal limits, and there were no neurological problems after surgery. The spinal cord monitorings ended successfully in these patients.

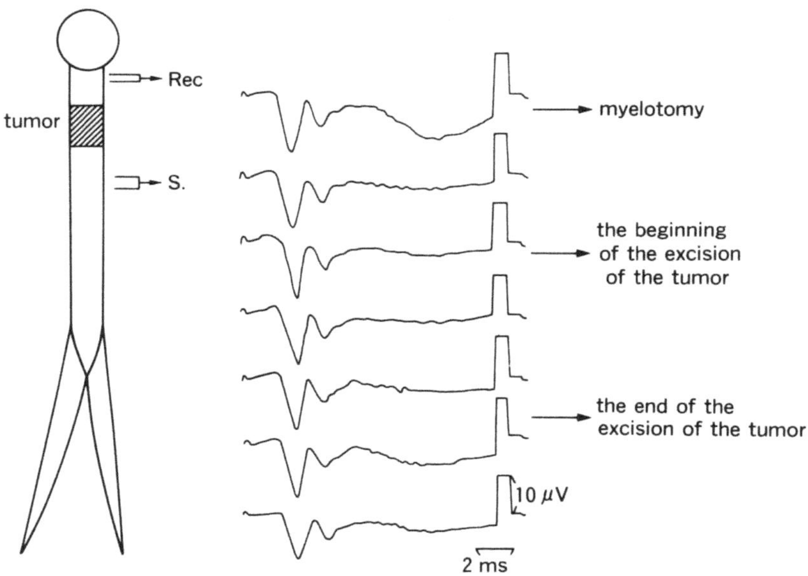

Fig. 3. Spinal cord monitoring using SC-AESCP recorded from a patient with intramedullary tumor (Case 10). The ESCP was initially a large positive deflection with small negative potentials, but the wave pattern was almost unchanged. *S*, stimulation; *Rec*, recording

Case 10. The patient was a 42-year-old man with an intramedullary tumor between C4 and C6. The ESCP was initially a large positive deflection with small negative potentials. The wave pattern of AESCP was almost unchanged during surgery. The patient was able to walk after surgery (Fig. 3).

Conductive ESCP were not recorded in two patients, but in one of them, spinal cord was monitored by MN-ESCP, which resulted in failure.

Case 9. The patient was a 52-year-old man with intramedullary tumor between T1 and T2. MN-ESCP was used as a monitoring wave because conductive ESCP was not recorded. The wave pattern was almost unchanged during the surgery (Fig. 4), but the patient remained paraplegic postoperatively. So we could not monitor the spinal cord by using MN-ESCP that did not conduct over the tumor.

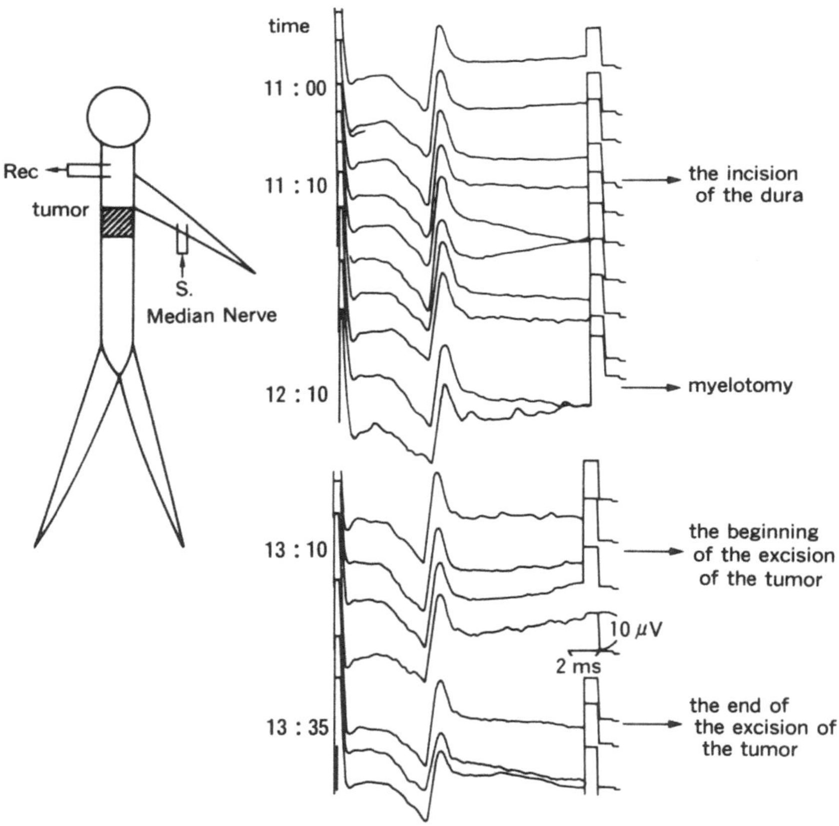

Fig. 4. Spinal cord monitoring using MN-ESCP recorded from a patient with intramedullary tumor (Case 9). The wave pattern of the MN-ESCP was almost unchanged during surgery, but the patient remained paraplegic postoperatively. It was impossible to monitor the spinal cord by using MN-ESCP that did not conduct over the tumor in this case. *S*, stimulation; *Rec*, recording

Dumbbell and Extradural Tumor

The ESCP were normal in all four patients with dumbbell and extra-dural tumors and there were no neurological problems after the surgery.

Arteriovenous Malformation

Spinal cord monitoring was performed successfully in one patient, but in another, the monitoring wave disappeared when the feeder artery was clipped.

Case 19. The patient was a 43-year-old man with AVM between T1 and T2. The potentials were completely flattened after clipping of the feeder artery

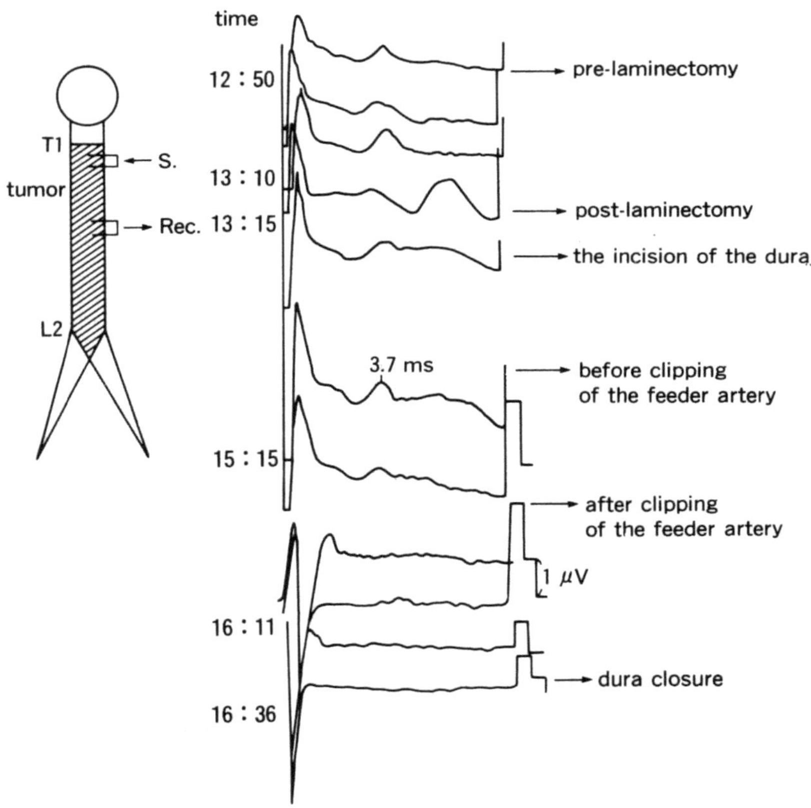

Fig. 5. Spinal cord monitoring using SC-DESCP recorded from a patient with AVM (Case 19). The potentials were completely flattened after clipping of the feeder artery. This patient woke up with paraplegia, but recovered 2 years after surgery with no neurological deficit. *S*, stimulation; *Rec*, recording

(Fig. 5). This patient woke up with paraplegia, but recovered 2 years after surgery with no neurological deficit except spasticity. This showed that the disappearance of the ESCP does not always imply complete paraplegia of the spinal cord, but neurological deficits for a short period after surgery.

Cauda Equina Tumor

Spinal cord was monitored successfully in two patients with cauda equina tumors. In one of them, SC-ETNP was used for monitoring instead of conductive ESCP.

Case 18. The patient was a 44-year-old man with cauda equina tumor at the level of L2. Potentials were recorded from bilateral peroneal (SC-EPNP) and tibial nerves (SC-ETNP) following stimulation of the spinal cord. Potentials were low in amplitude and atypical in wave pattern (Fig. 6), but they did not change during surgery and there were no neurological problems afterwards.

Discussion

Spinal cord monitoring using ESCP has been used for preventing postoperative paresis due to excessive distraction of a spinal cord which has been

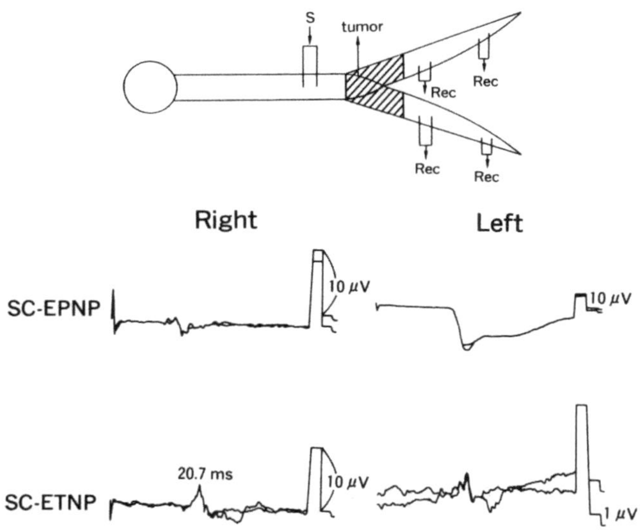

Fig. 6. SC-EPNP and SC-ETNP recorded from a patient with cauda equina tumor (Case 15). Both potentials generally consisted of multiphasic negative potentials and were used for spinal cord monitoring. *Left* SC-EPNP showed large positive potentials with unknown etiology in this case. *S*, stimulation; *Rec*, recording

functioning normally preoperatively in corrective surgery for scoliosis [3, 6], while monitoring the spinal cord which already has partial functional loss, such as spinal cord tumors and cervical myelopathy [7], has not been well established. In this study, we tried to record ESCP following stimulation of spinal cord, which traveled beyond the lesion in the spinal cord, associated with spinal cord tumor. However, conductive ESCP were not recorded in 4 of 19 patients, although the distance between the stimulation and recording electrodes was shortened. In these patients, a new method for monitoring should be established.

Recently, postoperative neurological deficits have been reported in association with unchanged SEP monitoring during surgery [1, 2]. One of the reasons for these false negative monitorings is that the injured site of the spinal cord might not be detected by an operative procedure. In fact, reliability in evaluating spinal cord function by means of ESCP still remains limited because the ESCP does not conduct in motor pathways, but rather in sensory pathways; in other words, in the dorsal and dorsolateral columns of the spinal cord. Therefore the wave that conducts in motor pathways must be obtained. However, we have had no experience in recording motor action potentials following transcranial stimulation of the motor cortex, either electrically or magnetically. So in the patient with cauda equina tumor, SC-EPNP and SC-ETNP were recorded. The origin and pathway of these waves are unknown, but there is a possibility that these waves can be used for spinal cord monitoring.

References

1. Ginsburg HH, Shetter AG, Raudzens DA (1985) Post-operative paraplegia with preserved intraoperative somatosensory evoked potentials: A case report. J Neurosurg 63:296–300
2. Lesser RP, Raudzens P, Luders H et al. (1986) Postoperative neurological deficits may occur despite unchanged intraoperative somatosensory evoked potentials. Ann Neurol 19:22–25
3. Machida M, Weinstein SL, Yamada T, Kimura I (1985) Spinal cord monitoring, electrophysiological measures of sensory and motor function during spinal surgery. Spine 10:407–413
4. Machida M, Weinstein SL, Yamada T et al. (1988) Dissociation of muscle action potentials and spinal somatosensory evoked potentials after ischemic damage of spinal cord. Spine 13:1119–1124
5. Owen JH, Laschinger J, Bridwell K et al. (1988) Sensitivity and specificity of somatosensory and neurogenic-motor evoked potentials in animals and humans. Spine 13:1111–1118
6. Ryan TP, Britt RH (1986) Spinal and cortical somatosensory evoked potential monitoring during corrective surgery with 108 patients. Spine 11:352–361
7. Satomi K, Okuma T, Kenmotsu K et al. (1988) Level diagnosis of cervical myelopathy using evoked spinal cord potentials. Spine 13:1217–1224

Clinical Correlations of Intraoperative Evoked Potential Monitoring in Spinal Cord Disorders

J.R. Cassidy[1] and T.B. Ducker

In the last decade, myelopathic patients or patients at high risk for spinal cord compression due to tumor, cyst, focal disc herniation, or degenerative disease have had evoked ascending sensory monitoring during decompressive operations. Personal operative experience (T.B.D.) for spinal cord pathology utilized the monitoring in 338 patients. A total of six patients are presented in whom changes in the monitored evoked potentials have either prompted the surgeon to alter his technique, or correctly predicted postoperative neurologic deficits. These cases are discussed with reference to the sensitivity and accuracy of somatosensory evoked potential (SSEP) monitoring.

Materials and Methods

Of 1700 patients who have undergone intraoperative SSEP monitoring at the Johns Hopkins Hospital and University of Maryland hospital in the last 10 years, 338 were operated upon by the senior author. The recordings were obtained with a Nicolet Med 80 monitor, and electrodes were routinely placed over the contralateral scalp and second cervical vertebra. Administered at a frequency of 5.9/s, 128 stimuli were usually acquired and then averaged to obtain reproducible waveforms.

A significant change in the SSEP was designated as a decrease in the peak amplitude of more than 50%. Changes in the latency, or loss of individual components of the waveform were not used as criteria.

In many instances, the immediate preoperative SSEP waveform was unobtainable. This was usually not due to a technical problem in applying the monitoring, but more commonly was a function of the patient's poor preoperative neurologic status. Nevertheless, with the administration of etomidate, acceptable tracings could be obtained in all but 14 patients [5].

[1] Department of Neurological Surgery, Johns Hopkins University School of Medicine, 600 North Wolfe Street, Baltimore, MD 21205, USA

K. Shimoji et al. (Eds.)
Spinal Cord Monitoring and Electrodiagnosis

Results

A decrease in the amplitude of the SSEP was detected intra-operatively on many occasions in which a clear etiology could be identified that was unrelated to the surgical technique. The most common causes were transient hypotension and oversedation from halogenated anesthetics. Occasionally, uncorrectable technical difficulties were discovered that accounted for the blunting of the tracings. However, in the six cases outlined below, the decrease in the amplitude of the tracings could be clearly temporally related to some aspect of the surgical procedure.

Case 1. P.M. is a 55-year-old woman with a 35-year history of rheumatoid arthritis. She is steroid dependent and has undergone multiple joint replacement surgeries and a C1–2 fusion for subluxation and basilar invagination. She presented with a 7-week history of numbness and tingling in all four extremities, but most severe in her right arm. She also reported weakness in her legs and difficulty walking. She denied bowel or bladder symptoms. Neurologic examination revealed signs of myelopathy, except for lower motor neuron involvement of the right arm. An MRI scan of the midcervical region clearly demonstrated a retrolisthesis of C5 on C6. The disc material appeared badly degenerated, with some loss of height of the C6 body, suggesting rheumatoid inflammatory involvement of that joint. An anterior cervical decompression and fusion was planned. Satisfactory SSEP tracings were maintained throughout the procedure, until the discectomy was com-

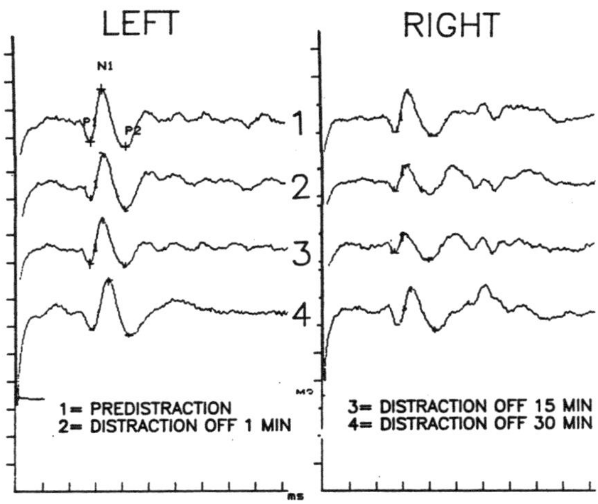

Fig. 1. Posterior tibial SSEP tracings of myelopathic rheumatoid patient undergoing anterior cervical discectomy and fusion with 8-mm iliac crest graft. SSEP decreased with distraction of the disc space. When smaller graft was fashioned and placed, SSEP returned to baseline

plete. At that time, the Cloward intervertebral spreader was placed and the disc space distracted while an appropriately sized tricortical iliac crest graft was fashioned. The amplitude of the SSEP began to decrease immediately after placement of the intervertebral spreader and continued to decrease to less than 50% of baseline after the graft was placed. In response to this change, the graft was removed, the distraction relieved and a shorter graft placed. The SSEP quickly returned to their original amplitude (Fig. 1). No appreciable worsening of the patient's preexisting deficit was detected in the postoperative neurologic examination.

Cases 2 and 3. In two patients with severe cervical spondylotic myelopathy, the amplitude of the SSEP tracings decreased just after the patient was turned and positioned in Mayfield three-point fixation. In neither patient was the neck excessively flexed, extended, or rotated during the process of turning the patient. Consequently, it was felt that their cervical spines must have been fixed in a slightly extended position and this caused cervical cord compression. In response to the changes in the SSEP, the patients were carefully repositioned such that their necks were slightly flexed with approximately 10 lbs of traction applied. In both instances the SSEP quickly recovered, and neither patient awakened with an exacerbation of their myelopathy.

Case 4. The patient is a 39-year-old woman who developed a progressive paraparesis in 1974, prompting surgical exploration and tube drainage of a spinal cord syrinx. To a large extent, the patient's symptoms resolved, until 1983 when she again developed a spastic gait, as well as bladder incontinence. CT myelopgraphy demonstrated an enlarged thoracic spinal cord, but no accumulation of intramedullary contrast material was seen, even on delayed films. The patient was taken to the operating room where a thoracolumbar laminectomy was performed. The dura was markedly thickened and scarred, and careful dissection was required to free it from the underlying spinal cord without compromising the neural tissue. A midline myelotomy was performed and the previously placed syringostomy tube removed. The tumor was identified and circumscribed using microsurgical technique. It extended from about L2 to T10 with smaller fragments positioned dorsally in the spinal cord at the most cephelad region of the tumor. The SSEP from the time of induction of anesthesia to the final stages of the tumor dissection remained unchanged. Unfortunately, the SSEP were lost as the last portions of tumor were removed from their rostral and dorsal location. Postoperatively, the patient's stength and spasticity slowly improved by one grade, although the neurogenic bladder remained. She has been left with profound loss of position sense in her lower extremities such that her gait is now broad based and she must constantly watch the floor to maintain her balance. An MRI of this region was recently obtained (Fig. 2) which clearly demonstrates atrophy of the dorsal region of the spinal cord, with relative preservation of the ventral horns.

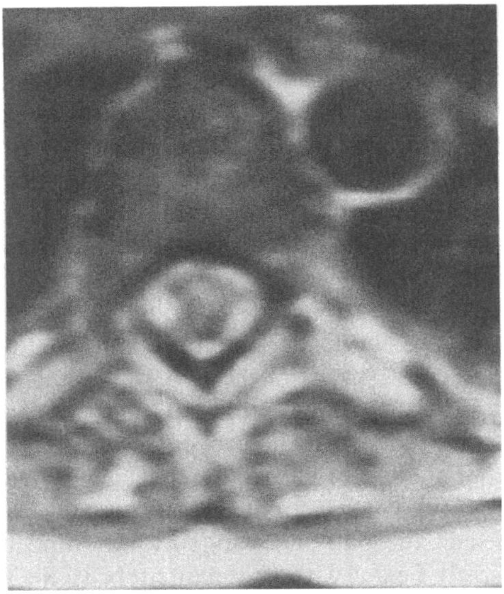

Fig. 2. MRI obtained 5 years after radical excision of thoracic hemangioblastoma. SSEP were unobtainable at the end of the procedure. The patient's strength improved postoperatively although she had profound loss of position sensation. MRI demonstrates marked atrophy of dorsal region of the spinal cord

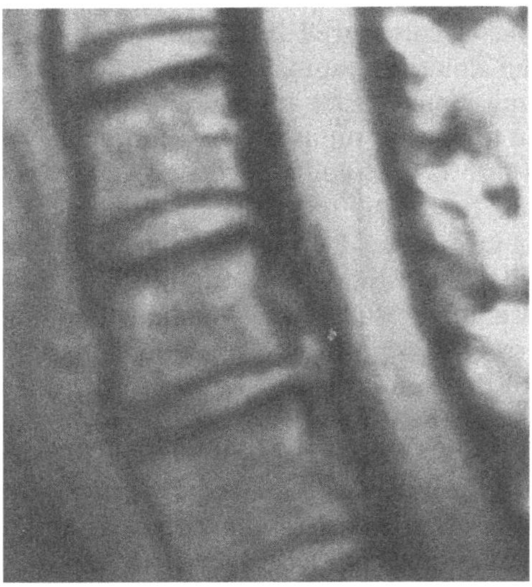

Fig. 3. Large soft central cervical disc herniation in a man with radiculopathy but no myelopathy. An attempt at removing the disc from a posterior approach resulted in a dramatic decrease in the amplitude of the SSEP every time the axilla of the nerve root was retracted superomedially. The disc was subsequently removed uneventfully via an anterior approach

Case 5. The patient is a 42-year-old, previously healthy laborer who presented with left arm and interscapular pain. He reported numbness and paresthesias in the second and third digits of his left hand. Physical examination revealed an absent triceps reflex and weakness of extension in that arm. No signs or symptoms of myelopathy were elicited. Diagnostic workup included an MRI of the cervical spine, and this showed a large soft central disc herniation at C6–7 (Fig. 3). The patient was taken to the operating room, and with the assistance of SSEP monitoring, the lesion was approached posteriorly via a keyhole foraminotomy. The nerve root was readily exposed and retracted upward to expose the disc space. The disc was not easily retrieved, however, and whenever any medial retraction was used to gain better exposure, the SSEP would decrease dramatically, with immediate recovery once the retraction was stopped. A decision was made to abort the posterior procedure and immediately to perform an anterior cervical discectomy and fusion. The anterior approach proceeded uneventfully, and the patient suffered no sequelae postoperatively. His arm pain was completely relieved, and his triceps strength improved to normal.

Case 6. The patient is a 55-year-old woman with a 1-year history of burning and numbness in her legs. She also complained of pain in her lower back. She denied weakness or bowel and bladder symptoms. Physical examination was significant for mild spasticity and hyperactive reflexes in the lower extremities. Plantars were downgoing. CT myelography demonstrated a large calcified T9–10 disc protrusion, with pronounced dorsal displacement of the spinal cord, and obliteration of the myelographic contrast column. A costotransversectomy was performed and a high-speed drill used to remove the mass. The surgery went smoothly, and no untoward intraoperative event was appreciated. The SSEP gradually decreased to approximately 50% of their baseline

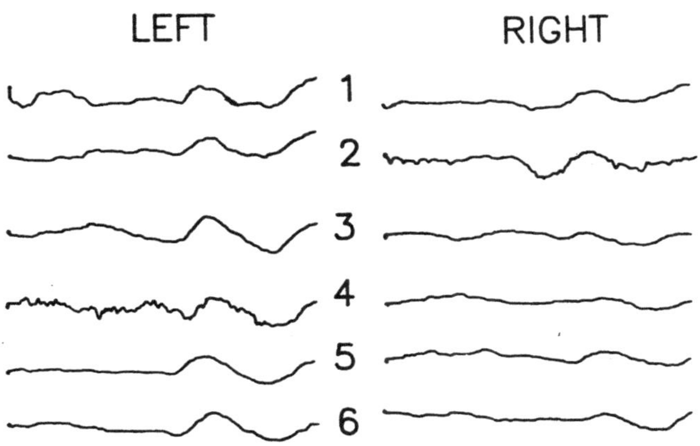

Fig. 4. These SSEP demonstrate only modest reduction in their amplitude on the *right* side only. This patient had an anterolateral approach to a thoracic disc. Post-operatively she had severe weakness of the ipsilateral leg, and no loss of posterior column function

amplitude, on the right side only, and then recovered to approximately 70% of baseline by the end of the operation (Fig. 4). The significance of these changes was not fully appreciated during the procedure, and profound weakness of the right leg was apparent postoperatively. There was mild loss of pin sensation, but no loss of proprioception at all. The patient eventually enjoyed near complete recovery of her motor deficit.

Discussion

The use of SSEP monitoring in neurosurgical spinal procedures has become commonplace. Nevertheless, it has been reported that as many as 62% of surgeries are interrupted by false positive decreases in the amplitude of the evoked potentials [2]. Furthermore, if at the end of the procedure, the amplitude remains decreased, only about 30% of these patients are reported to awaken with a neurological deficit [3, 7, 9]. False negative recordings, i.e., instances in which the SSEP fail to detect spinal cord injury, are well described [1, 4]. These findings have led some surgeons to conclude that the utility of SSEP monitoring is mainly limited to scoliosis procedures, because any changes detected while the rods are being manipulated are immediately reversible by reducing the correction. Others have argued that monitoring of intradural procedures yields limited information [8].

We have not had occasion to be concerned by the frequency of false positive decreases in SSEP amplitude. Usually, there is an obvious explanation, as in hypotension, increased anesthetic sedation, or use of cold irrigant. There are certain critical periods in spinal procedures in which the surgeon realizes that there is significant risk of spinal cord compromise. Careful attention to the SSEP during these times is crucial. Improper intraoperative positioning of the patient undegoing decompressive laminectomy for cervical spondylotic myelopathy has previously been assumed to be one cause of immediate post-operative neurologic deficit. We have described two cases in which, at least circumstantially, this is confirmed. Subsequently, we have preferred to position the patients prone, in slight flexion, with some degree of traction applied, and have had no further difficulties.

In our patient with rheumatoid arthritis, we discovered, as have our orthopaedic colleagues, that excessive distraction of the vertebral elements can result in spinal cord injury. Changes in the SSEP were immediately reversed by correcting the deformity that had been created. Similarly, efforts at retrieving a soft disc through a keyhole foraminotomy resulted in decreased SSEP amplitude coincident with retraction of the neural elements. Although the surgeon did not appreciate excessive intraoperative traction on the spinal cord, the procedure was terminated largely on the basis of the SSEP. Subsequent anterior cervical discectomy successfully allowed removal of the fragment, with complete neurologic recovery.

In our patient with thoracic disc herniation, the SSEP decreased by only 50%. Some authors would classify this as a false negative response of the test, whereas others consider a decrease of 30%–40% in the amplitude significant and would argue that the SSEP accurately predicted spinal cord injury [6]. Regardless, the conclusion we have reached is that when performing anterior procedures to decompress the spinal cord, smaller decreases in the amplitude of SSEP may be very significant in predicting motor impairment. Until motor or spinal evoked potentials are readily available, we feel somatosensory studies are still indicated, but smaller decreases in the SSEP should be heeded as a warning of spinal cord injury.

Lastly, we grant that SSEP may be overly sensitive to spinal cord manipulation during intradural procedures. In particular, changes may be frequently seen in excising intramedullary lesions. Furthermore, these are probably the surgical procedures that are most likely to leave the patient with at least a temporary postoperative neurologic deficit. In the operation for hemangioblastoma described above, dorsal column injury was predicted by the SSEP and confirmed by postoperative MRI. The patient accordingly suffered a loss of position sensation, although her strength ultimately improved. We feel that had the SSEP changes not alerted the surgeon to this problem, the post-operative deficit may have been worse.

The surgeon must exercise judgment in interpreting the SSEP. Decreases in amplitude that coincide with surgical manipulation of the spinal cord are particularly ominous, as compared to changes that are associated with hypotension or increased anesthetic sedation. In our experience, SSEP changes more accurately predict postoperative deficits in dorsal column function and may not accurately reflect injury to the motor elements. These are limitations of the testing that should be understood by its practitioners. Nevertheless, we believe that the consequences of disabling spinal paralysis are so great that the monitoring of all high-risk procedures is appropriate if only 1%–2% of the patients are truly helped.

References

1. Ben-David B, Haller GS, Taylor P (1987) Anterior spinal fusion complicated by paraplegia – a case report of a false negative somatosensory evoked potential. Spine 12:536–539
2. Dimitrijevic MR (1987) Neurophysiology in spinal cord injury. Paraplegia 25:205–217
3. Dinner DS, Luders H, Lesser RP, Morris HH, Barnett G, Klem G (1986) Intraoperative spinal somatosensory evoked potential monitoring. J Neurosurg 65:807–814
4. Ginsburg HH, Shetter AG, Raudzens PA (1985) Postoperative paraplegia with preserved intraoperative somatosensory evoked potentials – Case report. J Neurosurg 63:296–300
5. McPherson RW, Sell B, Traytsman RJ (1986) Effects of thiopental, fentanyl, and etomidate on upper extremity somatosensory evoked potentials in humans. Anesthesiology 65:584–590

6. More RC, Nuwer MR, Dawson EG (1988) True and false positive amplitude attenuations during cortical evoked potential spinal cord monitoring. In: Ducker TB, Brown RH (eds) Neurophysiology and standards of spinal cord monitoring. Springer-Verlag, Berlin Heidelberg New York Tokyo, pp 222–226
7. Raudzens PA (1983) Intraoperative somatosensory evoked potentials – correspondence. Anesthesiology 58:593–594
8. Whittle IR, Johnston IH, Besser M (1986) Recording of spinal somatosensory evoked potentials for intraoperative spinal cord monitoring. J Neurosurg 64:601–612
9. York DH, Chabot RJ, Gaines RW (1987) Response variability of somatosensory evoked potentials during scoliosis surgery. Spine 12:864–876

Intraoperative Evoked Potential Monitoring: Fringe Benefits in the Operating Room

R.H. Brown[1], C.L. Nash Jr., J.T. Stinson, and F.A. Bauman

Introduction

Somatosensory cortical evoked potential (SCEP) monitoring was first introduced in the early 1970s as a means of monitoring the integrity of the spinal cord during the correction of scoliosis. Since then it has become increasingly used in the direct diagnosis and treatment of a wide variety of spinal problems. As with most developing technology, increased use has also fostered a variety of associated beneficial applications or "fringe benefits" such as verification of safe patient positioning. The purpose of this paper is to share our intraoperative monitoring experience and report on a number of these secondary benefits or "spinoffs" derived from the primary monitoring process.

Background

Intraoperative spinal cord monitoring has been used routinely by the senior authors since 1977 in more than 1400 cases. During this time a number of fringe benefits have been identified as a part of the monitoring process. These cases were reviewed to identify the types and frequency of occurrence of these benefits and document their value to those who are using or considering using intraoperative spinal cord monitoring.

Methods

SCEP monitoring of all four extremities was the only method used over the entire period of the study. The stimulus was a constant current applied using percutaneous needle electrodes at both posterior tibial nerves at the ankle

[1] Division of Surgical Research, Saint Luke's Hospital, Shaker Boulevard 11311, Cleveland, OH 44104, USA

K. Shimoji et al. (Eds.)
Spinal Cord Monitoring and Electrodiagnosis

and both median nerves at the wrist. The stimulus rate was between 0.8 Hz and 1.2 Hz. Recording was obtained differentially over the appropriate somatosensory cortex using needle electrodes with a frontal indifferent and an ipsilateral reference behind the ear. Input bandwidth was 0.3–100 Hz, record length was 125 ms, and 64 successive records were averaged.

Our standard anesthetic protocol is described in the following: Premedication was secobarbital, nominally 2 mg/kg. Induction was accomplished using sodium pentobarbital, fentanyl, and a muscle relaxant. Anesthesia was maintained using a 60:40 mixture of nitrous oxide and oxygen, a continuous infusion of fentanyl, and muscle relaxants. A second continuous infusion of either nitroglycerin or sodium nitroprusside was employed to control blood pressure if necessary. The fentanyl drip was nominally started at a rate of 4 μg/kg h and adjusted as required according to the patient's physiologic responses and the monitoring records. If necessary, low-dose enflurane (Ethrane) (0.5% or less) is added to reach the desired depth of anesthesia or as an adjunct in controlling the blood pressure.

All patients had their initial monitoring in the operating room following premedication but prior to induction of general anesthesia. Any case in which an intraoperative change was noted in any of the monitoring records that could not be related to a spinal cord event, i.e., spinal distraction, were documented relative to the probable cause and the effects of remedial efforts.

Results

With time a number of fringe benefit experiences emerged and the following describes the most commonly encountered situations.

Positioning. One of the major areas where monitoring served an additional, significant role in the operative process was positioning. This is an especially critical procedure in the patient with spinal or neurologic instability given that normal muscle splinting is eliminated due to the muscle relaxants associated with general anesthesia. In these patients, immediate postpositioning monitoring proved to be very helpful. In a number of cases of spinal instability, the loss or deterioration of records soon after positioning signaled the need to change the alignment. This verification of conduction integrity also proved of value in cases of low positional tolerance (i.e., cervical spondylosis). In all cases, the monitoring of four extremities provided ongoing input regarding conduction, which may have been compromised during the course of surgery. Gradually increasing pressure on the axilla due to shifting position was noted in the upper extremity records on several occasions. Repositioning corrected this.

Early changes in upper extremity conduction occurred in 6% of the reviewed cases. All of these were resolved after changes were made in the patient's position. A similar phenomenon was noted due to positioning of the blood pressure cuff in 2% of cases. Significant lower extremity monitoring changes were noted post positioning in 2% of the cases which cleared shortly after repositioning was instituted.

Case 1344 was a 15-year-old boy who underwent a series of posterior laminectomies to drain multiple noncommunicating syrinxes. As a result of these procedures, his spine became increasingly unstable and developed a profound scoliosis requiring a two-stage anterior-posterior procedure to stabilize this spine. The anterior procedure was undertaken uneventfully with stable monitoring throughout. Ten days later, he returned to surgery for the posterior procedure. His preoperative, prepositioning monitoring records, obtained while he was awake in bed, were in good agreement with his prior procedure records. General anesthesia was then established while he was still in his bed. He was then positioned on the Hall-Reton frame and monitoring immediately reestablished which demonstrated almost total loss of conduction from both lower extremities. When an initial attempt at repositioning resulted in no improvement in the records, he was returned to the prone position in his bed and monitoring reestablished. Within 5 min, his potentials began to reappear, with good return by 20 min post turning. The records documentating this sequence of events are shown in Fig. 1. He was subsequently placed in a body cast and his posterior procedure successfully

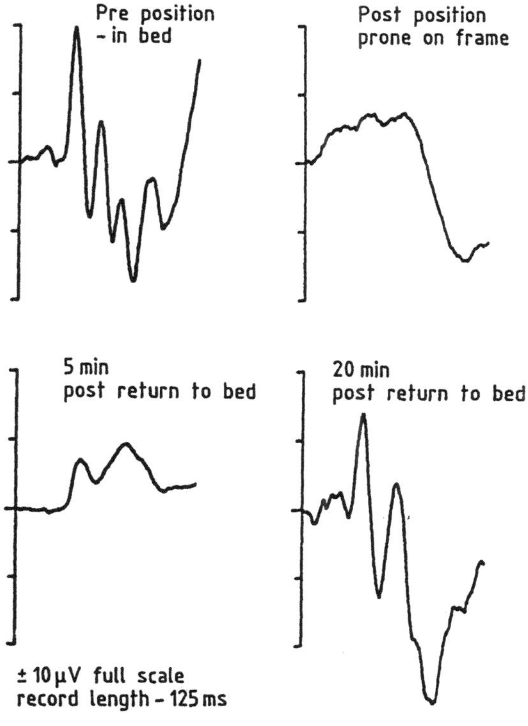

Fig. 1. Loss of conduction from lower extremities following positioning on Hall-Reton frame

performed a few days later using the front of the cast as an intraoperative support.

Case 1689 was a 37-year-old woman with cervical instability resulting from a motor vehicle accident. Her preoperative records indicated normal conduction patterns. Following induction of general anesthesia, she was placed in headhalter traction on a Mayfield headrest. The immediate post-positioning records demonstrated a virtual loss of lower extremity potentials. Based on these records, the head was repositioned in less extension with a rapid return of indicated conduction. By 10 minutes post repositioning, well-defined complexes were evident from both lower extremities. The records documentating this sequence of events are shown in Fig. 2. The remainder of the case proceeded uneventfully with a good outcome.

Case 1607 was a 52-year-old woman who underwent a posterior fusion on the Hall-Reton frame for progressive idiopathic adult scoliosis. A little over 4 h post positioning, her right median nerve potential indicated a decrease in amplitude and increase in latency but with preservation of the complex. However, by 5.5 h post positioning, the complex indicated a major loss of amplitude, prompting repositioning of the patient on the frame. Within 15 min post positioning, the complex had returned to its original latency and regained much of its original amplitude. The records demonstrating this sequence of events are shown in Fig. 3. Her postoperative findings were unremarkable.

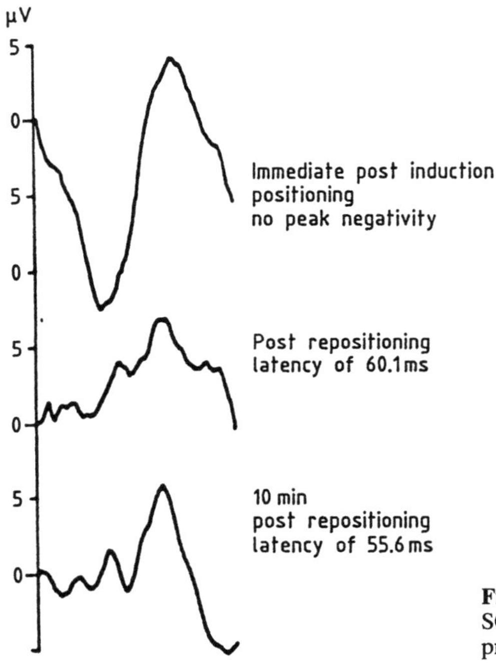

Fig. 2. Loss of posterior tibial nerve SCEP immediately post positioning prone in Mayfield headrest

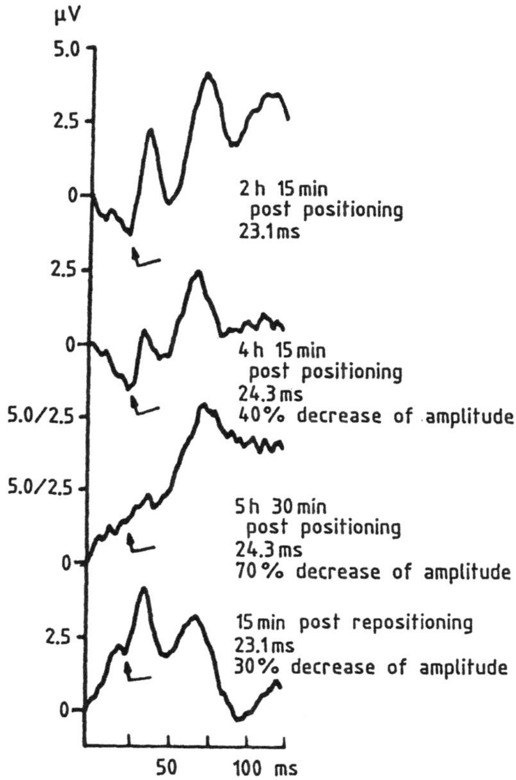

Fig. 3. Deterioration of median nerve evoked potential and restoration of its original latency and increase in amplitude following repositioning

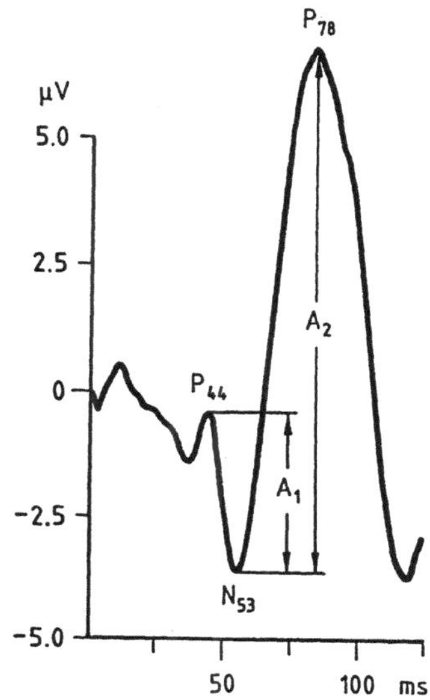

Fig. 4. Typical SCEP from stimulus of posterior tibial nerve

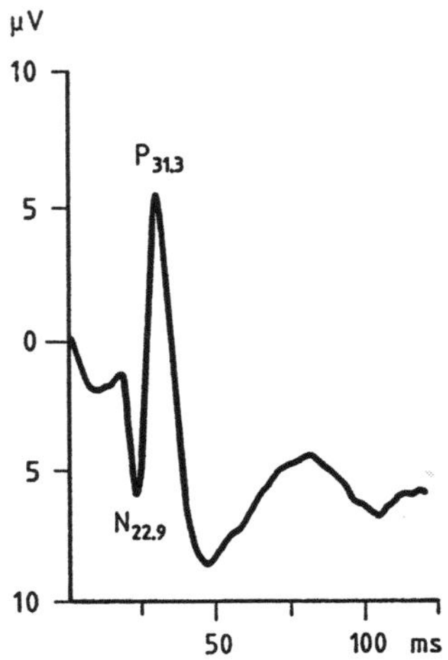

Fig. 5. Typical SCEP from stimulus of median nerve

Depth of Anesthesia. Monitoring depth of anesthesia is another, decidedly beneficial fringe benefit. While the basic evoked potential can be obtained using a record length of 75 ms, by taking 125 ms long records a cognitive after-potentials can be also be documented.

This cognitive after-potential is the cortical activity recorded following the nominal P60 inflection in the posterior tibial response (as shown in Fig. 4) and/or the nominal P30 in the median nerve response (as shown in Fig. 5) and has been shown in our hands to correlate extremely well with the patient's depth of anesthesia. When anesthesia is controlled such that only a slight negative inflection follows these peaks, the level of anesthesia is appropriate for the procedure, yet light enough to allow rapid arousal.

That this has allowed our patients to be carried at generally lighter, yet appropriate levels of anesthesia for the procedure is attested to by the fact that 98% of our patients undergoing posterior fusions were extubated in the operating room prior to going to the recovery room. This extubation generally occurred within 10 min of turning the patient off the frame into bed. This yielded the additional benefit of being able to verify voluntary, lower-extremity motor function to verbal command while the patient was still in the operating room. Another benefit from maintaining this somewhat lighter level of anesthesia was the ability to conduct rapid and repeatable wake-up tests whenever the monitoring became equivocal.

Case 1688 was a 34-year-old woman undergoing the first of a two-stage procedure for the correction of adult idiopathic scoliosis. Her immediate postinduction records showed her to still be somewhat lightly anesthetized, as demonstrated by her still active cognitive potential. Her level of anesthesia varied somewhat during the procedure, as demonstrated by the existence of a slight after-potential in midprocedure. Because of the need for a chest tube postoperatively, she was left intubated and her anesthesia was maintained at

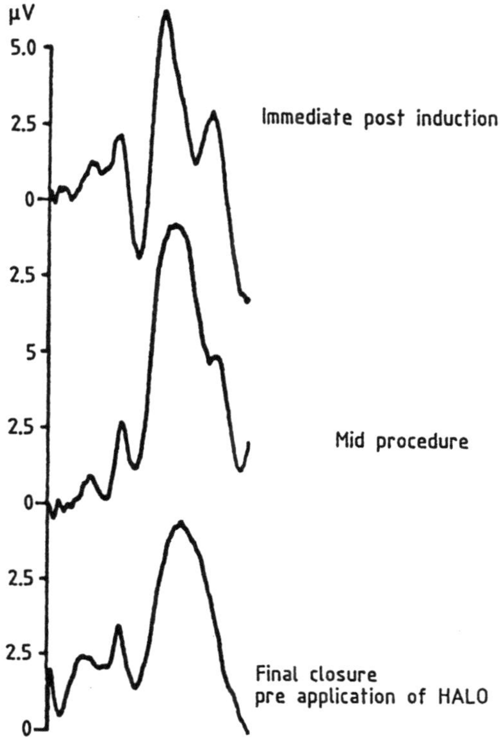

Fig. 6. Immediate postinduction, midprocedure, and final closure records showing appropriate anesthetic level in the case of transthoracic anterior diskectomy with interbody fusion. The patient was not extubated in the operating room

operative levels to the end of the case, as demonstrated by the absence of any after-potential at closure. The postinduction, mid-procedure, and closure records are shown in Fig. 6. She was transferred to the surgical intensive care unit and placed on a respirator.

Case 1695 was case 1688's second stage procedure 10 days later. Her postinduction and midprocedure records indicated an appropriate level of anesthesia, as demonstrated by the absence of any significant after-potential. Her final closure record indicated a lightening of anesthesia in preparation for return to bed, where she was successfully extubated. These records are shown in Fig. 7.

Case 1682 was a 30-year-old woman, who underwent anterior diskectomies and osteotomies for the correction of hyperkyphosis, and demonstrated the use of low-dose enflurane as a supplement to the balanced nitrous, narcotic, and muscle relaxant technique. While her initial postinduction records demonstrated a suitable level of anesthesia, as time progressed, she showed a tendency to lighten up under the stimulation of the procedure, as

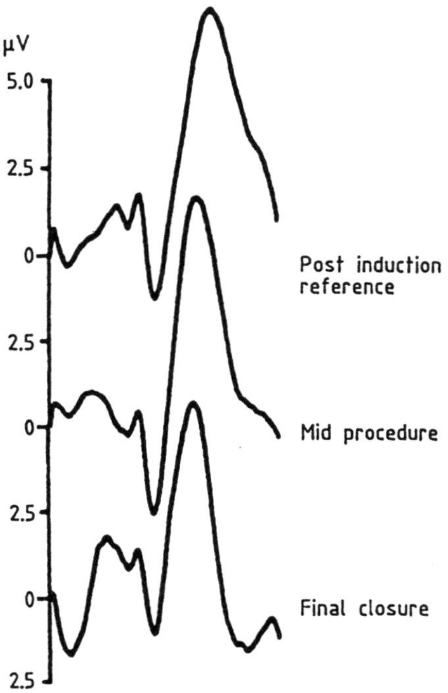

Fig. 7. Postinduction, midprocedure, and final closure records showing appropriate anesthetic level in the case of posterior fusion and using balanced nitrous-narcotic anesthesia. The patient was extubated in the operating room

shown by both the reemergence of an after-potential and an increase in blood pressure. For this reason, 0.3% enflurane was introduced with good results as shown in Fig. 8. However, during the lunch break of the certified registered nurse anesthetist, one of our new anesthesiologists felt the enflurane was not doing anything in such a low dose and turned it off. The result was that both the increased blood pressure and the after-potential reappeared in less than 0.5 h as shown in Fig. 9. The appropriate level of anesthesia was again attained 45 min following the reintroduction of the low dose enflurane when the CRNA returned as shown in Fig. 10, thus proving once again that it is easier to stay ahead of the patient than to play catch up.

Real Time EEG Monitoring. Another benefit derived from our system of intraoperative monitoring proved to be real time EEG monitoring. This was dramatically documented during a case in which the patient suffered a postinduction cardiac arrest. The patient was a 19-year-old man, who had suffered an L1 compression fracture resulting from a motor vehicle accident. He had undergone a decompressive laminectomy at another institution immediately post accident without incident. Due to increasing neurologic

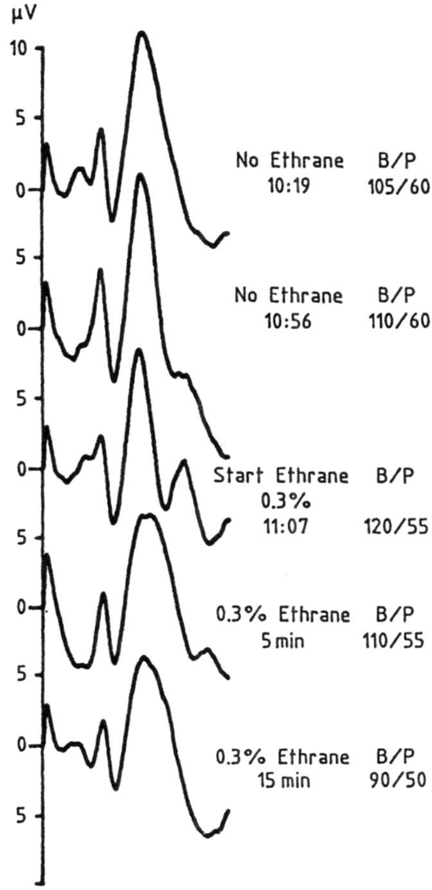

Fig. 8. Use of enflurane (Ethrane) as a supplement to the balanced nitrous, narcotic, and muscle relaxant form of anesthesia showing levels of anesthesia and the effect of enflurane on these

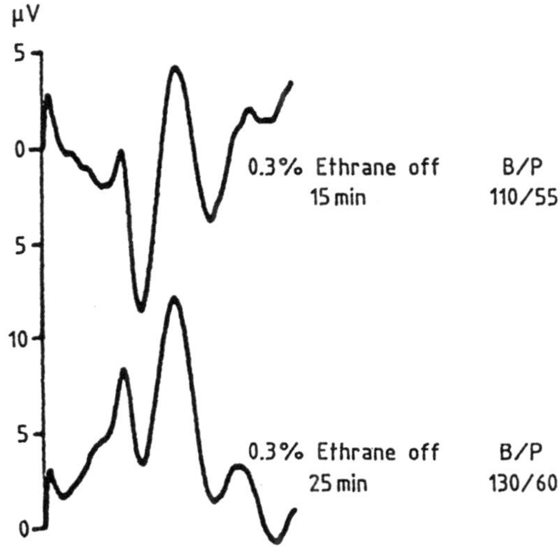

Fig. 9. Effect of turning off low dose of enflurane on the anesthetic level of the patient

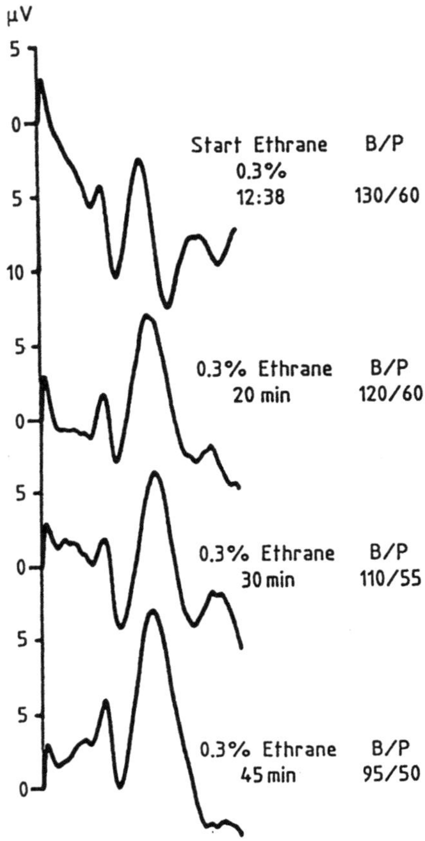

Fig. 10. Effect of reinstituting supply of low dose of enflurane on the anesthetic level of the patient

deficit, he was scheduled for further surgery at our institution. During induction, he experienced an episode of cardiac arrest requiring an extended period of resuscitation including four applications of defibrillation. Monitoring using both median nerve evoked potentials and real time EEG analysis was utilized throughout this period to ascertain the adequacy of cerebral perfusion. He was successfully resuscitated and underwent his stabilization procedure a few days later without further incident and made a full recovery.

Real time feedback is also used to monitor surgical technique by watching EEG activity during intraoperative procedures such as passage of sublaminar wires, anterior diskectomies, and decortication using the high-speed burr. Increased spikes on the EEG indicated irritation of the dura or meninges, which could be due to mechanical irritation or an irritable cord.

Case 1695 was a 34-year-old woman who underwent a posterior fusion with segmental instrumentation. Good, stable intraoperative records were being obtained, as demonstrated by the intraoperative reference record, up to passage of an apical sublaminar wire. During this procedure the real time

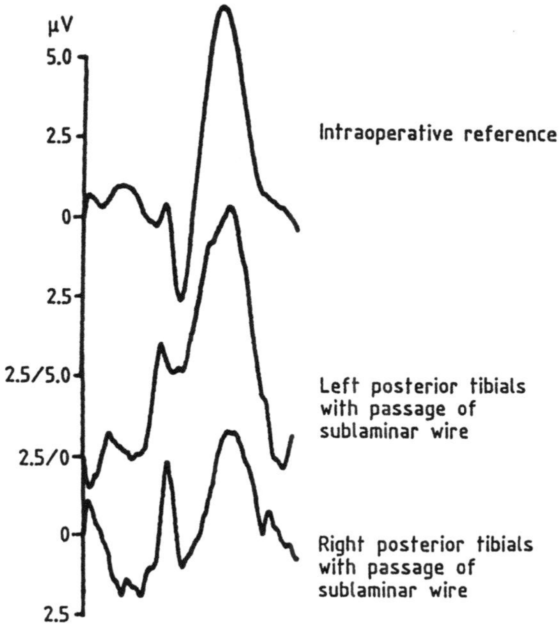

Fig. 11. EEG contamination in distal SCEP records coincident with significant EEG activity noted on real time monitor

EEG monitor showed a marked increase in cortical activity with a concomitant increase in high frequency contamination of the distal records, as shown in Fig. 11. These findings were communicated to the surgeon.

Use of real time EEG monitoring has allowed the surgeon to be kept continuously informed about the effect of the procedure on the cord. A not surprising finding of our real time EEG monitoring is that not only are all cords not alike but neither are all surgeons.

Long-Term Follow-up. In patients who require revision of prior procedures long-term follow-up is occasionally necessary. By using a standard protocol of monitoring and anesthesia, it is possible to utilize the patients' previous preoperative and intraoperative records to ascertain any changes in conduction that may have occurred in the interim.

Case 1508 was a 50-year-old woman who underwent a stabilization procedure in 1985, but developed a pseudoarthrosis and had documented progression of her deformity. By utilizing our standard protocol, it was possible to compare her original records with her records obtained at the time of her second procedure, which showed no change. These records are shown in Fig. 12.

Level of Muscle Relaxation. Induced muscle paralysis or level of muscle relaxation is another fringe benefit achieved through the routine application of intraoperative monitoring. The constant current stimulators used to elicit

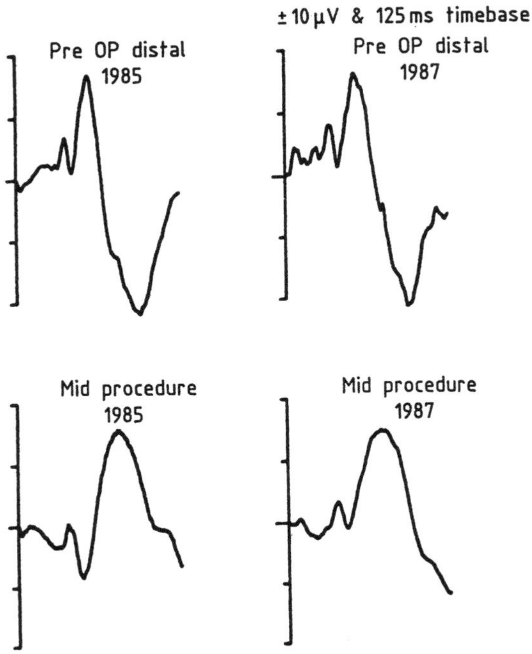

Fig. 12. Stability of standard monitoring protocol on second operative procedure 2 years after initial procedure

stimuli in the extremities are generally more sensitive than standard anesthesic stimulators and therefore provided earlier evidence of changes in the level of paralysis. Also, the use of a frontal indifferent recording electrode allowed us to pick up the initial onset of EMG activity in this area affording a second measure of the level of paralysis.

Summary

While the advent of evoked potential monitoring to monitor gross spinal cord function represented a major adjunct to the practice of spine surgery, its full potential to assess a patient's overall neurophysiologic status in the operating room is still evolving. Several additional areas of input from SCEP have been presented which can further add to the safe and efficient conduct of spinal surgery.

These additional benefits are based upon the use of SCEP monitoring utilizing all four extremities, which has the capacity to look at a wide variety of neurophysiologic systems. The versatility of this system has still not been thoroughly explored and users are encouraged to expand their thinking regarding the wider application of their monitoring technology.

Ketamine as a Major Component of an Anesthetic Plan for Spinal Cord Monitoring

W.T. FRAZIER[1], D.B. BIGGS, S.H. ODOM, and S.E. JENKINS

Introduction

Since our initial experience with spinal cord monitoring (SCM), we have searched for an anesthetic plan which would provide optimal conditions for the recording of somatosensory cortical evoked potentials (SSCEPs) during surgery involving the spine and/or spinal cord. As reported in a preliminary way at the Third International Symposium on Spinal Cord Monitoring [5], when faced with small baseline signals, we have (intraoperatively) switched from our "usual/best" anesthetic plan (50% N_2O + Fentanyl infusion + Thiopental infusion/or 0.2% Forane) [4] to a plan whereby ketamine is substituted for N_2O, leaving the other drugs the same. We have now expanded our series of ketamine cases to 33 and will summarize our data on 29 of the most complete cases done in a period from 1985 to April 1989.

Our patient population is approximately 40–50 cases per year, is almost all orthopedic/spine surgery, and is biased toward elderly patients with neurological deficits. Approximately 25% of our cases continue to have SSCEPs which are small enough to make it difficult to be sure that the baseline signals are clear enough to allow us to easily detect a diminution secondary to a surgical event (similar to case experience reported in [4]).

We chose to try ketamine (as a substitute for 50% nitrous oxide) on the basis of our general clinical experience with this drug and on the basis of cellular/animal work consistent with ketamine having a different mechanism of anesthetic action than typical anesthetics such as thiopental and halothane. Clinically, ketamine is one of the few drugs (if not the only one) which provides a general anesthetic state while still leaving airway/breathing reflexes relatively intact – implying a certain degree of specificity. On a cellular level, these differences were investigated by testing anesthetic agents on single invertebrate neurons [3]; thiopental (and in some instances halothane) was found to 'anesthetize" (i.e., silence) most cells by hyperpolarization and

[1] Department of Anesthesiology, Emory University School of Medicine 475 Woodruff Memorial Building, Atlanta, GA 30322, USA

K. Shimoji et al. (Eds.)
Spinal Cord Monitoring and Electrodiagnosis

by decreasing the size of post-synaptic potentials, both effects occurring in a manner consistent with increased potassium conductance. In contrast, the predominant effects of ketamine were selectivity (i.e., not effective on all cells), hyperpolarization, and a decreased size of postsynaptic potentials (all accompanied by an increased membrane resistance, consistent with a different biophysical mechanism than thiopental and possibly halothane).

In terms of clinical utilization, ketamine has *three potential advantages* other than its assistance with spinal cord monitoring: (1) its use can allow a very high oxygen concentration (i.e., no N_2O) in cases where this is advantageous (e.g., during anterior thoracic spine cases where "one-lung" anesthesia is sometimes indicated); (2) its use in cases needing high O_2% avoids using high concentrations of halogenated agents (e.g., forane) known to significantly depress signals; and (3) its use can allow (and we have occasionally used) a total intravenous anesthetic plan (e.g., a mixture of infusions of ketamine, fentanyl, and thiopental).

All of our experience with ketamine has been with a goal of using this drug as approximately ½ of the anesthetic, not as a total/single agent (i.e., use along with fentanyl/0.2% forane or with fentanyl/thiopental infusion). In all cases, the above anesthetic plan was complimented with a continuous infusion of a muscle relaxant (Atracurium), and the control of blood pressure and heart rate was achieved with intravenous agents (e.g., beta blockers and vasodilators, versus control with variable levels of an agent such as forane).

Summary of Ketamine Anesthetic Plan

There is one common component to all of our anesthetic plans – i.e., the continuous infusion of fentanyl (at an average rate of 2 μg/kg/h). We may increase this dose up to ~2.5–5 μg/kg/h (for patients who by history and by their response in the operating room (OR) clearly have a tolerance, primarily due to their preoperative use of analgesics). In the past, an essential part of our plan was thiopental at 2 mg/kg/h; on occasion, we used up to 3 mg/kg/h for patients who were tolerant to sedatives, primarily due to their use of ethanol or drugs such as Valium. We have now substituted 0.2% forane for thiopental, having noticed in our early experience that it seems to be the least offending of the halogenated agents [4], and does offer some advantage for blood pressure control and bronchodilitation.

The method of making the substitution of ketamine for nitrous oxide is to turn the N_2O off (i.e., "100%" O_2) and to monitor the gas levels with a time-shared mass spectrometer, documenting the washout of the nitrous oxide over 5–7 min. We load the ketamine over this same period in ~three divided doses. Typical ketamine doses (for a total ketamine anesthetic) are 2 mg/kg for an induction, with an onset of ~1 min and a duration of approximately

10–20 min. We have varied our loading dose from 1–2 mg/kg but we have settled on a 1 mg/kg dose, assuming that the forane and fentanyl cover approximately the other half of the needed level of anesthetic. During the loading period for the ketamine, we start an intravenous infusion at 3–3.5 mg/kg/h, having tried doses from 2.5–5 mg/kg/h. We must emphasize that all of these doses represent "intelligent guesses" since we have very little hard data to say exactly what these concentrations should be as part of a complex plan [1, 6, 15].

General Observations

No significant problems were encountered with the general management of the cases with the utilization of ketamine (e.g., either high or low blood pressure or heart rate). Perhaps the most significant change in our management of the patient is that there was consistently a prolonged awakening time. With our "usual/best" anesthesia plan, we are able to awaken our patients within 3–5 min after turning off the N_2O (having turned off the other drugs a few minutes before). We have considered this to be an advantage in that we can quickly assess the patient's motor function in the OR (e.g., during a "wake up" test and/or at the end of the case). With ketamine, the awakening times have been quite varied (i.e., with total infusion times ranging from approximately 10 min to approximately 3 h). Although not yet analyzed in detail, it is our clinical impression that (as expected) the awakening time is proportional to the total dose and the time since discontinuing the drug (i.e., in some cases, after passing the period of highest surgical risk, ketamine is terminated and N_2O reintroduced). Some patients have been extubated in the OR and all have been extubatable in the postanesthesia care unit within less than 2 h. We have no incidents of intraoperative recall, nor have we had any patient reports of postoperative "bad dreams." We have the definite clinical impression that we could use lower doses of ketamine, but until such time as we have parallel pharmacodynamic and pharmacokinetic evidence in animals and other patients (under more controlled and simpler conditions), we feel it is far better to err toward the higher ketamine doses.

In terms of onset of improvement, in some instances we have seen an effect on the SSCEPs within 2–5 min of making the substitution (even before the N_2O has been washed out). However, in other instances it has taken 5–10 min to achieve peak results.

With respect to the stability of the improvement, there is a tendency for the improvement to peak in 20–40 min and then begin to wane, possibly related to the accumulation of ketamine to a level which is unnecessarily high. However, there have been some cases where the improvement has lasted for more than an hour without any diminution.

Methods of Retrospective Analysis

In reviewing the cases we have done since 1985, we felt it would be helpful to create a grading scale for the SSCEPs as well as a grading scale for changes secondary to switching to Ketamine. In a preliminary way, we have also graded the preoperative neurological deficits.

For baseline grades, the grading scale is for the quality of the evoked potentials at the time we made the decision to switch from N_2O to ketamine. As seen in Fig. 1, this scale varies from 0 (essentially an absence of any signal) to 5 (a very large signal – i.e., among the best we had ever seen with such stimulation). For the improvement that we see when we switch from N_2O to ketamine, the scale was labeled O-III, as a ratio to the baseline signal. It is important to understand that our baseline grades were referred to an absolute "worst or best" that we have seen over the last 10 years, whereas the improvement scale is a ratio, with "I" being a slight increase (10%–15% or a ratio of 1.1 to 1.5), with II being a moderate increase (e.g., 1.5–2.0), and III being a marked increase (e.g., an amplitude change of some major component being a multiple of 2–3 of the baseline signal).

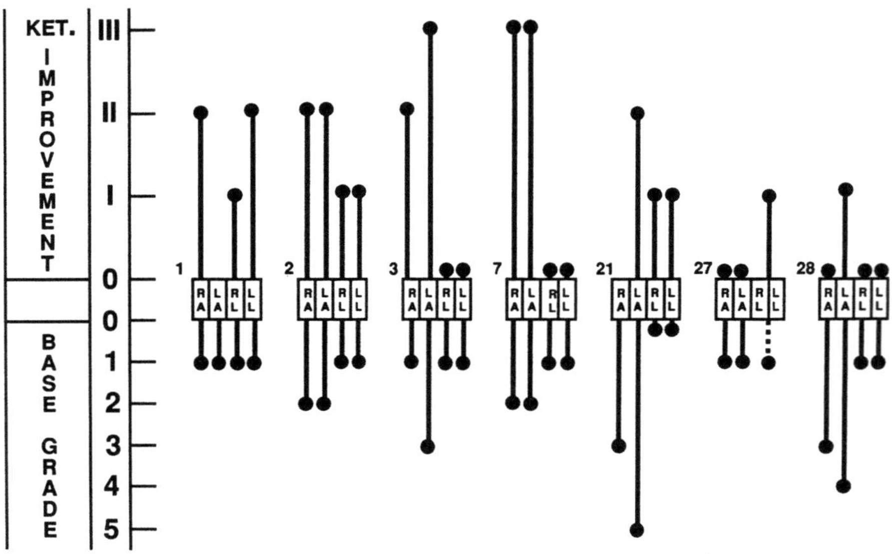

Fig. 1. Data survey. SSCEP improvement with ketamine substitution for N_2O. This data is from 7 patients (1, 2, 3, 7, 21, 27, 28), and illustrates interesting variants. For each patient, somatosensory cortical evoked potentials (SSCEPs) were elicited by stimulating at a site on each limb; e.g., *RA* = right arm (wrist), *LL* = left leg (ankle), etc. The baseline grade range was from 0–5 and the improvement in SSCEPs was ranked 0-III (see text for definitions). Absence of some data in cases 1 and 21 indicates no testing of this limb, and *dotted line* in case 27 indicates combined right and left leg stimulation

Results

The largest variable determining how much improvement was obtained is the grade for the baseline SSCEPs. That is (as we will demonstrate in some of our following analysis), if the signals were very poor, then there was not much improvement. Also, we are certain that we have clinical correlations that would indicate that those patients with very significant neurological deficits (and usually with the lowest baseline SSCEP scores) also generally received the least benefit. We also have the clinical impression that some of the more elderly patients (who may have age dependent decreased SSCEPs) may not receive as much benefit.

We have summarized the results we have obtained in our 29 cases in Fig. 1, which shows a sampling of seven of these cases illustrating some of the more interesting variations. The data is plotted such that we can see on one chart the baseline score as well as the improvement for each of the four limbs which we typically test. On occasion we may not test a particular limb either before or after the anesthetic change. That is, in some instances (e.g., cases 1 and 21 in Fig. 1), we are so busy checking the right and left leg signals that we only use one of the hands as a control signal during this period. In an occasional case (e.g., case 27), the signals from the legs have been so poor that we have resorted to an old technique to bring them out of the noise, i.e., stimulate both legs at the same time and then observe the change in the (larger) dual leg signal with the change from N_2O to ketamine (with the attendent loss of independent right and left leg evaluation).

In Fig. 1, note that the baseline SSCEP grades are 0–5 (as categories) whereas the improvement grade (estimated independently at a later time) is expressed as O-III, a ratio of the improved signal to the baseline signal acquired just before switching agents. Note that in some cases, no improvement was obtained (e.g., the right and left leg signals in patients 3, 7 and 28). Also, in one of the cases (21), very low grade baseline signals were improved to a barely usable level (grade I).

Another method of summarizing the data can be seen in Fig. 2 where we chart the improvement score obtained with the ketamine versus the baseline score (for the SSCEPs obtained with wrist stimulation). In our series, wrist SSCEPs were usually control signals, reflecting the fact that most of the surgery monitored was for problems below this level, potentially placing the legs at risk (and having mainly leg neurological deficits). It is of interest to note that when looking at the SSCEPs from the wrists, there were no zero scores and the peak concentrations of these scores were heavily in the grade 2-3 range. It was also of interest to see that within this grade 2-3 range, the improvement with ketamine is the most marked. It seems as though the most inadequate signals (grade zero) are inadequate secondary to a fundamental neurological problem, and therefore have limited capability of being enhanced by pharmacological manipulation (and also therefore have a low

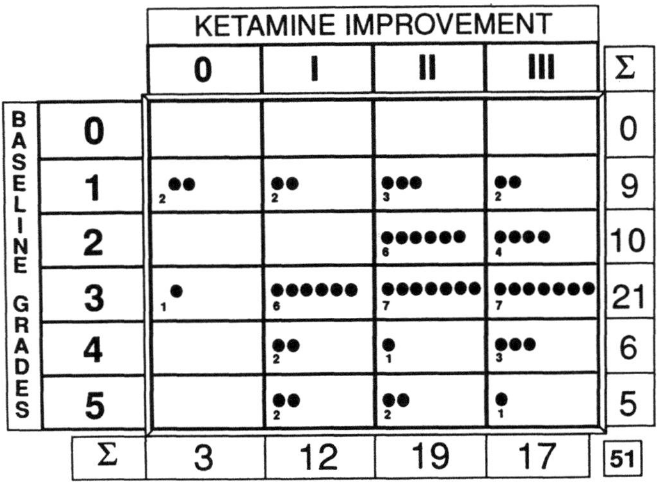

Fig. 2. As in Fig. 1, 0–5 is the grade scale for baseline wrist SSCEPs and 0-III is the scale for improvements with switch from 50% N_2O to ketamine infusion. Each *dot* is a grade for an SSCEP from one site (e.g., 2 *dots* may represent right and left wrist signals on the same patient or 2 different patients). Row and column sums indicate a peak concentration at cell 3/II. Note no baseline grades of 0 (wrist signals were largely used as controls for leg signals at risk)

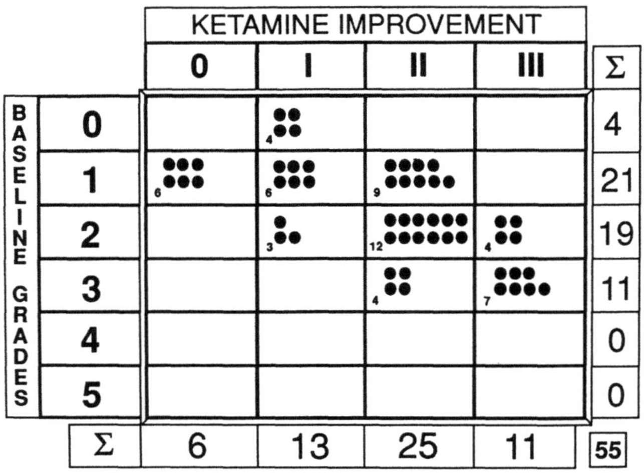

Fig. 3. Ankle SSCEPs. Grading scales are the same as Figs. 1 and 2. Note the row and column sums indicate a peak in cell 2/II, with the overall pattern being a diagonal distribution from cell 1/O to cell 2/III. Note also no cases with baseline grades in grades 4 and 5 (no improvement needed). The rationale for using ketamine was in the relative inadequacy of the signals in grades 0–3

improvement score). At the other extreme, the largest signals (i.e., grades 4–5) are already so large that perhaps they are near the maximum that is possible in the anesthetized state. Therefore, as one might anticipate, peak improvements would be in the intermediate baseline grade range where the patients have moderately good neurological function, but the signals have not reached their highest potential peak levels.

In contrast, in Fig. 3 it can be seen that with ankle stimulation the relationship between baseline scores and improvement scores was quite different from that with wrist stimulation. As we have noted before [4], it is much more difficult to get good leg signals than arm signals, even in healthy patients (but particularly in those who have neurological deficits). In this series, we did have two patients who had essentially zero baseline ankle scores, the majority had grade 1 scores, a large number had grade 2, a few had grade 3, and there were no cases with grades 4 and 5. Therefore, the patient population in which we chose to use ketamine exhibited SSCEPs in the lower ~60%–70% of the quality of signals that we have seen (i.e., ketamine not used/needed on patients with grade 4 and 5 ankle baseline scores). The pattern that you see in Fig. 3 is a rather diagonal one, stretching from the 1/0 corner down to the 3/III corner whereas the pattern with the

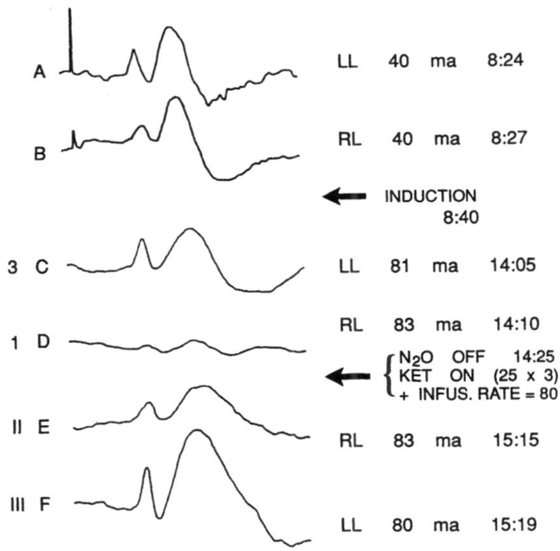

Fig. 4. Raw SSCEP data caused by stimulation of either the right ankle/leg (*RR*) or the left ankle/leg (*LL*). Traces *A* and *B* are preoperative controls. The numbers 3 and 1 and II and III represent baseline grades and improvement grades respectively (see text and previous figures for definitions and description of the case). The time-of-day is given as 8:24, etc. and the stimulation current is in milliamps (ma). The left-to-right time duration of the SSCEPs is 175 ms (positive up) with a 10 ms prestimulus delay (marked by the shock artifact in traces *A* and *B*). The infusion rate of 80 drops/min was for a solution yielding dose of 3 mg/kg/h. Note large improvements in SSCEPs in traces *E* and *F*

wrist (Fig. 2) was a left to right progression with a maximum number of cases in the grade 3 and no cases in the grade 0.

We have reviewed all of our cases and have graded the location and degree of neurological problems (e.g., pain, weakness, diminished sensation and mobility, etc.). Though beyond the scope of this presentation, it is relevant to note that our strong clinical impression is that high neurological problem scores are correlated with the poorest baseline SSCEP scores. For example, in Fig. 1, patient 27 was a patient who had severe cervical spine disease with sensory deficits in the arms and hands as well as long tract signs consistent with poor leg function.

Sample raw data from one of the patients in Fig. 1 is summarized in Fig. 4. This was a 50-year-old woman with a 12-year history of low back pain. She had undergone a lumbar laminectomy in 1984 and 1986 with some relief of her symptoms. Her pain had recently increased. Her diagnosis was a degenerative lumbar scoliosis with lumbar spinal stenosis, status postlumbar laminectomy. Her preoperative CT myelogram revealed foramenal compression at multiple lumbar levels and central canal compression at the L3–4 level. The initial control signals were obtained when the patient was supine and awake, lightly sedated with intravenous Midazolam. Traces A and B (in Fig. 4) were obtained at 8:24 and 8:27 before the patient was anesthetized. The induction took place at 8:40 and these signals diminished immediately (which is not unusual with the low current used for awake controls). The ankle stimulation currents were increased to the 80–83 mA range, which is typical (see [2] for the stimulation technique used). The preoperative control stimulation had been performed at a fairly low level that would be tolerable to the patient under these relatively awake conditions. The third and fourth traces (C and D) are the signals as they existed at 14:05 and 14:10, approximately 5.5 h after the control signals. During that whole period of time, the left leg signal remained essentially unchanged and the right leg signal drifted up and down, but generally was as seen here in trace D. No change in anesthesia plan was made until we and the surgeon estimated that the surgical risk was highest. The ketamine loading dose was 75 mg (her weight was 80 kg), and the ketamine infusion was at a rate of 3 mg/kg/h. Within a few minutes, we saw improvement in the previously good left leg signal and in the markedly diminished right leg signal. Note that we had rated the baseline quality as 3 and 1 for the left and right legs respectively. After the ketamine switch, the improvements were rated as II and III for the right and left legs respectively.

Discussion and Summary

It is clear that using ketamine as a major component of an anesthetic plan does have utility, but it is also clear that this is not a completely satisfactory anesthetic plan. The main limitations is that we do not feel comfortable in using it for extended periods, and therefore would generally want to limit it to approximately a 2-h period. As previously mentioned, prolonged awakening time seems to increase with longer use. Also, if the patient has a very significant neurological deficit, switching to ketamine (or perhaps any other drug) may not help. At the extreme, facilitation of synaptic transmission and cellular activity (e.g. by switching to a less depressing drug) cannot compensate for damaged axons and neurons.

Perhaps the most interesting question relates to the possible mechanism of action of ketamine. We mentioned at the time of the Annapolis meeting that it is intriguing to speculate that signal improvement with ketamine is related to the fact that it may have its anesthetic action by a more specific mechanism than drugs such as N_2O and forane, which probably act as much more general depressants of neurons [9, 13]. There is considerable evidence that ketamine does act as a noncompetitive blocker for the NMDA receptor (*N*-methyl-D-Aspartate, a subtype of the receptors of glutamate synaptic transmission) [10, 2, 11]. However, it is far from proven that this interaction in any species is clearly related to the anesthetic process.

Of course, our results with ketamine are not the only examples of the effort being made to find other anesthetics to cause improvements in evoked potential monitoring. There have now been several reports on the value of etomidate [7, 8, 12]. From these reports, it is probable that the improvements seen with etomidate are even larger than those seen with ketamine. However, we hesitate to use etomidate for any extended period of time since there is published evidence that this drug, at least in larger doses, results in adrenal suppression [13]. It is likely that the brief use of etomidate (e.g., 1 bolus) would be an acceptable risk under certain circumstances (such as neurosurgical patients already on steroids or other instances where an improvement in the signals even for a few minutes would be helpful). The studies indicating that there was adrenal suppression following the use of etomidate also indicated that there was increased morbidity in the postoperative intensive care units with patients who did receive etomidate (i.e., possibly related to adrenal suppression). In major spine surgery, one might not want to have to intervene with steroid therapy to support etomidate depressed adrenal function (given serious questions in regard to wound healing and infection in the postoperative period). Perhaps the most promising drug is propofol [8] which has been on the market in Europe for several years and will hopefully be available in the U.S. during 1990. Also, it is entirely possible that some combination of the drugs mentioned above (along with fentanyl and/or even lower doses of N_2O or forane) could give us better signals without exposing our patients to the full liabilities of the effects of having to use only one of these drugs at a time.

In summary, the message in 1990 is essentially the same as it was in 1986. That is, we are still at a very primitive state of our understanding of anesthetic drug action on the nervous sytem. A great deal of work is still to be done concerning the mechanisms of action of the anesthetics that we do have. It is still clear that we need new anesthetic agents. Hopefully, clinicians involved in SCM can provide some clues to the organic chemists and to the pharmacologists as to what directions we should go in order to discover or synthesize better anesthetics.

In conclusion, an anesthetic plan utilizing ketamine as a major component is very helpful in selected cases, but is not the final solution to the problem of having an anesthetic plan which provides near-ideal conditions for spinal cord monitoring as well as providing other features compatible with good total anesthetic case. For near-ideal SCM conditions, we await anesthetics with greater specificity.

References

1. Abin MS, Figallo E (1976) The use of intravenous Ketamine drip technique for fractured hip repair in the geriatric patient. In: Proceedings of the VI World Congress of Anesthesia, Mexico City, pp 277–279
2. Arnis NA, Berry SC, Burton NR et al. (1983) The dissociative anesthetics, Ketamine and Phencyclidine, selectively reduce excitation of central mammalian neurons. Br J Pharm 79:565–575
3. Frazier WT (1972) Effects of Thiopental and Ketamine on neurons in the isolated abdominal ganglion of aplysia. Fifth International Congress on Pharmacology, San Francisco, A428, p 72
4. Frazier WT, Odom SG, Biggs DB (1985) Anesthetic technique for spinal cord monitoring. In: Schram J, Jones SJ (eds) Spinal cord monitoring. Springer, Berlin Heidelberg New York, pp 69–81
5. Frazier WT (1986) Pharmacologic manipulation of evoked potentials, including changes in anesthesia. In: Ducker TB, Brown RH (eds) Neurophysiology and standards of spinal cord monitoring. Springer, Berlin Heidelberg New York, pp 125–139
6. Idval J, Ahlgren I, Aronson KF et al. (1979) Ketamine infusions: pharmacokinetics and clinical effects. Br J Anes 51:1167–1173
7. Kalkman CJ, van Rheineck AT, Hesslink EM, Bovil JG (1986) Effects of etomidate and midazolam on median nerve evoked potentials. Anes 65(3A):A356
8. Kochs E, et al. (1986) Alterations of somatosensory evoked potentials by etomidate and diprivan. Anes 65(3A):A353
9. Krnjevic K (1986) Cellular and synaptic effects of general anesthetics. In: Roth SH, Miller KW (eds) Molecular and cellular mechanisms of anesthetics. Plenum, New York, pp 1–16
10. Lodge D, Arnis NA (1982) Effects of Phencyclidine on excitatory amino acid activation of spinal interneurons in the cat. E J Pharm 77:203–204
11. Martine D, Lodge D (1985) Ketamine acts as a non-competitive n-methyl-d-aspartate antagonist on frog spinal cord in vitro. Neurophysiology 24(10):999–1003
12. McPherson RW, Ducker TB (1988) Effects of intravenous anesthetic induction agents on somatosensory evoked potentials: thiopental, fentanyl, and etomidate. In: Proceedings of the third international spinal cord symposium, Annapolis. Springer, Berlin Heidelberg New York, pp 163–167
13. Nicoll RA, Madison DV (1982) General anesthetics hyperpolarize neurons in vertebrate central nervous system. Science 217:1055–1057
14. Wanscher M, Tonneson E, Huttel M, Larson K (1985) Etomidate infusion and adrenocortical function – a study of elective surgery. Acta Anes Scan 29:483–485
15. White PF (1983) Use of continuous infusions versus intermittant bolus administration of fentanyl or ketamine during outpatient anesthesia. Anesthesiology 59:294–300

Sensory and Motor Thresholds to Electrical Stimulation of Ankles and Wrists During Spinal Cord Monitoring

W.T. Frazier[1], B.D. Biggs, S.H. Odom, and S.E. Jenkins

Introduction and Objectives

The following is an analysis of electrical stimulation threshold data from 181 cases in which spinal cord monitoring (SCM) was performed over the last 5 years. We have attempted to standardize the electrical stimulation of wrists and ankles in order to gain understanding and control of the stimulation process, leading to an effort to later relate apparent sensory deficits to the quality of the evoked potential (EP) monitoring performed on patients undergoing a full range of spine/spinal cord surgery.

Both sensory and motor threshold data were obtained by stimulating upper and lower limbs (i.e., wrist/arm and ankle/leg; right and left) in the preoperative/preanesthetic period. In most cases, data are available from each of the four limbs, and in others data from one or two limbs were not obtained due to design (e.g., only one wrist/hand used as a control for a potential lumbar area problem).

The initial two objectives were:

1. To obtain thresholds and then use the values to select the operating currents to be used at each site during the operative/anesthetic period (e.g., approximately 2.5 × motor threshold).
2. To confirm that the total stimulating system (e.g., electrodes, cables, etc.) was in working order before anesthetizing the patient.

As experience was gained, two additional objectives evolved:

1. To determine how the threshold currents related to the patient's size/shape. That is, general clinical experience (e.g., with SCM and with nerve stimulation used to monitor muscle relaxant levels) indicates that small and thin patients generally have lower thresholds than large and obese patients.
2. To determine how the threshold currents related to the patient's

[1] Department of Anesthesiology, Emory University School of Medicine, 475 Woodruff Memorial Building, Atlanta, GA 30322, USA

K. Shimoji et al. (Eds.)
Spinal Cord Monitoring and Electrodiagnosis

preoperative neurological status. That is, it became apparent that the thresholds generally reflected the pre-existing neurological deficit secondary to the underlying disease.

The first two objectives were basically procedural and usually easily met. The main drawback is that it takes approximately 15 min to obtain the initial data. Given the total time of the surgical procedure, the risks, and the value of the information, this seems to be a reasonable investment of time to us.

For the second two objectives:
1. Relative to the size/shape issues, we have derived a normalization formula and corrected all data to a standard size patient.
2. Relative to the patient's preoperative neurological status, we have collected extensive data from the preoperative history and physical examination and have organized this data in terms of region and severity (i.e., a numerical grading scale). We will only briefly comment on this area of our studies at this time. In general, as one would expect, many pre-existing neurological deficits are correlated with elevated thresholds (as well as with difficulties in quality of SCM signals), but the relationships are complex and await a more extensive analysis which goes beyond the scope and space of this presentation.

Patient Population

The population was biased toward patients with pre-existing neurological deficits and toward an older age group. The 181 patients were selected from a larger group (approximately 350) with acceptance criteria being only having performed sufficient preoperative/preanesthetic thresholds (e.g., at least one wrist/arm control for surgery placing leg nerves at risk). Some cases were anterior procedures, some posterior, and some both (i.e., done at the same or separate occasions). Some patients had multiple procedures (up to four).

Test Conditions

In almost all cases, there was some preanesthetic medication which could potentially influence the results. Early in this series, low-dose oral premedication consisting of benadryl and haldol was used, and was supplemented during the set-up time with in i.v. valium and fentanyl. With the latter part of the series, "sedation" was limited to a rapid/short acting agent (midazolam) given intravenously to assist with set-up (i.vs, etc.), replacing valium, benadryl and haldol. This i.v. medication was then allowed to "wear off" moderately during the testing period to facilitate patient communication. For airway drying agents and H_2-blockers, an attempt was made to use drugs which do not readily cross the blood-brain barrier (e.g., robinul and zantac).

Test Technique

In order to compensate for SCM changes secondary to anesthesia and temperature, in all cases at least one control site was used (e.g., one wrist/arm for legs at risk or one ankle/leg for arms at risk). In almost all cases, both right and left limbs in the region at risk were used as stimulation sites.

The stimulating electrode type and sites were as previously described elsewhere [1]. That is, relatively large layered pads were used, consisting of a conductive gel (in contact with the skin) overlaid with a conductive rubber pad connected to the usual wire cable. Subjective trials by the investigators found that such an electrode seems to give a less painful sensation (e.g., buzzing or tapping) than do small electrodes and/or needles (stinging and burning).

For wrist/arm stimulation, the size of the electrode was approximately 2×2.8 cm and the placement was over both the ulnar and median nerves at the wrist, with spacing approximately 0.5 cm apart. For ankle/leg stimulation, the size of the electrode was approximately 2×16 cm with the cathode placed circumferentially (just above the malleoli) and the anode on the dorsum of the foot. The intent of this arrangement was to stimulate all nerves innervating the foot. This arrangement would theoretically also give us a larger signal at the cortex (when compared to posterior tibial alone) and would also allow us to monitor for problems in multiple dermatomes (i.e., L4, L5, and S1).

For sensory thresholds, the technique was to note the current when the patient answered "yes" to the following questions (with increasing current): "Do you feel it?"; "Has it gone away?"; "Has it returned?"; an average of these three was recorded.

For motor thresholds, the technique was to observe (with increasing current) the first movement and then the disappearance of the movement as the current was lowered; an average of these two was recorded.

Although we observed general patterns, no systematic attempt was made to record the site of sensation (i.e., under the electrodes or distal, e.g., in the hand or foot), or the site of the first movement (e.g., which fingers or which part of the hand moved first).

Results

The primary method of analysis consisted of plotting histograms of each of eight sets of data: motor and sensory thresholds for each of the four limbs. Examples of these histograms are seen in Figs. 1A, 2A. The data was grouped in bins 1 mA wide (e.g., 0–0.99, 1–1.99, etc.), the height of the bars being the number of instances having a threshold value in the particular 1 mA bin. As we will discuss later, a secondary transformation of the data was to

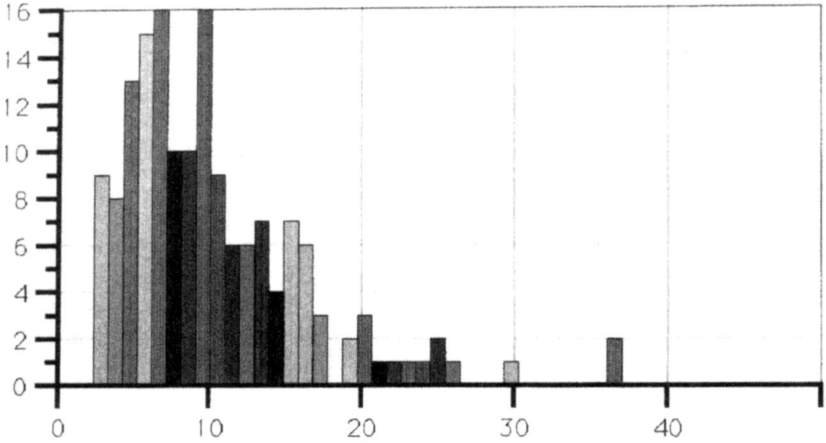

A

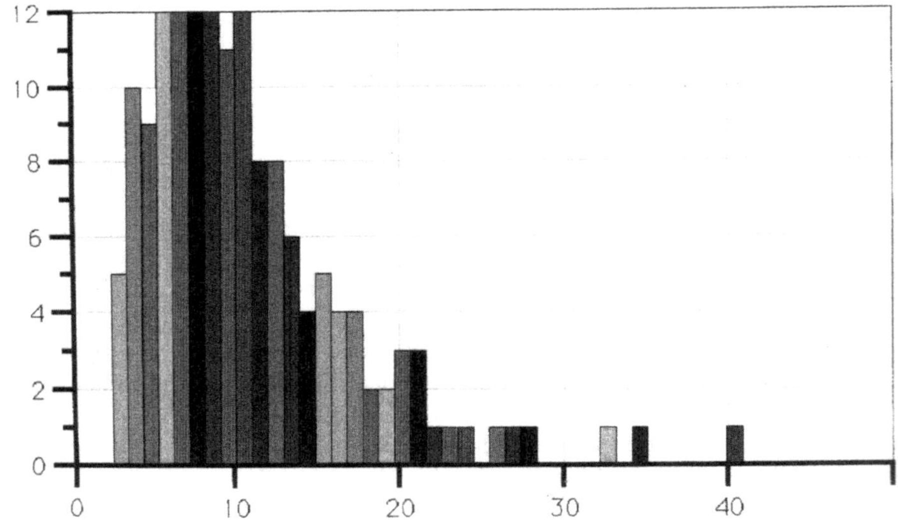

B

Fig. 1A,B. Electrically stimulated sensory thresholds. A histogram of the number of instances (on the *vertical axis*) versus stimulating current (grouped in 1 mA bins; e.g., 0–0.99, 1–1.99, etc.). Data for all right wrist/hand measurements for a population of 181 patients. **A** shows "raw data" and **B** shows data corrected for the patient's size. These data are the best obtained from all four limbs for illustrating the smoothing effect of the size correction and shows the very characteristic shape for sensory threshold histograms for all limbs. Note the relatively fast rise from 0 to a maximum, a maximum plateau, and then a slow decay to instances of 1 and 0

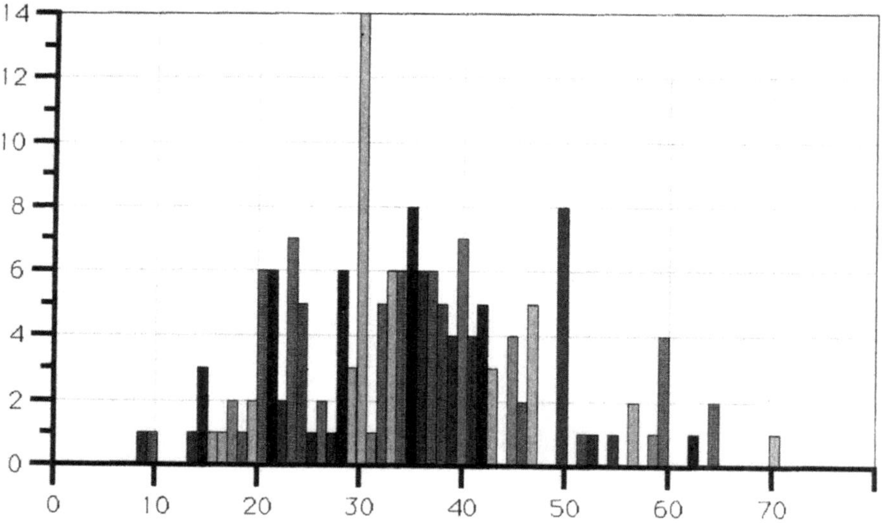

A

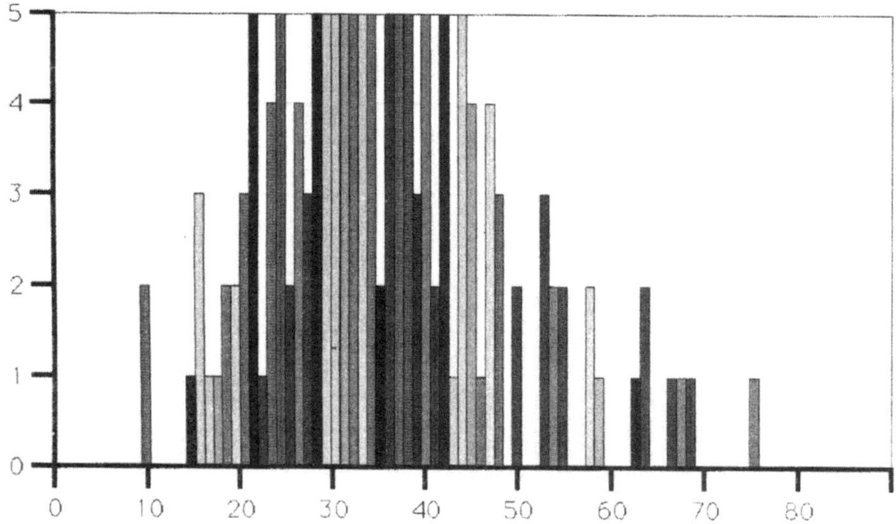

B

Fig. 2A,B. Electrically stimulated motor thresholds. A histogram of the number of instances (on the *vertical axis*) versus stimulating current (grouped in 1 mA bins; e.g., 0–0.99, 1–1.99, etc.). Data for all right ankle/leg measurements for a population of 181 patients. **A** shows "raw" data and **B** shows data corrected for the patient's size. These data are the best obtained from all four limbs for illustrating the "smoothing" effect of the size correction and shows the very characteristic shape for motor threshold histograms for all limbs (quite different from the sensory shape seen in Fig. 1). **B** Note the rather symmetrical distribution above and below a maximum cluster plateau (from approximately 21–44 mA), with a slightly slower decay at the higher currents

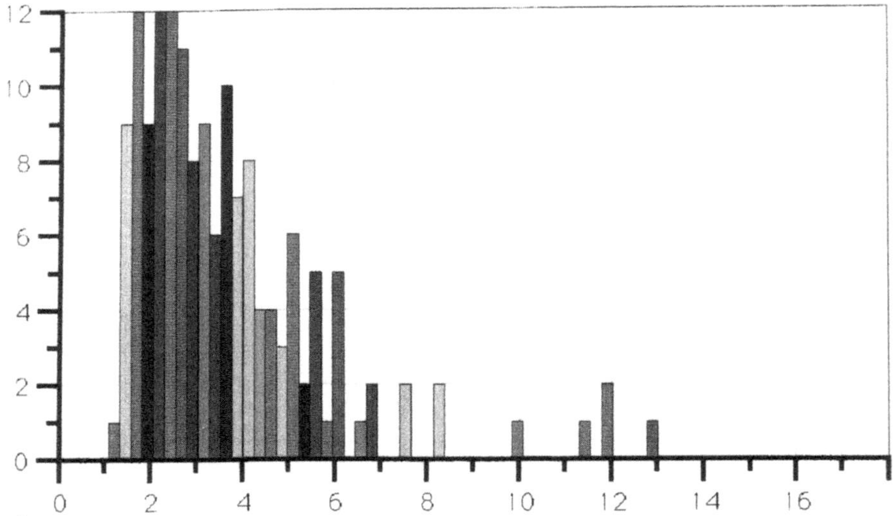

A

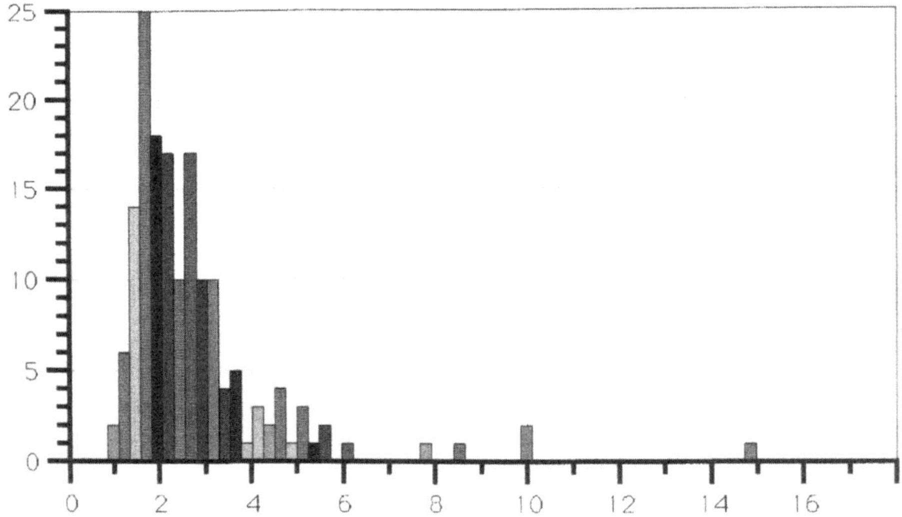

B

Fig. 3A,B. Motor to sensory ratios. Typical motor/sensory ratio histograms for **A** the right wrist/hand and **B** the right ankle/leg. Note the characteristic shape of such histograms (essentially the same for all limbs); i.e., a relatively fast rise from 0 to a maximum peak, and then a slow decay to 0

calculate the motor to sensory ratios, hopefully removing some of the individual size/shape differences by using the patient for his/her own control in the calculation of this dimensionless number (see Fig. 3).

In the example histograms of the sensory thresholds (Fig. 1), it can be seen that the sensory plot begins as a steep rise to a maximum plateau and then descends slowly to a few instances at higher currents, at least some of which may represent thresholds associated with preoperative sensory deficits. The range of currents is typically a ratio of 1:5 for approximately 90% of the population (a surprisingly large spread). In contrast, the example motor threshold histograms (Fig. 2) appear to be more symmetrical monotonic distributions (centered on a maximum plateau) with a range of currents which is a ratio of approximately 1:5 for about 90% of the population. The means and standard deviations of the above data are given in Table 1.

As mentioned above, it has been our general observation that wrist and ankle thresholds increase with wrist and ankle thickness. In turn, wrist thickness seems to vary with patient size – variability being secondary to just being symmetrically larger or to varying degrees of obesity or thinness. In retrospect, it would have been useful to have measured wrist and ankle dimensions – either circumstance and/or major or minor axes of an approximate ellipse – along with an estimate of subcutaneous soft tissue. However, not having these measurements, we have derived a formula for making size corrections based on the following assumptions:

1. The thresholds can be normalized to a standard size wrist or ankle.
2. The thresholds vary directly and linearly with thickness (diameter) of the wrist and ankles (i.e., related to distance between skin surface and deeper structures).
3. Congruent body enlargement means that all major dimensions change

Table 1. Simple statistical indices of electrical thresholds

Position of electrodes/ threshold measured		Measured data MEAN	SD	Corrected data MEAN	SD	Ratio M/S MEAN	SD
RH	M	23.85	8.33	23.85	8.36	3.38	2.18
	S	9.65	6.38	9.69	6.55		
LH	M	20.75	7.12	20.90	6.96	3.13	2.14
	S	9.27	6.66	9.33	6.77		
RL	M	34.54	11.63	34.45	12.09	2.44	1.79
	S	19.2	11.65	19.10	11.85		
LL	M	31.6	11.94	31.41	11.83	2.50	1.8
	S	16.52	10.06	16.46	10.24		

Means and standard deviations for data from all 181 patients included in the study. Measured data are raw data and corrected data are data recalculated to correct for body size (e.g., height, weight, thin, obese, etc.).

RH, right wrist/hand; *LH*, left wrist/hand; *RL*, right ankle/leg; *LL*, left ankle/leg; *M*, motor; *S*, sensory; *M/S*, motor to sensory ratio, *a* dimensionless number

equally and therefore changes in weight or volume are by a ratio that is related to the cube of any one dimension (e.g., height).

4. Non-congruent body size changes (i.e., thinness or obesity) result in diameter dimensional changes (e.g., a thicker wrist for a given height).

The formula for transforming the raw data (measured) into normalized data starts with a standard person with a height of 66 inches and a weight of 66 kg (these mixed units were chosen for numerical convenience). We then assume that size/shape relates to congruency. That is, one can have small, normal and large size people of the same shape. Also, one can vary shape independent of height (i.e., obese or thin). The formula derivation was carried out to correct for height (i.e., increase or decrease in height with normal shape) and then to correct for thinness or obesity, independent of height. The derivation assumes that all body parts can be approximated as cylinders or truncated cones. The derivation is quite simple but lengthy and beyond the scope of this paper. The final formula is as follows (where I_{cal} = corrected-calculated current, and I_{meas} = current measured):

$$I_{cal} = I_{meas} \times \frac{1}{\sqrt{\frac{\text{weight}}{\text{height}}}}$$

This formula was used for the calculations to make the transformation from raw data to corrected data (i.e., Figs. 1, 2).

It is important to emphasize certain relationships made apparent by plotting of the histograms. Only six examples of the 20 histograms generated (five for each wrist or ankle – sensory (S) and motor (M), corrected S or M, and M/S ratio) are included in this paper, but it was of considerable interest to note that all of the sensory plots (i.e., for all four limbs) had essentially the same shape (as seen in Fig. 1). It is also important to note that the shapes of the histograms are different for sensory and motor thresholds, and that all motor histograms are very similar. The sensory distributions are very unsymmetrical (a fast rise to a peak and then a slow decay), and the motor distributions are essentially symmetrical about a maximum plateau, as seen in Fig. 2. Of considerable importance is the fact that all the distributions are broad. That is, on a population basis there are no sharp thresholds and the range of thresholds is quite large (i.e., greater than 1:5). A secondary way of looking at the data is to convert the raw sensory and motor data for a particular site of stimulation to a ratio. This presumably would correct for at least some of the size of shape differences. Such derived data is shown in Fig. 3.

In terms of statistical manipulation (see Table 1), it is of interest to note that there is a great deal of uniformity, given the type of response (motor or sensory) for a particular limb stimulated (i.e., left and right resemble each other). Also it is to be noted that the standard deviations are similar and that the corrections secondary to the application of the size/shape formula do not make any clear differences in means and standard deviations. Also note that the ratios of motor to sensory currents for the hands are very similar and distinctly different from those for the legs (which are also very similar to each other).

As can be seen in Figs. 1 and 2, even though the corrections for size and shape do not change the means and standard deviations, there is a distinct pattern of smoothing of the data. It is very likely that data of this sort should be analyzed by a different type of statistical procedure which would indicate the efficacy of a smoothing function (i.e., the size/shape corrections).

Discussion

It is clear that the original objectives to obtain thresholds to allow calculating operating currents during the case were met and that in many cases inappropriately high or low intraoperative currents were avoided by knowing ahead of time what was appropriate for that particular patient. Secondly, on more than one occasion, flaws in the stimulating system (e.g., a bad electrode cable) were discovered before beginning the case.

In regard to being able to correct the raw data for the size/shape of the patient, the results are encouraging but obviously not definitive. The example histograms in Figs. 1 and 2 do indicate that there is some degree of smoothing of the data when it is corrected. However, we do not yet have a proven mathematical procedure for measuring what is apparent to the eye (i.e., smoother histograms) and what apparently makes good sense in terms of common clinical observations (that thicker wrists and ankles require a higher stimulating current).

Our general clinical experience and our preliminary look at the summaries of preoperative neurological deficits would definitely indicate that many of the higher current threshold values are associated with patients who have neurological deficits. However, it is also clear that some patients with neurological deficits do not have increased thresholds. In regard to sensory thresholds, it must be remembered that our electrode configurations span more than one dermatome (e.g., C7 and C8 for the hand and L4, L5, and S1 for the ankle). It is entirely possible that a deficit in one area would be obscured by a normal response in another area. This would be analogous to having a visual field defect, but still being able to detect a spot of light in that portion of the field remaining intact. In some instances, it would be desirable to actually use dermatomal stimulation [2] since in our experience most SCM problems seem to be nerve root related. For the motor systems, it is possible that once again a deficit would be missed by virtue of the fact that we stimulate more than one nerve. For example, it is possible that if there were a nerve fiber deficit in one of several nerves being stimulated, the threshold (of the observer) for being able to detect motor movement may not necessarily be changed if the nerve-muscle being stimulated/observed is not the deficient one.

It is important to point out that the sensory thresholds are probably dependent on the patient being able to communicate accurately what he or she senses. This certainly may be influenced to a certain extent by the presence of premedication. In regard to motor thresholds, the observer is "in the loop" and this person's own perception is part of the final result. It certainly might have been better to use an electromechanical detector of movement, although this would have been a much more complex arrangement. In a general way it should be noted that in the long run it is probably better to calculate the stimulating currents as milliamps per square centimeter and to look more closely at the volume of tissue through which the current is passing relative to the geometry of the individual nerves. Also, it is important to point out that it is not at all clear whether the first neural response is from stimulating deep trunk fibers (likely in the case of the motor responses), or whether with sensory thresholds the site of stimulation may be in the skin or at the deep nerve trunks or at any point in between. Regardless, it is fair to observe that there still seems to be a systematic pattern in that sensory results are very similar (for all limbs) and motor results are very similar (and different from sensory). Also, in all histograms a good correlation exists between the right and left sides.

In summary, certain clinical lessons and warnings can be emphasized:
1. Be suspicious of thresholds far out on the usual distribution and any unusual motor to sensory ratio.
2. Take the size and shape of the patient into account.
3. Be aware of the neurological history and physical examination.
4. Compare results from one side of the body to the other.
5. Modify expectations of SCM in reference to an abnormal history or physical examination, to abnormal thresholds, and to abnormal motor to sensory ratios.

In the long run, the electrical stimulation of wrist and ankles should be as systematic as would be the stimulation in any other sensory system. For example, in stimulating the visual system, it is certainly clear that the intensity of the signal, its spectral content, and its position on the retina as well as size, shape, and temporal pattern are important and would be related to results with electrical stimulation of any part of this system. It is likely that the control of somatosensory stimulation via electrical means (i.e., electrodes on the skin or in the subcutaneous tissue) is in a relatively primitive state regarding our knowledge of all of the variables such as site, type of electrodes, intensity, and current density. In our opinion, much work remains to be done in this area to help clarify and refine stimulus control so that we can be sure we have optimized peripheral stimulus conditions for obtaining optimal evoked potentials for spinal cord monitoring.

References

1. Frazier WT, Odom SH, Biggs DB (1985) Anesthetic technique for spinal cord monitoring. In: Schramm J, Jones SJ (eds) Spinal Cord Monitoring. Springer, Berlin Heidelberg New York Tokyo
2. Toleikis JR, Sloan T, Schrader S, Koht A (1985) Scalp distribution of dermatomal evoked potentials. In: Schramm J, Jones SJ (eds) Spinal Cord Monitoring. Springer, Berlin Heidelberg New York Tokyo

Somatosensory Evoked Potential Recordings for Decision Making on Instrumentation During Scoliosis Surgery

Y. Node[1] and S. Uematsu[2]

Abstract

Somatosensory evoked potential (SEP) recordings for 270 patients who underwent corrective surgery for kyphoscoliosis were reviewed. When the SEP was normal during the instrumentations, the chance that the patient would awaken from the anesthesia with paralysis of the limb was 0.4%. SEP abnormalities occured most frequently during instrumental manipulation of the spinal column. The surgeon could adjust the distraction rods or compression apparatus whenever the potential deteriorated. The results reconfirmed the validity of intraoperative SEP and provides important information for surgeons to use in decision making during critical maneuvering of the spinal cord and surrounding neurovascular structures.

Introduction

The main purpose of recording somatosensory evoked potentials (SEPs) during corrective surgery for kyphoscoliosis is detection of an abnormality before irreversible cord injury occurs. MacEwen et al. [4] reported that the incidence of neurological complication with scoliosis surgery is 0.72%. More recently, Wilber et al. [9] reported higher incidence, 17%, of complications with use of more complex instrumentation. As instrumentation has become more sophisticated, intraoperative monitoring of SEPs has become an essential part of spinal surgery [1–3, 5, 6]. This paper reviews the value of intraoperative SEPs for decision making during spinal surgery for scoliosis.

[1] Department of Neurosurgery, Nippon Medical School, 1-1-5 Sendagi, Bunkyo-ku, Tokyo 113, Japan
[2] Department of Neurosurgery, Johns Hopkins University, 600 North Wolfe Street, Baltimore, MD 21205, USA

K. Shimoji et al. (Eds.)
Spinal Cord Monitoring and Electrodiagnosis

Materials and Methods

The subjects of this study were 270 patients with scoliosis, aged 16 or above. All patient records of evoked potential studies over the 8 years prior to the study were reviewed by one of us (Y.N.). Of these patients 71 were male and 199 were female. The mean age was 30.8 years, with the oldest patient being 69 years old. The majority (239 cases) of patients had idiopathic or congenital scoliosis, and 31 had scoliosis due to neuromuscular diseases. The surgical procedures consisted of using various types of instrumentation, i.e., using Harrington distraction rods and/or Luque or Cotrel-Dubousset rods, or the Wisconsin compression apparatus.

Nitrous oxide, isoflurane, and nondepolarizing neuromuscular blocking agents were the most commonly used anesthetics. Morphine or a semisynthetic derivative (fentanyl) was used in some patients for induction and maintenance of anesthesia. Intraoperative blood pressure, rectal temperature, and blood gases were routinely recorded.

The SEPs were recorded with a TECA-TE42 (TECA Co., White Plains, New York) electromyograph, equipped with a DA6 digital averager and a nerve stimulator. Technical details are described elsewhere [8]. Major components of SEPs identified were P1, N1, P2, N2, P3, and N3, according to the polarity and sequence of occurrence. Downward polarity was considered positive and upward negative. When the amplitude dropped more than 50% from the baseline, it was called abnormal. When the latency was prolonged more than 20% for P1 or N1, and more than 30% for P2, N2, or P3 and N3, it was called abnormal.

Results

The study showed that when intraoperative SEPs were normal, the chance of surgical complications was less than 1%. Of 245 cases with normal intraoperative SEPs, all but one, i.e., 244 cases, woke up from anesthesia without surgical complications (Table 1). The exception, a 26-year-old achondroplastic

Table 1. Use of SEP in predicting complications during scoliosis surgery

Type of SEP recording	No of cases	%
Total SEP recording	270	
Normal SEP	245/270	
with complications	1/245	0.4
without complications	244/245	99.6
Abnormal SEP during surgery	13/270	4.8
Technically unsatisfactory records	12/270	4.8

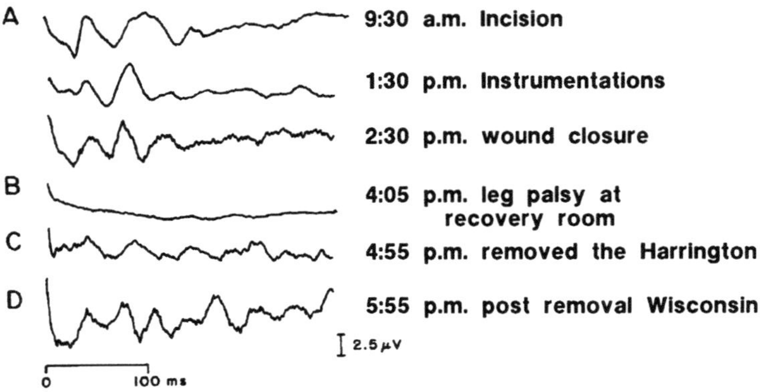

Fig. 1. SEPs of a 26-year-old achondroplastic man (*A*, baseline recordings; *B*, recording taken at recovery, SEP was absent; *C*, recording at time of removal of hematoma and Harrington distraction rods, the SEP recovered to 55% of the baseline amplitude, but remained poor quality; *D*, recording taken 25 min after removal of entire instrumentation including Wisconsin compression apparatus, the amplitude returned to baseline values)

patient with severe kyphoscoliosis at the thoracolumbar junction, had normal SEPs throughout the procedure. The SEPs did not change during the application of Harrington rods or Wisconsin compression apparatus to correct spinal curvature. However, upon the patient's awakening from anesthesia in the recovery room, it became apparent that he could not move his toes, quadriceps, and hamstring muscles. Postoperative SEP recordings confirmed that the evoked potential was absent (Fig. 1). Exploration of the wound uncovered a hematoma. With irrigation of the hematoma and removal of the Harrington distraction rod, the SEP recovered to 55% of baseline amplitude; however, once wound closure had been performed, the amplitude began to drop again, and therefore, decision was made to remove the Wisconsin compression apparatus. The SEP promptly return to baseline. The time until removal of the Harrington rod was 2 h and 20 min after final distraction of the spine and the Wisconsin apparatus was removed 30 min after the removal of the Harrington rod. The wound was once again closed, and SEP remained

Table 2. Duration of abnormal SEP and surgical decision

Duration of abnormality (min)	No of cases	Effect on decision making and on instrumentation
< 5	5	Did not affect
5 - 10	3	Adjustment of distraction, followed by "wake-up" test
> 20	3	Removed all instruments
25	1	BP 50/25 mm Hg
360 (6 h)	1	A case with myelopathy

stable. The patient was awakened and returned to his bed and was able to move both legs to command at this time. Careful neurological examination later found that he had moderate sensory and motor deficits limited to the lower extremities. His deficit completely returned to normal at his postoperative clinic visit 4 months later.

In 13 of 270 patients (48%), the intraoperative SEPs were interpreted as abnormal (Table 2). In all 13, the abnormality was encountered during placement of the distraction rods or compression apparatus. The P3 and N3 components of the SEP dropped to half (50%) or less of the baseline amplitude in all 13 patients. A drop of P2 and N2 was noted in 11 of 13 patients, and in P2 and N2 in 9 of 13 patients. Prolonged conduction time (latency) was apparent in only half or less of the 13 patients.

A transient abnormality of the SEP (less than 5 min) was observed in 5 cases, but baseline values were recovered spontaneously without direct influence on surgical decisions (Fig. 2). However, in 6 cases the SEP abnormality persisted for more than 5 minutes, requiring adjustment of the distraction or compression apparatus or complete removal of the instrumentation. The SEP recovered to the baseline level in all cases before closure of the surgical wound. All 13 patients had no neurological deficit upon recovery from general anesthesia (Table 2). In one case, the SEP disappeared during the instrumentation; however, it was promptly restored when the blood pressure was normalized from extreme hypotension: 50 mm Hg systolic, 25 mm Hg

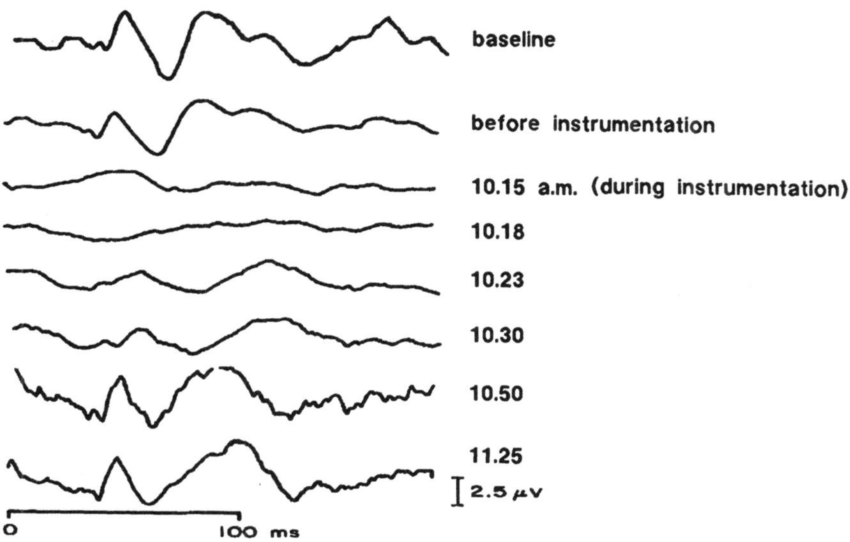

Fig. 2. SCEP of a 38-year-old woman during instrumentation. The SEP disappears transiently for 3 min as is apparent on the recordings taken at 10:15 a.m. during application of the instrumentations. The SEP recovered to 50% of the baseline level 8 min later at 10:23 a.m. Application of the instrumentation was completed using the Harrington rod and Wisconsin apparatus without complications

diastolic (mean BP 34 mm Hg). The last case showed poor SEPs from the beginning of surgery to the end. The patient had involvment of an undefined spinal cord degenerative disease associated with kyphoscoliosis.

Twelve patients (4.4%) had unsatisfactory SEPs. The recordings were not interpretable due to poor quality with apparent contamination of electromyographic activity, 60-cycle or other high-frequency electrical noises.

Discussion

An experimental spinal cord compression study in cats indicated that the SEP drops instantly, within 250 ms, upon compression of the dorsal aspect of the spinal cord [7]. The SEP is less disturbed by hypotension (hypovolemia), hypoxia, or hypothermia until a critical level is reached for some length of time [7].

Intraoperative use of SEP recording during kyphoscoliosis surgery, where the cord is apt to be compressed, is therefore effective. For this reason, SEP monitoring is essential to avoid catastrophic neurological complications during this procedure [1–3, 5, 6].

Our study indicated that normal SEP gives a "green light" with a confidence level of over 99% to the surgeon during surgical manipulation of the spine. SEP recording should continue until the completion of wound closure, and it may be worthwhile to repeat SEP recording during the recovery period from general anesthesia after transfer from the operating table to the hospital bed and until the patient is fully awake. In our series, in only one patient was the SEP discontinued after the patient was moved from the operating table to his hospital bed. An apparent neurological deficit was noted upon postoperative neurological examination in the recovery room. The SEP recovered to baseline upon complete removal of the instruments but it took 4 months for the patient to recover completely from the neurological deficit.

The deterioration of the SEP after closure of the wound could be distortion or malalignment of the spine, compressing the neurovascular structures around and within the spinal canal. Other possibility is hematoma or edema of the cord and soft tissue around the cord. It also could be due to restoration of the paraspinal muscle tone as the muscle recovers from the neuromuscular junction blocking agents used during anesthesia. Another possibility is that simply transferring patient from the fixed body position used on the operating table to the rather soft hospital bed at the end of the surgery distorts the spine. Based on the review of our experience, we recommend the guidelines shown in Table 3.

Artifacts were not studied, although the problem occurred in 4.4% of our cases and continued to be troublesome. Preoperative SEP recording, including particularly peripheral nerve conduction study, is worthwhile particularly when dealing with a patient with known peripheral neuropathy

Table 3. Guidelines for surgeon if SEP becomes abnormal

Duration of abnormal SEP	Recommended procedure
< 5 min	Check recording system electrode contact, grounding, etc. SEP on median nerve stimulation notify surgeon and anesthesiologist, check cardiovascular and respiratory system, etc.
5 – 10 min	Reduce distraction or compression
15 – 20 min	Remove all instruments
> 20 min	Prepare "wake-up" test

or myelopathy. SEPs from these patients may not be reliable and it would be better to prepare for the "wake-up" test. If the patient has not been instructed beforehand, even the "wake-up" test on the operating table may be less reliable.

Deterioration of SEPs should alert not only the surgeon but also the anesthesiologist to the cardiovascular and respiratory status of the patient. In one case, the SEP change coincided with a drop in blood pressure to 50/25 mm Hg.

Our records indicated that the later components of SEP, N3 and P3, were more helpful indicators. One should not be misled by the relatively intact appearance of the early components N1 and P1. We have the impression that N3 or P3 tend to show deterioration before the earlier components do. Study of latency is not helpful in monitoring spinal surgery.

We conclude that abnormal SEPs, even transient ones, indicate that the cord, particularly the dorsal column, the large myelinated ascending sensory tract of the cord, could be in serious trouble. Surgeons should take decisive action to avoid catastrophic neurological deficit without delay. The SEP should be recorded throughout the procedure from the beginning of the induction of anesthesia to the patient's awakening from general anesthesia.

References

1. Brown RH, Nash CL Jr, Berilla JA, Amaddio MD (1984) Cortical evoked potential monitoring – A system for intraoperative monitoring of spinal cord function. Spine 9:256–261
2. Dinner DS, Luders H, Lesser RP, Morris HH, Barnett G, Klem CT (1986) Intraoperative spinal somatosensory evoked potential monitoring. J Neurosurg 65:807–814
3. Lesser RP, Randzens PA, Luders H, Nuwer MR, Goldie WD, Morris HH, Dinner DS et al. (1986) Postoperative neurological deficits may occur despite unchanged intraoperative somatosensory evoked potentials. Ann Neurol 19:22–25
4. MacEwen GD, Bunnell WP, Shiram K (1975) Acute neurological complications in the treatment of scoliosis. J Bone Joint Surg 57[A]:404–408
5. Mostegl A, Bauer R, Eichenauer M (1988) Intraoperative somatosensory potential monitoring. A clinical analysis of 127 surgical procedures. Spine 13:396–400

6. Roy EP 3rd, Gutmann L, Riggs JE, Jones ET, Byrd A, Ringel RA (1988) Intraoperative somatosensory evoked potential monitoring in scoliosis. Clin Orthop 229: 121–124
7. Uematsu S, Rocca U (1981) Effect of compression, hypoxia, hypothermia and hyporelemia on the evoked potentials of the spinal cord. Electromyogr Clin Neurophysiol 21:229–252
8. Uematsu S, Tolo V (1981) Recording of the somtosensory, evoked potentials during surgery for scoliosis and midline myelotomy to monitor spinal cord function. Electromyogr Clin Neurophysiol 21:253–266
9. Wilber RG, Thompson GH, Shaffer JW, Brown RH, Nash CL (1984) Postoperative neurological deficits in segmental spinal instrumentations: A study using spinal cord monitoring. J Bone Joint Surg 66[A]:1178–1187

Incidence of Spinal Cord Impairment Related to Extent of SEP Decline and Recovery, Aetiology and Instrumentation in Operations for Scoliosis

C.S. Waller[1], J.M.H. Paterson, M.A. Edgar, and S.J. Jones

The generators and conduction properties of the spinal somatosensory evoked potentials (SEPs) recorded from the epidural space following stimulation of nerve trunks in the lower limbs are now well understood [2–3]. On account of their relative stability and speed of acquisition, spinal SEPs are often preferred to cortical SEPs for monitoring the spinal cord during both orthopaedic and neurosurgical procedures [1, 4, 6–10]. Here we present a retrospective analysis of a large number of scoliosis operations, to examine the incidence of neurological complications in relation to: (a) the maximum degree of SEP deterioration, (b) the extent to which the SEPs subsequently recovered, (c) the surgical manoeuvres which may have been responsible and (d) the type of patients who were most at risk.

Clinical and Surgical Methods

During the period 1981–1987, 754 consecutive patients undergoing posterior spinal fusion and instrumentation were monitored. There were 511 females and 243 males, aged 6–48 years (mean 15.1 years). Of these cases, 68% were idiopathic scoliosis, most of the remainder being congenital osteogenic deformities or neuromuscular scoliosis. The majority of patients were instrumented with a single Harrington distraction rod or a combination of compression and distraction rods. Neuromuscular cases (Duchenne dystrophy or spinal muscular atrophy) were instrumented with Luque apparatus. The 152 most recent idiopathic cases were treated with Harrington rods and multiple level sublaminar wires, secured to the rod for more rigid fixation. In all except the muscular dystrophies and atrophies, hypotensive anaesthetic techniques were used. Post-operative neurological sequelae were classified as follows:

1. No impairment (no symptoms or abnormal neurological signs)
2. Mild impairment (numbness or paraesthesiae in the limbs or trunk, abnor-

[1] Scoliosis Unit, Royal National Orthopaedic Hospital, London, UK

K. Shimoji et al. (Eds.)
Spinal Cord Monitoring and Electrodiagnosis

malities on sensory testing or change in tendon reflexes without motor weakness)
3. Severe impairment (demonstrable motor weakness ranging from foot drop to paraplegia, with or without sensory abnormalities)

Monitoring Methods

The recording electrode was a 3 mm length of stainless steel tubing, 1 mm in diameter, at the end of an insulated lead of the same diameter. The reference electrode was either a stainless steel needle inserted in paraspinal muscles (monopolar configuration) or, in more recent cases, a second length of stainless steel tubing mounted on the same lead with an inter-electrode distance of 2 cm (bipolar configuration). The lead was introduced to the dorsal epidural space between the exposed vertebrae at the rostral end of the exposure and advanced cephalad by approximately two levels. The stimulus was a square-wave impulse 0.1 ms in duration, delivered to the posterior tibial nerve at the popliteal fossa with an intensity sufficient to cause a muscle twitch (prior to administration of muscle relaxants). The stimulus frequency was 20/s. Amplification was with a band-pass of 200–2000 Hz and typically 300–400 responses were averaged over an epoch of 30 ms with a sampling frequency of 17 kHz. The left and right legs were stimulated alternately throughout the operation and recordings were made until the electrode was removed after insertion of the final sutures.

Results

Eighteen patients (2.4%) were excluded on account of technically inadequate recordings (Table 1). Some of these had a pre-existing neurological problem

Table 1. Incidence of SEP deterioration and neurological complications in 754 operations for scoliosis

SEP deterioration category of deterioration	No of cases		Neurological complications (No of cases)
Undetermined (technical failure)	18		0
Less than 50%	679		0
More than 50%:			
Transient	27		0
Persistent	30	Mild	9
		Severe	6
Total	754		14

but none suffered any further impairment of spinal cord function as a result of the operation. In 679 cases, identifiable responses were stably recorded throughout the procedure, stability being defined as no more than a 50% drop in amplitude compared with the initial baseline recordings and no marked change in the SEP waveform. Experience with the first 410 cases had established that a 60% attenuation was associated with a tenfold increase in the risk of neurological complications [4], whereas 30% was considered to be the approximate limit of non-pathological variability [5]. None of the 679 patients with stable SEPs suffered any post-operative neurological complications attributable to a spinal cord lesion, although two showed minor evidence of sacral root involvement which would not have been detected by tibial nerve SEPs. A gradual increase in the latency of the response by up to 1.0 ms was commonplace and probably due to cooling; latency increase without significant loss of SEP amplitude was not found to be associated with neurological complications.

In 57 patients (7.6%) the SEP dropped in amplitude by more than 50% on one or both sides at some stage during the operation. In 27 cases the decrement was rapidly reversed, either spontaneously or after re-siting the recording electrode; none of these suffered neurological complications. The remaining 30 patients (4.0%) with persistent SEP falls of more than 50% included all 14 (1.85%) who suffered post-operative complications, and were therefore analysed in detail.

The group comprised 22 females and 8 males, aged 6–46 years (mean 15.6 years). There were 19 patients with idiopathic scoliosis, four with congenital osteogenic deformities, six with neuromuscular scoliosis and one with a post-traumatic kyphosis. Mild post-operative neurological complications occurred of nine cases and severe complications in five, while the remaining 16 had no deficits. In every case there had been a significant SEP decrement with stimulation of the leg ipsilateral to the side of neurological impairment. The incidence of complications was particularly high in those patients for whom the lowest recorded SEP amplitude was less than 20% of the baseline value (Table 2). The incidence of complications was also greater in the 15

Table 2. Incidence of neurological complications in relations to degree of SEP attenuation and recovery (% of initial amplitude)

	Neurological complications		
	None	Mild	Severe
Lowest amplitude:			
20–50%	14	5	2
Less than 20%	2	4	3
Final amplitude:			
More than 50%	10	4	1
30–50%	4	2	2
Less than 30%	2	3	2

Table 3. Incidence of neurological complications in relations to surgical events provoking SEP attenuation and recovery

Surgical event	Number of occasions implicated	Neurological complications		
		None	Mild	Severe
Decortication	3	1	2	0
Hook placement	0	0	0	0
Distraction	11	8	2	1
Wire insertion	4	2	0	2
Wire tightening	10	4	4	2
Other	2	1	1	0
Distraction released	(11)	6	3	2
Distraction sustained	(19)	10	6	3

patients for whom the final SEP amplitude remained below 50% of baseline, as compared with the 15 for whom there was improvement to above 50%.

The same 30 cases were examined to determine at what stage of the operation the SEP reduction occurred (Table 3). Most frequently (14 cases) this was during the insertion or tightening of sublaminar wires, and more than half of these showed post-operative complications. The significance of this finding is heightened by the fact that sublaminar wires were used in less than on quarter of all operations. In 11 patients, the SEP reduction occurred during distraction of the spine, although neurological complications occurred in only three of these. No significant difference in the incidence of complications was found according to whether or not the distraction force was released.

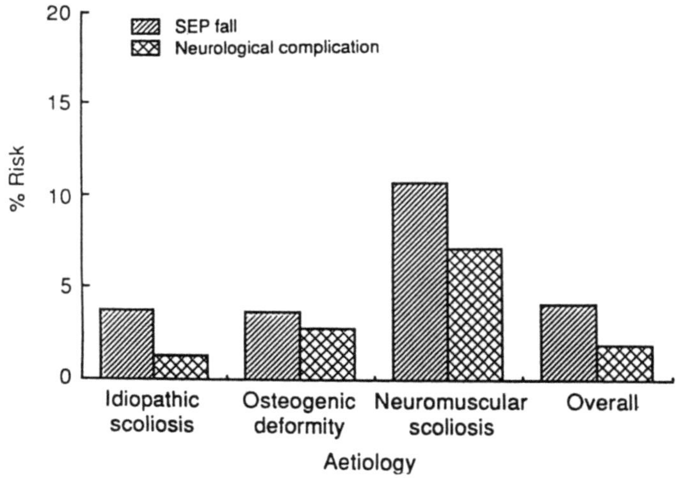

Fig. 1. Percentage incidence of persistent SEP attenuation by more than 50% and post-operative neurological complications in patients with idiopathic, osteogenic, and neuromuscular scoliosis

Sublaminar wires were always slackened when significant SEP deterioration occurred, although it is not possible to conclude that this was directly responsible for the SEP recovery which often ensued.

The incidence of SEP reduction was greatest in patients with neuromuscular scoliosis (10.5%), as was the risk of neurological impairment (7.1%; Fig. 1). The incidence of SEP loss was similar in osteogenic and idiopathic cases (3.6% and 3.8% respectively) but the former had a much higher incidence of neurological complications (2.7% compared with 1.3%). Neither the incidence of significant SEP attenuation nor the development of postoperative complications was related to the age or sex of the patient, the severity, direction or stiffness of the curve or the degree of correction achieved.

In three cases the SEP reduction occurred when the systolic blood pressure fell below 50 mm Hg. In one of these these was an associated episode of ventricular tachycardia during final wire tightening. In two out of three cases, the SEP failed to recover to above 50% of baseline and the patients suffered post-operative complications, one mild and one severe. In the third case the SEP returned to normal when the blood pressure was increased and no neurological abnormalities ensured.

All patients who sustained neurological complications recovered to some degree. Of the five with severe complications, two recovered completely and three partially over a period of months. No patient was wheelchair-dependent in the long term. Of the nine patients with mild complications, seven recovered completely and two continued to report localised truncal numbness months or years after surgery, suggesting involvement of the dorsal roots or their associated ganglia. In both of the latter, there were no clinical signs of a spinal cord lesion at any time, in spite of the SEP reduction. This reinforces the belief that, in these as well as others whose SEP reduction was associated with no neurological complications, there may have been a real cord impairment which was reversed before becoming clinically significant.

Discussion and Conclusions

In the previous Symposium, we presented data relating to SEP amplitude and latency variations occurring an uneventful operations, and therefore considered non-pathologic [5]. It was concluded that amplitude fluctuations exceeding 30% were very unlikely to occur spontaneously. Here our object was to evaluate an empirically chosen criterion of 50% amplitude loss, considerably beyond the range of "normal" variability, by retrospective analysis of all 754 patients, who were monitored using similar techniques in one hospital over a 7-year period. We also wished to determine what other factors might contribute to morbidity, such as the aetiology of the deformity and the surgical procedures employed. The main conclusions were as follows:

1. SEP amplitude reduction exceeding 50%, and not rapidly reversed either spontaneously or by technical manipulation, is an appropriate criterion with which to identify a very large proportion of patients who subsequently demonstrate mild or severe neurological complications. In this series the incidence of correct detections was 1.9%, of technical failures 2.4%, of false positives 2.1%, and of false negatives 0%, although some sensory abnormalities of a minor nature may have gone undetected. The incidence of mild and severe complications was 1.2% and 0.7% respectively, but a zero incidence of paraplegia compares favourably with previously published figures. The possibility that ischaemic damage in the distribution of the anterior spinal artery may be missed when only the sensory pathways are monitored remains a real one, but we have encountered no such cases.
2. The danger of neurological impairment is reduced when the SEP can be restored to an amplitude above 50% of its initial baseline value, although this appears not to depend to any large degree on whether or not the distraction force is removed.
3. Although many SEP reductions occurred soon after distraction, proportionately many more were apparently provoked by insertion and tightening of sublaminar wires, and the occurrence of neurological complications was still more strongly associated with this procedure. From the available data, it is not possible to discover whether removal of the wires alone (without reduction of the distraction force) reduces the danger of postoperative complications.
4. The patients at greatest risk of SEP attenuation and (particularly) neurological complications were those with neuromuscular scoliosis. Clearly, in such cases any additional motor impairment is liable to be clinically significant and should be avoided if at all possible.
5. SEP reduction occurred in two patients, for whom the only postoperative defects were sensory loss in the trunk, i.e., there was not clinical evidence of a spinal cord lesion. This provides support for our belief that SEP reductions which have no neurological sequelae should not necessarily be regarded as false positives. Rather, a certain proportion (the exact number of which can never be determined) may be precisely those patients who benefit most from monitoring, having defects which are detected early enough to be fully reversible before any clinically significant neuronal damage can ensue.

References

1. Bradshaw K, Webb JK, Fraser AM (1984) Clinical evaluation of spinal cord monitoring in scoliosis surgery. Spine 9:636–643
2. Halonen J-P, Jones SJ, Edgar MA, Ransford AO (1989) Conduction properties of epidurally-recorded spinal cord potentials following lower limb stimulation in man. Electroencephalogr Clin Neurophysiol

3. Jones SJ, Edgar MA, Ransford AO (1982) Sensory nerve conduction in the human spinal cord: epidural recordings made during scoliosis surgery. J Neurol Neurosurg Psychiatry 43:495–503
4. Jones SJ, Carter L, Edgar MA, Morley T, Ransford AO, Webb PJ (1985) Experience of epidural spinal cord monitoring in 410 cases. In: Schramm J, Jones SJ (eds) Spinal cord monitoring. Springer, Berlin Heidelberg New York, pp 215–220
5. Jones SJ, Howard L, Shawkat F (1988) Criteria for detection and pathological significance of response decrement during spinal cord monitoring. In: Ducker TB, Brown RH (eds) Neurophysiology and standards and spinal cord monitoring. Springer, Berlin Heidelberg New York Tokyo, pp 201–206
6. Macon JB, Poletti CE, Sweet WH, Ojemann RG, Zervas NT (1982) Conducted somatosensory evoked potentials during spinal surgery. II Clinical applications. J Neurosurg 57:354–359
7. Nainzadeh, NK, Neuwirth MG, Bernstein R, Cohen LS (1988) Direct recording of spinal evoked potentials to peripheral nerve stimulation by a specially modified electrode. In: Ducker TB, Brown RH (eds) Neurophysiology and standards of spinal cord monitoring. Springer, Berlin Heidelberg New York Tokyo pp 234–244
8. Romstock J, Watanabe E, Schramm J (1988) Variability of epidural SEP from below and above spinal cord lesions – the effect of the lesion on spinal SEP. In: Ducker TB, Brown RH (eds) Neurophysiology and standards of spinal cord monitoring. Springer, Berlin Heidelberg New York Tokyo, pp 261–267
9. Ryan TP, Britt RH (1986) Spinal and cortical somatosensory evoked potential monitoring during corrective spinal surgery with 108 patients. Spine 11:352–361
10. Whittle IR, Johnson IH, Besser M (1986) Recording of spinal somatosensory evoked potentials for intraoperative spinal cord monitoring. J Neurosurg 64:601–612

Pre- and Intraoperative Electrodiagnosis of Compressive Lumbosacral Nerve Root Lesions

Y. Miyawaki[1], T. Doi, K. Amano, I. Rai, H. Kagiyama, S. Nishiue, Y. Inoue, and T. Takenaka

Summary

Thirty-four patients with radiologically undetectable compressive lumbosacral radiculopathy, who had been surgically treated, were evaluated preoperatively according to electromyography (EMG), somatosensory evoked potential obtained by single root stimulation (root SEP), and elicited pain produced by root stimulation. Comparing the diagnostic accuracy of each preoperative test, the elicited pain test showed the highest accuracy rate (87.5%) and root SEP revealed abnormality with a rate of 80.8%. On the other hand, EMG detected nerve root lesions at a rate of 71.7%. However, a combination of different electrodiagnostic tests is necessary to improve the detection rate of lumbar radiculopathy. In addition, intraoperative electrodiagnosis was also conducted in 15 cases. Action potentials of the cauda equina (CEAP) and the nerve root (ERP) recorded directly from the operative site provided more precise information about the site of nerve root lesions.

Introduction

Compressive lumbosacral nerve root lesions are clinically diagnosed on the basis of symptoms and radiological examinations, such as plain radiograph, myelogram, myelographic computed tomography, and MRI. If the result of the radiograph is negative in spite of the presence of intense clinical symptoms or too many abnormal findings, it is difficult to determine which root must be decompressed. In such cases electrophysiological evaluations are of great value in diagnosis.

While there have been many reports about the application of electrodiagnosis to lumbosacral radiculopathy [1, 2, 6], in which only one or two electro-

[1] Department of Orthopedic Surgery, Osaka Rosai Hospital, 1179–3, Nagasone-cho, Sakai, Osaka, 591 Japan

K. Shimoji et al. (Eds.)
Spinal Cord Monitoring and Electrodiagnosis

diagnostic methods for the evaluation are described, it is difficult to explain the origin of symptoms in which at least three evaluation tests must be used, i.e., when muscle weakness, sensory disturbance, and pain must be tested. In addition, as various electrophysiological tests have their own limitations of diagnosis potential, a combination of various tests is essential for correct diagnosis. In this study, three electrophysiological tests of three symptoms were made for evaluation of preoperative status. Intraoperative electrodiagnosis of lumbosacral radiculopathy is not as commonly applied as spinal cord electrodiagnosis. There have been two major methods: recording action potentials of the cauda equina and the nerve root directly from the site of operation [3] and obtaining dermatomal somatosensory evoked potentials (SEPs) by electrical stimulation of the appropriate dermatome [4]. Our methods in this study are based on the former technique for the reason that diagnosis was employed not only to determine lesion segment but also to detect the accurate location of nerve root involvement.

Materials and Methods

Preoperative Electrodiagnosis

Thirty-four patients, 26 men and 8 women, aged 18–77 years were investigated (Table 1). All cases were radiologically examined before operation, by plain radiograph, myelography, myelographic computed tomography, and MRI, but unfortunately we could not determine which root had to be decompressed from radiological findings alone. Therefore, three different electrophysiological examinations were conducted before surgical decompression. For electrophysiological tests we used a DANTEC 1500 system.

Conventional needle electromyography (EMG) was performed in lower extremities. EMG was considered abnormal when it showed pathologic spontaneous activity (fibrillation, positive sharp wave) and/or neurogenic change (polyphasic motor unit, giant spike) of segmental distribution in corresponding muscles.

Table 1. Cases

	Number of patients			
	Disc herniation	Lumbar spinal canal stenosis Degenerative	 Spondylolisthesis	Combined
Preoperative Electrodiagnosis	18	7	3	6
Intraoperative Electrodiagnosis	7	3	1	4

Recording and stimulation of SEPs were done according to Matsuda's [7] and Hashimoto's [3] techniques. An electrical pulse (duration, 0.2 ms; frequency, 2 Hz) was applied to the needle electrode (DANTEC) inserted percutaneously. For stimulation of the L4 or L5 root, the electrode tip was placed just lateral to the intervertebral foramen. In order to stimulate the S1 root, the electrode was inserted into the dorsal sacral foramen. Reference electrode for stimulation was placed on the lower part of the sacrum. In some cases, other root stimulations (L2, L3, and S2) were added to aid diagnosis. SEPs induced by root stimulation (root SEPs) were recorded from a corresponding contralateral scalp location using amplification with bandwidth 2 Hz to 1 kHz. A reference electrode was placed on Fz. Root SEP was considered abnormal when it showed low amplitude, increased latency of the first component (P1 or N1) as compared to the response obtained by the contralateral intact root or just above or below root stimulation.

When weak electrical stimulation was applied to the affected nerve root, patients felt radiating pain similar to sciatica. For evaluation of electrically elicited pain, the pain drawing method [8] carried out by patients themselves was used. Patients had drawn their sciatic pain area on the human body map before stimulation. When electrical stimulation was applied to a nerve root, patients drew the radiating pain area on another map. By comparing the elicitied pain area with their spontaneous pain area, it was decided which nerve roots were affected. The stimulation frequency used for the evaluation of elicited pain was 2 Hz or 20 Hz.

Preoperative diagnosis of affected roots was determined in consideration of clinical, radiological, and electrophysiological findings. However, if the diagnosis was uncertain in cases of discrepancy among preoperative tests, or if further information was necessary to decide on decompression area, we performed intraoperative electrodiagnosis for the final diagnosis.

Intraoperative Electrodiagnosis

Intraoperative electrodiagnosis was performed in 15 patients (Table 1). Electrical stimulation was applied to the peripheral nerves of the lower extremities, i.e., peroneal, tibial, and femoral nerves, by means of a surface electrode. An active needle electrode for recording was inserted intraoperatively into the exposed yellow ligament at several segmental levels and a reference electrode was placed in the subcutaneous tissue. Before the surgical procedure of laminectomy began, cauda equina action potentials (CEAPs) and evoked root potentials (ERPs) were recorded from a needle electrode placed on the midline of the dural sac and the lateral portion of the yellow ligament, respectively, using amplification with bandwidth 20 Hz–2 kHz. ERPs recorded directly from the exposed nerve root after laminectomy were also evaluated. CEAPs and ERPs were considered abnormal when they showed a positive wave, defined as initial positive deflections with an amplitude exceeding 50% of the whole response amplitude, desyncronization, and low amplitude.

Clinical Evaluations

Clinical evaluations of the surgical results were conducted following the standards of the Japanese Orthopedic Association (JOA) [5].

Results

Preoperative Electrodiagnosis

In this study three different electrodiagnostic methods, i.e., EMG, root SEPs and elicited pain, were used for the evaluation of muscle weakness, sensory disturbance, and sciatica, respectively. Table 2 lists the relationship of clinical findings to electrophysiological findings. All electrophysiological tests revealed abnormal findings at a rate of above 80% in cases with corresponding clinical findings. Of these tests, the elicited pain method is considered most useful; it showed positive findings in 92.3% of the patients with sciatica. However, EMG and root SEPs indicated abnormal findings in 50% and 28.6% of the patients without motor and sensory loss, respectively.

The preoperative decision as to which segment should undergo surgical decompression was made on the basis of the combination of clinical findings, radiological examinations, and three electrodiagnostic tests. Diagnosis of lumbar radiculopathy was easy when all five findings indicated the same nerve root lesions. However, some discrepancies among five testings usually occurred. Figure 1 shows the relationships between the postoperative improvement of the JOA score and number of matched testing findings. Postoperative improvement was best in patients with five matched test data and good when at least three findings were matched.

Table 2. Relationships between clinical and electrophysiological findings

Clinical findings		No of patients	Electrophysiological finding (%)	
			Positive	Negative
			EMG	
Muscle weakness	(+)	19	100	0
	(−)	14	50	50
			Root SEP	
Sensory disturbance	(+)	24	83.3	16.7
	(−)	7	28.6	71.4
			Elicited pain	
Sciatica	(+)	26	92.3	7.7

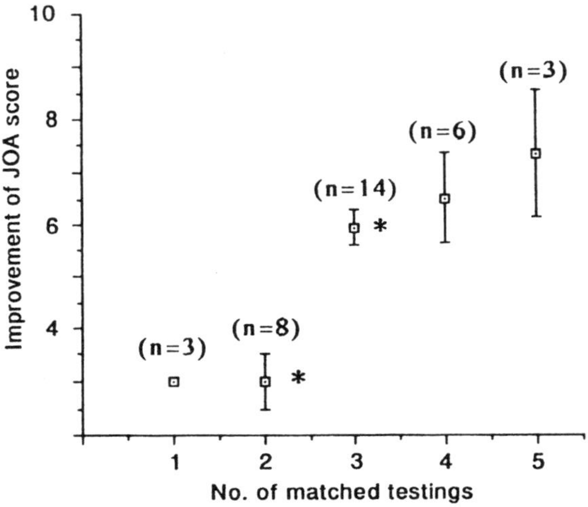

Fig. 1. Relationship of postoperative improvement of JOA score and the number of matched test results, among clinical examinations, radiological evaluations, EMG, root SEP and elicited pain. Data are expressed as mean ± standard error of mean (SEM). Statistical analysis was performed using Wilcoxon's rank test to represent the significant different between groups (*, $p < 0.01$) Accompanying parenthesis to each data indicates number of cases

Table 3. Preoperative diagnosis of affected roots

No of affected roots[a]		No of diagnosed roots (detection rate: %)
Radiology	50	27 (54)
EMG	46	33 (71.7)
Root SEP	44	35 (80.8)
Elicited pain	40	35 (87.5)

[a] Affected roots were decided by operative findings

Table 3 shows the diagnostic accuracy of each test which was based on the operative findings. As described in Materials and Methods, all cases in this study were not fully diagnosed by radiological methods, therefore, detection rate by radiology was worst. When three electrodiagnostic methods were compared, elicited pain had the highest diagnostic potential of all.

Intraoperative Electrodiagnosis

Intraoperative electrodiagnosis was performed in 15 patients. Abnormalities of CEAPs were seen in 14 cases with positive waves (11 segments), desynchronization (4 segments), and low amplitude (5 segments), as listed in Table 4. ERPs showed abnormalities in 13 cases. A positive wave of the ERP was

Table 4. Abnormal findings in CEAP and ERP

Electrodiagnosis	No of patients	No of affected segments or roots		
		Positive wave	Desynchronization	Low amplitude
CEAP	14	11	4	5
ERP	13	16	1	0

most frequently observed finding, in 16 roots, and another finding was the desynchronization of ERP in one root (Table 4). However, there was one patient who showed no abnormalities in his CEAPs and ERPs. He had a severe symptom of sciatica without sensory and motor loss. This illustrates that care must be taken making a diagnosis in patients without neurological deficits.

Comparing pre- and intraoperative electrodiagnoses, it is found that both tests revealed the same nerve root lesions in nine cases; however, five cases showed a discrepancy between pre- and intraoperative test results. In four of these five cases, the nerve root lesions detected using preoperative tests implied intraoperatively detected ones; the operative findings were disc herniation (three cases) and sequestrated type disc herniation (one case). Only one case with disc herniation at L2–3 and L5–S1 showed complete discrepancy between both tests.

Case Reports

The first case was a 58-year-old man with complaints of left leg pain and numbness. Clinical examinations of his sensory and motor loss indicated on L5 nerve root lesion. Radiographs of the lumbar area of the spine demonstrated spondylolisthesis at L4–5; however, there was myelographic evidence of cauda equina compression from the asymptomatic right side. Therefore, electrophysiological tests were conducted to diagnose the true lesion. Denervation potentials, which were found in the EMG recorded from muscles of the L5 segment, and disappearance of an early component of the root SEP suggested a left L5 nerve root lesion. Elicited pain, which was similar to his leg pain, was observed when electrical stimulation was applied to L5 and S1 roots (Fig. 2). From these findings, the preoperative diagnosis was considered as left L5 and S1 nerve root lesions produced by spondylolisthesis at L4–5.

The second case was a 44-year-old man who had low back pain and left side sciatica. Neurological examinations revealed muscle weakness of his left toe extensors and flexers as well as sensory disturbance in the left S1 dermatome. The myelogram showed a root anomaly of L5 and S1 on the left side (conjoined nerve roots) without providing evidence of disc herniation or stenosis (Fig. 3). Myelographic computed tomography also supported

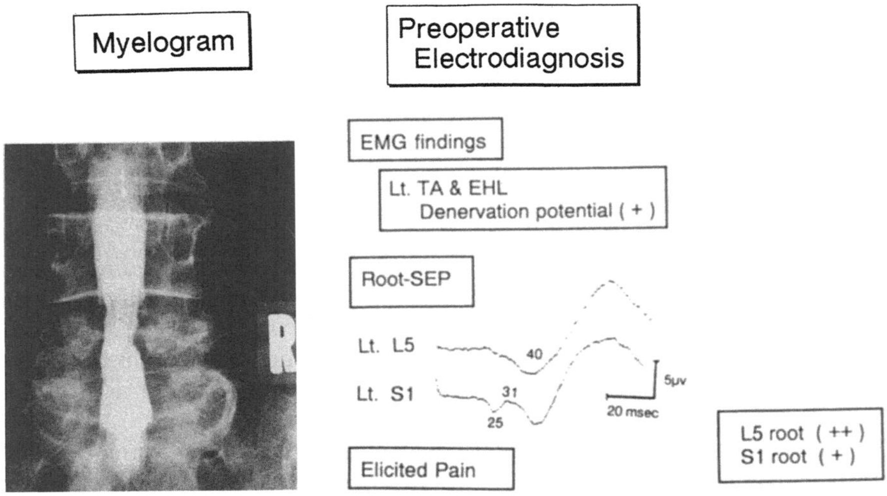

Fig. 2. Case 1.58-year-old man. Spondylolisthesis at L4–5

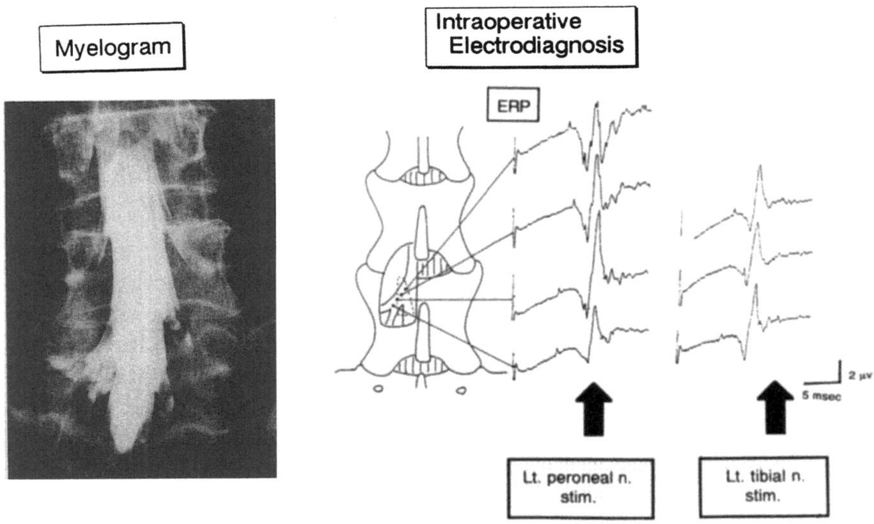

Fig. 3. Case 2.44-year-old man. Root anomaly of L5 and S1

these findings. EMG indicated evidence of an S1 nerve root lesion with the presence of denervation potentials and an L5 root lesion with neurogenic change. Root SEPs, obtained by stimulating L5 and S1 roots, showed abnormalities with the absence of an early component and increased latency. The elicited pain method was also positive when L5 and S1 nerve roots were stimulated. From these abnormal findings, preoperative diagnosis was decided as L5 and S1 nerve root lesions. Intraoperative electrodiagnosis was conducted to detect the location of the nerve root lesion. Exploring the conjoined L5 and S1 roots operatively, compression of the conjoined roots produced by the edge of superior facet was found ERPs recorded from various points of the conjoined roots are shown in Fig. 3. ERPs obtained by stimulating the peroneal nerve showed positive wave at the entrapment point and increased amplitude at more lateral recording points. There was no abnormality in ERPs produced by stimulating the tibial nerve. Foraminotomy was performed to decompress, the nerve root until recorded ERPs showed normal amplitude. The patients clinical symptom completely disappeared after the operation.

Discussion

An important feature of this study was to clarify the nerve root lesions with absent or equivocal radiological evidence. All of our cases had presented some difficulty in radiological investigations and all patients had multilevel involvement, lateral disc herniation, sequestrated herniation, lateral stenosis, or a previous operation or a combination of these. As a rule, patients with manifest radiological and clinical findings were excluded from investigation. Many electrodiagnostic methods, such as EMG, H reflex, F response, SEPs, and CEAPs, are available for the evaluation of these difficult cases. However, from a neurophysiological point of view, each examination provided only limited information about neural damages which produce a patient's symptoms. For example EMG and F response reveal the abnormalities of motorneurons and SEP represents the conduction failure of an afferent pathway. Therefore, a combination of various electrodiagnostic tests to diagnose true nerve root lesions may be justified on the grounds that there are no useful methods which enable us to examine all neural functions. In the present study, three electrodiagnostic tests for clinical symptoms were used preoperatively to determine which nerve roots had to be decompressed surgically.

Comparing the diagnostic accuracy of each preoperative test (Table 3), the elicited pain method showed the highest accuracy rate of diagnosis (87.5%). This method was considered very useful in the evaluation of sciatic pain. Although there is another method to investigate sciatic pain, i.e., when selective infiltration of the nerve root with local anesthesia or a contrast medium is performed, the inserted needle mechanically stimulates the nerve root and reproduces sciatic pain. However, electrical stimulation is preferable than mechanical stimulation because of changeable stimulus intensity. In addition, if one intends to evaluate the radiating pain area precisely, repeated stimulation is necessary, and this is more easily applicable using electrical stimulation. On the other hand, root SEPs also revealed abnormalities with a high rate of accuracy (80.8%) in the nerve root which really was affected, as was proved by surgery. However, EMG showed abnormal findings at a rate of 71.7%. These findings may be interpreted by the different sensitivity of picking up abnormalities, i.e., EMG has very sensitive potentials for picking up subclinical nerve root damage compared with root SEPs. Actually, positive findings of EMG were found in 50% of patients without muscle weakness and those of root SEPs in 28.6% of patients without sensory disturbance. Thus, careful investigation is necessary to diagnose true lesions on the basis of the presence of subclinical nerve root lesions.

If the decompression area was not accurately determined by the preoperative tests, intraoperative electrodiagnosis was conducted for the final diagnosis and determination of the extent of the precise decompression area by searching the conduction block. Positive waves recorded at the site of the nerve indicate the presence of a conduction block in the nerve fibers. When positive wave of the CEAP is recorded directly from the operative field, a conduction block of the nerve fiber may be located in the recorded segment; however, because possible contamination from another segment by volume conduction must be considered here, the wave form of the CEAP recorded from an adjacent segment must be referred to, to prevent misjudgement. On the other hand, the wave form of ERPs recorded from the exposed nerve root is relatively uninfluenced by the volume conductor; therefore, positive wave of ERPs indicates the location of the conduction block. Decompression procedure must be conducted along the nerve root as far as the site of conduction block. While the amplitude of the initial positive deflections which became 50% of whole amplitude of the response was defined as a positive wave, the criteria to determine the response as a positive wave is controversial. In cases where abnormal fingings of low amplitude and desynchronization were observed in CEAPs, the nerve root lesion may be located more peripherally, and elaborate recording of ERPs along the nerve root enables us to detect the nerve root lesions.

Cauda equina lesions of L3–4, L4–5 and L5–S1 were detectable using those methods. However, when the lesions are located in upper lumbar segments, L1–2 and L2–3, diagnosis is difficult. CEAPs recorded at L1–2 or L2–3 did not usually reveal abnormality of positive waves, but increased latency was found in some cases.

References

1. Haldeman S (1984) The electrodiagnostic evaluation of nerve root function. Spine 9:42–48
2. Haldeman S, Shouka M, Robby S (1988) Computed tomography, electrodiagnostic and clinical findings in chronic workers compensation patients with back and leg pain. Spine 13:345–350
3. Hashimoto T, Matsuda H, Hirose T et al. (1984) Cauda equina action potential in lumbosacral compression radiculopathy. In: Honma S, Tamaki T (eds) Fundamentals and clinical application of spinal cord monitoring. Saikon Pub, Tokyo, pp 26

4. Herron LD, Tripp AC, Gonyeau M (1987) Intraoperative use of somatosensory-evoked potential in lumbar stenosis surgery. Spine 12:379–383
5. Inoue S (1986) Assessment of treatment for lowback pain. Japanese Orthopaedic Association. J Jpn Orthop Ass 60:905–911
6. Leyshon A, Kirwan EOG, Parry CBW (1981) Electrical studies in the diagnosis of compression of the lumbar root. J Bone Joint Surg (Br) 63[B]:71–75
7. Matsuda H, Hirose T, Hashimoto T et al. (1979) Electrodiagnosis of compression of individual nerve roots of the cauda equina. Int Orthop 3:121–131
8. Udén A, Åstrom M, Bergenudd H (1988) Pain drawings in chronic back pain. Spine 13:389–392

Intraoperative Monitoring of Segmental Evoked Spinal Cord Potentials in Cervical Lesions

T. GOYA[1], S. MIYAHARA, T. MIKURA, and K. KINOSHITA

Introduction

Evoked spinal cord potentials (ESCPs) are divided into segmental and conductive potentials [1–8]. The segmental potentials are obtained from the segmental level of the spinal cord by stimulating the peripheral nerves [1–3]. The conductive potentials are recorded at the rostral portion of the spinal cord far from the stimulated segment by stimulating the peripheral nerves or the spinal cord directly [5, 7]. In this paper, we report on segmental ESCPs directly obtained from the exposed dural surface intraoperatively in various cervical diseases. Although it seemed difficult to obtain completely normal waveforms in our operative cases, the evaluation of waveforms was tried by comparing them through pre- and postdecompression periods and by comparing the difference between affected and nonaffected sides. We also tried to evaluate whether monitoring of segmental ESCPs was useful as an indicator for sufficient decompression to the spinal cord and the nerve root. We obtained some interesting findings during these intraoperative monitorings.

Subjects and Methods

Recordings from the posterior dural surface through the posterior approach to the cervical spine were taken in 14 cases, and recordings from the anterior dural surface through the anterior approach in six cases. In cases of posterior approach, recordings were taken from electrodes introduced into the epidural space through the midline cleft of the ligamentum flavum, as a control of the predecompression stage before laminectomy was performed. In some cases, electrodes were placed on the lateral side of the dural surface or inserted into the intervertebral foramen. After decompression procedures, record-

[1] Department of Neurosurgery, Miyazaki Medical College, 5200 Kihara, Kiyotake, Miyazaki, 889-16 Japan

K. Shimoji et al. (Eds.)
Spinal Cord Monitoring and Electrodiagnosis

ings were once again taken from the same points. A reference electrode was placed in the subcutaneous tissue of the operative field, the contralateral shoulder of the stimulated limb, or on the forehead (Fpz).

Electric stimuli were applied to the median or ulnar nerve at the wrist or elbow. Square-wave electrical pulse stimuli with 0.2 ms duration were applied at supramaximal intensity. The evoked potentials were recorded and averaged with a Dantec Evomatic 8000 apparatus and the recording bandwidth was 20 Hz–2 kHz. The stimulus rate was 3–5 Hz. We averaged 50–100 responses and the analysis time was 40 ms. All recordings were displayed with the negative potentials upward.

Results

Recordings from the Posterior Dural Surface

Waveform Characteristics

Waveforms of the segmental ESCPs obtained by stimulating the peripheral nerve and recorded from the posterior dural surface are variably reported. The difference in waveforms almost certainly seems to depend on the location of the recording electrodes in the posterior epidural space. Some findings could be obtained from our recorded waveforms by comparing them.

Figure 1a shows one of the examples of the segmental ESCPs recorded from the posterior dural surface in the case of ossification of the posterior longitudinal ligament in the cervical region. These recordings were done just after decompressive laminoplasty. The lower trace shows a response recorded from the dural surface of the C7 root near the intervertebral foramen. Initial sharp and deep positive P1 (9.92 ms) and the following negative component (10.7 ms) were observed. The upper trace reveals a response obtained from the midline dural surface of the spinal cord at the level of C6–7. In this trace, P1 is 10.3 ms in latency, becomes shallower than P1 in the lower trace, and the following wide negative wave (N1) is observed. This N1 wave is roughly divided into two components, N1a and N1b. N1a is 11.9 ms in latency and N1b seems to have two peaks. P2 follows this N1 potential. From these recordings, it is confirmed that waveforms recorded from the posterior dural surface are considerably variable according to the location of the recording electrodes. On evaluation of the recorded waveforms of segmental ESCPs, attention should be paid to the location of the electrodes.

Figure 1b is another pattern of the segmental ESCPs obtained from the posterior dural surface in a case of cervical spondylosis. These recordings were performed after decompressive laminoplasty. Electrodes 1 and 2 are ipsilateral to the stimulated side, and electrodes 3 and 4 on the contralateral. Latencies of P1 are slighly different according to the location of

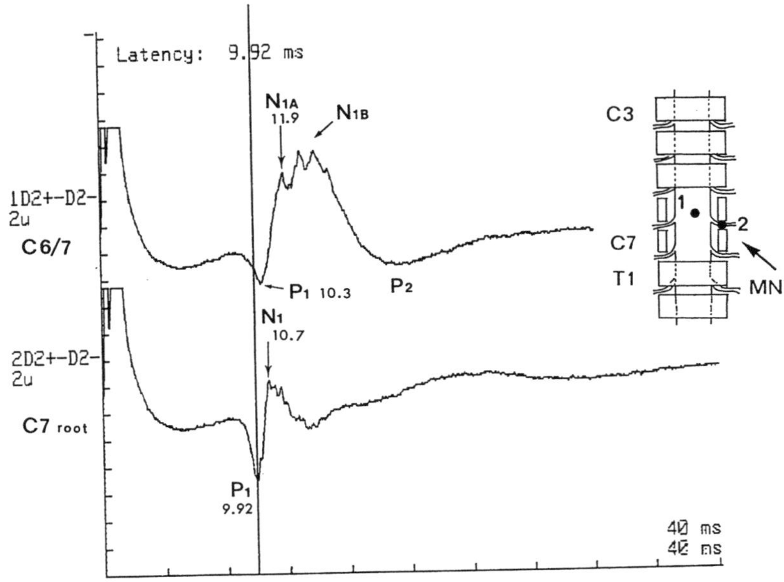

a

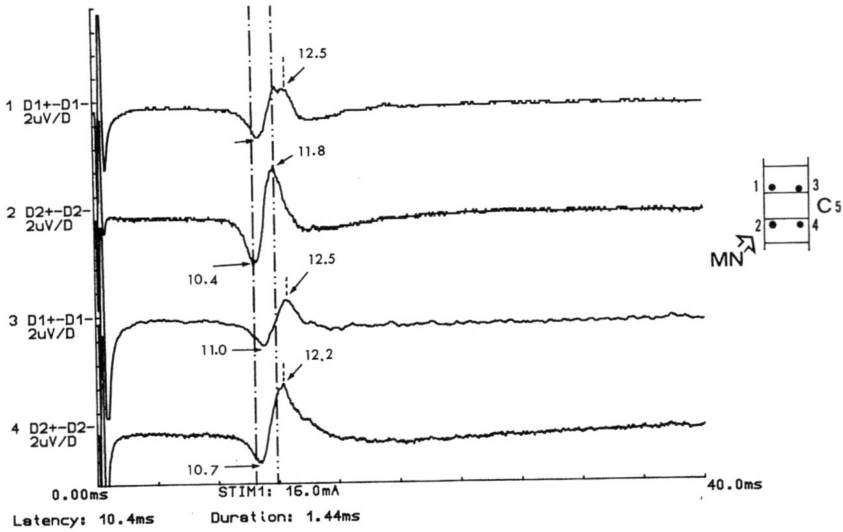

b

Fig. 1a,b. Segmental evoked spinal cord potentials (ESCPs) **a** recorded at the dorsal midline of the exposed dura (upper trace) and at the C7 root (lower trace), and **b** obtained from the ipsilateral and contralateral sides after median nerve stimulation

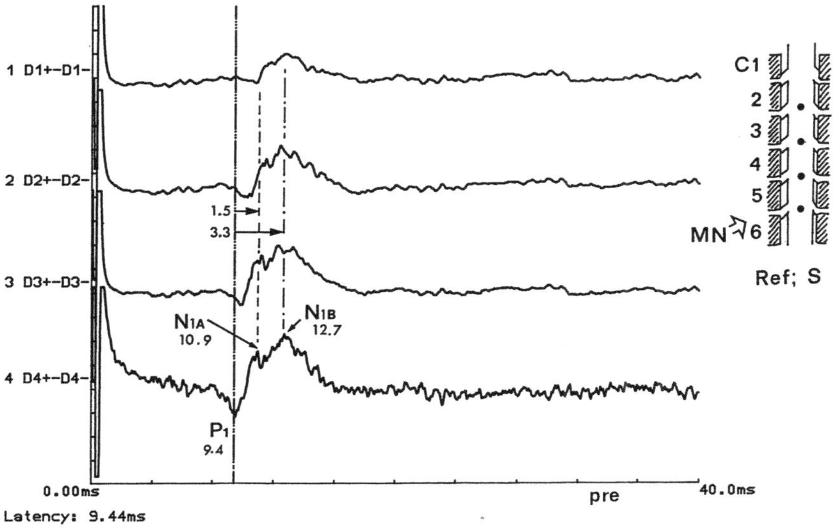

a

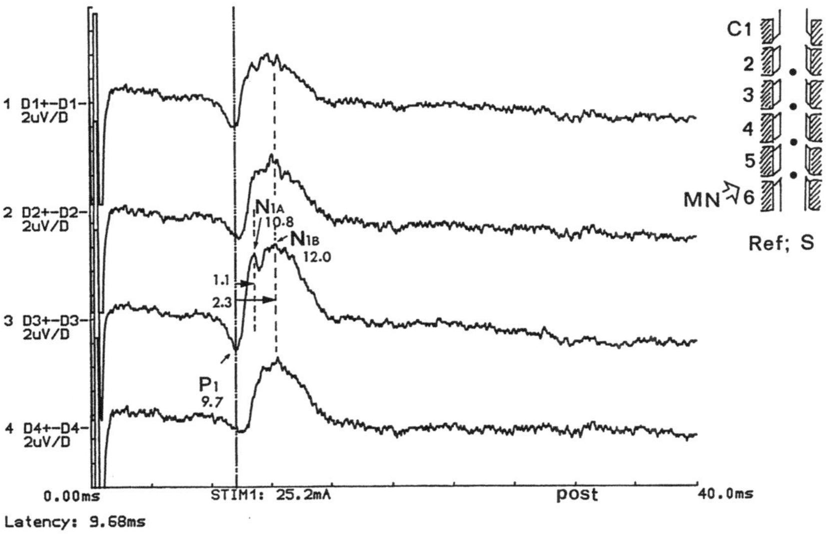

b

Fig. 2a,b. Case 1, ESCP recorded **a** in predecompression stage and **b** in postdecompression stage

the electrodes. The shortest point of P1 latency is at the electrode 2, and in this trace, a solitary N1 peak with 11.8 ms in latency followed. In trace 1, two small N1 peaks are faintly observed. These are assumed to be N1a and N1b. In traces 3 and 4, P1 and N1 peaks are more delayed than in trace 2. Therefore, N1 had a solitary peak in some cases, as seen in this instance, and in other cases, two peaks, N1a and N1b.

Pathological Findings

Case 1 was a 70-year-old man with old atlantoaxial dislocation and a narrow cervical spinal canal. Removal of the posterior arch of C1 and decompressive laminoplasty from C2 down to C6 were performed. Figures 2a and 2b show the recorded waveforms during pre- and postdecompression stages. Location of the electrodes is shown on the right side of these figures. In trace 4 of Fig. 2a, P1 (9.4 ms), N1a (10.9 ms), and N1b (12.7 ms) components are noted. The intervals of P1–N1a and P1–N1b were 1.5 ms and 3.3 ms respectively.

After sufficient decompression procedure via the posterior route had been performed, evoked potentials were recorded (Figure 2b). Shortening of P1-N1a and P1-N1b intervals and an increase in amplitude were noted, in comparison with recordings taken in the predecompression stage (Fig. 2a).

Case 2 was a 56-year-old woman with cervical syringomyelia. She had segmental sensory disturbance from C4 down to T4 levels on the left side, marked scoliosis, and muscle atrophy in the left hand. Decompression of the foramen magnum together with plugging to the obex, and syringosubarachnoid shunt were performed. Evoked potentials were recorded from the posterior dural surface after operative procedures. Figure 3a was recorded from the less affected right side. Recording electrodes were placed as shown on the right side of this figure. In trace 4, P1 and N1a can be clearly observed. N1a delays in latency toward rostral levels, but N1b is bifid, which may be abnormal, although these recordings are from the less affected side. Figure 3b is recorded from the affected left side. N1a is preserved but the following negative component (N1b) is not observed. As the gray matter of the spinal cord is involved in the early period in syringomyelia, the suppression of the later component of N1 (N1b) would suggest disturbance of interneurons in the posterior horn.

Recordings from the Ventral Dural Surface

Waveform Characteristics

Segmental ESCPs were also recorded from the ventral dural surface after anterior removal of a soft disk hernia, an osteophyte, and ossification of the posterior longitudinal ligament.

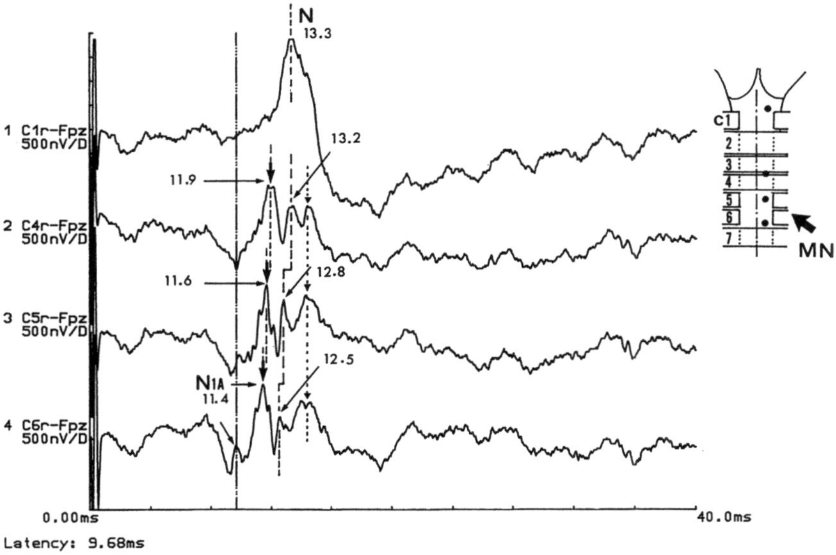

a

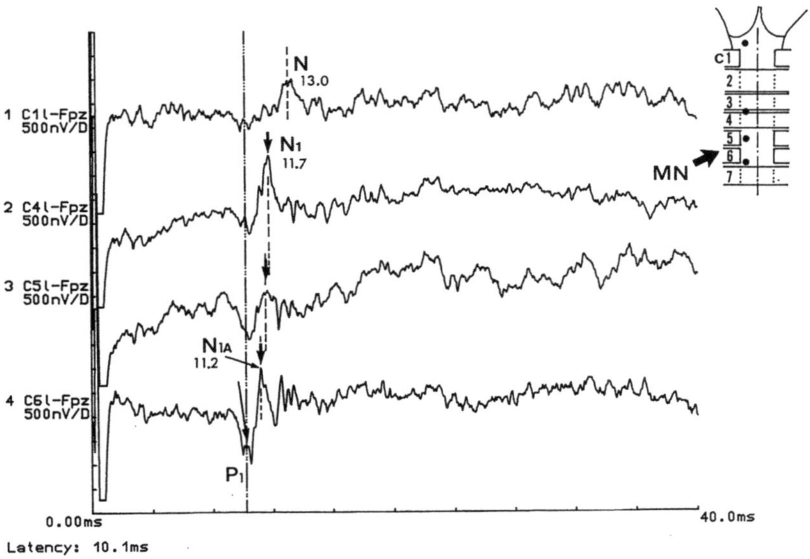

b

Fig. 3a,b. Case 2. Recorded evoked potentials obtained by stimulation of **a** the less affected and **b** the affected sides in a case of cervical syringomyelia

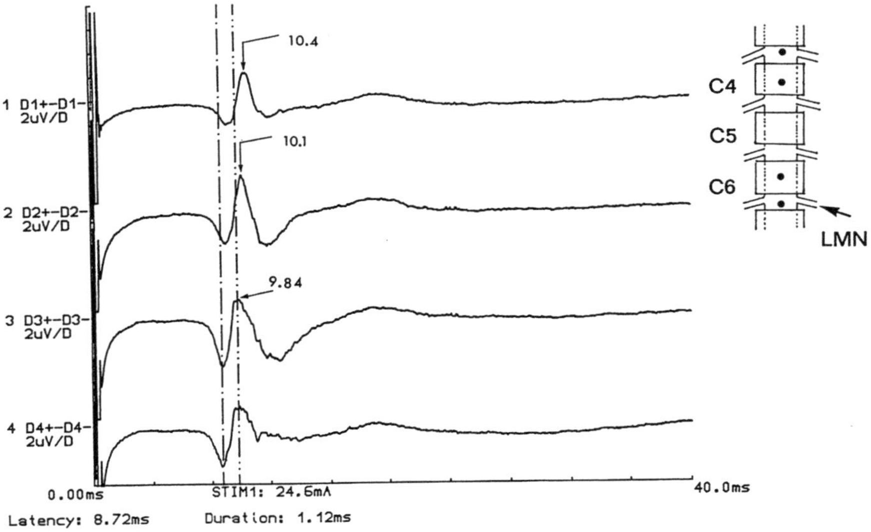

a

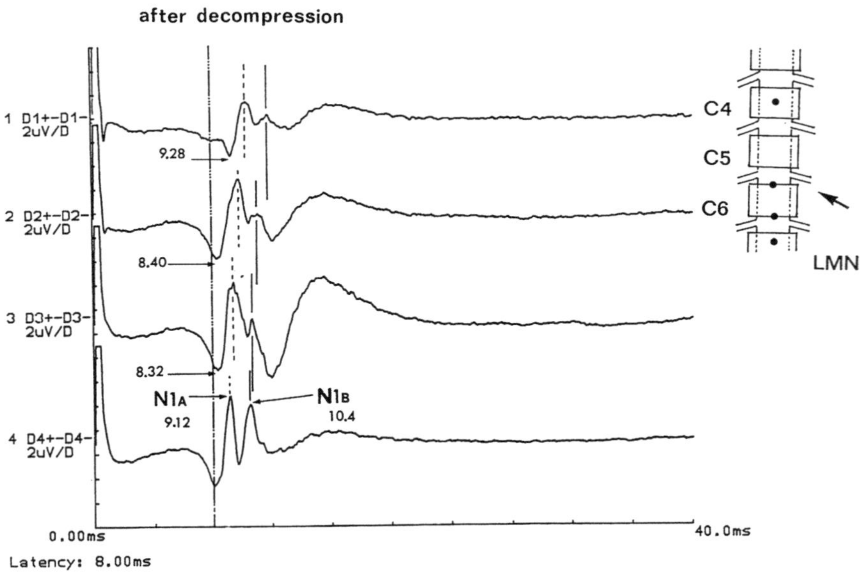

b

Fig. 4a,b. ESCP recorded from the anterior dural surface after removal of osteophyte (**a**), and **b** another pattern of waveforms from the ventral dural surface. P1, N1a, N1b and N2 components can be clearly observed in trace 3

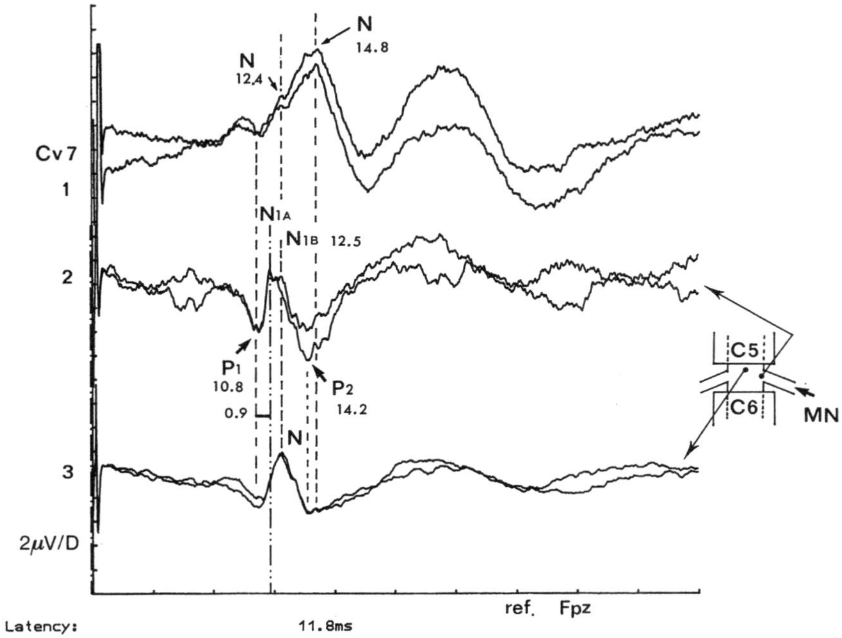

a

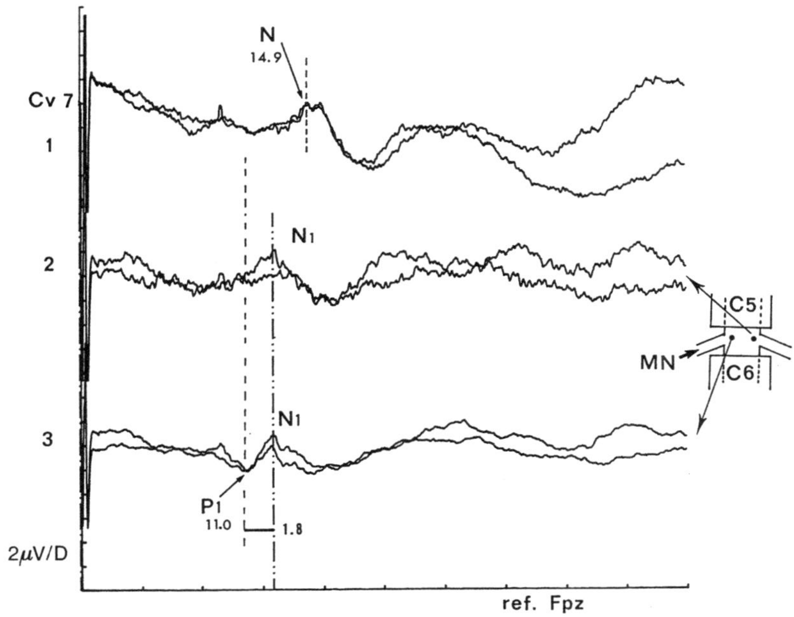

b

Fig. 5a,b. Case 3. Segmental ESCP recorded from the ventral dural surface by stimulation of **a** the less affected and **b** the affected sides in a case of cervical spondylosis

Figure 4a shows the waveforms from the ventral electrodes, demonstrated on the right side of the figure. This is a case of the soft disk hernia between C5 and C6. After removal of the protruding disk, recordings were taken. In trace 3, P1 and N1 are 8.72 ms and 9.84 ms in latency, respectively. The P1–N1 interval is 1.12 ms. P1 in traces 3 and 4 is sharp and deep. These levels are supposed to be the entry portions of impulses on stimulation of the median nerve. N1 delays in latency when the recording electrode went up toward the higher levels.

Figure 4b shows another pattern of segmental ESCPs recorded from the ventral cord after removal of a disk hernia at the levels of C4–5 and C5–6. P1, N1a, and N1b were recorded. The latency of P1 and N1a in trace 4 is 8.00 ms and 9.12 ms respectively, and the P1–N1a interval is 1.12 ms in each trace. When recording electrodes were placed vertically, N1a and N1b delayed in latency and N1b rapidly decreased in amplitude, as the electrodes went up toward the rostral levels.

Pathological Findings

Case 3 was a 50-year-old man with cervical spondylosis at the levels of C4–5 and C5–6. Recordings were taken from the ventral dural surface after anterior removal of osteophytes. Figure 5a shows the responses obtained by nonaffected left side stimulation. In trace 2, P1, N1a, N1b, and P2 peaks are observed. There is a small negative notch within P1. The intervals between P1–N1a and P1–N1b are 0.9 ms and 1.7 ms respectively. From the midline electrode (trace 3), only the N1 component is observed and this N1 peak latency corresponds to N1b in trace 2. The N1a component is not observed from the midline electrode. Figure 5b shows waveforms obtained by median nerve stimulation on the affected right side. As compared with N1 in trace 2 of Fig. 5a, the amplitude of N1 is suppressed in trace 3 of Fig. 5b. A solitary negative peak (N1) is observed. In this trace, the P1–N1 interval is 1.8 ms and this interval approximately corresponds to the P1–N1b intervals in trace 2 of Fig. 5a (1.7 ms). Therefore, the N1a component is markedly inhibited and N1 (N1b), per se, is suppressed in amplitude.

These findings suggest disturbance not only in the nerve root but also in the interneuron pool in the posterior horn.

Discussion

In cases of cervical lesion segmental ESCPs were recorded from the exposed dural surface directly when there was stimulation of the median or ulnar nerve intraoperatively [3, 6]. Recordings were taken from the anterior or posterior epidural space of the cervical spinal cord, according to the lesions.

We tried to evaluate waveforms as an indicator of the function of the spinal cord and the nerve root.

Segmental ESCPs are basically composed of an initial P1, followed by N1 and P2 peaks [5–8]. N1 is sometimes, not always, divided into two components, N1a and N1b. P1 consists of afferent impulses arising from the nerve root and N1a possibly originates from nerve terminals within the spinal cord or from the posterior column [3]. N1b is assumed to arise from interneurons in the posterior horn and to be a postsynaptic component [7, 8]. With this background, we would like to discuss our recorded waveforms.

First of all, the position of the recording electrode is the most important factor when we evaluate the waveforms of segmental ESCPs [2]. As shown in Fig. 1, if the electrode is placed at the root entry portion or in the intervertebral foramen, deep positive and negative potentials are recorded. This P1 shows impulses arriving along the nerve root. From the electrode placed on the posterior dural surface, P1, N1a, and N1b components are recorded, This P1 is not so deep as the positive wave obtained from the electrode on the nerve root, and it has a sightly delayed latency. N1a is seen on the rising phase of this N1 and is followed by N1b. The P1–N1a interval is usually within 1.5 ms, and the P1–N1b interval is also constant. In case 1 (Figs. 2a,b), an increase in amplitude of P1–N1a and P1–N1b seems to have occurred after the decompression.

In case 2 (Fig. 3) the later component of N1 (N1b) was markedly suppressed by stimulation of the affected side. This means disturbance of the postsynaptic component arising from the dorsal horn. At the same time N1a was considerably well preserved, and so it is suspected that N1a does not arise mainly from the posterior horn. Even when stimulation of the nonaffected side was used, N1b had two peaks and this may be abnormal. The latency of N1a is gradually delayed as recording electrodes went up to the higher levels and therefore, N1a is partly acting as a conducting potential. This suggests that N1a arises not only from the afferent terminals, but also from the dorsal column.

Waveforms recorded from the ventral epidural space are somewhat different from those recorded from the posterior epidural space. As shown in Fig. 4a, N1 is seen as a solitary negative peak. This N1 also delays in latency toward the rostral position of the spinal cord and corresponds to the N1a component recorded from the posterior epidural space.

Figure 4b shows other waveforms recorded from the ventral surface. There are two negative peaks, N1a and N1b, in this case. N1a is also assumed to be a conductive potentials. However, N1b is most prominent at the level of C5–6 and rapidly decreases in amplitude at higher levels of the spinal cord. This suggests that N1b may arise from a localized region and the interneurons in the gray matter of the cord.

In pathological recordings shown in Fig. 5, P1–N1b (N1) has exactly the same intervals in affected and nonaffected sides. But, on the affected side, the amplitude of N1 (N1b) is markedly suppressed and N1a could not be detected. These findings may demonstrate disturbances not only of afferent impulses but also of interneurons of the dorsal horn.

Summary

We recorded segmental ESCPs directly from the dural surface in the form of intraoperative monitoring. Regarding evaluation of waveforms obtained, attention must be taken as to the positioning of recording electrodes. In the midline electrode, typical P1, N1a, N1b, and P2 components are recorded, but when electrodes shift laterally toward the stimulated side, P1 and N1a become prominent and N1b is decreased. In many cases, only P1 and N1b are observed on the contralateral dural surface, and N1a becomes unclear. N1a is assumed to arise not only from afferent nerve terminals but also from the dorsal column. In addition, N1a, partly, may be a conducting potential through the dorsal column. N1b is suppressed in gray matter lesions such as syringomyelia. This component may arise from the interneuron pool in the dorsal horn. If a solitary N1 peak is present, N1a and N1b can be differentiated by measuring the inter-peak latency between P1 and the following negative peak.

Although it may be difficult to observe immediate change of segmental ESCPs just after an operative procedure, abnormal evoked potentials correspond to neurological disturbances in some cases.

References

1. Austin GM, McCouch GP (1955) Presynaptic component of intermediary cord potential. J Neurophysiol 18:441–451
2. Ertekin C (1976) Studies on the human evoked electrospinogram I. The origin of the segmental evoked potentials. Acta Neurol Scandinav 53:3–20
3. Kaneda A, Yamaura I, Kamikozuru M, Shinomiya K, Sato H, Yokoyama M (1985) An analysis of spinal cord potentials evoked by median nerve stimulation. In: Schramm J, Jones SJ (eds) Spinal cord monitoring. Springer, Berlin Heidelberg New York, pp 35–42
4. Kobayashi M (1985) On the origin of spinal cord potentials evoked by peripheral nerve stimulation and its clinical meaning Part 1. Clinical study. J Jpn Orthop Ass 59:27–38 (in Japanese)
5. Maruyama Y, Shimoji K, Shimizu H, Kuribayashi H, Fujioka H (1982) Human spinal cord potentials evoked by different sources of stimulation and conduction velocities along the cord. J Neurophysiol 48:1098–1107
6. Nashold BS, Ovelmen-Levitt J, Sharpe R, Higgins AC (1985) Intraoperative evoked potentials recorded in man directly from the dorsal roots and spinal cord. J Neurosurg 62:680–693
7. Shimoji K (1985) Clinical application of spinal cord evoked potentials recorded from the epidural space in man. Asian Med J 28:349–362
8. Yates BJ, Thompson FJ, Mickle JP (1982) Origin and properties of spinal cord field potentials. Neurosurgery 11:439–450

Intraoperative Evoked Potential Monitoring in Patients with Poliomyelitis

A. KOHT[1], H. WATTS, and R. ELLIS

Introduction

Poliomyelitis is a rare disease in the Western world and most industrialized countries; however, almost half of the world population is still at risk and many patients continue to seek medical help. Most physicians trained at present time lack experience in poliomyelitis and hence have only limited ability in the treatment of such patients. Poliomyelitis strikes the motor rather than the sensory tracts creating weakness in certain parts of the body and imbalance in others, and thus generates more deformities. Such impairments are common in the spinal column and many require surgical instrumentation [5]. During surgery, special attention is needed to prevent adding insult to already limited function. This is normally achieved by careful intraoperative monitoring, utilizing either the wake-up test or evoked potentials [1]. The value of the wake-up test may be limited by the baseline weakness of the lower extremities. The value of somatosensory evoked potentials (SSEPs), which are being used more frequently in the diagnosis and monitoring of spinal and intracranial operations [2–4, 7], has not been reported in patients with poliomyelitis. In this paper, we describe our experience with monitoring SSEPs in addition to the wake-up test in 16 patients with poliomyelitis who underwent spinal instrumentation.

Method

Sixteen poliomyelitis patients (10 men and 6 women) with spinal column deformities who underwent spinal instrumentation (Harrington rod) at King Faisal Specialist Hospital and Research Centre in Riyadh, Saudi Arabia were

[1] King Faisal Specialist Hospital and Research Centre, P.O. Box 3354, Riyadh 11211, Saudi Arabia

K. Shimoji et al. (Eds.)
Spinal Cord Monitoring and Electrodiagnosis

included in this study. Their ages ranged from 8–20 years with an average weight of 34.2 kg (18–76 kg). These patients had had no immunization and had had poliomyelitis at an earlier age and were suffering at the time of treatment from weakness, deformities, and limited functional life. They underwent surgery by either or the two pediatric orthopedic surgeons (H.W., R.E.) to stabilize the spinal column in hope of better functional life.

In addition to fentanyl, thiopental sodium or midazolam was used for induction of anesthesia, and atracurium besilate or pancuronium bromide was used to achieve relaxation. During the maintenance of anesthesia, moderate doses of narcotics and low concentrations of inhalation agents, without nitrous oxide, were used. After the start of anesthesia with the patient in the supine position, standard subdermal EEG recording needles were placed on the scalp at Cz′ (2 cm behind Cz) and Fz (according to the international EEG 10–20 system). An electrode needle was placed in the subdermal area posterior to the second cervical spinal vertebra (C2).

Impedances were kept below 5 kΩ. Two subdermal needles were used to stimulate the posterior tibial nerve at the ankle using 5.1 Hz, 0.3 ms constant current pulses set at 19.9 mA. Three hundred and fifty responses were filtered (30–3000 Hz), amplified, and averaged on a Nicolet CA 1000/DC-2000 signal averager. Each test was done in duplicate to ensure reproducibility. Special attention was paid to SSEP signals before the start of surgery. A comparison between right and left responses in relation to spinal curvature was made. After obtaining these baseline tracings, patients were positioned in the prone position in preparation for surgery. During surgery SSEPs were recorded continuously alternating between right and left. Changes were called significant if the amplitude decreased by more than 50% or the latency increased by more than 10%. A differential diagnosis was made to evaluate the cause of the change and the appropriate responses were taken. All tracings were stored on floppy disks for future analysis.

In addition to the SSEPs the neurological integrity of the spinal cord was tested intraoperatively using the wake-up test which was performed after the placement of the rod. Before the wake-up test, inhalation agents were stopped, intravenous agents withheld, and the degree of muscle relaxation assessed. A reversal of the muscle relaxants was given if needed. In previous cases because the two surgeons did not speak the patient's language a third person, either an orderly or a technician, had been asked to perform the wake-up test. During this study all wake-up tests were performed by a doctor who spoke the patient's language and was familiar with the wake-up test. In each patient, we tried to predict the neurological outcome first by evaluating the SSEPs, then by performing the wake-up test and a correlation between the two methods was made. Due to technical problems and shortage of manpower, patients were not tested preoperatively.

In all patients the surgeon evaluated the curvature of the spinal column, identified the convex of the curve, and evaluated the motor strength before and after surgery.

Results

In general the patients had below average weight and baseline muscle weakness in different parts of the body. The motor strength of their lower extremities ranged from 0% to about 50% of normal. Reliable SSEP signals were present in all patients which permitted intraoperative monitoring. The baseline signals, taken before incision, showed differences between right and left side stimulation correlating with the direction of the curvature. Latencies of the SSEP signals on the convex side were increased in both cervical and cortical leads (Fig. 1). In patients where two curvatures were present, the convex side with the worst curve was associated with the longer latency. If the two curvatures were equal then there were no differences. In nine patients, 11 changes of SSEPs were seen. These included five technically, three anesthetically, and three surgically related changes. Some of the patients had more than one cause or one episode of SSEP changes. Most of the technical causes (four cases) were related to the wake-up test. In one patient, his movement dislodged the Fz lead and in three others the muscular activity affected the SSEP responses recorded from the cervical region (Fig. 2). Two of the three anesthesia associated changes were related to fluctuation in the depth of anesthesia around the time of the wake-up test. One of the surgically related changes was the result of tightening the rod (Fig. 3) and two were related to the placement of the hooks. One of the changes related to

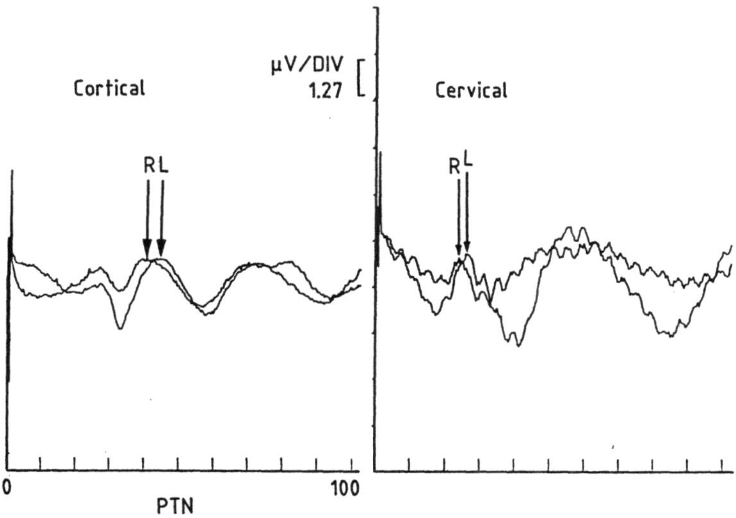

Fig. 1. Tracing of cortical responses to right and left posterior tibial nerve stimulation in a patient with poliomyelitis whose convexity of the spinal curvature pointed to the left. It demonstrates that the side of the convexity is the side with the later latency (*R*, the cortical responses to right posterior tibial nerve stimulation; *L*, the cortical responeses to left posterior tibial nerve stimulation)

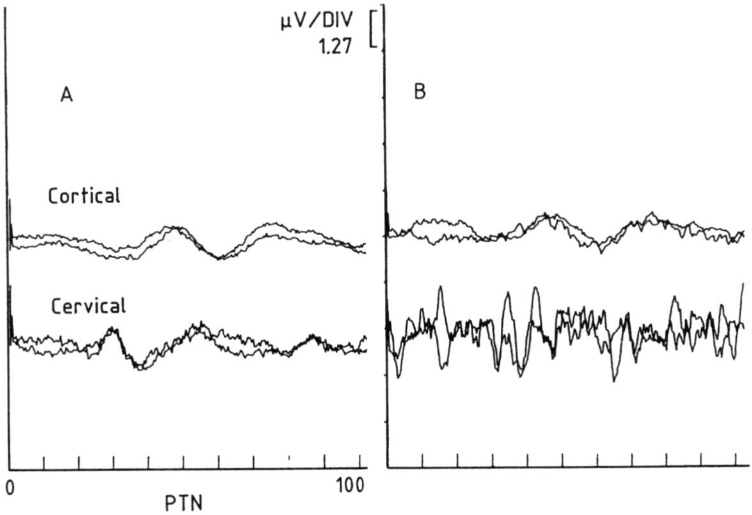

Fig. 2A,B. The effects of the wake-up test on SSEP recordings. **A** The SSEP tracing of cortical and cervical responses at the start of anesthesia; **B** The SSEP tracing of cortical and cervical responses at the time of the wake-up test

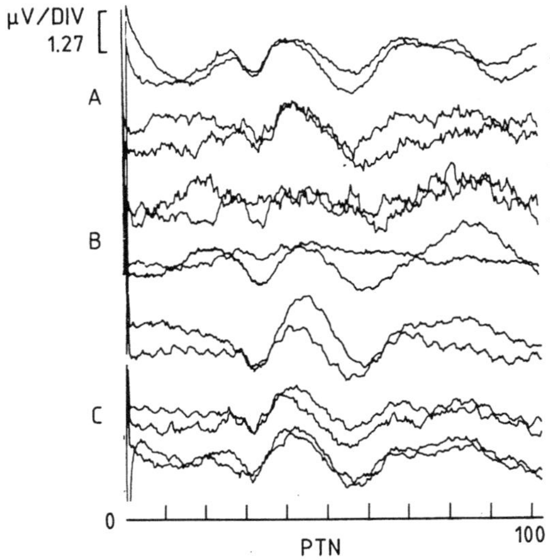

Fig. 3A–C. Cortical responses in patient with poliomyelitis during Harrington rod instrumentation (**A** Tracing at the start of anesthesia and surgery; **B** Tracing after the placement of hooks and rods; **C** Tracing after the removal of the rods)

Table 1. Evaluation of SSEP and wake-up test in 16 patients with poliomyelitis

	SSEP	Wake-up test
Patients tested	16	11
Outcome predicted correctly	16	7
Outcome predicted incorrectly	0	4
Surgical course altered	3	0
Surgeon made comfortable	16	7
Complication	1	4

the placement of the hook recovered spontaneously while the second change persisted during the tightening of the rod and recovered only after its removal. In all cases, SSEPs predicted the neurological outcome correctly, and three times altered the surgical course. The surgeon removed the rods in two patients, and proceeded with surgery in the third in spite of a delay in the wake-up test. One complication, a broken needle, occurred during SSEP monitoring and was removed without problem.

Due to profound weakness in the lower extremities before surgery, the wake-up test was not performed in five patients. And in the remaining patients where the test was performed, it was definitive only in seven, where this test correctly predicted the outcome. In the other four, the wake-up test was either slow, not definitive, or did not correlate with the outcome (one false positive). In this patient, SSEPs were severely changed during the tightening of the rod and the wake-up test was positive (no movement of lower extremities). The rod was removed and the SSEP signals recovered within 2–3 min; however, the patient could not move her legs. The surgery was then halted, anesthesia deepened (0.5% isoflurane plus thiopental sodium for amnesia), and the operative field flooded with warm saline. Half an hour later, a second wake-up test was attempted without success; the patient moved the upper but not the lower extremities. At that point surgery was terminated without placement of the rod. At the end of surgery the SSEP signals were as good as at baseline (Fig. 3C); however, the patient was still not able to move her legs. One hour after the initial SSEP changes, the patient started to move her legs and within 90 min she regained her previous motor strength. The complications associated with the wake-up test included one dislodged SSEP lead, one disconnected intravenous line and one patient moved before the wake-up test, which worried the surgeon. Table 1 presents a comparison between the two intraoperative monitors.

Discussion

The wake-up test, which has been used for some time in idiopathic scoliosis [6, 9], has more limitations in patients with poliomyelitis. These limitations stem from the baseline motor weakness which makes the assessment of motor function during anesthesia more difficult. The small residual anesthetic effect is sufficient to limit the success of the wake up test. In our study, five patients did not have the wake-up test because of profound muscle weakness before surgery. Such anesthetic effects were seen in three patients in whom the wake-up test was delayed in comparison with the more normal motor functions of the upper extremities. The uncertainty regarding the cause of the delayed or weak response during the wake-up test is troublesome. While the delay is not so crucial if it is caused by anesthesia, it could be detrimental if surgically related and there was a need to remove the rod. A delay in removing the rod means prolongation of the cord insult, which could decrease the chances for recovery. The second problem observed in this institution is communication. This problem could lead to false results and uncertainty about the test. While this condition is taken for granted in most areas of the world, it should be taken into consideration if such a procedure is to be carried out by a traveling team in a third world country. During the present study one of the authors (AK) who spoke the language evaluated the SSEPs and performed the wake-up test.

SSEP monitoring, which utilizes stimulation of a nerve, and recording the signals from over the neural tract are not affected by the above-mentioned two major limitations of the wake-up test. This is evidence from our ability to use SSEPs in all sixteen patients, compared with only eleven, who underwent the wake-up test. SSEPs were obtained in all patients after the induction of anesthesia in spite of the absence of preanesthetic baseline. The choice of anesthetic agents with minimum effects on SSEP monitoring may have contributed to our success rate [8]. The three changes related to anesthesia were the result of a bolus injection, an increase in isoflurane concentration, and starting nitrous oxide around the time of the wake-up test. The high frequency of technical problems we observed (five cases) were also related to the wake-up test. These technical changes were the result of muscle activities, light anesthesia, and a dislodged needle (Fz) and were seen mainly in the cervical leads. Most of the technical and anesthetic problems can be overcome by a careful planning of anesthesia, better timing of the wake-up test, and could be minimized if the wake-up test were not used. The surgical causes of SSEP changes started when the surgeon placed the hook in two of the three cases and once when the rod was tightened. Changes in amplitude at the time of placing the hooks should be carefully watched.

In these 16 cases of patients with poliomyelitis we observed a better correlation between SSEP responses and the outcome than between the wake-up test and the outcome. This was due to the impossibility of utilizing the wake-up test in all patients and the failure of the test in others. The success rate for the use of the SSEPs and for their use to predict the outcome was 100%, while the wake-up test was used only in 68% of the patients and predicted the neurological outcome only in 45% of the cases. The low correlation of the wake-up test to the postoperative neurological outcome as compared to idiopathic scoliosis may be due to the baseline motor deficit. In the critical moments after tightening the rod, SSEP monitoring gave us faster and more reliable results than the wake-up test.

Lastly, the discrepancy between the latency of right and left posterior tibial nerve stimulation, as related to the direction and degree of convexity of the spinal column, may be related to the length of the neural tract. It may indicate a longer tract on the side of the convexity. Meanwhile the similar values of amplitude may imply an equal number of transmitting axons. These differences in SSEP responses need further evaluation to see if any correlation between such differences and intraoperative SSEP changes do exist.

In conclusion, SSEP monitoring is a useful indicator in patients with poliomyelitis. It is superior to the wake-up test and can be reliably used during spinal instrumentation.

References

1. Ben-David B (1988) Spinal cord monitoring. Orthop Clin North 19: No 2.
2. Bradshaw K, Webb JK, Fraser AM (1984) Clinical Evaluation of spinal cord monitoring in scoliosis surgery. Spine 9:636–635
3. Chiappa KH (1985) Evoked Potentials in Clinical Medicine. Raven Press, New York, pp 176–240
4. Dinner DS, Luders H, Lesser RP et al. (1986) Intraoperative spinal somatosensory evoked potential monitoring. J Neurosurg 65:807–814
5. James JIP (1987) Poliomyelitis, Essentials of surgical management. Edward Arnold Press, London pp 127–157
6. Jones ET, Mathews LS, Hensinger RN (1982) The wake-up technique as a dual protector of spinal cord function during spine fusion. Clin Orthop 168:113–118
7. Kidooka M, Nakasu Y, Watanabe K, Matsuda M, Handa J (1987) Monitoring of somatosensory evoked potentials during aneurysm surgery. Surg Neurol 27:69–76
8. Koht A, Schuetz A, Schmidt G, Schramm J, Watanabe E (1988) Effects of etomidate, midazolam and thiopental on median nerve somatosensory evoked potentials and the additive effects of fentanyl and nitrous oxide. Anesth Analg 67:435–441
9. Vauzwelle C, Stagnara P, Jourinroux P (1973) Functional monitoring of the spinal cord activity during spinal surgery. Clin Orthop 93:173

Motor Evoked Potentials Monitoring During Neurosurgical Operations on the Spinal Cord

J. Zentner[1]

Summary

In order to monitor descending pathways during neurosurgical operations on the spinal cord, motor evoked potentials (MEPs) were recorded from the thenar and anterior tibial muscles and from the epidural space along the spinal cord and the cauda equina in a total of 83 patients following transcranial electrical cortex stimulation. It was the aim of our study to find out whether MEP monitoring is a reliable and sensitive method for the detection of impending motor deficits. Interaoperative recordability of potentials was 79.7%. On the basis of acceptable changes in amplitudes of up to 50% at the end of surgery, 82.7% of the recordings correlated correctly with the postoperative motor status; there were false positive results in 17.3%. We did not observe false negative correlation. In conclusion, intraoperative MEP monitoring is a sensitive method for the detection of impending motor deficits. Major problems concern the influence of anesthesia, the definition of acceptable limits for changes in potentials, and the most suitable stimulation and recording techniques.

Introduction

Electrophysiological assessment of the descending pathways has received more interest now that noninvasive motor cortex stimulation is available [5]. Animal studies of spinal cord trauma and ischemia confirmed close relationships between changes in motor potentials and the neurological condition [2–4, 6–8]. These results encouraged us to use motor evoked potentials (MEPs) in neurosurgical operations on the spinal cord. It was the aim of our

[1] Department of Neurosurgery, University Medical School, Hoppe-Seyler-Str. 3, W-7400 Tübingen, FRG

K. Shimoji et al. (Eds.)
Spinal Cord Monitoring and Electrodiagnosis

Table 1. Patients and diagnoses

Diagnosis	Patients No	%
Metastasis	18	21.7
Neurinoma	12	14.5
Meningioma	9	10.8
Ependymoma	8	9.6
Syringomyelia	7	8.4
Glioma	6	7.2
AV-Malformation	5	6.0
Cavernoma	3	3.6
Chordoma	2	2.4
Germinoma	2	2.4
Nontumorous	11	13.4
Total	83	100.0

work to find out whether MEP monitoring is a reliable and sensitive method for detection of impending motor deficits.

Patients and Methods

Intraoperative MEP monitoring was performed in a total of 83 patients with ages ranging from 9 to 82 years. Table 1 provides an overview of the diagnoses. Disease was located within the medulla oblongata and cervical spine in 43 patients (51.8%), thoracically in 33 (39.8%), and lumbar in seven (8.4%) cases.

The motor cortex was stimulated transcranially with voltage-constant condenser discharges using a Digitimer D 180 stimulator which provides a maximum output of 750 V. Only single stimuli were applied with a time interval of at least 3 s. Stimulation was performed anodally using Ag/AgCl electrocardiogram electrodes. For stimulation of the motor hand area, the anode was placed at the foot of the precentral gyrus and the cathode at the bregma. When the motor leg area was stimulated, the anode was placed at the bregma and the cathode 6 cm behind it at the midline. We recorded the electromyographic responses from both the thenar and anterior tibial muscles. Furthermore, potentials were recorded from the epidural space of the cauda equina and along the spinal cord above and below the lesion, using bipolar cardiac pacing electrodes.

The patients were fully informed about the procedure and gave their consent. They received 20 mg trifluorpromazine thiopentone, 50 mg pethidine, and 0.5 mg atropine as premedication. After an anesthetizing dose of 250–300 mg thiopental sodium, 1 mg/kg succinylcholine was given before intubation. No muscle relaxants were given thereafter. Usually, anesthesia was maintained with an oxygen and nitrous oxide mixture (50%–60% nitrous oxide) and fentanyl. In addition, flunitrazepam (single dose 0.02 mg/kg) and thiopental sodium (single dose 2–3 mg/kg) were administered as required. In some patients, continous analgesia and sedation with fentanyl and midazolam were maintained while the patient breathed an oxygen/air mixture, thus avoiding the use of nitrous oxide. Baselines were recorded after the relaxation necessary for intubation had waned. Further recordings were performed during the essential steps of surgery and at the end of the operation. Stimulus strength was gradually increased until a clear response was obtained or the absence of any response was documented despite a maximum stimulus strength of 750 V. Once the stimulus strength was determined, it was kept constant throughout the monitoring procedure. Usually the averages of 5–15 signals were evaluated. The time base was 100 ms with a gain ranging from 100 μV to 1 mV/div. Filter settings ranged from 20 Hz–3 kHz (Tesy II, Toennies; Compact 4, Nicolet).

Onset latencies of the initial negative wave and peak-to-peak amplitudes were measured. For evaluation of potentials, we decided upon an acceptable limit for changes in amplitudes of up to 50% at the end of the operation as compared with the baseline values. This calculation was arbitrary and resulted from practice with intraoperative somatosensory potential monitoring [1]. On that basis, MEP findings were correlated with the postoperative motor status taking three categories into consideration: correct correlation was defined as either unchanged postoperative motor status and difference in amplitudes not more than 50%, or additional motor deficit coinciding with difference in amplitudes of more than 50% at the end of surgery. Correlation was considered to be false positive in cases with unchanged postoperative motor status and permanent reduction in amplitudes of more than 50%. False negative correlation was defined as deteriorated postoperative motor status despite no permanent difference in amplitudes of more than 50%.

Results

Intraoperatively, MEPs were recordable from the spinal cord in 11 out of 19 patients (57.9%) and from the cauda equina in 23 out of 30 patients (76.7%).

Table 2. Intraoperative recordability of potentials

Recording site	No of patients	MEP recordable	
		No	%
Spinal cord	19	11	57.9
Cauda equina	30	23	76.7
Thenar muscle	24	21	87.5
Tibial muscle	50	43	86.0
Total	123	98	79.7

With recording from the thenar and anterior tibial muscles, in 21 out of 24 patients (87.5%) and 43 out of 50 patients (86.0%) intraoperative potentials were obtained. On the whole, 98 out of 123 recordings (79.7%) produced useful potentials. In the remaining patients, no signals were obtained despite a maximum stimulus strength of 750 V (Table 2).

On the basis of acceptable changes in amplitudes of up to 50% at the end of surgery, the expected postoperative motor status was correctly predicted in 9 (81.8%) recordings from the spinal cord and in 21 (91.3) from the cauda equina, while there were false positive findings in two cases in each category (18.2% and 8.7% respectively). Recordings from the thenar and anterior tibial muscles showed correct results in 16 (76.2%) and 35 (81.4%) cases respectively. At the same time, correlation was false positive in 5 (23.8%) and 8 (18.6%) cases. However, there were no false negative results in our patients (Table 3; Figs. 1–4).

In seven out of 83 patients, the motor status deteriorated postoperatively. Five of them showed a decrease in amplitudes of 60%–90% compared with the baseline values at the end of the operation. In two other cases, potentials were lost intraoperatively. One of them showed a complete, the other an incomplete postoperative motor deficit.

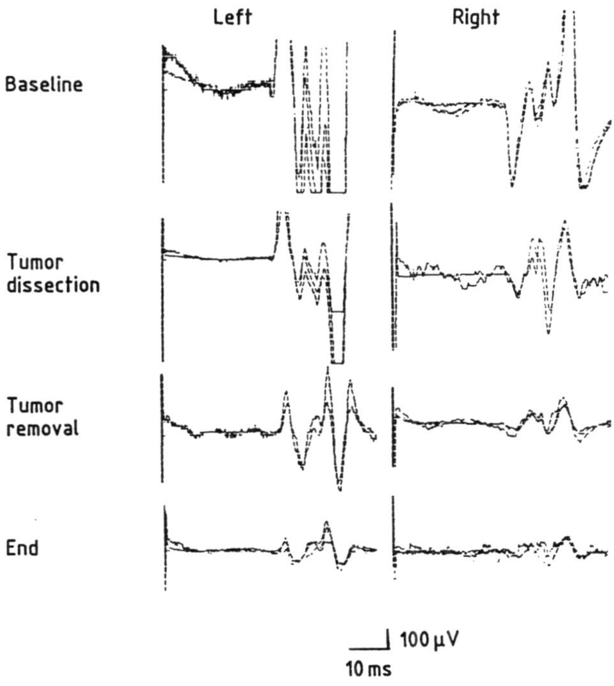

Fig. 1. Anterior muscle MEP of a 64-year-old male patient with a metastatic tumor at C3–4. During tumor dissection and removal, amplitudes decreased on both sides with differences of more than 50% at the end of surgery. The postoperative motor status had deteriorated

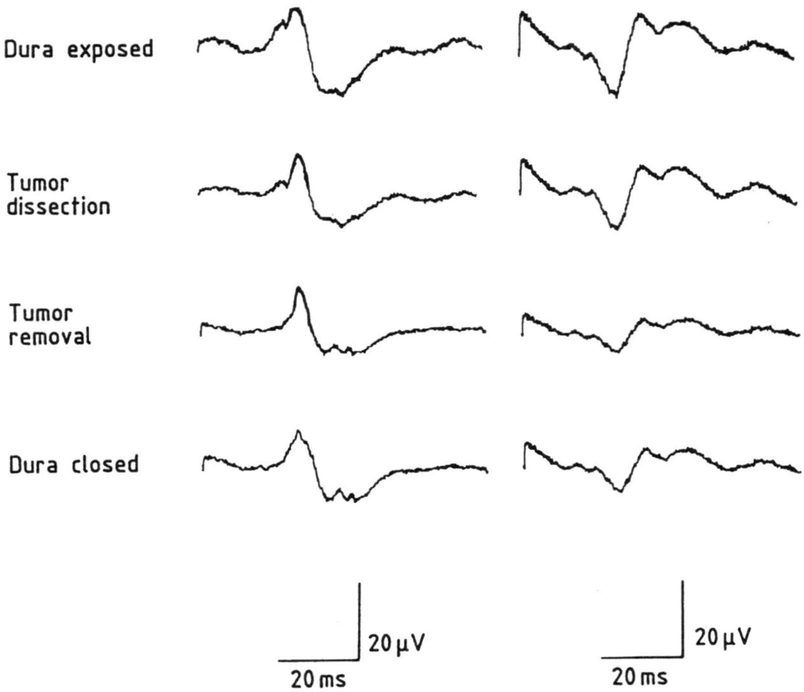

Fig. 2. MEP recorded epidurally above and below a metastasis at L1 in a 67-year-old male patient. During tumor dissection and removal, potentials recorded above the lesion were stable, whereas those obtained below the tumor showed a noticeable reduction in amplitudes. At the end of the operation, no differences more than 50% of the baselines were found. There was no additional postoperative motor deficit

Table 3. Correlation of intraoperative changes in potentials with the postoperative motor status

Recording site	Correct		False positive		False negative	
	No	%	No	%	No	%
Spinal cord	9	81.8	2	18.2	–	–
Cauda equina	21	91.3	2	8.7	–	–
Thenar muscle	16	76.2	5	23.8	–	–
Tibial muscle	35	81.4	8	18.6	–	–
Total	81	82.7	17	17.3	–	–

Discussion

Our results indicate MEPs to be highly sensitive to impending motor deficits. On the basis of acceptable changes in amplitudes of up to 50% at the end of

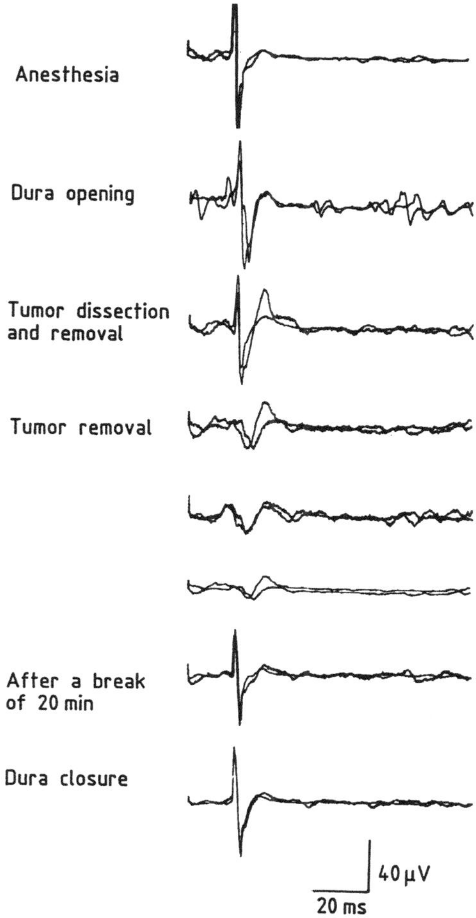

Fig. 3. MEP recorded epidurally from the cauda equina in a 46-year-old female patient with an ependymoma of the 4th ventricle. During removal of tumor remnants which had infiltrated the medulla oblongata, we found a temporary reduction in amplitudes of 75% of the baselines. Deterioration of potentials led the surgeon to leave a small tumor strip at the medulla. After a break of 20 min, potentials had recovered completely. There was no motor deficit postoperatively

surgery as compared with the baseline values, we found no false negative results in our series. Postoperative deterioration of motor function coincided in every case with permanent reduction in amplitudes of more than 50%. However, only intraoperative loss of potentials, as observed in two patients, accurately predicted poor neurological outcome while there were many false positive results in cases with preserved although deteriorated potentials. Recordings from the cauda equina predicted slightly more accurately, showing correct results in 91.3% and false positive results in 8.7%, as opposed to cor-

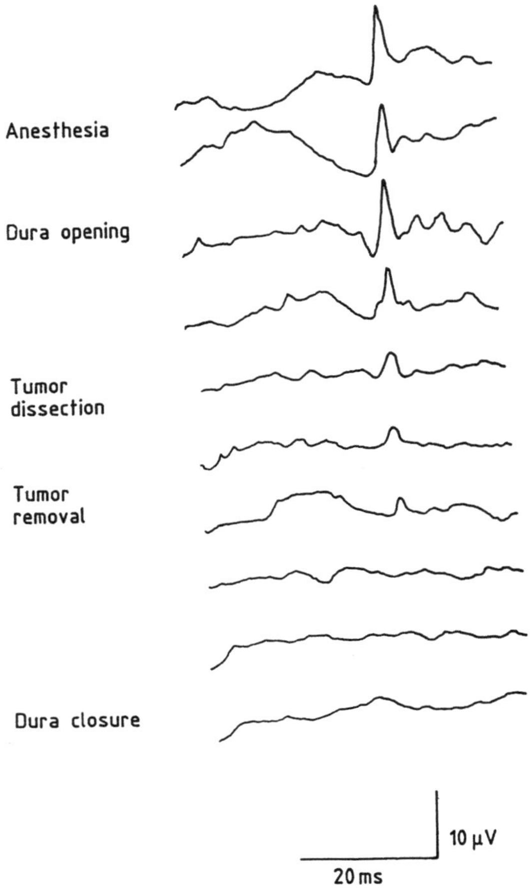

Fig. 4. MEP recorded epidurally from the cauda equina in a 66-year-old female patient who was operated under suspicion of an intramedullary tumor at C3–6. During "tumor" dissection and removal, amplitudes decreased continuously and were finally lost. In this case, cauda equina potentials showed an unusually long latency of 26 ms. Note also that latencies did not change essentially during successive reduction in amplitudes. Postoperatively, the patient had an incomplete tetraparesis. Histological examination of the specimen showed only a gliosis; tumor tissue was not detected

responding values ranging from 76.2% to 81.8% and from 18.2% to 23.8% for the other recording sites.

We did not observe close correlation between intraoperative changes in the latencies of the potentials and the postoperative motor status. Increase in latencies, as observed in some of our patients with and without neurological deterioration, may be due to physiological parameters such as a drop in body temperature rather than to damage of neural structures. It is worth mentioning that in the two patients with intraoperative loss of potentials, latencies essentially did not change during successive reduction in amplitudes.

Therefore, amplitudes were superior to latencies for the evaluation of intraoperative changes in potentials.

Although preoperative potentials were obtained in all of our patients, intraoperative recordability was only 79.8% on average. This indicates noticeable anesthesia-related suppression of MEP amplitudes. Investigation of anesthesiological influences on MEP showed a major suppressive effect of nitrous oxide [9]. In consequence, we tried to avoid this agent and found that continous analgesia and sedation with narcotics administered intravenously such as midazolam and fentanyl with the patient breathing an oxygen/air mixture is the most suitable mode of anesthesia for intraoperative MEP monitoring. Differences in recordability of potentials comparing invasive and noninvasive recording sites cannot be evaluated in our series since analgesia and sedation showing improved recordability of potentials were more often applied to recordings from the distal extremity muscles than to the epidural recording sites.

In conclusion, MEP monitoring during neurosurgical operations on the spinal cord is a sensitive method for detection of impending neurological deficits. Major problems concern the influence of anesthesia and the definition of acceptable limits for changes in amplitudes. Further work must be done to improve the intraoperative recordability of potentials. Thus, MEP monitoring should prove to be a method that is easy to implement and reliable in its interpretation. Nevertheless, the decisive question remains: how many patients will actually benefit from this procedure?

References

1. Dinner S, Lueders H, Lesser RP, Morris HH (1986) Invasive methods of somatosensory evoked potential monitoring. J Clin Neurophysiol 3:113–130
2. Ducker TB, Saleman M, Lucas JT, Garrison WB, Perot PL (1978) Experimental spinal cord trauma. II: Blood flow, tissue oxygen, evoked potentials in both paretic and plegic monkeys. Surg Neurol 10:66–72
3. Fehlings MG, Tator CH, Linden RD, Piper IR (1987) Motor evoked potentials recorded from normal and spinal cord-injured rats. Neurosurgery 20:125–130
4. Konrad PE, Tacker WA, Levy WJ, Reedy DP, Cook JR, Geddes LA (1987) Motor evoked potentials in the dog: effects of global ischemia on the spinal cord and peripheral nerve signals. Neurosurgery 20:117–124
5. Merton PA, Morton HB (1980) Stimulation of the cerebral cortex in the intact human subject. Nature 285:227
6. Oro J, Levy WJ (1987) Motor evoked potential as a monitor of middle cerebral artery ischemia and stroke. Neurosurgery 20:192–193
7. Patil AA, Nagaray MP, Mehta R (1985) Cortically evoked motor action potentials in spinal cord injury research. Neurosurgery 16:473–476
8. Simpson RK, Baskin DS (1987) Corticomotor evoked potentials in acute and chronic blunt spinal cord injury in the rat: correlation with neurological outcome and histological damage. Neurosurgery 20:131–137
9. Zentner J, Ebner A (1989) Nitrous oxide suppresses the electromyographic response evoked by electrical stimulation of the motor cortex. Neurosurgery 24:60–62

Analysis of Conductive Spinal Cord Function Using Spinal Cord Evoked Potentials During Surgery of Spinal Cord Tumors

K. YOKOGUSHI[1], G. KATAHIRA, H. YOKOZAWA, H. KAWAMURA, T. YAMASHITA, and N. OGIWARA

Introduction

The monitoring of conductive spinal cord evoked potentials (conductive SpEPs) has been widely used for the purpose of preventing damage to the spinal cord during operation [5, 7–9]. It has not been resolved whether changes in the conductive SpEPs correlate to the pathological condition of the spinal cord under chronic pressure or not. Excision of the tumor usually ensures recovery from paralysis caused by chronic pressure and circulatory disturbance due to extramedullary spinal cord tumors. Therefore patients with spinal cord tumors are useful models for studying changes of the conductive SpEPs in the spinal cord under chronic pressure. The purpose of this study is to investigate a correlation between changes in the conductive SpEPs and prognosis of the patients.

Clinical Materials and Methods

The 14 patients available for this study had spinal cord tumors; six patients were men and eight patients were women. The age of the patients ranged from 22 to 61 years (average, 44.8 years). Eight patients had neurinomas, three had metastatic tumors, two had sarcomas, and one had meningioma. The site of the tumors was the cervical spine in four cases and the thoracic spine in ten cases. Six tumors were intradural-extramedullary and eight epidural. Two of the eight epidural tumors were neurinomas of type III according to Eden's classification. The other six epidural tumors were malignant tumors associated with destruction of the vertebral bodies and pedicles. Therapeutically, tumorectomy in combination with laminectomy or laminoplasty was performed on the patients with a benign tumor. Intra-

[1] Department of Orthopedic Surgery, Sapporo Medical College, Minami-1 Nishi-16, Sapporo, 060 Japan

K. Shimoji et al. (Eds.)
Spinal Cord Monitoring and Electrodiagnosis

regional excision was performed for malignant tumor, and reconstruction of the spine was carried out using Luque instrumentation for the cervical spine or Cotrel-Dubousset instrumentation for the thoracic and thoracolumbar spine.

Record of Conductive SpEPs

After exposure of the laminas, the bipolar catheter electrodes were visually inserted on the dorsum of the dura along the midline from the intact interlaminar space. A cranial electrode was used for recording and a caudal one for stimulating. Stimulation was given using rectangular potentials of 10 Hz, with a duration of 0.2 ms and an intensity about 5 times as large as the motor threshold. The band pass filter was set at 3 Hz–2 kHz. Each recording was averaged from 50 to 200 stimulations to eliminate electric noise. The DISA N-2000 electromyograph was used for stimulation and recording. A conductive SpEP was continuously recorded during the operation (Fig. 1). It was classified as a normal or an abnormal potential. The abnormal potentials were subdivided into three types: electrically silent, low amplitude potentials, and slow sustained potential. Neurological symptoms were evaluated by using Frankel's grades.

Results

Analysis of Conductive SpEPs Before Decompression of the Spinal Cord

Neurological symptoms before operation showed Frankel grade A in one case, grade B in one case, grade C in six cases, grade D in five cases, and grade E in one case. Before decompression of the spinal cord, six of 14 patients showed normal SpEPs (Frankel grade E in one case, Frankel grade D in five cases). The period from the onset of symptoms to the operation ranged from 1 to 6 months (average, 3.6 months). The tumors were located anteriorly in one case, laterally in two cases and circumferentially in three cases. Abnormal SpEPs were shown in eight of 14 cases (Frankel grade C in six cases, Frankel grade B in one case, and Frankel grade C in one case). The period from the onset of symptoms to the operation ranged from 1 week to 4 months (average, 2.4 months). The tumors were located anteriorly in two cases, laterally in three cases, posteriorly or posterolaterally in two cases, and circumferentially in one case. Three of eight cases showed electrically silent potentials, three cases showed low amplitude potentials, and two cases showed slow sustained potentials. The patients with Frankel grades E and D showed normal SpEPs, the patients with Frankel grades C, B, and A showed

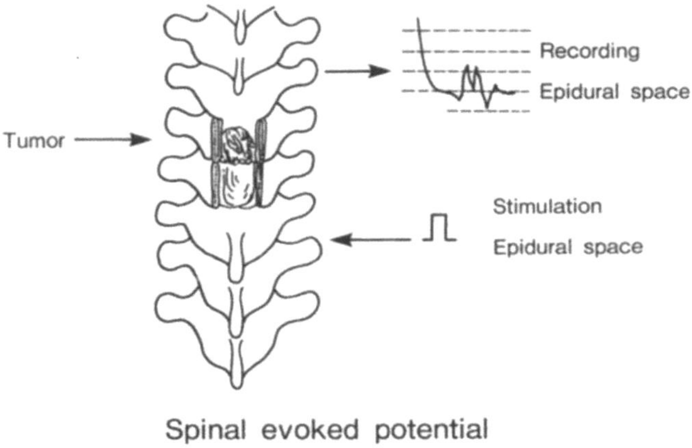

Fig. 1. Schematic illustration of monitoring technique of spinal cord functions during the surgery of the spinal cord tumor. Conductive SpEP recorded on the dorsal surface of the dura is elicited through epidural stimulation

abnormal SpEPs. The period from the onset of symptoms to the operation was shorter in the patients with normal SpEPs than with abnormal SpEPs. The patients with tumors located anteriorly, posterolaterally, or posteriorly often showed abnormal SpEPs. There was no correlation between the histological features of the tumors and the conductive SpEPs.

Analysis of Conductive SpEPs after Decompression of the Spinal Cord

The changes in the conductive SpEPs after decompression of the spinal cord were noted in six of the eight cases which showed abnormal SpEPs before decompression. Three cases showed an appearance of a conductive SpEP after decompression. The amplitudes of the SpEPs were 1.2 μV, 5 μV, and 10 μV, respectively. Augmentation was noted in three cases which showed

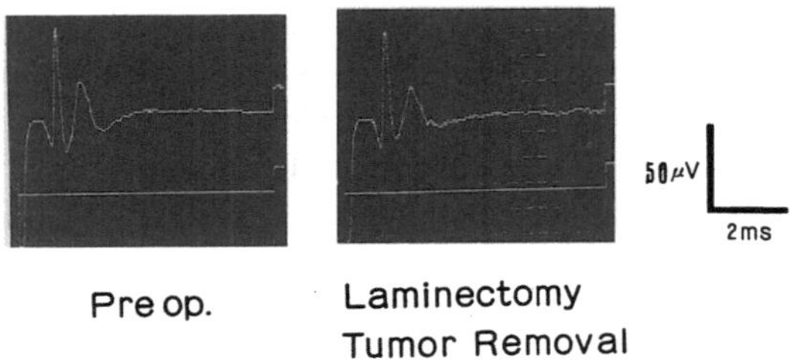

Fig. 2. Case 1. 60-year-old woman who had numbness in her trunk and both legs and showed spastic gait for 2 months before admission. The conductive SpEP was a typical potential which had two negative peaks with amplitude of 46 μV. It remained unchanged during operation, The patient improved one Frankel grade from D to E

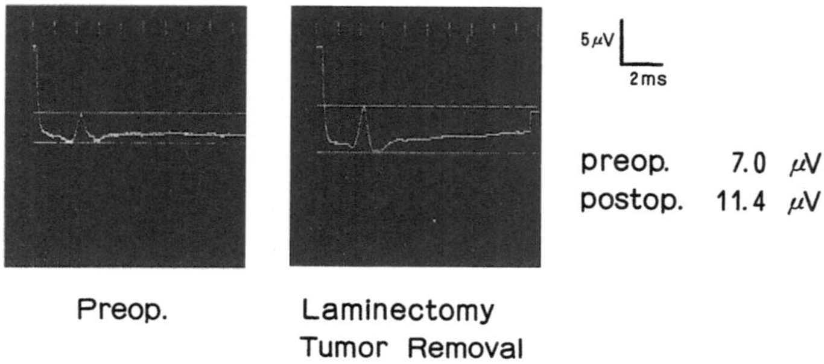

Fig. 3. Case 7. 61-year-old woman who stated she had had paraparesis for 2 months before admission and could not ambulate without crutches. Intradural-extramedullary tumor at T9. The conductive SpEP was a low amplitude potential of 7.2 μV before decompression. It was potentiated to 11.4 μV after decompression. The patient improved two Frankel grades from C to E

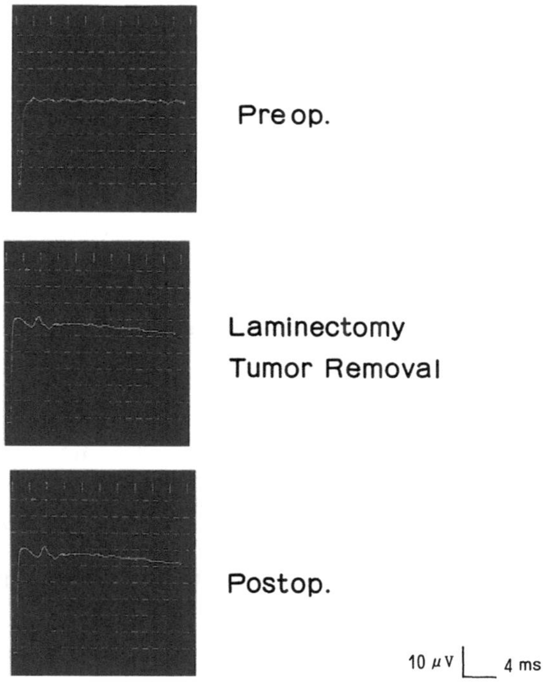

Fig. 4. Case 12. 27-year-old woman who had had back pain for 3 months and could not ambulate on admission. Extradural neurinoma of Eden's type III at T7–8. The conductive SpEP was electrically silent before decompression. SpEPs gradually appeared after removal of the tumor, the amplitude was low at 4.8 μV after decompression. The patient improved two Frankel grades from C to E

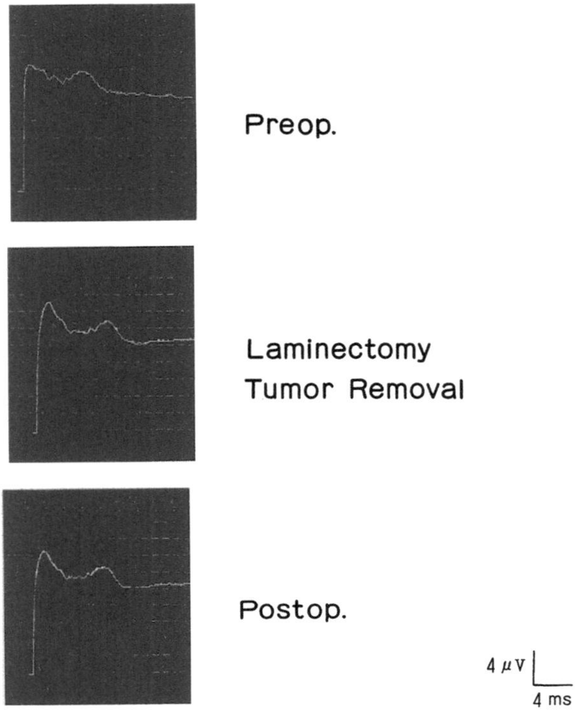

Fig. 5. Case 13. 22-year-old man who had numbness of his left shoulder 6 months prior to admission and suddenly started to have tetraparesis at 1 month before admission. Extradural neurinoma of Eden's type III at C4–5. The conductive SpEP was a slow sustained negative potential with amplitude of 3 μV before decompression. There was no change in the waveform and amplitude during the operation. The patient improved two Frankel grades from C to E

low amplitude potentials before decompression. The increases in amplitude of the conductive SpEPs in these three cases were from 6.0 μV to 9.2 μV (15.3%), from 7.2 μV to 10.2 μV (14.2%), and from 10 μV to 25 μV (150%). No changes in the conductive SpEPs after decompression were noted in eight cases. Six of these showed normal SpEPs and two showed the slow sustained potentials (Figs. 2–5).

Correlation Between Changes in Conductive SpEPs and Prognosis

Neurological symptoms improved in 12 out of 14 cases. Two cases remained unchanged after decompression. Patients showing potentiation or appearance of the conductive SpEPs after decompression showed symptomatic improvement. The conductive SpEPs did not change in two cases after decompression. Neurological symptoms remained unchanged in one case and the other improved from Frankel grade C to grade E (Figs. 2–4). This suggests that neurological deficits will recover in the patients whose conductive SpEPs remain normal or are improved after decompression.

Discussion

It has been confirmed that conductive SpEPs were evoked on the superficial layer of the posterior part of the lateral and posterior column of spinal cord, and that changes in the conductive SpEPs reflects mainly the pathological conditions in white matter [6]. SpEPs are sensitive enough to reflect events which might be hazardous to the normal spinal cord during operation. The peak amplitude of the potential is used as an indicator of spinal cord function. However, it is difficult to evaluate how changes in the waveforms and fluctuation of the amplitude of conductive SpEPs reflect conditions of the spinal cord under chronic pressure. Some kinds of waveforms of the conductive SpEPs were observed in the patients with spinal cord tumors in this study. Abnormal SpEPs can be classified into three types, (a) electrically silent, (b) low amplitude potentials, and (c) slow sustained potentials. The patients with moderately or severely impaired neurological states showed abnormal SpEPs. The patients with a spinal cord tumor located anteriorly, posterolaterally, or posteriorly showed abnormal SpEPs. Eight of the 14 patients (57%) had an abnormal SpEP before decompression. However, no case showed amplitude attenuation during the operation. It may be considered that there should be new criteria for evaluating conductive SpEPs indicating impending danger to the spinal cord under chronic pressure which are different from those applicable to the normal spinal cord. We have to pay attention to the patients whose amplitude attenuations are greater than 50%, because any changes in abnormal SpEPs may reflect damage to the spinal cord. Six (75%) of eight patients with abnormal SpEPs showed potentiation or augmentation. The augmentation of SpEPs was observed in the following situations (a) after excision of the spinal cord tumor, (b) on correction of scoliosis, (c) after removal of chronic lesions squeezing the spinal cord, (d) immediately after a sudden drop of the blood pressure in aortic aneurysm surgery, (e) on cooling of the spinal cord [1, 3, 5, 10]. We also emphasize that false positive potentiation may be caused by technical problems during monitoring [2, 8]. The findings in this study suggested that symptomatic improvement did not always reflect neurophysiological recovery in conductive SpEPs. In fact we have experienced that the conductive SpEP may remain unchanged even if symptoms show marked recovery after the operation. However we cannot completely deny the value of monitoring conductive SpEPs, because the change in abnormal SpEPs during operations gives us information about the reversibility of damaged spinal cord and prognosis of spinal function.

References

1. Akimoto H, Tohno S, Harada S, Ohmi Y, Nakano K, Suetsuna F, Arai T (1987) The augmentation of spinal cord evoked potential at the caudal site of a lesion in the spinal cord pathway. J of Electrodiagnosis of the Spinal Cord 10:15–17
2. Brown RH, Nash CL (1988) Standardization of evoked potential recording. In: Ducker TB, Brown RB (eds) Neurophysiology and standards of spinal cord monitoring. Springer, Berlin Heidelberg New York Tokyo, pp 1–10
3. Harada Y (1983) Study on experimental spinal cord injuries by distraction using spinal cord evoked potentials. J Jpn Orthop Ass 57:685–701
4. Iizuka T, Kurokawa T (1982) Transient augmentation of the evoked spinal cord and peripheral nerve action potentials to ischemia. J Jpn Orthop Ass 56:163–170
5. Imai K (1988) A clinical study on intra-operative spinal cord monitoring with spinal evoked potential for scoliosis. J Jpn Orthop Ass 62:511–521
6. Imai T (1976) Human electrospinogram evoked by direct stimulation on the spinal cord through epidural space. J Jpn Orthop Ass 50:1037–1056
7. Kurokawa T (1976) Spinal cord action potentials evoked by epidural stimulation of cord – A report of human and animal record. Jpn J Electroencephalogr Electromyogr Clin Neurophysiol 1:64–66
8. Shinomiya K, Sato Y, Sato M (1983) Monitoring of spinal cord function using evoked spinal cord potentials during surgery of the cervical spine. Orthopaedic Surgery and Traumatology 26:1961–1968
9. Tamaki T, Noguchi T, Takano H, Tsuji H, Nakagawa T, Imai K, Inoue S (1984) Spinal cord monitoring as a clinical utilization of the spinal evoked potential. Clin Orthop 184:58–64
10. Yasukawa K (1980) Experimental and clinical studies of the evoked electrospinogram for monitoring spinal cord function. J Jpn Orthop Ass 54:1661–1677

Anorectal Pressure Monitoring During Surgery for Reuntethering of the Spinal Cord in Cases of Lumbosacral Lipomyelomeningocele

H. SAKAMOTO[1], A. HAKUBA, K. FUJITANI, H. HAYASHI, M. KIYAMA, and S. NISHIMURA

Introduction

Tethered cord syndrome is most often seen in children with a lumbosacral lipomyelomeningocele associated with occult spinal dysraphism, and often manifests as urinary bladder dysfunction, sensory disturbance, and motor weakness or deformity of the lower extremities. The aim of surgical treatment of a lipomyelomeningocele is to release the tethered spinal cord from the lipomatous tissue. Despite the satisfactory results obtained through surgical treatment for lumbosacral lipomyelomeningoceles without anorectal pressure monitoring [2], we found that untethering of the spinal cord was quite difficult to perform in some patients in which the lipoma involved the ventral part of the spinal cord [1, 4, 9]. We also found that appropriate repair of a lipomyelomeningocele at the time of initial surgery sometimes failed to prevent retethering of the spinal cord due to postoperative adhesion between the cord and the surrounding structures [1, 5], and that dissection of the neural structure deeply embedded in dense scar tissue was much more difficult to perform in a repeat operation for reuntethering of the spinal cord than in the initial operation. In such cases, intraoperative monitoring for identification of the nerve root and the conus medullaris is necessary to untether the spinal cord completely and preserve the neural structures. During the operation, the lower lumbar and first sacral ventral roots were easily identified by applying electrical nerve stimulation and palpating contractions of the segmental muscle groups. Identification of the second, third, and fourth sacral roots, which have neurogenic control of the urinary bladder and also the anal sphincter, requires measurement of sphincter function [3, 7, 8].

Reported here are the usefulness and limitation of intraoperative anorectal pressure monitoring, especially for repeat operations of patients with a lipomyelomeningocele who had postoperative retethered spinal cord.

---

[1] Department of Neurosurgery, Osaka City University Medical School, 1-5-7 Asahi-machi, Abeno-ku, Osaka, Japan

K. Shimoji et al. (Eds.)
Spinal Cord Monitoring and Electrodiagnosis

Clinical Materials and Methods

Out of 40 patients with lumbosacral lipomyelomeningoceles who were treated surgically in our hospital, six patients showed signs of a retethered spinal cord in the late postoperative period after initial repair. Appropriate repair of the lipomyelomeningocele had been carried out at the time of initial surgery, i.e., the intradural lipoma had been sufficiently removed, the thickened film terminale had been divided, if present, and the spinal cord had been completely released from lipomatous tissue with a dural plasty [2].

We found that reuntethering of the spinal cord was quite difficult in the first two patients with retethering and thus decided to employ anorectal pressure monitoring while operating on the last four patients with retethered cords. Table 1 summarizes the clinical data on these four patients. Initial repair was performed in patients aged 1 month–4 years. Four to 12 years after initial surgery, dysfunction of the lumbosacral cord such as urinary bladder dysfunction and motor weakness in the lower extremities appeared. Preoperative neuroradiological studies [6], such as computed tomography and magnetic resonance imaging, showed that in all cases a small residual lipoma was attached to the terminal portion of the spinal cord and the low-lying conus medullaris was shifted dorsally with no space for the cerebrospinal fluid. These studies indicated that the spinal cord was retethered by a postoperative adhesion around the conus medullaris. A repeat operation for reuntethering of the cord was performed in patients aged 4–14 years. Case 1 already showed atonic bladder and required intermittent catheterization of the bladder 1 year before the repeat surgery. Cases 2 and 3 had foot deformity with minimum abnormality or mild uninhibited bladder on cystometry. In case 4, gait disturbance and occasional urinary incontinence developed 1 month before repeat surgery, and this patient had mild atonic bladder on cystometry. The patients were classified into two groups according to the severity of the urinary bladder dysfunction before the repeat surgery: a severe dysfunction group and a minimum or mild dysfunction group. Case 1 belonged to the severe dysfunction group and the other three cases fell into the minimum or mild dysfunction group.

We prepared an intraoperative electrical stimulation system (block monitor. Sanyoudenshi Co., Japan) with a bipolar nerve stimulator and an elongated rubber balloon as a pressure-sensing device [3, 8, 9]. The balloon was placed just into the anal canal, so that contraction of the external anal sphincter could be detected. The tip of the balloon was connected to a pressure transducer which was then filled with water. The changes in anorectal pressure were continuously monitored and recorded on a paper-chart recorder. Electrical nerve stimulation was started with 0.5 ms duration and 1 count/s frequency. Nerve stimulation was tested on an apparent normal nerve root with a low-voltage amplitude. All questionable structures were then stimulated. If no response was observed to electrical stimulation at a voltage ranging from the threshold up to 10 times higher than the threshold amplitude, the stimulated structure was determined to be a nonfunctioning or non-neural structure. Administration of a muscle relaxant was stopped at least 2 h prior to nerve stimulation.

Complete untethering of the spinal cord with intraoperative monitoring was accomplished in the following manner. Intraoperative nerve stimulation was used for identification of the roots, while the spinal cord was dissected circumferentially, leaving some lipomatous tissue around the conus medullaris. When the volume of the conus medullaris containing the lipomatous tissue seemed too large to fit back into the shallow spinal canal associated with the bifid laminae, stimulation was applied to the dorsal surface of the conus medullaris for identification of the liponeural interface. The lipomatous tissue was then detached from the dorsal surface of the conus medullaris, and duraplasty was subsequently performed.

Results

Figure 1 shows intraoperative recordings of the anorectal pressure. The upper graph demonstrates spontaneous changes in the pressure before electrical nerve stimulation. These rhythmic changes possibly resulted from spontaneous contractions of the internal anal sphincter. In the lower graph, stimulation of the lower sacral root caused a group of abrupt increases in

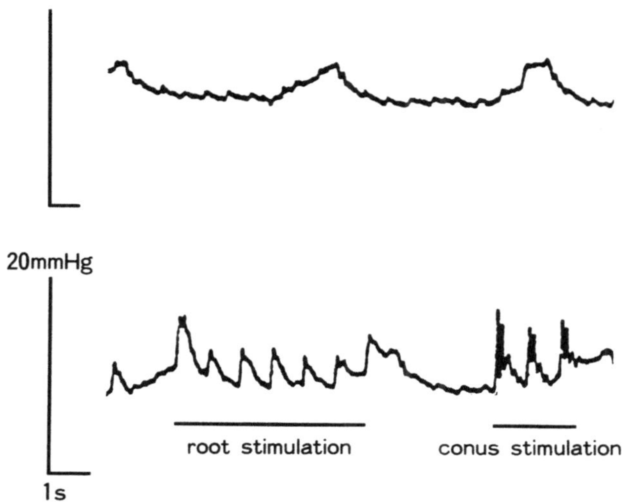

Fig. 1. Intraoperative recordings of the anorectal pressure. The upper graph shows spontaneous rhythmic changes in pressure. The lower graph shows the pressure changes during electrical stimulation of the apparently normal S3 root and the conus medullaris

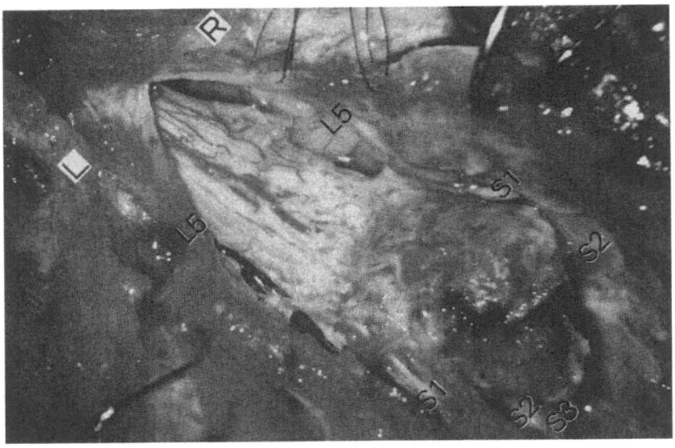

Fig. 2. Intraoperative photograph of case 1. The spinal cord was untethered and the right S3 and bilateral S4 roots could not be identified by electrical stimulation. A small amount of lipomatous tissue was left attached to the conus medullaris to avoid injury to it by further removal of the lipomatous tissue. *R*, right side; *L*, left side; *L5*, L5 nerve root; *S1*, S1 nerve root; *S2*, S2 nerve root; *S3*, S3 nerve root

pressure at 1 Hz frequency. Stimulation of the conus medullaris resulted in abrupt pressure increases with multiple peaks.

In all four cases, the safely dissectable plane between the spinal cord, the lipoma, and the leptomeninges was quite narrow at the time of repeated surgery because a large amount of the lipoma in the intradural space had been removed at the time of initial surgery. In addition, the nerve roots were intermingled with lipomatous tissue or dense scar tissue between the dura mater and the neural tissue.

Figure 2 shows an intraoperative view of case 1 after untethering the spinal cord. The bilateral S1 and S2 roots and the left S3 root were identified and preserved according to their morphological appearance under an operative microscope. Electrical stimulation of these roots or the dorsal surface of the conus medullaris containing lipomatous tissue produced no change in anorectal pressure. The right S3 and bilateral S4 roots were not identified either by electrical stimlation or by careful dissection under an operative microscope. Debulking of the lipomatous tissue attached to the dorsal surface of the conus medullaris was performed mainly by careful dissection under an operative microscope. A small amount of lipomatous tissue was left behind on the dorsal surface of the conus medullaris because further removal of the lipomatous tissue might have resulted in injury of the conus. Significant urinary incontinence remained unchanged postoperatively in this case.

Figure 3 shows intraoperative recordings of the anorectal pressure changes in case 4. Abrupt increases in pressure were seen during electrical stimulation of the right S3 root. Similar pressure changes were observed during stimulation of the other lower sacral roots. When the lipomatous tissue attached to the dorsal surface of the conus was removed, no response

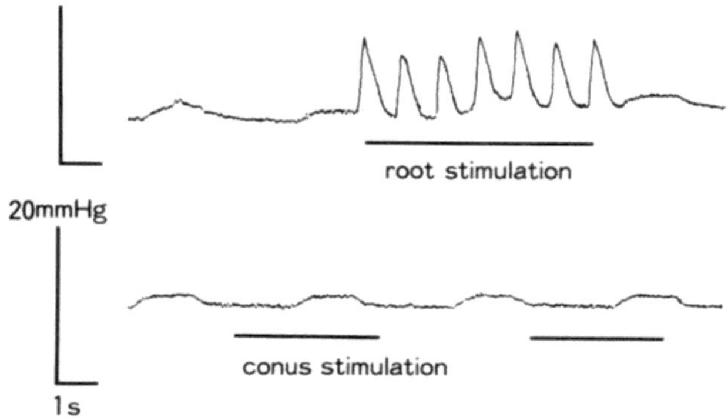

Fig. 3. Intraoperative recordings of the anorectal pressure in case 4. The upper graph shows prominent changes in pressure during electrical stimulation of the S3 root. No significant change in pressure was seen during repeated stimulation of the dorsal surface of the conus medullaris while attempting to locate the liponeural interface

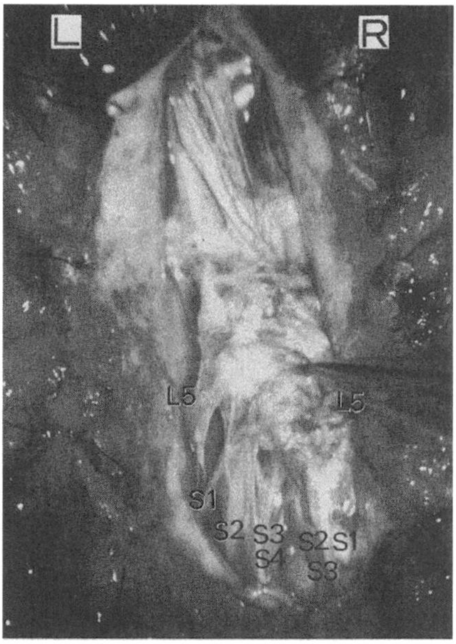

Fig. 4. Intraoperative photograph of case 4. The spinal cord was untethered and the bilateral S1, S2, and S3 roots and the left S4 root were identified. A small amount of lipomatous tissue was left attached to the conus medullaris to avoid injury to it by further removal of the lipomatous tissue, *R*, right side; *L*, left side; *L5*, L5 nerve root; *S1*, S1 nerve root; *S2*, S2 nerve root; *S3*, S3 nerve root; *S4*, S4 nerve root

Table 1. Case summaries

Case	Present age (years)	Sex	Age at time of initial operation	Interval between initial and repeat operation (years)	Degree of preoperative bladder dysfunction	Age at time of repeat operation (years)	Response to stimulation: Sacral roots	Response to stimulation: Lipo-neural interface	Preservation of sacral roots	S2	S3	S4	Degree of postoperative bladder dysfunction	Follow-up period
1	10	F	4 months	8	severe (atonic)	8	–	–	R	+	–	–	atonic (unchanged)	2 years
									L	+	+	–		
2	16	F	6 months	12	minimum	14	+	–	R	+	+	+	transient spastic	1.5 years
									L	+	+	+		
3	12	F	4 years	6	minimum	11	+	–	R	+	+	+	transient spastic	10 months
									L	+	+	+		
4	5	F	1 month	4	mild	4.5	+	–	R	+	+	+	transient spastic	6 months
									L	+	+	–		

was found during repeated stimulation of the dorsal surface of the conus medullaris. Debulking of the lipomatous tissue was then performed by careful dissection under an operative microscope. Figure 4 is an intraoperative photograph of case 4 after untethering the cord. The bilateral S1, S2, and S3 roots and the left S4 root were preserved. A small amount of lipomatous tissue was left behind the conus medullaris in an attempt to avoid injury to the conus during its dissection from the lipomatous tissue. This patient had transient spastic bladder after surgery.

Table 1 shows the results of the operation and the preservation of the S2, S3, and S4 nerve roots, which have primary neurogenic control over the urinary bladder and anal sphincter. The spinal cord was sufficiently untethered in all cases. In the severe bladder dysfunction group, case 1, the lack of response to electrical stimulation of the lower sacral roots indicated that these nerve roots had no significant control over the bladder function. Dissection and identification of the neural structure was performed by careful observation of the tissue under an operative microscope. There was no change in this patient's bladder dysfunction after surgery. In the three cases in the minimum or mild bladder dysfunction group (cases 2, 3, and 4), the lower sacral roots responded to electrical stimulation and could be preserved. However, removal of the residual lipomatous tissue in the conus could be carried out by careful observation of the tissue contiguous to the lipoma under an operative microscope, because intraoperative electrical stimulation of the possible liponeural interface caused no response in anorectal pressure. This group of patients showed transient spastic bladder postoperatively. Urinary bladder function in case 2 returned to the preoperative neurological condition in 6 months and showed no further deterioration in the follow-up period of 1.5 years. Case 3, who had had no complaint about urinary voiding and minimum abnormality on cystometry, returned to the preoperative neurological condition within 1 month. In case 4, preoperative bladder dysfunction with a mild atonic pattern on cystometry changed to a spastic pattern after surgery, followed by return to an atonic pattern with occasional urinary incontinence in 1 month. No further deterioration was observed in this case.

Discussion

The operative results in the minimum or mild bladder dysfunction group suggest that intraoperative anorectal pressure monitoring is useful for preserving the function of the lower sacral nerve roots. However, it is of little help in finding the liponeural interface before injury to the conus medullaris, because no response was found during conus stimulation. Moreover transient spastic bladder after surgery possibly resulted from mild injury of the conus medullaris during its dissection. The results obtained in the severe bladder dysfunction group suggest that anorectal pressure monitoring is of little benefit in identifying the lower sacral roots or the liponeural interface in a patient with long-standing dysfunction of a severe atonic bladder.

We made no attempt to employ intraoperative external anal sphincter electromyography [5]. This technique requires bulky equipment and very accurate insertion of electrodes. Recording evoked spinal cord action potentials in tethered cord syndrome presents difficulties with regard to interpreting the results and there is also some question about the origin of the recorded potentials [10]. The anorectal pressure monitoring system used is simple and non-invasive, and requires no special skills. Also, this pressure recording system is not affected by electromagnetic interference generated in the operating room.

Identification of the liponeural interface by anorectal pressure monitoring is sometimes difficult [8]. There are a couple of possible reasons why we failed to apply anorectal pressure monitoring successfully to identification of the liponeural interface: the electrical stimulation might have been too small in amplitude or such a large volume of lipomatous tissue could have been left behind on the dorsal surface of the conus medullaris that it blocked the electrical stimulation of the conus. Stimulation with too large an amplitude may damage the neural tissue. We had to leave some lipomatous tissue attached to the dorsal surface of the conus medullaris to avoid injury to conus medullaris. The large volume of a conus containing lipomatous tissue may increase the likelihood of postoperative adhesion between the conus and the dura mater, which may cause retethering of the spinal cord.

Preoperative evaluation of urinary bladder function can provide useful information about the value of intraoperative anorectal pressure monitoring. In patients without severe atonic bladder dysfunction, intraoperative anorectal pressure monitoring is highly beneficial for preserving the sacral nerve roots, which have neurological control over the urinary bladder function. On the other hand, in patients with long-standing dysfunction of a severe atonic bladder, such monitoring is unable to identify even the lower sacral nerve roots. We feel, however, intraoperative anorectal monitoring can be effectively applied to patients with a lipomyelomeningocele and various degrees of bladder dysfunction after due consideration of its usefulness and limitations.

All four patients with retethering of the spinal cord returned to their preoperative condition after surgery. The operative results in the severe bladder dysfunction group indicate that the lower sacral roots have no significant control over urinary bladder function. The neural structure controlling urinary bladder function was irreversibly damaged before the reuntethering operation. Reuntethering of the spinal cord must be performed as soon as possible, if signs of a retethered cord appear after initial surgery.

In the minimum or mild bladder dysfunction group, transient spastic bladder developed after surgery, even though removal of the lipomatous tissue was halted before identification of the liponeural interface by electrical stimulation. This suggests that the retethered conus is too vulnerable to be surgically dissected. Highly skillful microsurgical techniques are necessary to operate on patients with a lipomyelomeningocele and retethered spinal cord.

Conclusion

When performing a repeat operation on a lipomyelomeningocele for re-untethering a spinal cord deeply embedded in scar tissue and a residual lipoma, intraoperative anorectal pressure monitoring is useful for preserving the functioning sacral nerves, but is of little help in dissecting the conus medullaris from the lipomatous tissue attached to the dorsal surface of the conus.

References

1. Chapman PH (1982) Congenital intraspinal lipomas. Anatomic considerations and surgical treatment. Child's Brain 9:37–47
2. Hakuba A, Fujitani K, Hoda K, Inoue Y, Nishimura S (1986) Lumbo-sacral lipoma, the timing of the operation and morphological classification. Neuro-Orthopedics 2:34–42
3. Ikeda K, Kubota T, Kashihara K, Yamamoto S (1986) Anorectal pressure monitoring during surgery on sacral lipomeningocele. Case report. J Neurosurg 64:155–156
4. James CCM, Lassman LP (1981) Spina bifida occulta. Orthopaedic, radiological and neurosurgical aspects. Grune & Stratton, New York, San Francisco, pp 160–179
5. James HE, Mulcahy JJ, Walsh JW, Kaplan GW (1979) Use of anal sphincter electromyography during operations on the conus medullaris and sacral nerve roots. Neurosurgery 4:521–523
6. Komiyama M, Hakuba A, Inoue Y, Yasui T, Yagura H, Baba M. Nishimura S (1987) Magnetic resonance imaging: Lumbosacral lipoma. Surg Neurol 28:259–264
7. Pang D, Casey K (1983) Use of an anal sphincter pressure monitor during operations on the sacral spinal cord and nerve roots. Neurosurgery 13:562–568
8. Pang D (1985) Use of an anal sphincter pressure monitor for identification of sacral nerve roots and conus. In: Holtzman RN, Stein BM (eds) The tethered spinal cord. Thieme-Stratton, New York, pp 74–84
9. Pierre-Kahn A, Lacombe J, Pichon J, Giudicelli Y, Renier D, Sainte-Rose C, Perrigot M, Hirsch J (1986) Intraspinal lipomas with spina bifida. Prognosis and treatment in 73 cases. J Neurosurg 65:756–761
10. Satomi K, Fukui Y, Kenmotsu K, Nakamura Y, Okuma T, Hirabayashi K (1988) Evoked spinal cord action potentials on tethered cord syndrome. In: Seko K (eds) The journal of electrodiagnosis of the spinal cord. Vol. 11, (Proceeding of the 11th meeting of Japanese society of spinal cord electrodiagnosis), pp 107–110

The Effect of Hypothermia on SEPs Recorded from the Posterior Pharynx and the Scalp During Cardiopulmonary Bypass in Man

T. TAKADA[1], H. FUJIOKA, H. FUJIHARA, Y. MARUYAMA, N. URUSHIBARA, and K. SHIMOJI

Summary

Somatosensory evoked potentials (SEPs) from the posterior pharynx (C5–7 vertebral level) and scalp (C3′, C4′) were simultaneously recorded with nerve action potentials (NAPs) (ulnar sulcus and Erb's point) in response to both ulnar and median nerve stimulation in 10 adult patients undergoing cardiac surgery. In profound hypotherma – esophageal temperature (Te) of 22.0°C during cardiopulmonary bypass (CPB) – peak latencies of P13 from the posterior pharynx and N20 from the scalp increased more than that of P9 from the posterior pharynx. Amplitudes of P13 and N20 significantly decreased to 87.5 ± 6.7 (mean ± S.D.) and 63.8 ± 14.8% of the control value (before CPB), respectively, during profound hypothermia (Te, 20.0°–28.0°C). By contrast, amplitudes of NAPs from the ulnar sulcus and Erb's point significantly increased to 133.3 ± 13.3 and 138.0 ± 10.5%, respectively, in comparison to those before CPB. The results suggest that in profound hypothermia by CPB, latency prolongation of SEPs from the spinal cord or upper central pathways are largely due to the synaptic delay by cooling. SEPs recorded from the posterior pharynx were stable and reproducible in profound hypothermia even below Te of 20°C, while SEPs from the scalp were unrecordable in some cases. Thus, recording from the posterior pharynx during CPB was considered to be a useful method to monitor spinal function, since it is noninvasive and reproducible.

Introduction

As a result of the development of intraoperative anesthetic management during cardiac surgery, the number of high risk surgical cases, including aortic surgery, has increased. Paraplegia due to spinal ischemia is among the severe neurological complications of cardiac surgery [4, 12]. Spinal ischemia is one

[1] Department of Anesthesia, Niigata Cancer Center Hospital, Niigata, Japan

K. Shimoji et al. (Eds.)
Spinal Cord Monitoring and Electrodiagnosis

of the major complications encountered during surgery of aortic aneurysm [2, 7, 9, 14]. In order to monitor spinal ischemia by recording of spinal SEPs, it might be necessary to accumulate basic data recording the influence of hypothermia on SEPs during cardiac surgery.

We previously demonstrated noninvasive and reproducible prevertebral recording from the posterior pharynx (C5–7 vertebral level) with comparison to segmental spinal cord potentials simulteneously recorded from the posterior epidural space during spine surgery (13).

We have now applied the methods to open heart surgery and recorded spinal SEPs during CPB, while simultaneously recording cerebral SEPs from the scalp and nerve action potentials (NAPs) from the ulnar sulcus and Erb's point. We also investigated the relationship between body temperature and SEPs during CPB, and assessed the utility of the methods for intraoperative monitoring during CPB.

Materials and Methods

Ten patients (25–70 years; 4 males, 6 females) with acquired heart diseases undergoing cardiac surgery were included in this study. Premedication was carried out with intramuscular injection of scopolamine (0.01 mg/kg) and morphine (0.1 mg/kg) 1 h before induction of anesthesia. A high dose of fentanyl (0.1 mg/kg) was given intravenously under inhalation of 100% oxygen for maintenance of anesthesia. Following induction of anesthesia, two stimulating needle electrodes each were inserted into the skin close to both the median and ulnar nerves at the wrist 10 mm apart. Recording electrodes were inserted at the ulnar sulcus and Erb's point for recording NAPs. Spinal SEPs were recorded from the posterior pharynx at the C5–7 vertebral level and cerebral SEPs from the scalp (C3′, C4′). Reference electrodes for spinal and cerebral SEPs were placed at the dorsal neck skin surface and bilateral earlobes, respectively. Square-wave pulses of 0.2-ms duration with a frequency of 2 Hz and an intensity of 10–15 mA were delivered to the median or ulnar nerve. Bandpass, 0.5–5000 Hz was used. The reproducibility of evoked potentials was confirmed by superimposing two or three traces at each period. Sequential recording of both spinal and cerebral SEPs was made using Neuropack-II (Nihon Kohden KK).

Esophageal temperature (Te), arterial blood pressure, pulmonary arterial pressure, electrocardiogram, and electroencephalogram (C3′, C4′) were continuously monitored throughout the procedures. Arterial blood gas analysis was carried out frequently. Mean arterial pressure was kept at more than 60 torr before CPB (Te: 35.0°–37.0°C) and more than 45 torr during CPB (Te:35.0°–20.0°C). Ventilation volume was set at 100–120 ml/kg/min to keep arterial pCO2 as close to 40 torr as possible. Arterial pCO2 was maintained at 30–40 torr by CPB during hypothermia.

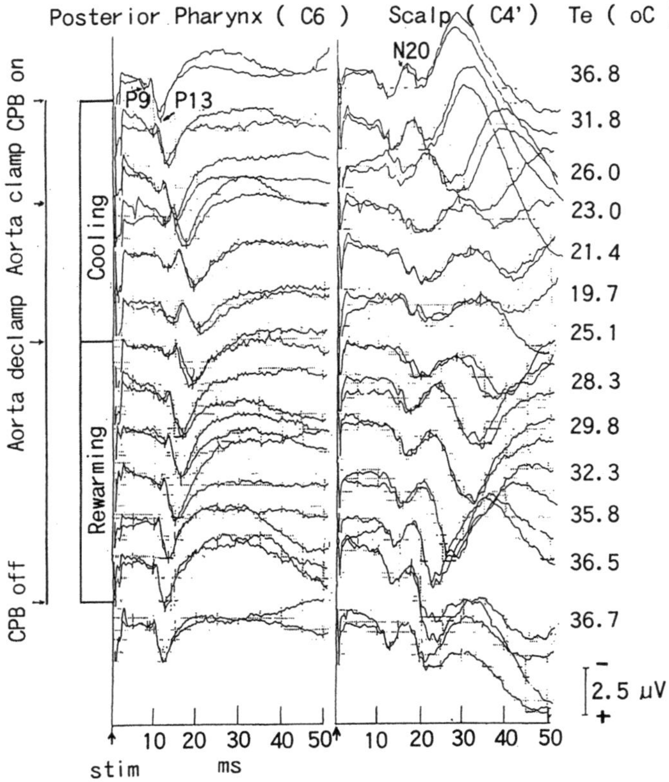

Fig. 1. An example of sequential recording of the somatosensory evoked potentials recorded from the posterior pharynx (C6 vertebral level) and the scalp (C4′) in response to left ulnar nerve stimulation at the wrist before, during and after CPB. When the esophageal temperature (Te) fell from 36.8° to 21.4°C, peak latencies of P9, P13 and N20 increased. Peak latency of P9 increased from 8.7 to 13.3 ms, P13 from 11.6 to 18.4 ms, and N20 from 18.4 to 30.6 ms. Note that the amplitude of P9 increased with cooling, while those of P13 and N20 decreased

Results

Figure 1 shows sequential recordings of SEPs from the posterior phrynx (C6 vertebral level) and the scalp (C4′) in response to left ulnar nerve stimulation. When Te fell from 37.0°C to 21.0°C, all peak latencies of P9, P13 and N20 increased. Peak latency of N20 increased more than those of P13 and P9, and P9 showing the smallest increase.

Figure 2 shows the changes of peak latencies of P9, P13 and N20 of SEPs recorded from the posterior pharynx and the scalp during cooling and rewarming. Before CPB, when Te was 35°C, peak latencies of P9, P13 and N20 were 9.29 ± 0.83 ms (mean ± S.D.), 12.93 ± 0.88 ms, and 18.55 ± 1.29 ms,

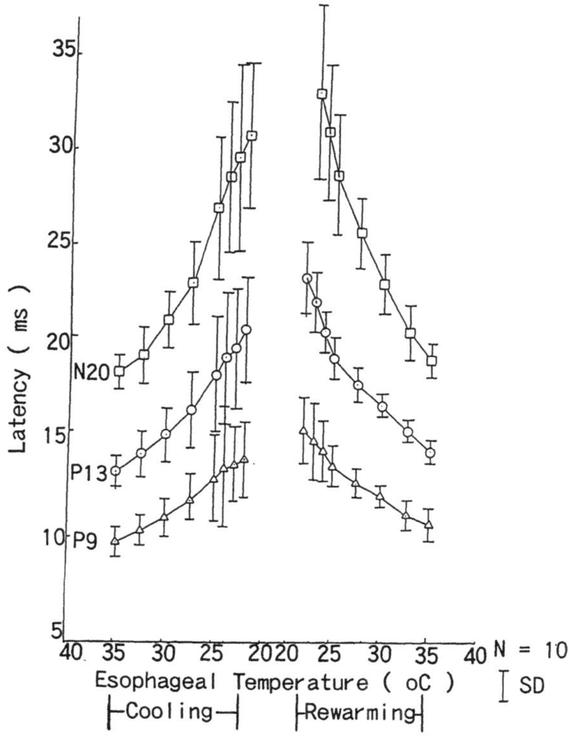

Fig. 2. Changes in peak latencies of P9, P13 recorded from the posterior pharynx and N20 recorded from the scalp by cooling and rewarming during CPB

respectively. In profound hypothermia, when Te fell to 22.0°C, the peak latency of P9 increased to 13.31 ± 1.36 ms (+43.3%), while that of P13 increased more steeply to 19.33 ± 1.90 ms (+49.5%) and that of N20 increased most steeply to 30.51 ± 2.46 ms (+64.5%). Table 1 shows the re-

Table 1. Regression lines and correlation coefficients (*r*) for peak latencies of P9, P13 in SEP from the posterior pharynx and N20 in SEP from the scalp during moderate (esophageal temperature; Te = 35.0°–28.0°C) and profound hypothermia (Te = 28.0°–22.0°C) on colling by CPB

Peak latency	Te (°C)	Regression lines of peak latency as a function of Te	
P9	35.0–28.0	−0.26Te + 18.28	−0.35
	28.0–22.0	−0.34Te + 20.56	−0.51
P13	35.0–22.0	−0.36Te + 25.03	−0.67
	28.0–22.0	−0.64Te + 33.15	−0.59
N20	35.0–28.0	−0.65Te + 41.37	−0.65
	28.0–22.0	−0.96Te + 49.22	−0.68

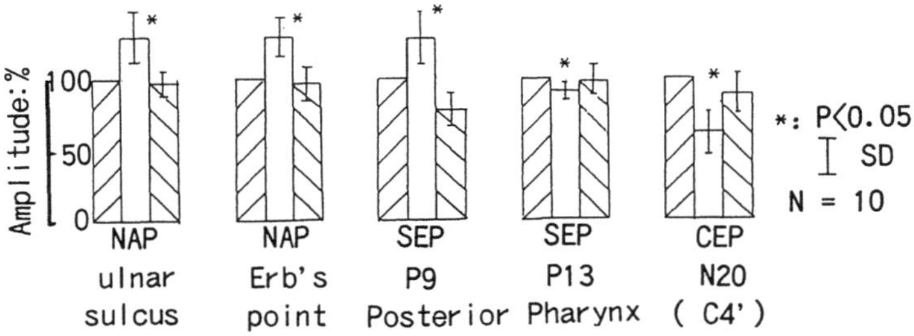

Fig. 3. Percent changes (with standard deviation) in amplitudes of nerve action potentials (NAPs) (ulnar sulcus and Erb's point) P9 and P13 recorded from the posterior pharynx at the C5–7 vertebral level and N20 recorded from the scalp (C4′) before (the *left bar*) (Te = 35.0°–37.0°C), during (the *middle bars*) (Te = 22.0°–26.0°C) and after (the *right bar*) (Te = 35.0°–37.0°C) CPB. *Asteriks* denote a significant difference ($p < 0.05$) ($N = 10$) from the control value (before CPB)

gression line slopes with correlation coefficients (r) for peak latencies of P9, P13 and N20 as a function of Te during CPB. In profound hypothermia, when Te fell to 22.0°C, the P9 peak latency slope was 0.34 ms/°C (+3.66%). In contrast, that of P13 was steeper (0.64 ms/°C (+4.95%)) and that of N20 was steepest (0.94 ms/°C (+4.96%)).

Figure 3 show the changes in amplitudes of NAPs (ulnar sulcus, Erb's point), P9 and P13 of spinal SEP, and N20 of cerebral SEP before (Te = 35.0°C), during (Te = 22.0°–26.0°C), and after (Te = 35.0°–37.0°C) CPB. During CPB the amplitudes of both NAPs and P9 significantly increased, as compared to before CPB, to 133.3 ± 13.3 (mean ± S.D.), and 138.5 ± 10.5%, respectively. In contrast, those of P13 and N20 significantly decreased to 87.5 ± 6.3 and 63.8 ± 14.1% of the control values (before CPB), respectively.

Discussion

The present results may show that SEP recorded from the posterior pharynx is useful for monitoring spinal cord function during cardiac surgery due to its noninvasive nature and reproducibility. Epidural recording is a highly reliable and useful method for monitoring and estimating spinal functions during operative procedures [10, 11]. However, this method may be difficult to use in cardiac surgery under heparinized conditions.

Prevertebral recording of SEPs using an esophageal probe in conscious volunteers was reported by Desmedt et al. [5]. We applied this method to anesthetic patients undergoing spine surgery, and found that peak latencies of P9 and P13 coincided with those of the initial positive spike (P1) and sharp negative wave (N1) of the segmental spinal cord potential recorded from the posterior epidural space [13].

This study was undertaken to clarify the effect of hypothermia on cerebrospinal functions during CPB. Several authors reported the influences of hypothermia on SEPs in cardiac surgery. Hume and Durkin [6] reported that both central conduction time (CCT; N13–N20 interpeak interval) and spinal conduction time (SCT; N10–N13 interpeak interval) increased exponentially by cooling during CPB in man. The CCT increased most steeply and the SCT increased more steeply than the N10 peak latency. They suggested that conduction delay by hypothermia was more pronounced centrally than peripherally [6]. Coles et al. [3] reported that prolongation of peak latencies of N18 and N13, reflecting a slowing of neural transmission, occurred as a function of decreasing systemic temperature during profound hypothermia. Markand et al. [8] studied the relationships of peak latencies of the N10, N14, and N19 components of SEPs to nasopharyngeal temperature during CPB. They found that all the peak latencies increased linearly as nasopharyngeal temperature decreased, and that the increasing slopes of the regression lines for the peak latencies of later components were more prominent. None of these authors, however, noted the differences in the increasing delays of peak latencies of P13 and N20, both of which are thought to be postsynaptic, from that of P9 peak latency in profound hypothermia during CPB.

In animal experiments, Brooks et al. [1] found that excitatory processes at synaptic junctions were more retarded during hypothermia than non-synaptic components especially below 30°C.

Our data in human support their finding. In profound hypothermia, when Te fell from 28.0° to 22.0°C during CPB, the peak latencies of P13 and N20 increased more steeply than that of P9; the slopes of the peak latencies of P13 and N20 were 0.64 (+4.95%) and 0.94 (+4.96%) ms/°C, respectively, which were larger than that of P9 (Fig. 2, Table 1). These conduction delays were attributed to synaptic delays resulting from cooling, as well as delays of axonal conduction to the somatosensory pathways during profound cooling. On the other hand, amplitudes of both P13 and N20 significantly decreased in comparison to those before CPB, while that of P9 significantly increased in profound hypothermia.

SEPs from the posterior pharynx were clearly and reproducibly recorded in all cases during CPB. They were also recordable below Te of 20°C in some cases. These findings indicate that spinal function can be monitored even during profound hypothermia. By contrast, cerebral function could not always be determined from cerebral SEP during profound hypothermia below Te of 20°C. Intraoperative monitoring of SEPs from the scalp are now mainly carried out during cardiac surgery in many institutes. However, such monitoring can not completely prevent neurological complications during cardiac surgery. SEPs from the scalp or skin surface may not be an index of motor function. It is necessary to increase the averaging number to record stable potentials from the skin surface while body temperature decreases rapidly during CPB. Thus it becomes difficult to investigate the exact relationship between SEP and body temperature when SEPs from the skin surface are recorded. Further, SEPs from the skin surface are highly vulnerable to mechanical artifacts during operation.

Recording with posterior pharyngeal electrodes has several advantages. First, it can be applied in cardiac surgery under heparinized conditions because it is noninvasive. Second, it is very stable and highly reproducible in comparison to recording from the skin surface. Third, by using this method, we can record spinal SEP even during profound hypothermia below Te of 20°C during CPB. The method, performed together with recording of scalp SEP, might allow us to determine more precisely cerebrospinal function during cardiac surgery, and provide us with a tool to differentiate cerebral and spinal ischemia during cardiac surgery.

References

1. Brooks C McC, Koizumi K, Malcolm JL (1955) Effects of changes in temperature on reactions of spinal cord. J Neurophysiol 18:205–216
2. Cole JG, Wiloson GJ, Sima AG, Kelmant P, Tait GA (1982) Intraoperative detection of spinal cord ischemia using somatosensory cortical evoked potentials during thoracic aortic occlusion. Ann Thorac Surg 34:299–306
3. Coles JG, Taylor MJ, Pearce JM, Lowry NJ, Stewart DJ, Trusler GA, Williams WG (1984) Cerebral monitoring of somatosensory evoked potentials during profoundly hypothermic circulatory arrest. Circulation 70 (Suppl I):96–102
4. Cunnigham JN, Laschinger JC, Merkin HA, Nathan IM, Colvin S, Ransohoff J, Spencer FC (1982) Measurement of spinal cord ischemia during operations upon the thoracic aorta. Ann Surg 196:285–296
5. Desmedt JE, Cheron G (1981) Prevertebral (oesophageal) recording of subcortical somatosensory evoked potentials in man: the spinal P13 component and the dual nature of the spinal generators. Electroencephalogr Clin Neurophysiol 52:257–275
6. Hume AN, Durkin MA (1986) Central and spinal somatosensory conduction times during hypothermic cardiopulmonary bypass and some observations on the effects of fentanyl and isoflurane anesthesia. Electroencephalogr Clin Neurophysiol 65:46–58
7. Laschinger JC, Cunningham JN, Nahtan IM, Knopp EA, Cooper MM, Spencer FG (1983) Experimental and clinical assessment of the adequacy of partial bypass in maintenance of spinal cord blood flow during operations on the thoracic aorta. Ann Thorac Surg 36:417–426
8. Markand ON, Warren CH, Moorthy SS, Stoelting RK, King RD (1984) Monitoring of multimodality evoked potentials during open heart surgery under hypothermia. Electroenceph Clin Neurophysiol 59:432–440
9. Maruyama Y, Shimoji K, Fujioka H, Takada T, Endoh H (1988) Brain and spinal cord monitoring by multispatial and multimodal evoked potentials during aortic surgery. In: Ducker TB, Brown RH (eds) Neurophysiology and standards of spinal cord monitoring. Springer, Berlin Heidelberg New York Tokyo, pp 177–187
10. Shimoji K, Higashi H and Kono T (1971) Epidural recording of spinal electrogram in man. Electroencephalogr Clin Neurophysiol 30:236f–239
11. Shimoji K, Kano T, Higashi H, Morioka T, Henschel EO (1972) Evoked spinal electrograms recorded from epidural space in man. J Appl Physiol 33:468–471
12. Szilagyi DE, Hageman JE, Smith RF, Elliott JP (1978) Spinal cord damage in surgery of abdominal aorta. Surgery 83:38–56

13. Takada T, Fujioka H, Fujihara H, Maruyama Y, Shimoji K (1988) Somatosensory evoked potentials recorded from the retropharyngoesophageal region and its relation to segmental spinal cord potential. Jap J EEG EMG 16(2):167 (abstr)
14. Takagi T, Okumura F (1985) Application and limitation of somatosensory evoked potential monitoring during thoracic aortic aneurysm surgery. Anesthesiology 63:700–703

Spinal Cord Monitoring in Aortic Aneurysm Surgery

H. Takano[1], K. Takakuwa, H. Kitagawa, N. Yamamoto, H. Tsuji, and T. Ueyama

Introduction

Spinal cord damage caused by spinal cord ischemia during surgery for aortic aneurysm is a most serious complication. The prevalence of this complication has remained between 0.5% and 15% [6]. The sites most vulnerable to spinal cord ischemia are the upper thoracic and lumbosacral spinal cord. It is difficult to predict the sites of ischemia. Spinal cord potentials elicited by spinal cord stimulation (spinal cord evoked potential, SCEP) and by peripheral nerve stimulation (peripheral NEP) are therefore recorded simultaneously to detect spinal cord ischemia and the level of the ischemia. By recording different potentials, we can detect the ischemic site and can decrease the incidence of false negative cases.

Materials and Methods

Intraoperative spinal cord monitoring during surgery for aortic aneurysm was performed on 23 patients. The diagnoses were descending thoracic aortic aneurysm in 19 patients and thoracoabdominal aneurysm in four patients. The average age of the patients was 63 years (range, 39–72 years). There were 16 men and seven women. All the patients gave informed consent prior to surgery. Distal aortic perfusion was utilized in all patients during cross clamping of the aorta.

In ten patients who underwent surgical treatment in 1985–1986, two tube-type electrodes were introduced into the epidural space at the cervical and thoracolumbar vertebral levels on the day prior to surgery. We recorded SCEP to monitor spinal cord function. In the 13 patients seen after 1987, we

[1] Department of Orthopedic Surgery, Faculty of Medicine, Toyama Medical and Pharmaceutical University, 2630 Sugitani, Toyama, 930-01 Japan

K. Shimoji et al. (Eds.)
Spinal Cord Monitoring and Electrodiagnosis

Table 1. Data of the patients whose evoked potentials changed

Case no.	Age (years)	Sex	Aortic cross clamp placement		Duration (min)	Distal aortic perfusion pressure (mmHg)	BT during cross clamping (°C)	Evoked potential change	
			Proximal	Distal				Peripheral NEP	Conductive SCEP
1	64	m	T5	T12	85	above 60	35.1–30.8	/	increase
2	70	f	T5	L1	73	above 70	36	/	decrease[a]
3	72	m	presubclavian	postsubclavian	60	30–40	36	/	decrease
4	66	m	T5	T10	80	above 70	35.5–34.0	increase	increase followed by decrease
5[b]	69	f	T11	iliac	120	above 70	33.8–32.5	decrease[c]	increase
6	62	m	presubclavian	T5	30	above 70	36.2–33	decrease	increase
7	39	f	postsubclavian	T8	85	40–60	35.5–33.7	decrease	decrease
8	69	m	T5	T10	60	20–30	36	decrease	no change
9	72	m	presubclavian	postsubclavian	56	above 70	34.2–34.8	decrease[d]	increase
10[e]	69	m	T5	T11	54	above 70	36.2–36.7	decrease[c]	decrease[c]

[a] evoked potential recovered after reimplantation of critical intercostal arteries
[b] showed paraplegia after the surgery
[c] evoked potential changed and did not return to the control level
[d] evoked potential changed after transient cardiac arrest which occurred after reperfusion
[e] intraoperative death after graft replacement

added a thoracic epidural electrode and recorded not only SCEPs but also peripheral NEPs elicited by tibial nerve stimulation.

A constant current pulse of 0.3 ms duration with 10 c/s frequency was delivered. The intensity of stimulation ranged from 3 to 10 mA to the epidural space and from 30 to 50 mA to the tibial nerve. Action potentials, recorded using a DISA 1500 EMG system (DISA, Denmark), were averaged from 16 to 64 responses. The frequency response of the recording system ranged from 10 to 2000 c/s.

Results

Of the 23 patients, 13 showed no changes in evoked potential and no neurological deterioration. In ten of the 23 patients, evoked potentials changed during surgery. In two of the 10 patients (cases 5 and 10 in Table 1), the evoked potentials decreased and did not return to the control level. One

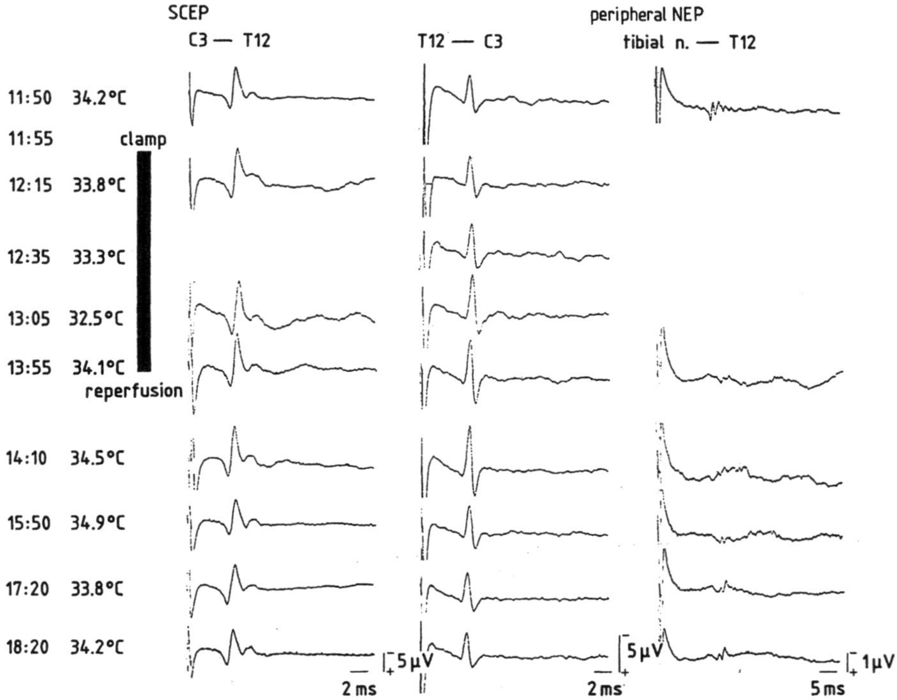

Fig. 1. Changes in evoked potentials in case 5. Peripheral NEP disappeared and SCEP showed transient increase at the time of cross clamping the aorta. Peripheral NEP reappeared after reperfusion but the fast potential showed more marked changes than the late potential

patient showed postoperative paraplegia and the other died during surgery. The remaining eight patients showed transient changes and returned to the control level. They showed no neurological deficit after surgery.

Representative Case Reports

Case 5

A 69-year-old woman underwent graft replacement for thoracoabdominal aneurysm. Aortic cross clamps were placed at the levels of T11 and the iliac artery for 120 min. Distal aortic perfusion pressure was maintained above 70 mmHg throughout the cross clamp interval. She suffered flaccid paraplegia below the L1 segment with preserved light touch and deep sensation after surgery.

Both descending and ascending SCEPs showed transient increases in amplitude during cross clamping of the aorta. Transient increase of the potentials was more markedly recognized in the ascending SCEP. Peripheral NEP, which could not be recorded because of noise during cross clamping of the aorta, was abolished just before reperfusion. This potential did not return to the control level (Fig. 1).

Case 8

A 69-year-old man underwent graft replacement. Aortic cross clamping was performed between T5 and T10 for 60 min. Body temperature was kept at about 36°C throughout the cross clamp period. After the distal aortic perfusion pressure had fallen below 60 mmHg (at 1305 h; Fig. 2a), peripheral NEPs and the late component of the SCEP recorded at L1 decreased in amplitude, while the SCEP recorded at T5 and the fast component of the SCEP recorded at L1 did not change. All of the evoked potentials recovered after reperfusion (Fig. 2). The patient did not show any neurological symptoms after surgery.

Case 10

A 69-year-old man died during surgery because of massive bleeding. Both SCEP and peripheral NEP decreased in amplitude and were abolished with decrement of arterial pressure. Peripheral NEP was more sensitive than SCEP (Fig. 3).

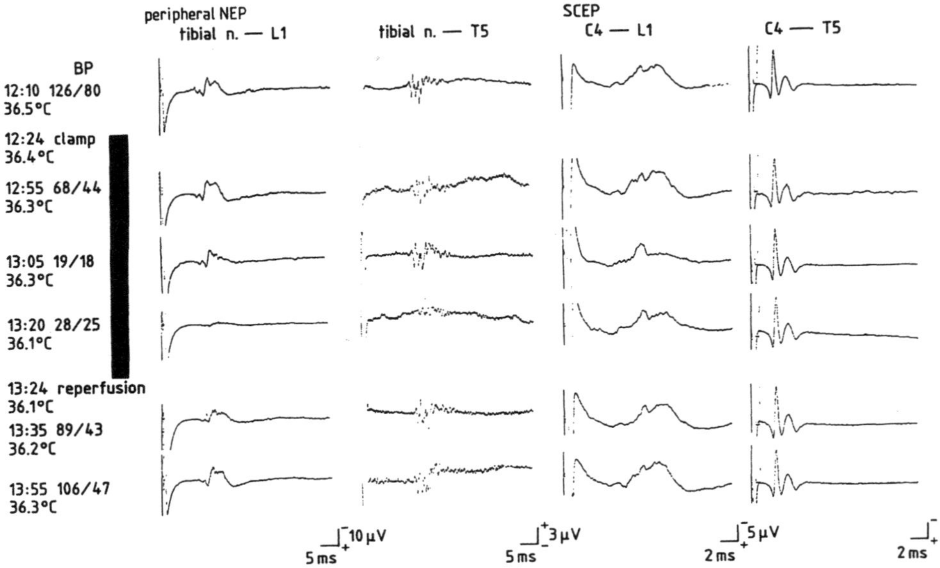

a

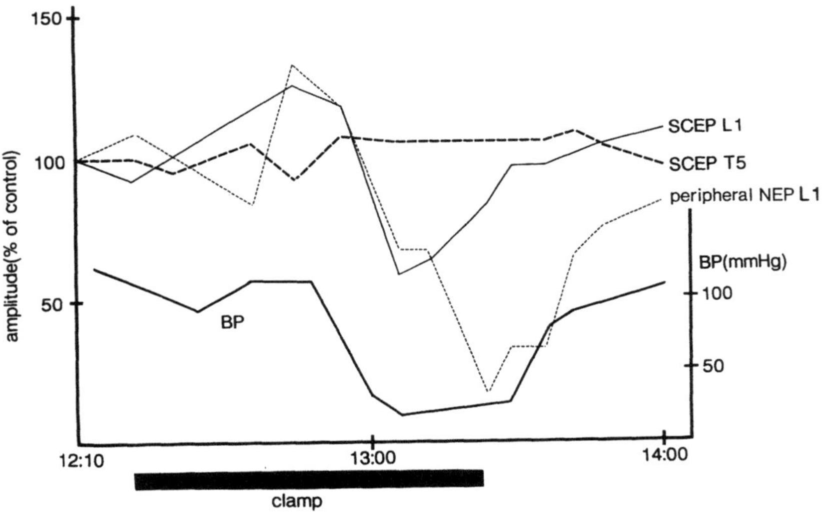

b

Fig. 2a,b. Evoked potential changes in case 8. **a** After distal perfusion pressure fell below 60 mmHg, peripheral NEP and the late component of SCEP (L1) decreased in amplitude without any change in SCEP (T5). The blood pressure values recorded during aortic cross clamping are the distal aortic perfusion pressure and **b** amplitude changes during aorta cross clamping in case 8

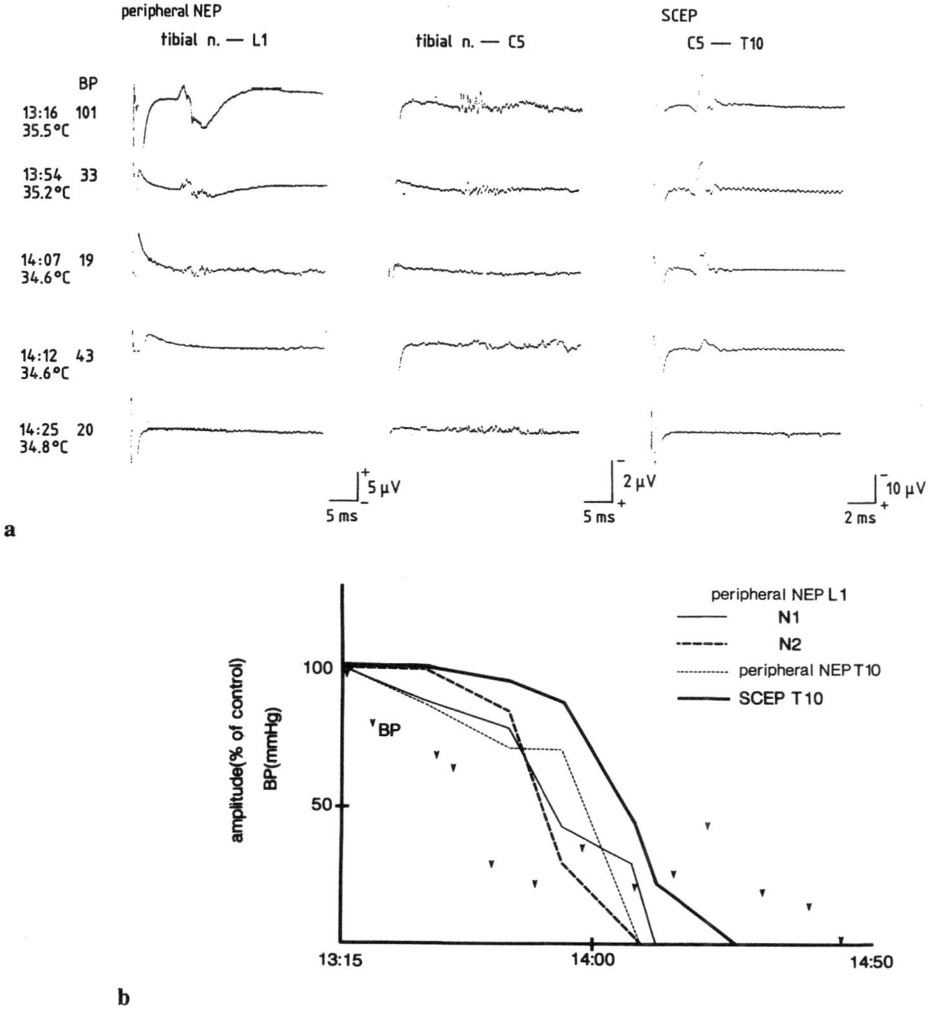

Fig. 3a,b. Evoked potential changes in case 10. **a** Each evoked potential decreased in amplitude and disappeared in proportion to the decrease in blood pressure and **b** graphical representation of the sequential changes of each potential in case 10

Discussion

Changes in evoked potentials during aortic aneurysm surgery may be due not only to spinal cord ischemia but also to hypothermia. In three (cases 2, 3, and 8) of the eight patients (cases 1–8) whose evoked potentials changed during clamping, the body temperature was kept constant. The postoperative neurological deficit of one patient (case 5) suggested that spinal cord ischemia occurred during surgery. These findings suggest that the evoked potentials in four patients (cases 2, 3, 5, and 8) changed because of spinal cord ischemia. In another four (cases 1, 4, 6, and 7) of the eight patients, changes in body temperature were marked. Therefore there is the possibility that the effect of hypothermia may also cause evoked potential change. Although the effect of hypothermia on evoked potentials is controversial [5], it is interesting that in four of five patients whose body temperature was not maintained during clamping of the aorta (cases 1, 4, 5, and 6), the SCEP increased in amplitude.

Ischemic regions suggested by combined recordings were the lumbosacral spinal cord in two patients (cases 5 and 8) and the entire region from the peripheral nerve to the spinal cord in one patient (case 10). In one patient (case 5), peripheral NEP disappeared, while SCEP showed transient augmentation. The ischemic level diagnosed in this case was the peripheral nerve and lumbosacral cord. Although peripheral NEP was recorded at T12, it was not typical in this patient. This potential is thought to be not a segmental potential but an ascending cord potential. Cracco [1] explained that this potential is thought to arise largely in postsynaptic fibers whose synapses are more resistant to asphyxia than those which produce the late potential recorded over caudal spinal cord segments, and partly in nonsynapsing first order sensory axons which traverse the dorsal column. Halonen [3] explained that components 1 and 2 are often found to be more vulnerable to cord ischemia than component 3, which is conducted through the dorsal column. In this patient, peripheral NEP was abolished during cross clamping of the aorta and partially recovered. The patient showed paraplegia with spared dorsal column function. These findings suggest the spinal SEP recorded at the end of surgery reflects the dorsal column function. SCEP showed a transient increase in amplitude. Increase in the amplitude of SCEP was more marked in the ascending potential than in the descending potential. The increase in the potentials is explained by the fact that the threshold for stimulation of nervous tissue transiently decreases in response to ischemia [4]. But the effect of hypothermia was undeniable in this patient. This might have been a false negative case missing the spinal cord ischemia if only SCEP monitoring had been performed. In the other patient (case 8), SCEP recorded at L1 and peripheral NEP decreased in amplitude after the distal perfusion pressure fell below 60 mmHg, while SCEP recorded at T5 did not change. These findings suggest that the peripheral nerve and lumbosacral cord were ischemic. Autoregulation of the spinal cord blood flow was not maintained below 60 mmHg [2]. The late potential of SCEP recorded at L1 decreased, but the fast potential did not change. The fast potential is conducted through descending and ascending tract fibers. The late potential may include both the action potential conducted through the dorsal column and the potentials produced by the activity of neurons in the spinal gray matter [7].

Changes in the evoked potentials, which are assumed to reflect the activity of neurons, were obvious within 10 min, while the evoked potentials produced by propagated action potentials changed slowly (cases 2, 3, 8, and 10). Various evoked potentials must be recorded in order to monitor the whole spinal cord because it is difficult to predict the sites of ischemia prior to surgery, and because each evoked potential has a different origin and different conducting tracts. However, the following problems still remain:

1. The effect of hypothermia on evoked potentials is not yet clarified.
2. The critical point at which neurological deficits develop is not clearly defined.
3. There is a possibility of overlooking localized upper thoracic cord ischemia by the potentials which are conducted through fibers.
4. We can not detect the function of the gray matter of the upper thoracic cord.

Further investigation is needed.

Summary

Recording evoked spinal cord potentials from various stimulation sites can be used to detect spinal cord ischemia and its level. Spinal cord monitoring using this method will decrease the number of the false negative cases.

References

1. Cracco RQ, Evans B (1978) Spinal evoked potential in the cat: effect of asphyxia, strychnine, cord section and compression. Electroencephalogr Clin Neurophysiol 44:187–201
2. Griffiths IR (1973) Spinal cord blood flow in dogs: the effect of blood pressure. J Neurol Neurosurg Psychiatry 36:914–920
3. Halonen JP, Edgar MA, Jones SJ, Ransford AD (1988) Interpretation of anterior and posterior spinal SEPs during scoliosis surgery. In: Ducker TB, Brown RH (eds) Neurophysiology and standards of spinal cord monitoring. Springer, New York Berlin Heidelberg, pp 40–45
4. Iizuka T, Kurokawa T (1982) Transient augmentation of the evoked spinal cord and peripheral nerve action potentials through ischemia (in Japanese). J Jpn Orthop Assoc 56:163–170
5. Maruyama Y, Shimoji K, Fujioka H, Takada T, Endoh H (1988) Brain and spinal cord monitoring by multispatial and multimodel evoked potentials during aortic surgery. In: Ducker TB, Brown RH (eds) Neurophysiology and standards of spinal cord monitoring. Springer, New York Berlin Heidelberg, pp 177–187
6. Szilagyi DE, Hageman JH, Smith RF, Elliott JP (1978) Spinal cord damage in surgery of the abdominal aorta. Surgery 83:38–56
7. Toyoda A, Kanda K (1984) Origins of spinal cord potentials evoked by stimulation of the cat spinal cord. In: Homma S, Tamaki T (eds) Fundamentals and clinical application of spinal cord monitoring. Saikon, Tokyo, pp 99–111

B. Electrodiagnosis

Somatosensory Evoked Potentials and Spinal Evoked Potentials in Patients with Cervical Myelopathy and Cerebrovascular Disease

T. HASHIMOTO[1], M. NARUSHIMA, S. KATSUYAMA, Y. KITAMURA, H. MATSUDA, A. SHIMAZU, M. YASUI, and K. OOTA

Introduction

Subtle changes in somatosensory evoked potential (SEP) are useful to monitor the disturbance of somatosensory pathways. We attempted to clarify the lesion responsible for the disturbance of sensory tracts and to evaluate whether the level of the lesion could be determined. We investigated cerebrovascular disorders with hemiplegia and cervical myelopathy which had abnormal waves and delayed latency in SEPs.

Material and Methods

The subjects were 24 patients with cerebral vascular disorders (12 men and 12 women aged between 46 and 83 years), including 16 patients with cerebral infarction, eight patients with brain bleeding, and 18 patients with cervical myelopathy (12 men and six women aged between 39 and 81 years).

SEPs were induced by stimulating the median and ulnar nerves at the wrist. Electrical stimuli were 0.2 ms duration pulses and 2 times the intensity of the motor threshold at 3 Hz frequency. Electrodes were placed at F3′, F4′, C3′, C4′ and CV7 in the 7th cervical spinous process and at Erb's point. The tied ear reference (A1 + A2) was used. A total of 500 responses were counted, using an EVOMATIC 8000. The input of electrical impulse from the scalp, neck, and Erb's point was filtered with a 10 Hz low cut and 2 kHz high cut. Table 1 shows the mean peak latencies measured in our institute.

The negative peaks in latency were designated N9, N11, N13, N16, N19, N31, and N66. To compensate for the variation in latency due to variation in length of arm, the average latency N9 was subtracted from subsequent peaks and calculated as mean plus 3 standard deviation (SD).

[1] Department of Rehabilitation Medicine, Fuchuu Hospital, 1-10-17 Hiko-cho, Izumi City, Osaka, 594 Japan

K. Shimoji et al. (Eds.)
Spinal Cord Monitoring and Electrodiagnosis

Table 1. SSEP and SEP after stimulation of the median nerve in 46 normal subjects

Components	Latency (left and right combined)		Latency difference between left and right	
	mean ± SD(ms)	mean + 3SD	mean ± SD(ms)	mean + 3SD
N9	9.4 ± 0.6	11.2	0.0 ± 0.3	0.8
N11	10.8 ± 0.6	12.7	0.0 ± 0.4	1.1
N13	12.6 ± 0.8	14.9	0.0 ± 0.4	1.3
N16	15.8 ± 0.7	17.8	0.0 ± 0.3	1.0
N19	18.7 ± 1.0	21.7	0.0 ± 0.5	1.4
N31	31.6 ± 3.3	41.4	0.7 ± 4.7	14.7
N66	66.0 ± 8.6	91.9	0.9 ± 3.9	12.5

Table 2. Latency difference between N9 and the subsequent peaks in 46 normal subjects

Components	Latency difference	
	mean ± SD(ms)	mean + 3SD
N11–N9	1.3 ± 0.4	2.5
N13–N9	3.0 ± 0.5	4.5
N16–N9	5.4 ± 1.0	8.4
N19–N9	9.3 ± 0.7	11.3
N31–N9	21.9 ± 4.5	35.5
N66–N9	56.6 ± 8.6	82.5

The evaluation of the SEP abnormality was defined as the value more than mean plus 3 SD in latency and to lower than half a peak in amplitude on healthy side as shown in Table 2.

Results

Typical cases examined are summarized here. Case 1 was a 70-year-old woman who had right-sided hemiplegia and loss of deep sensation due to bleeding in the left thalamus. Figure 1 shows the image obtained by computed tomography 5 years after the attack. A low density area was observed in the thalamus over the lateral side of left lateral cerebral ventricle. Normal findings in the left median and ulnar nerves but diminished waves after N16 in the right median and ulnar nerves were observed in the SEP (Fig. 1).

Case 2 was a 74-year-old man with myelopathy due to cervical spondylosis. He had been suffering from paresthesia in the lower regions from the left side at C5. Incomplete block at the C3–4 level was seen on the myelogram and CTM. The SEP had delayed latency and low amplitude in N13, N16, and N19 bilaterally, evoked by electrical stimulation of the median nerve (Fig. 2).

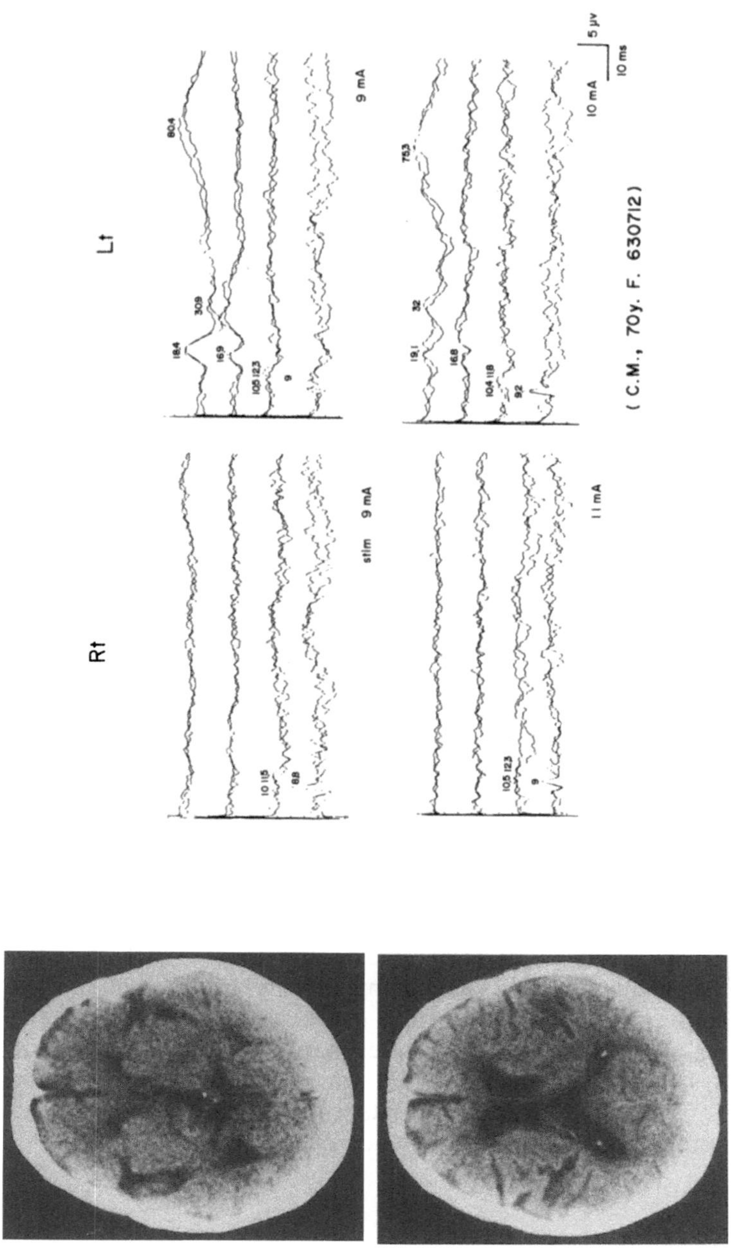

Fig. 1. SEP findings and computed tomogram of the brain in case 1, a 70-year-old woman who had bleeding in left side of the thalamus

SEPs evoked by electrical stimulation of the median nerve were investigated in patients with cerebrovascular disorders and the abnormal findings in components after N16 in SEPs were noted in all patients with central brain stem disturbance. The frequency of abnormal findings in 26 extremities of the 23 patients with hemispheric lesions in the thalamus and putamen was 38% in N16, 85% in N19, and 88% in N31 and N66. The ratio of low or diminished amplitude to delayed latency was the same in N16, and three times higher in N19, whereas components after N31 in all patients were of low or no amplitude. Delayed latency in N19 on the healthy side was seen in three patients with hemispheric lesions, and a high amplitude in N66 was noted in one patient.

As summarized in Fig. 3, the abnormal findings in N16 were seen in about half of all the patients and the low or no amplitude in components after N19 appeared in almost all paients with hemispheric lesions. The abnormal findings in components after N19 were frequently noted in the patients with infarction, presumably caused by occlusion of the middle cereberal artery. Low amplitude in components after N31 appeared in one patient with infarction in the thalamus and disappearance of components after N16 was noted in two patients with bleeding in the thalamus. Low amplitudes in N16 only were seen in two patients with infarction in the putamen, and abnormal findings in components after N19 were observed in other patients.

In patients with cervical myelopathy, SEPs were induced by stimulating the median and ulnar nerves. These patients showed abnormal findings from N11 or N13. The abnormal findings in N11, N13, N16, and N19 were seen in half of the cases upon stimulation of the median nerve and in more than half of them upon stimulation of the ulnar nerve. The appearance rate of the abnormal findings in components after N31 was 35% upon stimulation of the median and ulnar nerves. Delayed latency with low or diminished amplitude, low amplititude in N11 and N13, delayed latency in N16 and N19, and low amplitude in N31 and N66 were often noted upon stimulation of the median nerve. Stimulation of the ulnar nerve, frequently produced low amplitude of SEP components.

In the cases of cervical myelopathy (Fig. 4), SEP showed low or diminished amplitude in components after N11 in the patients with multiple disk lesion, including those at C3–4. Delayed latency in N11 and abnormal findings in N13 to N19 were often seen, but the frequency of abnormal findings in components after N31 was low in the patients with lesions at C4–5, C5–6, and C5–6. In patients with lesions at C6–7, abnormal findings were only observed upon stimulation of the ulnar nerve.

Discussion

Computed tomography and magnetic resonance imaging provide useful information for the diagnosis of cerebrovascular disorders and cervical myelo-

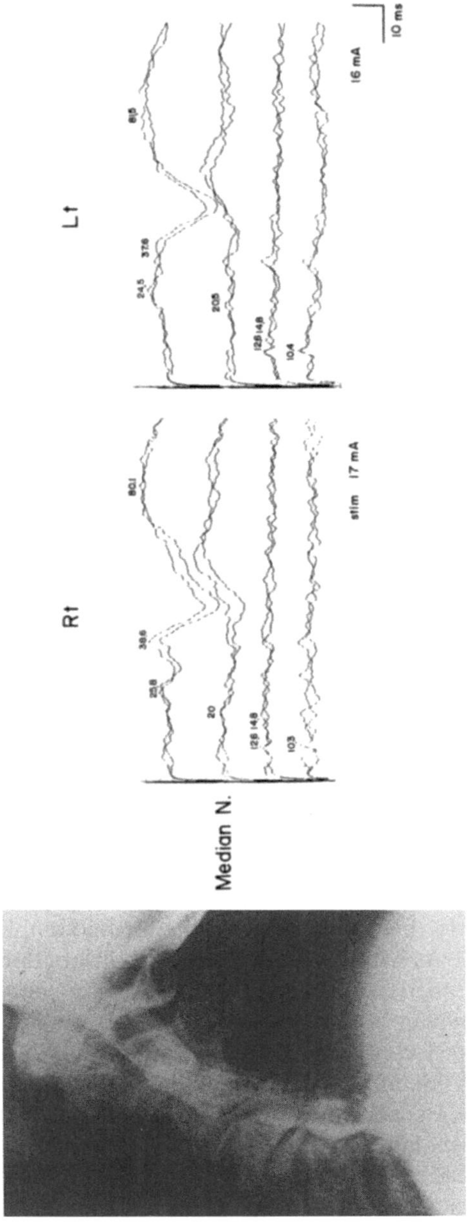

Fig. 2. SEP findings and myelogram in Case 2, a 74-year-old man who had cervical spondylotic myelopathy

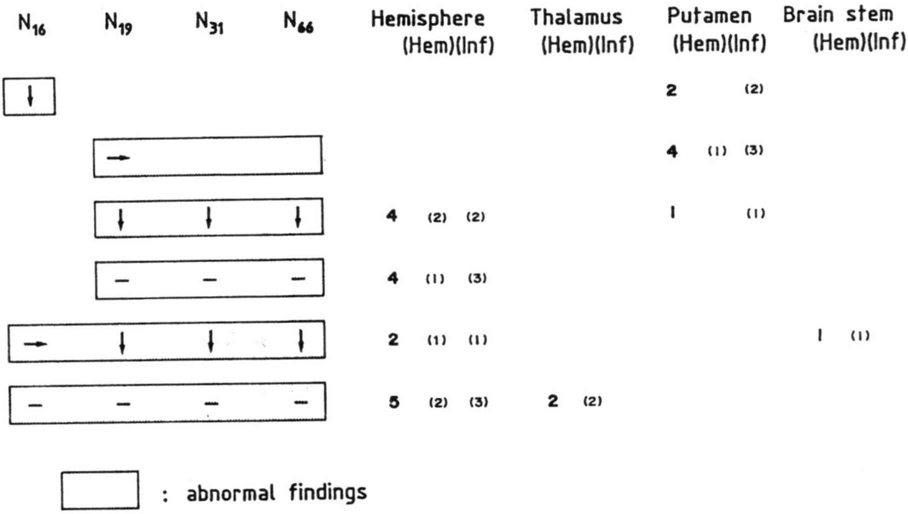

Fig. 3. Abnormal SSEP and SEP findings in the patients with cerebrovascular disorders (→ delayed latency; ↓, low amplitude; −, diminished wave; Hem., hemorrhage; Inf., infarction)

pathy. Based on the assumption that SEPs sensitively reflect the focus in disturbance of the medial lemniscus, even in the absence of definite motor disturbances, SEPs have been utilized to diagnose the lesion responsible for somatosensory tract disturbances and to evaluate the degree of their tract disturbance.

N9 has been reported to be mainly generated in the vicinity of the plexus brachialis [6], N11 in the dorsal tract of the upper cervical spine [1] over the orifice of the spinal cord in the cervical spinal root, N13 in the dorsal tract of the cervical spinal cord or nucleus cuneatus, P14 in the medial lemniscus [3], N17 in the vicinity of the thalamus [4], and N19 in the specific thalamocortical projection system or sensory area [3, 7].

In this study, low and diminished amplitudes from N16 were observed in patients with cerebrovascular disorders due to infarction and bleeding which extended over the hemisphere. These changes in SEP may have resulted from occlusion of the internal carotid artery due to infarction and extension in the posterior portion of the capsula interna due to bleeding per se. Two patients with bleeding in the thalamus had no sensation and had absence of components after N16. This suggests that N16 originates from the vicinity of the thalamus (8) and the SEP enables further identification of the range of damage. Half of the patients with the focus in the putamen showed delayed latency in N19. This indicates mild damage of the fibers in the thalamocortical projection, in the light of anatomical location supplied to the posterior portion of the capsula interna by blood vessels. Abnormal findings in N11 and N13 were noted in half of the patients with cervical myelopathy, and N31 and

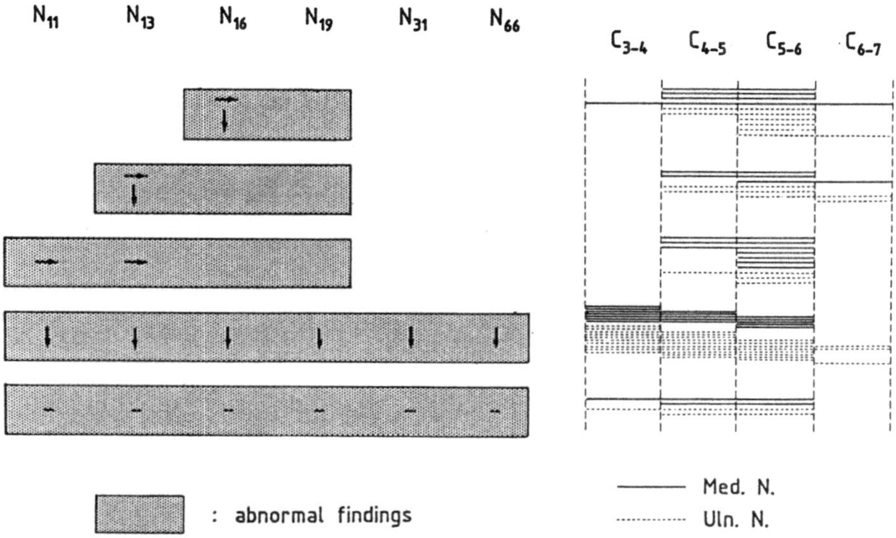

Fig. 4. Abnormal SSEP and SEP findings in the patients with cervical myelopathy. The *horizontal lines* in the right-hand panel represent the disk levels at which these findings were recorded (→, delayed latency; ↓, low amplitude; –, diminished wave)

N66 were detected in these patients. These findings enable easy differential diagnosis from the cases of cerebrovascular disorders. The low and diminished amplitudes in components after N11 may result from damages extending over the multiple disk levels in the cases of cervical myelopathy, including C3–4. A different diagnosis based on the damage at the sites of C4–5 and C5–6 may be difficult. In the cases of damage at C6–7, abnormal findings were noted in the ulnar nerve due to disturbance in the lower cervical spinal cord. The diagnosis of the lesion level in cervical myelopathy requires measurement of evoked potentials from near the cervical spinal cord in order to obtain definite information [5]. However, SEPs may be useful for the diagnosis of damage to the cervical spinal cord.

In conclusion, SEPs give useful information for evaluating disturbances in the sensory pathway and its vicinity, as well as the site of the damage in somatosensory tracts.

Summary

1. Patients with hemispheric lesions due to cerebrovascular disorders, were divided into two groups, one with abnormal findings in components after N16 and the other with abnormal findings after N19. The diminished components, mainly those after N16, appeared in the patients with thalamic

bleeding, and frequency of delayed latency in components after N19 occurred in half of the patients with putamen lesions.

2. More than half of the patients with cervical myelopathy had abnormal findings from N11 and N13. The components in N31 and N66 were detected in these patients. These findings were useful to provide a diagnosis different from that made in cases of cerebrovascular disorders. Although the level of the lesion in cervical myelopathy could not be determined, SEPs were useful to verify whether the cervical spinal cord was damaged or not.
3. Determination of SEPs may be useful for the evaluation of structural damage to the medial lemniscus.

References

1. Desmedt JE, Cheron G (1980) Central somatosensory conduction in man. neural generators and interpeak latencies of the far-field components recorded from neck and right or left scalp and earlobes. Electroencephalogr Clin Neurophysiol 50:382–403
2. Jones SJ (1977) Short latency potentials recorded from the neck and scalp following median nerve stimulation in man. Electroencephalogr Clin Neurophysiol 43:853–863
3. Kimura J (1983) Somatosensory evoked potentials. In: Kimura J (ed) Electrodiagnosis in diseases of nerve and muscle. Davis, Philadelphia, pp 399–427
4. Maccabee PJ, Pinkharov EI, Cracci RQ (1983) Short latency somatosensory evoked potentials to median nerve stimulation. Effect of low frequency filter. Electroencephalogr Clin Neurophysiol 55:34–44
5. Matsuda H, Kondo M, Kureya S (1987) Clinical application evoked spinal cord potentials. Orthop Traum Surg 30:281–293 (in Japanese)
6. Small DG, Beauchamp M, Mattews WB (1980) Subcortical somatosensory evoked potentials in normal man and in patients with central nervous system lesion. In: Desmedt JE (ed) Clinical Uses of Cerebral, Brainstem and Spinal Somatosensory Evoked Potentials, Progress in Clinical Neurophysiology, vol 7. Karger, New York Basel, pp 190–204
7. Yamada T, Machida M, Tippin J (1985) Somatosensory evoked potentials. In: Owen JM, Davis H (eds) Evoked potential testing. Clinical application. Grune & Stratton, Orlanado FL, pp 109–158
8. Yamada T, Kayamori R, Kimura J, Beck DO (1984) Topography of somatosensory evoked potentials after stimulation of the median nerve. Electroencephalogr Clin Neurophysiol 59:29–43

Spinal Evoked Potentials After Stimulation of Ventral Roots: Do Ventral Root Afferents Exist in Man?

V. DELETIS[1], R. ABBOTT, and W. YOUNG

Introduction

One of the approaches to understanding the complexity of the nervous system's function is to dissect subparts of the system functionally. This strategy sometimes leads to the wrong conclusion due to the oversimplification of the very complex relationships in the system. As a result we operate with theories of the function of the nervous system which might be incorrect, or are not able to explain some clinical findings. Thus, our intervention in the nervous system does not yield the expected results.

As an example, we offer the statement that the dorsal roots of the spinal cord contain sensory fibers and the ventral roots contain motor fibers. Starting from this accepted knowledge, our prediction is that the dorsal root rhizotomy will relieve patients of intractable pain. Clinical practice has shown that this is not always the case. Dorsal rhizotomy as a method in pain treatment sometimes fails, but an additional dorsal gangliectomy can be successful [7]. One possible explanation for this is that there are other fibers for pain transmission, and it is highly probable that these are ventral root afferents [3].

This "heretic idea," that the ventral root has both efferent and afferent fibers is as old as the classification of the sensory and motor roots, known as "Law of Bell-Magendie." An entire book has even been published about this problem [4].

Sherrington, in his article in the *Journal of Physiology* in 1894, showed that in the distal stump of completely cut ventral roots, some fibers did not degenerate [8]. This experimental data strongly suggests that the cell bodies of these fibers do not originate from the anterior horn of the spinal cord. Further, pathohistological examination of the human ventral roots demonstrated that almost 30% of the ventral root fibers are unmyelinated [3, 7]. Similar data has been obtained from cats [2].

[1] Department of Anesthesiology, New York University Medical Center, 550 First Avenue, New York, NY 10016, USA

K. Shimoji et al. (Eds.)
Spinal Cord Monitoring and Electrodiagnosis

Methods and Materials

All records were obtained with the patient lying in a quiet room. Square electrical pulses of 0.2 ms were delivered at a rate of 3.3 Hz to the posterior tibial nerve at the ankle, at motor threshold for the musculus abductor hallucis. SEPs were recorded using a 2-channel montage of disk skin electrodes situated at the vertex (Cz) and referred to the midfrontal site (Fpz) (Fig. 1). The signal was amplified and filtered with a bandpass of 10–1000 Hz. Each record was the average of 1000 responses over an analysis time of 100 ms. Nicolet-Viking equipment was used. Two runs were performed for each recording. Traces displayed on a terminal screen were hardcopied without smoothing for illustrations. P1, N1, P2, and N2 amplitudes were measured from baseline to each peak (Fig. 1).

The subjects were 17 patients with diabetic neuropathy, three with incomplete spinal cord injury (C5 and T6–7), and seven hemiplegics (15 men aged 63.5 ± 10.5 years, 12 women aged 70.5 ± 7.2 years). In these patients, 51 legs were examined, some unilaterally. Thermal and vibratory sensation were scored on a scale from 1 to 5 as follows:

Thermal sensation. Test tubes filled with water heated to 22°C, 24°C, 26°C, 28°C, 30°C, 32°C, 34°C, 36°C, 38°C, and 40°C were applied to the patient's ankle. The patient's sensitivity to the difference of temperature was classified as: Grade 5, (normal) sensitive to 2°C or less difference, Grade 4, sensitive to 4°C difference; Grade 3, sensitive to 6°C difference; Grade 2, sensitive to 8°C

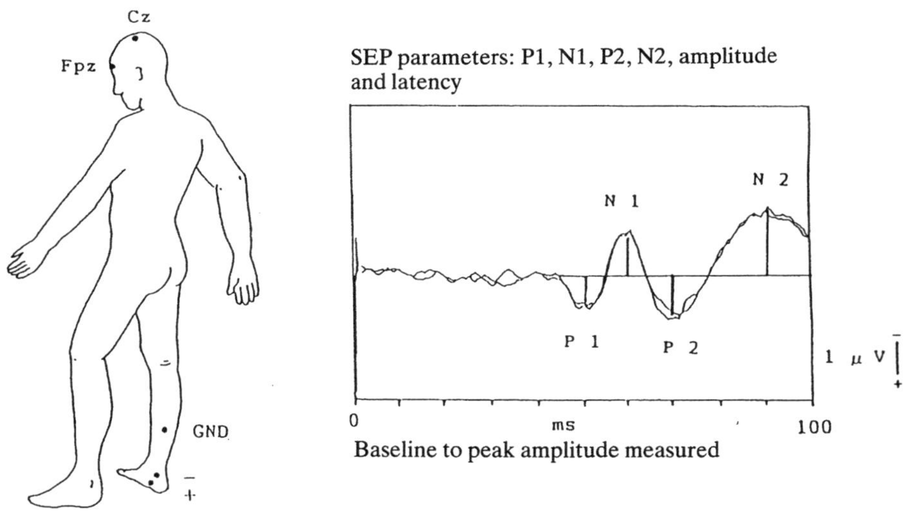

Fig. 1. Recording condition of SEP and SEP parameters following stimulation of the posterior tibial nerve at the ankle. The amplitudes of P1(P36), N1(N48), P2(P60), N2(N75) measured from baseline to each peak. Analysis time: 100 ms; Bandpass filter: 10–1000 Hz; Averages: 1000 times; Delivered stimuli: 0.2 ms square wave pulse, 3.3 Hz with an intensity sufficient to obtain a minimal twitch of the great toe

difference; Grade 1, sensitive or insensitive to 10°C or more difference. These grades were decided through three trials.
Vibratory sensation. A 128 Hz tuning fork was placed at the same ankle site used for stimulation in recording SEPs. The grades were estimated by measuring the sensation time (seconds) to one light stroke of the vibrating tuning fork (normal 18–20 s) as; Grade 5, 16–20 s; Grade 4, 12–15 s; Grade 3, 8–11 s; Grade 2, 4–7s; Grade 1, 0–3 s.

Results

Figures 2 and 3 show one case of spinal cord injury and another case of diabetic neuropathy. Table 1 shows the mean SEP parameters of 51 records following stimulation of 51 legs in 27 patients. Table 2 shows the distribution of the 51 legs according to the grade of thermal and vibratory sensation. Tables 3 and 4 show the result of multiple regression analysis of P1, N1, P2, and N2 amplitudes following thermal and vibratory sensation. P1 and P2 were positively well correlated to thermal and vibratory sensation. Tables 5 and 6 show the results of multiple regression analysis in the 20 cases of diabetic neuropathy and incomplete spinal cord injury, excluding 7 cases of hemiplegia, since in the case of hemiplegia, the deranged brain might distort the SEP components themselves. The grade of thermal sensation disturbance was correlated with P1(a) and P2(c). In addition, P1(a) was correlated with the grade of vibratory sensation disturbance.

Discussion

SEP components have been thought to be dependent upon the integrity of the dorsal column medial lemniscus [2, 4]. Later components of SEP can be

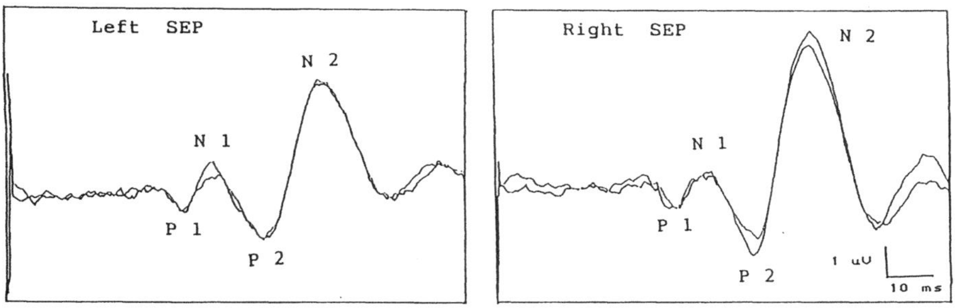

	Amp. (μV)	Lat. (ms)		Amp. (μV)	Lat. (ms)
P 1	0.63	38.8	P 1	0.78	39.2
N 1	0.57	45.0	N 1	0.21	46.0
P 2	1.72	56.0	P 2	1.66	56.6
N 2	3.63	68.8	N 2	4.63	69.0

Thermal sensation 1
Vibratory sensation 2

Thermal sensation 3
Vibratory sensation 1

Fig. 2. SEP parameters and the grade of sensory disturbance in a 63-year-old man with incomplete C5 spinal cord injury

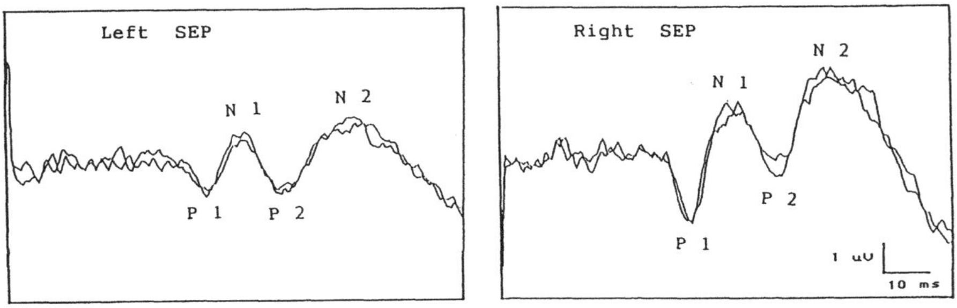

	Amp. (μV)	Lat. (ms)
P 1	0.87	43.2
N 1	1.20	51.2
P 2	0.51	59.4
N 2	1.62	75.6

Thermal sensation 1
Vibratory sensation 2

	Amp. (μV)	Lat. (ms)
P 1	2.08	42.0
N 1	1.60	52.0
P 2	0.57	61.2
N 2	2.70	72.2

Thermal sensation 1
Vibratory sensation 3

Fig. 3. SEP parameters and the grade of sensory disturbance in a 62-year-old woman with diabetic neuropathy

Table 1. Mean SEP parameters of 51 records following stimulation of 51 legs from 27 patients

Component	Amplitude (μV)	Latency (ms)
P1	2.0 ± 1.3	41.7 ± 4.5
N1	1.7 ± 1.3	49.4 ± 5.1
P2	1.8 ± 1.5	61.9 ± 5.8
N2	3.3 ± 1.7	78.8 ± 7.7

Table 2. Distribution of 51 legs according to the grade of thermal and vibratory sensation

Grade	1	2	3	4	5
Thermal sensation	5	16	12	8	10
Vibratory sensation	7	14	18	5	7

processed on the basis of the earlier component. However, in some pathologic cases, the later components have been detected without the earlier ones [5]. Large diameter fibers produced an elevation in P1/N1 amplitude, while A δ fibers produced an elevation in P2/N2 amplitude [1].

In the present study, the grades of sensory disturbance in thermal and vibratory sensation were not significantly related to the amplitude of each SEP component following stimulation of the posterior tibial nerve at the ankle. However, thermal sensation tended to have a closer relationship to both P1 and P2. Also vibratory sensation was correlated positively to P1 and negatively to N2. These results are different from those obtained by Ziganow. According to his conclusion, vibratory sensation should be related more

Table 3. Multiple regression analysis of SEP components following thermal sensation test

	Partial regression coefficient	*F* value	probability
a	0.366	3.21	0.08
b	−0.211	1.00	0.32
c	0.390	4.78	0.03
d	0.002	0.00	0.99

Multiple correlation coefficient 0.478
F value (probability) 3.58 (0.01)
$Y = aX_1 + bX_3 + cX_3 + dX_4 + E_0$, where
X_1 = P1 amplitude
X_2 = N1 amplitude
X_3 = P2 amplitude
X_4 = N2 amplitude
Y = Thermal sensation

Table 4. Multiple regression analysis of SEP components following vibratory sensation test

	Partial regression coefficient	*F* value	probability
a	0.333	2.64	0.11
b	−0.302	2.03	0.16
c	0.427	5.71	0.02
d	−0.290	4.23	0.05

Multiple correlation coefficient 0.352
F value (probability) 1.63 (0.18)
$Y = aX_1 + bX_3 + cX_3 + dX_4 + E_0$, where
X_1 = P1 amplitude
X_2 = N1 amplitude
X_3 = P2 amplitude
X_4 = N2 amplitude
Y = Vibratory sensation

Table 5. Multiple regression analysis of thermal sensation in patients with diabetic neuropathy and incomplete spinal cord injury

	Partial regression coefficient	*F* value	probability
a	0.585	7.98	0.01
b	−0.259	1.52	0.23
c	0.367	2.93	0.09
d	−0.108	0.63	0.43

Multiple correlation coefficient 0.567
F value (probability) 4.03 (0.01)
$Y = aX_1 + bX_3 + cX_3 + dX_4 + E_0$, where
X_1 = P1 amplitude
X_2 = N1 amplitude
X_3 = P2 amplitude
X_4 = N2 amplitude
Y = Thermal sensation

Table 6. Multiple regression analysis of vibratory sensation in patients with diabetic neuropathy and incomplete spinal cord injury

	Partial regression coefficient	*F* value	probability
a	0.437	3.86	0.05
b	−0.190	0.72	0.40
c	0.491	0.05	0.83
d	−0.321	4.85	0.03

Multiple correlation coefficient 0.383
F value (probability) 1.46 (0.23)
$Y = aX_1 + bX_3 + cX_3 + dX_4 + E_0$, where
X_1 = P1 amplitude
X_2 = N1 amplitude
X_3 = P2 amplitude
X_4 = N2 amplitude
Y = Vibratory sensation

closely to P1/N2 and thermal sensation should be related more closely to P2/N2. This difference might come from the difference in method: Ziganow examined only incomplete spinal cord injuries. His patient group was more homogeneous than ours. In his study, P1/N1 and P2/N2 were analyzed as a group, whereas in our study, P1, N1, P2, and N2 were analyzed separately. However, Ziganow did not exclude the possible influences of two sensory modalities to both P1/N1 and P2/N2. In that sense, our conclusions do not completely contradict those of Ziganow. There is also experimental evidence that both large and small diameter fibers project to both dorsal and anterolateral columns in the 35–70 ms range [6].

References

1. Alpsan D (1981) The effect of the selective activation of different peripheral nerve groups on the somatosensory evoked potential in the cat. Electroencephalogr Clin Neurophysiol 51:589–598
2. Halliday AM (1967) Changes in the form of cerebral evoked responses in man associated with various lesions of the nervous system. Electroencephalor Clin Neurophysiol [Suppl] 24:178–192
3. Kimura J (1983) Electrodiagnosis in diseases of nerve and muscle: principle and practice. Davis, Philadelphia, pp 409–415
4. Noel P, Desmedt JE (1975) Somatosensory cerebral evoked potentials after vascular lesions of the brainstem and diencephalon. Brain 98:113–128
5. Seko K, Fujiwara T (1984) The spinal cord conduction velocity with body surface electrodes. In: Honma S, Tamaki T, Shimoji K, Kurokawa T (eds) Fundamentals and clinical application of spinal cord monitoring. Saikon Publ, Tokyo, pp 319–330
6. Simpson RK, Blackburn JG, Martin HF, Katz S (1981) Peripheral nerve fiber and spinal cord pathways contributions to the somatosensory evoked potential. Exp Neurol 73: 700–715
7. Ziganow S (1986) Neurometric evaluation of the cortical somatosensory evoked potential in acute incomplete spinal cord injuries. Electoencephalogr Clin Neurophysiol 65:86–93

Dermatomal Somatosensory Evoked Potentials (DSSEPs) in Patients with Cervical Lesions and Brachial Plexus Injuries in Comparison with Somatosensory Evoked Potentials (SSEPs)

T. Izumi[1], Y, Nishijima, M. Okada, Y. Yamazaki, N. Toda, and M. Yamaguchi

Introduction

Somatosensory evoked potentials (SSEPs) have been used as a method of diagnosing the level of cervical lesions and brachial plexus injuries [1, 3, 4, 6]. It seems, however, to be difficult to diagnose damage occuring to a single nerve root at the level of each medullary segment by using SSEPs, which are recorded following stimulation of the nerve trunk.

Therefore, dermatomal somatosensory evoked potentials (DSSEPs) produced by stimulation of the sensory nerves of each skin segment from C5 to C8 are considered an appropriate method to diagnose the level of cervical lesion and brachial plexus injuries. In this study, the DSSEPs were compared with SSEPs recorded after stimulation of the median and ulnar nerves. Recordings were taken from the scalp over the contralateral cortex, and then the level and the degree of damage to the nerve root at each medullary segment were determined. Furthermore, both potentials were recorded chronologically, and it was investigated which was more related to recovery of clinical symptoms.

Subjects and Methods

Healthy Control Subjects

The number of healthy adults tested was 12 (six men and six women). Their mean age was 40.5 ± 13.2 years (range 20–61 years), and their mean height was 162.6 ± 10.9 cm (range 145–178 cm).

[1] Department of Orthopedic Surgery and Rehabilitation, Kanazawa Medical University, 1-1 Daigaku, Uchinada-machi, Kahoku-gun, Ishikawa, 920-02 Japan

K. Shimoji et al. (Eds.)
Spinal Cord Monitoring and Electrodiagnosis

Patients

There were six patients with cervical spondylotic radiculopathy (CSR), two patients with cervical spondylotic myelopathy (CSM), one patient with ossification of the posterior longitudinal ligament (OPLL), and three patients with brachial plexus injuries (BPI). They were seven men and five women with a mean age of 41.9 ± 13.8 years (range 21–64 years), and a mean height of 161.3 ± 10.3 cm (range 142–174 cm).

Measurements were made with the subjects in a quiet room kept at a fixed temperature. Subjects kept a supine position at rest with their eyes closed. The method of stimulation was as follows: Using Keegan's Dermatome (1947) [5], C5 (one third of the lower part of the anterolateral region of the upper arm, one third LAUA), C6 (the thumb), C7 (the middle finger), C8 (the little finger) (Fig. 1), and the median and ulnar nerves at the wrist were stimulated individually on both the left and right sides. The stimulation of C5 were performed using two saucerlike electrodes set 3 cm apart (silver–silver chloride disk), while stimulation of C6, C7, and C8 was performed using a pair of ring electrodes. The supramaximal stimulation intensity for SSEPs was also used for DSSEPs. The frequency of stimulation was determined by using a square wave with a duration of 0.1 ms at 3 Hz. For recording, the 1500 electromyograpm, manufactured by the DISA Corporation, was used, attached to a needle electrode with the filter band width set from 20 Hz–20 kHz, and the sensitivity set at 5 μV. The averaging times were 128 for SSEPs, and 512 or 1024 for DSSEPs. The indifferent electrode was set at Fpz, and the reference electrode at Shagass C3′ (C4′). Both DSSEPs and SSEPs were recorded from the scalp over the contralateral cortex. In order to confirm reproducibility, 3–5 recordings were made. Sensory nerve action potentials (SNAPs) and needle electromyograms (EMGs) were recorded in some cases.

As evaluation for sensory disturbance, tactile and deep sensation and senses of pain and temperature were selected.

The period of measurement spanned from before to after surgical or conservative treatment had been applied (a mean of 8 months, range from 1 to 15 months).

Results

Group of Healthy Volunteers

The wave patterns of DSSEPs were similar to those of SSEPs, in that they were composed of three negative and positive components (P1, N1, P11, N11, P111, N111). Wave pattern analysis at latency of each negative and positive peak was performed. For the amplitude, the peak to peak value of each negative-positive wave was measured (Fig. 2). Table 1 shows the normal

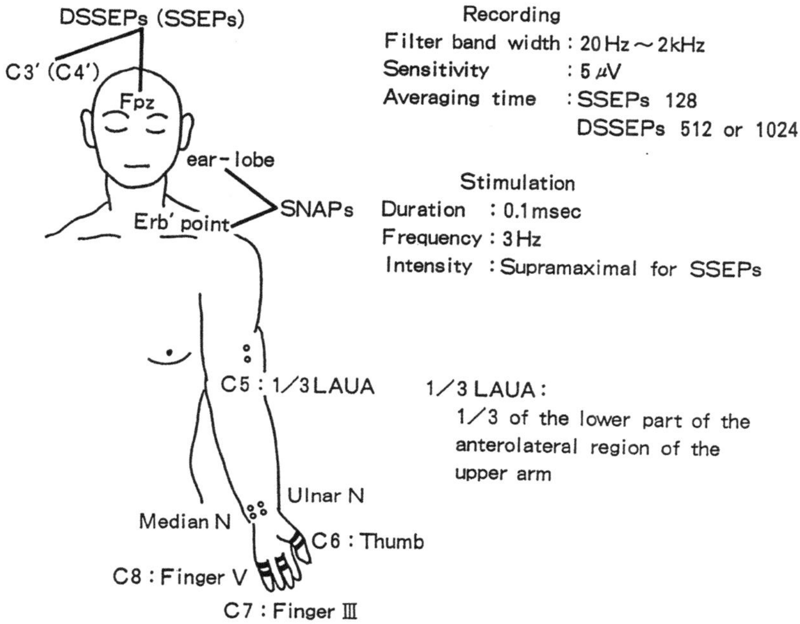

Fig. 1. Stimulus site and recording method

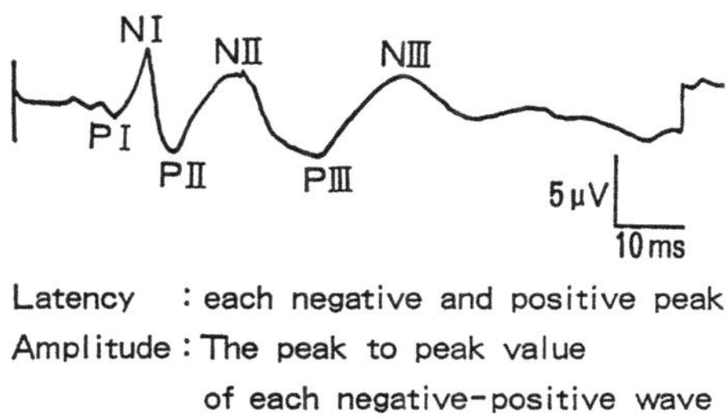

Fig. 2. The normal wave pattern and measurement method of latency and amplitude

values determined in healthy adults. Wave pattern components showed no difference between the left and right sides. The standard deviation (SD) of latencies was small, but the SD of amplitudes was large.

Patients

For the definition of an abnormal wave pattern, a delay of more than 2 SD at latency (N1, N11) and concerning amplitude (P1–N1, P11–N11), a decrease

Table 1. Mean latency and mean amplitude of DSSEPs and SSEPs in healthy volunteers (12 patients, 24 upper extremities)

Mean latency Stimulus site	Root level	NI mean ± SD	NII mean ± SD	NIII mean ± SD
⅓ LAUA	C5	15.6 ± 1.3	28.8 ± 2.3	60.1 ± 6.1
Thumb	C6	21.7 ± 1.1	34.6 ± 2.1	63.1 ± 5.5
Finger III	C7	21.9 ± 1.2	34.8 ± 2.0	63.5 ± 7.0
Finger V	C8	21.8 ± 1.1	34.8 ± 2.1	63.5 ± 6.6
Wrist	Median. N	18.7 ± 1.2	32.0 ± 2.3	62.4 ± 6.3
Wrist	Ulnar. N	18.7 ± 1.3	32.0 ± 2.5	62.7 ± 6.1
Mean amplitude (μV) Stimulus site	Root level	PI–NI mean ± SD	PII–NII mean ± SD	PIII–NIII mean ± SD
⅓ LAUA	C5	1.8 ± 0.9	2.3 ± 1.6	3.2 ± 2.4
Thumb	C6	2.0 ± 0.7	2.4 ± 1.3	3.6 ± 2.0
Finger III	C7	1.9 ± 0.8	2.5 ± 1.5	3.4 ± 2.1
Finger V	C8	1.9 ± 0.8	2.4 ± 1.5	3.5 ± 2.1
Wrist	Median. N	4.0 ± 1.3	7.3 ± 4.5	9.4 ± 3.1
Wrist	Ulnar. N	3.9 ± 1.4	7.4 ± 4.1	8.8 ± 4.2

Table 2. The results of wave pattern analysis before treatment. Latency (NI, NII) is prolonged more than 2 SD, amplitude (PI–NI, PII–NII) is decreased more than 50%

Case	DSSEPs and SSEPs C5	C6	C7	C8	Med.	Uln.	Clinical diagnosis
1	N	AB	N	N	N	N	CSR (C6)
2	N	AB	AB	N	N	N	CSR (C5, 6)
3	N	N	N	N	N	N	CSR (C5, 6)
4	N	AB	AB	N	N	N	CSR (C6, 7)
5	N	AB	AB	AB	N	AB	CSR (C6, 7, 8)
6	AB	N	N	N	N	N	CSR (C5)
7	AB	AB	AB	AB	AB	AB	OPLL (C3/4 ~ C6/7)
8	N	AB	AB	AB	N	AB	CSM (C4/5)
9	N	AB	AB	AB	N	AB	CSM (C4/5)
10	AB	AB	AB	AB	AB	AB	BPI (C5 ~ C8)
11	(–)	(–)	AB	AB	AB	AB	BPI (C5 ~ C8)
12	AB	AB	(–)	(–)	AB	AB	BPI (C5 ~ C8)

N: normal, *AB* abnormal, (–): absent

to below 50% in comparison with the unaffected sides was considered to indicate abnormality. Table 2 shows the results of pattern analysis before treatment. The level of lesions determined by neurophysiological and neuroradiological findings and the level of abnormal wave patterns obtained from DSSEPs almost corresponded with each other. DSSEPs also showed the

Table 3. The results of wave pattern analysis after treatment. Latency (NI, NII) is prolonged more than 2 SD, amplitude (PI–NI, PII–NII) is decreased more than 50%

Case	DSSEPs and SSEPs						Clinical diagnosis
	C5	C6	C7	C8	Med.	Uln.	
1	N	N	N	N	N	N	CSR (C6)
2	N	N	N	N	N	N	CSR (C6, 7)
3	N	N	N	N	N	N	CSR (C5, 6)
4	N	N	N	N	N	N	CSR (C6, 7)
5	N	N	N	AB	N	AB	CSR (C6, 7, 8)
6	N	N	N	N	N	N	CSR (C5)
7	N	N	AB	AB	N	AB	OPLL (C3/4 ~ C6/7)
8	N	N	N	N	N	N	CSM (C3/4)
9	N	N	N	N	N	N	CSM (C3/4)
10	N	N	AB	AB	N	AB	BPI (C5 ~ C8)
11	(−)	(−)	N	N	N	N	BPI (C5 ~ C8)
12	N	N	(−)	(−)	N	AB	BPI (C5 ~ C8)

N: normal, *AB* abnormal, (−): absent

degree of damage to the nerve. In other words, when the degree of damage was more serious, DSSEPs wave patterns showed abnormality or absence. SSEPs showed normal wave patterns when DSSEPs showed abnormal wave patterns following stimulation at one to three vertebral levels. Furthermore, when all DSSEPs showed abnormal wave patterns, SSEPs also showed abnormal wave patterns. Table 3 showed the results of wave pattern analysis after treatment. Abnormal DSSEP wave patterns at each level normalized simultaneously with improvement of clinical symptoms. For SSEPs, normal wave patterns were shown, similar to before treatment, even though abnormal wave patterns existed for DSSEPs produced by stimulation between one or two vertebrae. When all DSSEPs showed abnormal wave patterns, SSEPs also showed abnormal wave patterns.

Case 4 was a 51-year-old male patient with CSR whose main symptoms were sensory disturbance and a decrease of muscle power in the region innervated by the medullary segments of the left C6 and C7 spinal nerves. These major complaints began to appear in about June 1987 and were improved by conservative treatment. DSSEPs before treatment showed abnormal wave patterns after stimulation of C6 and C7, while SSEPs were normal following stimulation of the median and ulnar nerves. SNAPs also showed that stimulation of C6 and C7 caused abnormal wave patterns. After treatment, all DSSEPs, SSEPs, and SNAPs showed normal wave patterns (Fig. 3).

Case 11 was a 28-year-old male patient fell off a motorcycle on November 26, 1986, and suffered from C5 and C6 root avulsions. Palsy was complete at C5 and C6 and was partial at C7 and C8. An intercostal nerve was transplanted to the musculocutaneous nerve on June 9, 1987.

Before the operation, DSSEPs did not produce wave patterns following stimulation of C5 and C6, while stimulation of C7 and C8 caused abnormal

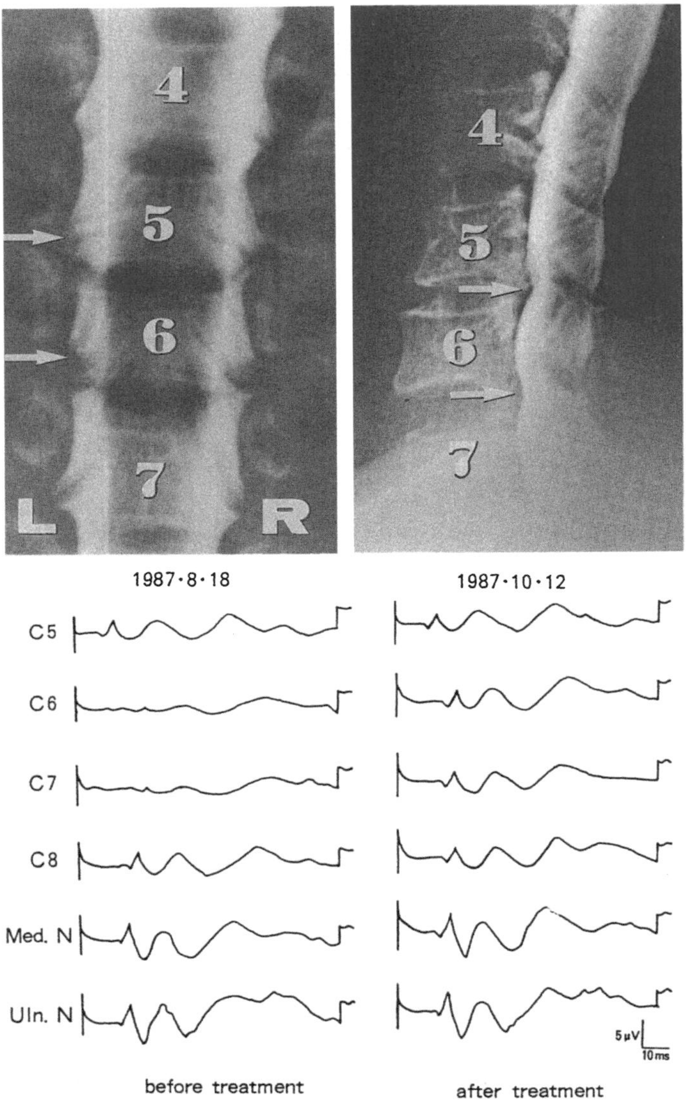

Tactile	Mild ~ Middle	Normal
Pain	Mild ~ Middle	Normal

Fig. 3. The myelogram and changes of DSSEP and SSEP wave patterns in a 51-year-old male patient with cervical spondylotic radiculopathy. The left C6 and C7 spinal nerve roots were defect in the myelogram. Before treatment, DSSEPs showed abnormal wave patterns following stimulation of the left C6 and C7, while both SSEPs showed normal wave patterns

wave patterns to appear. SNAPs showed that stimulation of the levels from C5 to C8 induced abnormal wave patterns. SSEPs produced abnormal wave patterns after stimulation of both median and ulnar nerves.

At 7–12 months postoperatively, DSSEP stimulation of C5 and C6 still did not induce wave patterns, whereas stimulation of C7 and C8 induced normal wave patterns after 12 months. SNAPs showed that stimulation of C7 and C8 produced normal wave patterns after 12 months. Normal wave patterns were also seen after SSEP stimulation of both median and ulnar nerves after 12 months. At this time, tactile and deep sensation were improved, however, no improvement of senses of pain and temperature was observed. As for the needle EMG, muscle contraction was noted in the biceps 6 months after the operation and after 1 year it became possible for the patient to contract the muscle voluntarily (Fig. 4).

Discussion

It was found that in cases where cervical nerve root lesions were clearly indicated by neurophysiological and neuroradiological findings, SSEPs still showed normal wave patterns. This is considered to be due to the masking of the damaged nerve root by the nondamaged nerve root, suggesting that

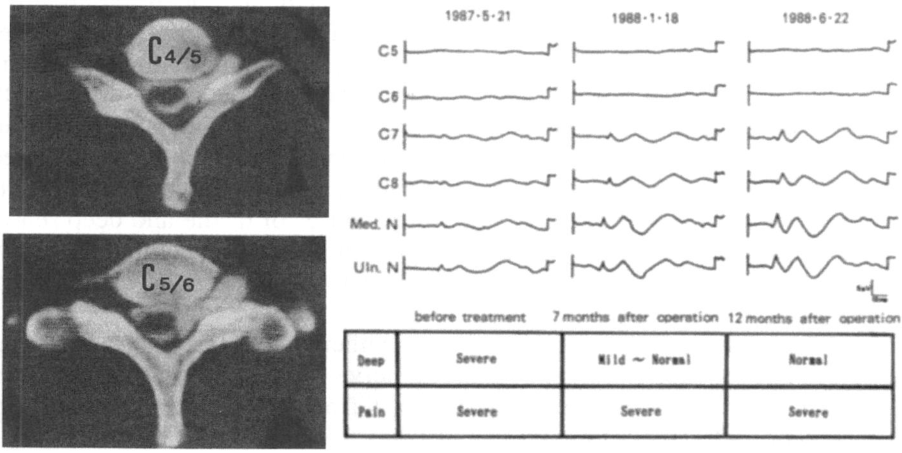

Deep	Severe	Mild ~ Normal	Normal
Pain	Severe	Severe	Severe

Fig. 4. The myelogram and changes of DSSEP and SSEP wave patterns in a 28-year-old male patient with brachial plexus injuries. There were two root avulsions of the left C5 and C6 in the myelogram. Before treatment, DSSEP wave patterns were absent after stimulation of C5 and C6. DSSEP wave patterns following stimulation of C7 and C8 and both SSEP wave patterns showed abnormality. At 7 and 12 months postoperatively, DSSEP wave patterns following stimulation of C5 and C6 were still absent, whereas DSSEP wave patterns following stimulation of C7 and C8 and both SSEP wave patterns showed normality after 12 months

Table 4. The rate of appearance of abnormal wave patterns in patients

	SSEPs (med. and uln.)	DSSEPs (C5 ~ C8)
Before treatment	11 (n = 24) 45.8%	29 (n = 31) 93.5%
After treatment	4 (n = 10) 40.0%	12 (n = 12) 100%

diagnosis of damage of a single nerve root is not easy using SSEPs [10].

EL-Negamy (1979) [3] and Ganes (1980) [4] have reported that abnormal SSEP wave patterns in cervical spondylosis were shown, in only about 50% of all their cases. It was found that in our cases, similar results were obtained for SSEPs both before and after treatment (Table 4).

On the other hand, Eisen et al. (1983) [2] reported that they stimulated the sensory nerve in each skin segment individually to record DSSEPs, and found it possible to diagnose damage to a single nerve root. Lajoire (1983) [7], Machida et al. (1986) [8], and Maeda et al. (1988) [9] also reported on the usefulness of DSSEPs. It was found in the present report that DSSEPs could delinate quite well the level and the degree of damage for lesions of each medullary segment. It was also found that the processes of recovery of the damage nerve root and normalization of abnormal DSSEP wave patterns almost corresponded with each other.

In Case 11, it was noted that recovery of tactile and deep sensation occurred with normalization of abnormal DSSEP wave patterns, while senses of pain and temperature did not show recovery. From this finding, it appears that the tract mediating DSSEPs is the dorsal column-medial lemniscus tract. However, since senses of pain and temperature are also stimulated in recording DSSEPs as a mass response, DSSEPs may include the potentials mediated by the spinothalamic tract. We were unable to define this from the results of the present research. At any rate, disturbances of tactile and deep sensation and those of senses of pain and temperature often occur stimultaneously, as shown in case 4, and the process of their recovery must be correlated in many cases.

The following points must also be remembered as potential disadvantages of DSSEPs: (a) the dermatome and the nerve root level do not always correspond, (b) all lesions in the nerve tract from the periphery to cerebral cortex are reflected, (c) measurement takes a considerable amount of time.

Conclusion

The following findings were obtained by serially recording DSSEPs and SSEPs from patients with cervical lesions and brachial plexus injuries:

1. DSSEP wave patterns denoted accurately the level and the degree of damage to each medullary segment. It was also noted that improvement of clinical symptoms and normalization of abnormal DSSEP wave patterns at each damaged level corresponded almost with each other.
2. Abnormal SSEP wave patterns in cervical lesions and brachial plexus injuries occurred in only about 50% of cases both before and after treatment.

References

1. Eisen A, Elleker G (1980) Sensory nerve stimulation and evoked cerebral potentials. Neurology 30:1097–1105
2. Eisen A, Hoirch M (1983) The Electrodiagnostic Evaluation of Spinal Root Lesions. Spine 8:98–106
3. El-Negamy E, Sedgwick EM (1979) Delayed cervical somatosensory potentials in cervical spondylosis. J Neurol Neurosurg Psychiatry 42:238–241
4. Ganes T (1980) Somatosensory conduction times and peripheral, cervical and cortical evoked potentials in patients with cervical spondylosis. J Neurol Neurosug Psychiatry 43:683–689
5. Keegan JJ (1947) Dermatome hypalgesia with posterolateral herniation of lower cervical intervertebral disk. J Neurosurg 4:115–125
6. Kondo M (1976) Cervical and Experimental Studies on Neurological Disease by means of Somatosensory Evoked Potentials in Orthopaedic Surgery. J Jap Orthop Ass: 569–585
7. Lajoie WJ, Melvin JL (1983) Somato-sensory evoked potentials elicited from individual cervical dermatomes represented by different fingers. Electromyogr Clin Neurophysiol 23:403–411
8. Machida M, Asai A, Okada A (1986) Dermatomal somatosensory evoked potentials (DSSEPs) in patients with cervical lesions and brachial plexus injuries. Orthop Surg 37:319–325
9. Maeda A, Fujiwara M (1988) Clinical application of dermatomal somatosensory evoked potentials in patients with cervical lesions. Rinsho Noha 30:29–33
10. Synek VM (1986) Validity of median nerve somatosensory evoked potentials in the diagnosis of supraclavicular brachial plexus lesions. Electroencephalogr Clin Neurophysiol 65:27–35

Level Diagnosis Using Spinal Cord Evoked Potentials in Cervical Myelopathy

Y. Ohmi[1], S. Harata, K. Ueyama, Y. Okamura, H. Sasaki, and D. Iwaya

Spinal cord evoked potentials could offer some functional information needed for the diagnosis of the level of spinal cord lesions. We have recorded the ascending potential evoked by direct spinal cord stimulation in the thoracic spine since 1979. The potential was recorded at each intervertebral disk level. However, we noticed that the positivization and the augmentation were recognized at different positions of the recording electrode at the same disk level. We were convinced that for precise diagnosis it was necessary to take recordings at multiple levels in the cervical spine. Therefore, recordings at multiple levels have been performed in the cervical spine using personal computer since 1985. The purpose of this paper is to report the changes in the wave form of the spinal cord evoked potentials, and the cases in which this examination method was applied usefully.

Materials and Methods

Fifty cases of cervical myelopathy were examined. There were 17 cases of cervical spondylotic myelopathy, nine cases of cervical disk hernia, eight cases of ossification of the posterior longitudinal ligament, four cases of cervical canal stenosis, four cases of syringomyelia, two cases of motor neuron disease, two cases of spinal cord tumor and four other cases.

Bipolar wire electrodes were used for both stimulation and recording. Stimulation was given by the electrode inserted into the thoracic epidural space with a 0.1 ms pulse duration at a stimulus frequency of 30 Hz. Recording was done at 7–60 levels, with an average of twenty levels in the cervical epidural space. Stimulation, recording, and averaging were all done using the Dantec N2000. The NEC PC9801 personal computer system received the wave data from the N2000, then displayed the wave pattern during the examination, and printed the wave pattern after the examination. The recording

[1] School of Allied Medical Science, Hirosaki University, 66-1 Honcho, Hirosaki-shi, Aomori-ken, 036 Japan

K. Shimoji et al. (Eds.)
Spinal Cord Monitoring and Electrodiagnosis

Table 1. The relation between the diagnosis and the wave patterns showing that the incidence of the abnormal wave patterns became higher as the lesion was localized

	CSM	CDH	OPLL	CCS	S-M	MND	Tumor
Decrease	14/17 (82%)	8/9 (89)	6/8 (76)	1/4 (25)	1/4 (25)	0/2	0/2
Augmentation	12/17 (70)	8/9 (89)	4/8 (50)	0/4	0/4	0/2	0/2
Polyphasia	10/17 (59)	2/9 (22)	4/8 (50)	1/4 (25)	2/4 (50)	0/2	1/2 (50)
Positivization	7/17 (41)	4/9 (44)	3/8 (38)	0/4	0/4	0/2	1/2 (50)
Normal	0/17	1/9 (11)	0/8	2/4 (50)	2/4 (50)	2/2 (100)	1/2 (50)

CSM, cervical spondylotic myelopathy; *CDH*, cervical disc hernia; *OPLL*, ossification of the posterior longitudinal ligament; *CCS*, cervical canal stenosis; *S-M*, syringomyelia; *MND*, motor neuron disease

electrode was introduced into the epidural space through C7–T1 to C2, under X-ray control, then moved backward manually or by puller, while recording the ascending volley.

Results

The abnormal wave patterns were classified into four types. Plural abnormalities were recognized in a case. Abnormal decrease in amplitude occurred in 66% of the cases, augmentation at the caudal site of the lesion in 52%, polyphasia in 42%, and positivization in 30%. The relation between the diagnosis and the wave pattern showed that the incidence of an abnormal wave pattern became higher as the lesion was localized (Table 1). The relation between the neurological findings in the lower limbs and the wave pattern showed that there were no abnormal wave patterns which had specific relations to a single neurological sign (Table 2).

Cases

Case 1 was a C4–5 cervical disk hernia. Normal potentials were recognized between C4–5 and C7. Slight polyphasic patterns were recognized on the rostral side of C4–5 (Fig. 1).

Case 2 was a patient with cervical spondylotic myelopathy. He complained of numbness in the upper limbs, muscle weakness in the left upper and lower limb, and atactic gait. There was no sensory disturbance, and no hyperreflexia in the lower limbs. His myelogram showed defects at C3–4, C4–5, and C6–7. Normal wave pattern was seen between C5 and T1. Augmentation was recognized at the C4 vertebral body level. Positivization and polyphasic patterns of the second component were seen on the rostral side of C3 (Fig. 2). It was considered that the conduction block occurred at C3–4.

Case 3 was a patient with a cervical disk hernia and ossification of the

Table 2. The relation between the neurological findings in the lower limbs and the wave patterns

	Sensory disturbance		Muscle weakness		Hyperreflexia	
	+	−	+	−	+	−
Decrease (33 cases)	11 (33%)	22 (67)	7 (21)	6 (79)	30 (91)	3 (9)
Augmentation (26)	15 (58)	11 (42)	7 (27)	9 (63)	23 (88)	3 (12)
Polyphasia (21)	10 (48)	11 (52)	6 (29)	15 (71)	18 (86)	3 (14)
Positivization (15)	8 (53)	7 (47)	4 (27)	11 (73)	13 (87)	2 (13)
Normal (8)	3 (38)	5 (62)	1 (13)	7 (87)	5 (63)	3 (37)

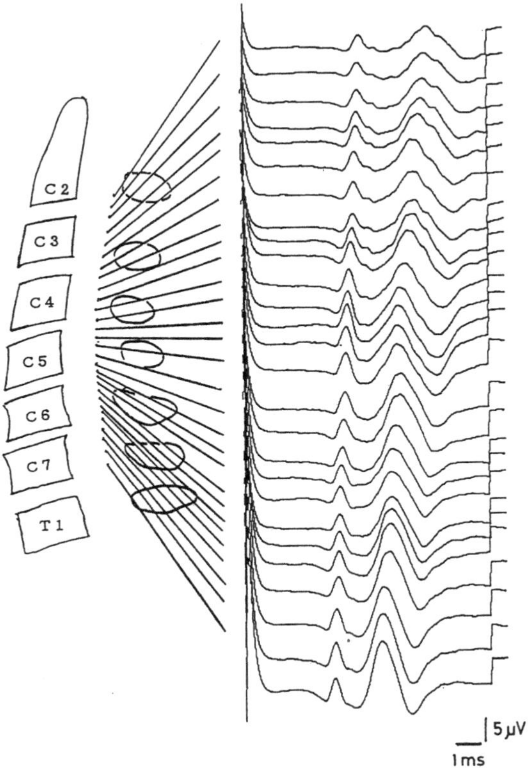

Fig. 1. Case 1, a 53-year-old patient with C4–5 cervical disk hernia. Normal wave patterns were recognized between C4–5 and C7. Slight polyphasic patterns were recognized on the rostral side of C4–5

posterior longitudinal ligament (OPLL). He complained of sensory disturbance and muscle weakness in the right upper and lower limbs. His myelogram showed defects between C3–4 and C5–6. Computed tomography showed compression by herniation at C3–4 and by OPLL at C5.

We recognized augmentation at two vertebral body levels, C4 and C6, and positivization at two disk levels, C3–4 and C5–6 (Fig. 3). Double lesions were considered in this case.

Discussion

Augmentation at the caudal site of the lesion is assumed to occur when the reference electrode exists in the lesion area (Fig. 4). This finding suggested that there was a lesion on the rostral side. If the augmentation was recognized at C4 vertebral body level, there was a decrease in the amplitude and the positivization at the C3–4 disk level in the majority of cases. The same phenomenon was seen at C5 and C4–5, and also at C6 and C5–6. Experimental study in the peripheral nerve showed the change in the wave form from a

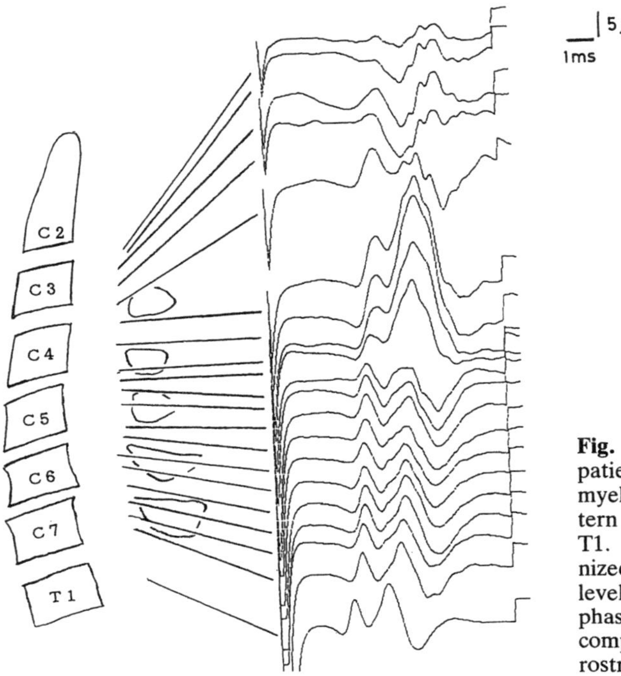

Fig. 2. Case 2, a 53-year-old patient with cervical spondylotic myelopathy. Normal wave pattern was seen between C5 and T1. Augmentation was recognized at the C4 vertebral body level. Positivization and polyphasic patterns of the second component were seen on the rostral side of C3

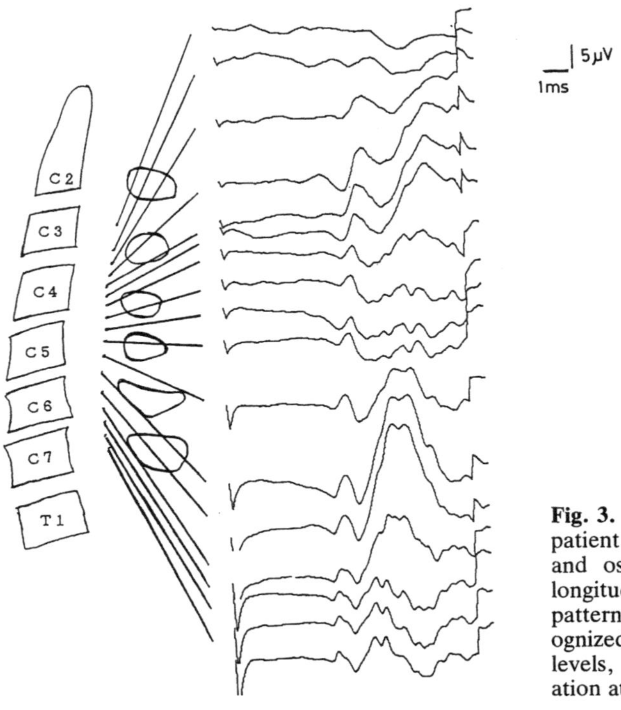

Fig. 3. Case 3, a 59-year-old patient with cervical disk hernia and ossification of the posterior longitudinal ligament. Polyphasic pattern and augmentation were recognized at two vertebral body levels, C4 and C6, and positivization at disc levels C3–4 and C5–6

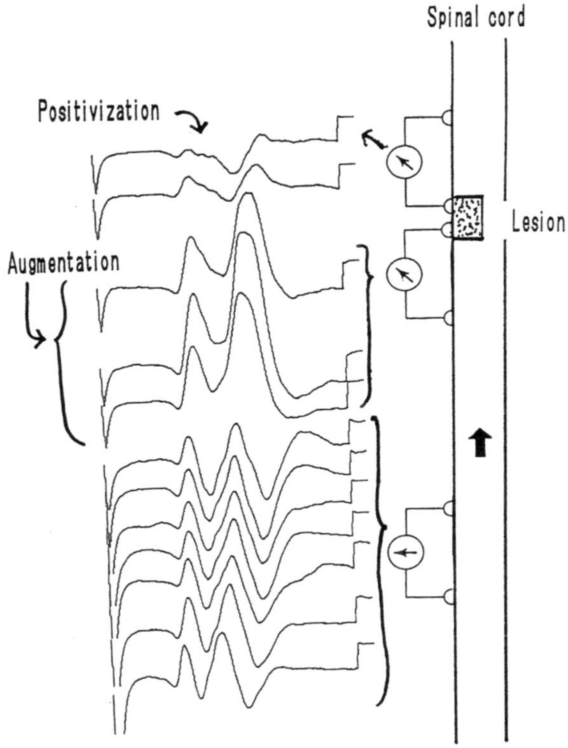

Fig. 4. Augmentation at the caudal site of the lesion is assumed to occur when the reference electrode is placed in the lesion area. This finding suggested that there was a lesion on the rostral side

biphasic to a monophasic wave and augmentation when the reference electrode is placed at the damaged end [2].

This phenomenon was recognized in monopolar recording in our experimental study of the spinal cord. We concluded that this phenomenon in monopolar recording might be produced by the change of the local current at the site of the lesion [1]. It was not clear whether the wave originated from the sensory fibers only. Abnormal findings in the wave were recognized in about 50% of the cases, which showed no sensory disturbance. This seemed to suggest that the abnormal wave pattern may, in part, reflect disturbance of the motor conduction. But there was the possibility of subclinical sensory disturbances in these cases.

Summary

1. The abnormal wave pattern could be classified into four types.
2. The incidence of abnormal wave patterns became higher as the lesion was localized.

3. An abnormal wave pattern was recognized in about 50% of the cases, which had no sensory disturbance in the lower extremities. There were no abnormal wave patterns which had specific relations to a single neurological sign.
4. The appearance of augmentation suggested that the lesion existed at the rostral level. It was considered that augmentation occurred when the reference electrode was placed in the lesion area.

References

1. Akimoto H, Tohno S, Harata S, Ohmi Y, Nakano K, Suetsuna F, Araki T (1987) The augmentation of spinal cord evoked potential at the caudal site of a lesion in the spinal cord pathway. The Journal of Electrodiagnois of the Spinal Cord 10:15–18
2. Katz B (1966) Nerve, Muscle and Synapse. McGraw-Hill, New York

Experience of Diagnosis of Thoracic Myelopathy Using Spinal Cord Evoked Potentials

Y. Kikuchi[1], H. Baba, N. Kawahara, S. Nagata, K. Tomita, S. Nomura, and H. Yugami

Summary

Conducted spinal cord evoked potentials (SEPs) were elicited preoperatively from the thoracic epidural space following spinal cord stimulation in 22 patients with thoracic myelopathy. The waveform changes in SEPs, namely, decrease of amplitude or positive-going potentials were observed and proved to be a successful method for determining the main lesions in all cases. In 20 patients with compressive myelopathy, waveform changes were recorded at the most compressed site in each case. A decrease in amplitude of more than 70% was demonstrated frequently in the moderately involved patients, whereas positive-going potentials were elicited predominantly in the patients with severe myelopathy. The waveform changes reflected to some extent the severity of the neurological symptoms; however, it was not possible to estimate the prognosis of the postoperative neurological symptoms by studying the preoperative SEP waveform. Conducted SEPs provide important information regarding determination of the surgical procedure and range of surgical decompression.

Introduction

It is sometimes difficult to diagnose accurately the main lesions in thoracic myelopathy by routine radiographic procedures, and electromyograms (EMGs) and/or somatosensory cortical evoked potentials are even less useful in the level-specific diagnosis of thoracic myelopathy. Diagnosis is complicated by the fact that myelopathy patients having OPLL and/or OYL often have multiple involvement. For the precise diagnosis of the damaged thoracic cord, we have utilized conducted spinal cord evoked potentials (SEPs) elic-

[1] Department of Orthopedic Surgery, School of Medicine Kanazawa University, 13-1 Takara-machi, Kanazawa, 920 Japan

K. Shimoji et al. (Eds.)
Spinal Cord Monitoring and Electrodiagnosis

ited by epidural stimulation since 1982 [1, 3]. In most cases, we used SEPs to diagnose the level of the main lesion and the neurophysiological properties of the thoracic myelopathy and to determine the extent of the indicated surgical decompression. This report describes our clinical experience with conducted SEP in thoracic myelopathy.

Materials and Methods

Conducted SEPs were recorded in a total of 22 patients with thoracic myelopathy. There was an equal number of male and female patients with a mean age of 55 years, ranging from 25 to 79 years. There were 18 patients with chronic compressive myelopathy caused by ossification of the posterior longitudinal ligaments (OPLL), ossification of the yellow ligaments (OYL), or spondylosis, and two patients with spinal cord tumors complicated by spondylosis or OPLL (Table 1). Each of these cases had at least two lesions of some importance and some of them also had multilevel lesions in up to 12 vertebrae, with a mean involvement of 5.3 vertebrae, often complicated by cervical OPLL or spondylosis. The remaining two patients suffered from hematomyelia and spinal cord infarction, both of which were difficult to identify radiographically. Nineteen of 22 patients received operation and 14 patients could be followed-up for more than 12 months. The clinical condition of thoracic myelopathy was assessed using the Japanese Orthopedic Association's (JOA) score system, in which the categories of upper extremities are excluded from the system for cervical myelopathy (Table 2). A full score is 11 points. In our patients, the preoperative score ranged from 0 to 9 points, with a mean score of 4.5 points. Myelographic findings at the most stenotic site were classified into three types according to the authors' criteria.

Table 1. Etiology of thoracic myelopathy. The number of involved segments and the grading of SEP

Diagnosis	No. of cases	No. of involved segments in thoracic spine	No. of cases with cervical lesions	Grading of the SEP waveform	
				Grade 3	Grade 4
OPLL	4	2 ~ 10; mean 5.5	2	1	3
OYL	6	2 ~ 6 mean 3.7	1	6	0
OPLL with OYL	7	3 ~ 12 mean 7.1	5	0	7
Spondylosis	1	2	0	0	1
Neurinoma with OPLL	1	4	0	0	1
Hemangioma with Spondylosis	1	3	1	0	1
Hematomyelia	1	1	0	1	0
Spinal Cord Infarction	1	1	0	0	1
Total	22 cases	1 ~ 12; mean 4.9 segments	9 cases	8 cases	14 cases

Table 2. The JOA score system for assessment of thoracic myelopathy

I. Motor function	
0 Impossible to walk	
1 Needs cane or aid on flat ground	
2 Needs cane or aid only on stairs	
3 Possible to walk without cane or aid, but slow	
4 Normal	
II. Sensory	
A. Lower extremity	
0 Apparent sensory loss	
1 Minimal sensory loss	
2 Normal	
B. Trunk	same as A
III. Bladder function	
0 Complete retention	
1 Severe disturbance (staining, dribbling)	
2 Mild disturbance (pollakiuria, hesitation)	
3 Normal	

In the first type, identations were seen in six patients, in the second type, incomplete blocks were seen in five patients, and in the third type, complete blocks were seen in nine patients.

We elicited conducted SEPs at the epidural site of the lesion using epidural stimulation. When a lesion spread from the upper thoracic spine to the middle thoracic spine, an ascending SEP was elicited by stimulation of the lower thoracic spinal cord. Also, a descending SEP was elicited for the whole thoracic spine including conus medullaris lesions by stimulation of the cervical or upper thoracic spinal cord. By this method, both the lower and the upper margins of the lesion were clearly identified. The stimulus was a 0.2 ms duration pulse at three times the sensory threshold, usually 40–100 V. The reference electrode was inserted into the subcutaneous tissue of the thoracic spine. The band–filter width of the recording system was applied from 10 Hz to 2 kHz. Thirty-two evoked signals were amplified and averaged in a DISA-1500 digital EMG system.

We classified the waveforms of SEP into four grades for evaluation (Fig. 1): normal, moderate decrease in amplitude, marked decrease, and positive-going killed-end potential were named grades 1, 2, 3, and 4, respectively. We investigated the relationship between the SEP findings, the myelographic findings and the JOA scores.

Results

In our all patients, abnormal waveform changes were recognized. Abnormal SEPs of grade 4 were recorded in 14 patients and abnormal SEPs of grade 3

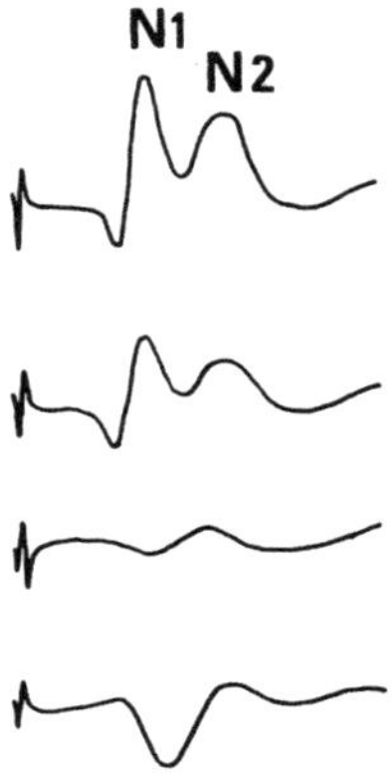

Fig. 1. Grading of the conducted SEP. Normal evoked potentials consist of two negative deflections as Grade 1, including less than 30% decrease in amplitude. Moderate (30%–70%) and marked (more than 70%) decrease in amplitude are classified into Grades 2 and 3, respectively. Grade 4 abnormality represents positive-going potentials

Grade of SEP	Myelographic Findings		
	Indentation 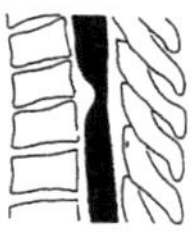	Incomplete Block 	Complete Block 
Grade 3	4 (50%)	3 (38%)	1 (12%)
Grade 4	2 (17%)	2 (17%)	8 (66%)

Fig. 2. Comparison between the grading of SEP and the myelographic findings

were recorded in eight patients (Table 1). In 20 patients with compressive myelopathy, an abnormal waveform was recorded at or beyond the site of maximum stenosis on the myelogram. In two cases with spinal cord vascular disorders, the spinal cord lesions showing SEP abnormalities were identified by magnetic resonance imaging, not myelography.

Comparison of the waveform abnormalities of SEP with the type of myelogram indicated that positive-going potentials tended to appear at the level of complete block, since 66% of the grade 4 abnormalities showed myelographic complete block, but only 12% of the grade 3 abnormalities (Fig. 2).

Comparing the SEP abnormlities with the preoperative JOA socres, the patients with grades 3 and 4 abnormalities were found to have an average score of 6.8 points and 3.5 points, respectively (Fig. 3). This difference was

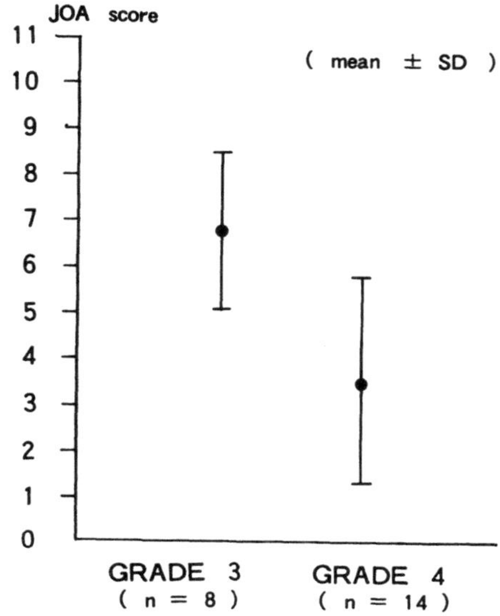

Fig. 3. Comparison between preoperative SEP and preoperative JOA score. The difference was significant by Wilcoxon testing ($p < 0.01$)

defined as significant by Wilcoxon testing. It is clear that positive-going potentials indicated more severe spinal cord damage than that indicated by a marked decrease of amplitude.

The average rate of recovery of the five patients showing grade 3 abnormalities on SEP was 59%, which is equal to that of the nine patients with grade 4 abnormalities. Consequently, it was impossible to predict postoperative neurological recovery only on the basis of preoperative SEP.

Case Presentations

Case 18 was a 37-year-old woman with OPLL at T2–3 and T7–8, who became unable to ambulate without aid, and had pollakiuria and apparent sensory loss below the T4 dermatome (Fig. 4). Her preoperative JOA score was only 3 points. The tomogram after ascending myelography showed complete block at T2–3 and anterior indentation at T7–8. On magnetic resonance imaging, the spinal cord also appeared to be compressed more seriously at T2–3 than at T7–8. Both physical and radiographic examinations revealed that the OPLL at T2–3 was responsible for her paraparesis. However, it was questionable whether the middle thoracic lesion at T7–8 played any role in the paraparesis or not. The ascending SEP showed positive-going potentials at T2–3 as expected, and was normal around T7–8. She received decompression surgery for T2–3, and 4 weeks after surgery, she could ambulate without any assistance.

Case 17 was a 37-year-old man with OPLL from T5 to T8 and OYL at T7–8, who had apparent sensory loss below the T7 dermatome and received

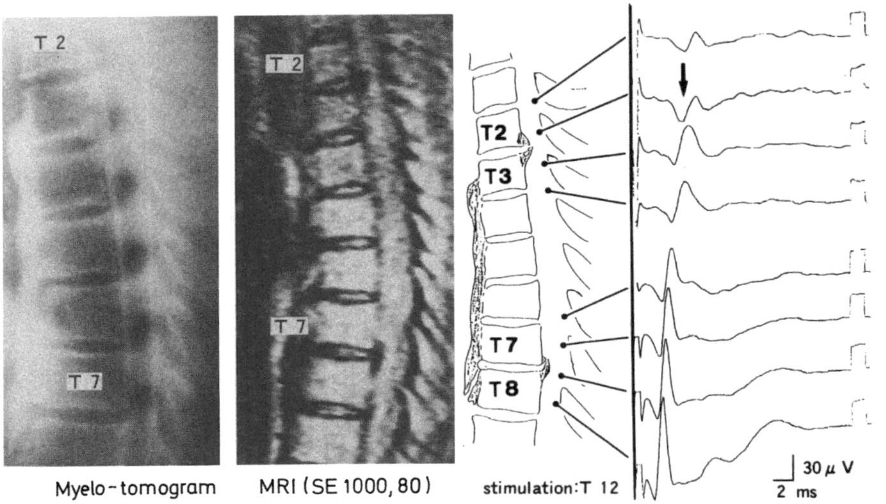

Fig. 4. Case 18. OPLL at T2–3 and T7–8. Myelotomogram showed complete block at T2–3 and anterior indentation at T7–8. Spinal cord was compressed more severely at T2–3 than at T7–8 on MRI (SE 1000,80). The ascending SEPs showed a positive-going potentials at T2–3 (*black arrow*) and were normal around T7–8

a JOA score of 5 points (Fig. 5). The myelogram showed anterior indentation at T5 and complete block at T7–8. Positive-going potentials were elicited at T7–8 on both descending and ascending SEPs, and amplitude augmentation of SEP was seen around the T6 level on descending SEP. Consequently, the T7–8 segment was diagnosed as the level of the main lesion, while it was suggested that the spinal cord around T6 might also be damaged. Total decompression [11] of T7–8 was performed, while T5 and T6 were decompressed posteriorly. One year later, he had obtained 11 points and was able to return to his original work.

Case 3 shows one of the most useful applications of recording SEPs (Figs. 6a,b). A 52-year-old woman had thoracic OPLL and OYL with cervical OPLL spread over 19 vertebrae. She could not ambulate, had apparent sensory loss below the T5 dermatome, and was assigned a JOA score of only 1 point, while her upper extremities were not involved at all. Complete block was seen at C3–4 on the descending myelogram and at T4–5 on the ascending myelogram. The entire eight vertebrae did not show at all on the myelogram. Magnetic resonance imaging showed a gap in the spinal cord image at the T4–5 level. Positive-going potentials emerged at T4 and T5 on ascending and descending SEPs, respectively. The ascending SEP in the cervical spinal cord was normal. We performed wide laminectomy from T2 to T8 as the first step and anterior decompression and fusion from T3 to T6. The rate of recovery at 2 years after surgery was 70%.

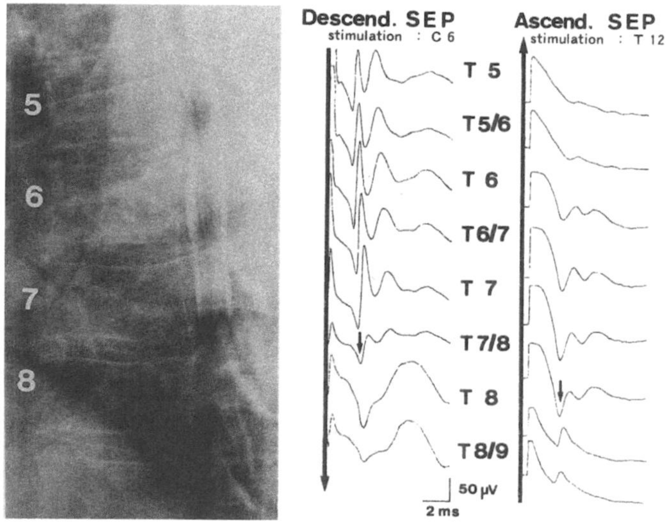

Fig. 5. Case 17. OPLL from T5 to T8 with OYL at T7–8. The descending myelogram showed anterior indentation at T5 and complete block at T7–8. Positive-going potentials were elicited at T7–8 on both descending and ascending SEPs

Case 10 was 71-year-old man with OYL from T7–8 to T12 who developed flaccid paraparesis below L2, with a preoperative JOA score of 5 points (Fig. 7). The myelotomogram indicated posterior indentations at T9–10 and T10–11 and incomplete block at T11–12. On the descending SEP, the amplitude decreased gradually below T10–11. Laminectomy of T11 and T12 gave a 67% improvement 3 years after surgery.

Case 16 was a 40-year-old man, who suffered from subacute paraplegia below the T12 level caused by hematomyelia (Fig. 8). His myelogram did not show any abnormality. Subsequently descending SEP was elicited along the lower thoracic spine. The descending SEP showed positive-going potentials at the T11 level suggesting the presence of possible spinal cord damage. A high signal area was detected in the same region on magnetic resonance image, and at last definite operative diagnosis of hematomyelia could be made.

Discussion

It is essential to make a functional diagnosis of the main lesion of myelopathy, especially in cases involving multiple segments. For this reason, we have utilized various electrodiagnostic methods such as EMG, somatosensory cortical evoked potentials [2, 4, 5], and segmental and conducted SEPs [1, 3, 6, 8, 10].

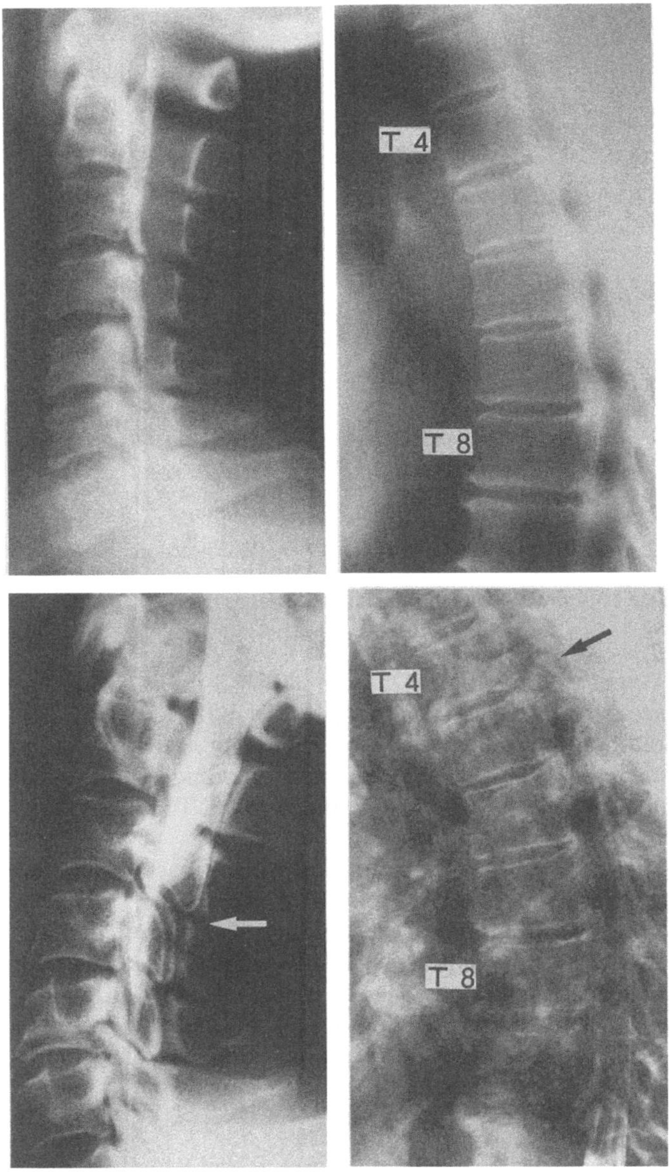

Fig. 6a,b. Case 3. Thoracic OPLL and OYL with cervical OPLL. **a** Mixed type of OPLI in the cervical spine and continuous type of OPLL in the whole thoracic spine were seei on the tomograms. The descending myelogram showed complete block at C3–4 (*whit arrow*). The ascending myelogram also showed complete block at T4–5 and vague imag of filling from T5 to T9, and **b** Cervical ascending SEP was normal. Positive-going poten tials were elicited at T4 on thoracic ascending SEP and at T4–5 on thoracic descending SEF

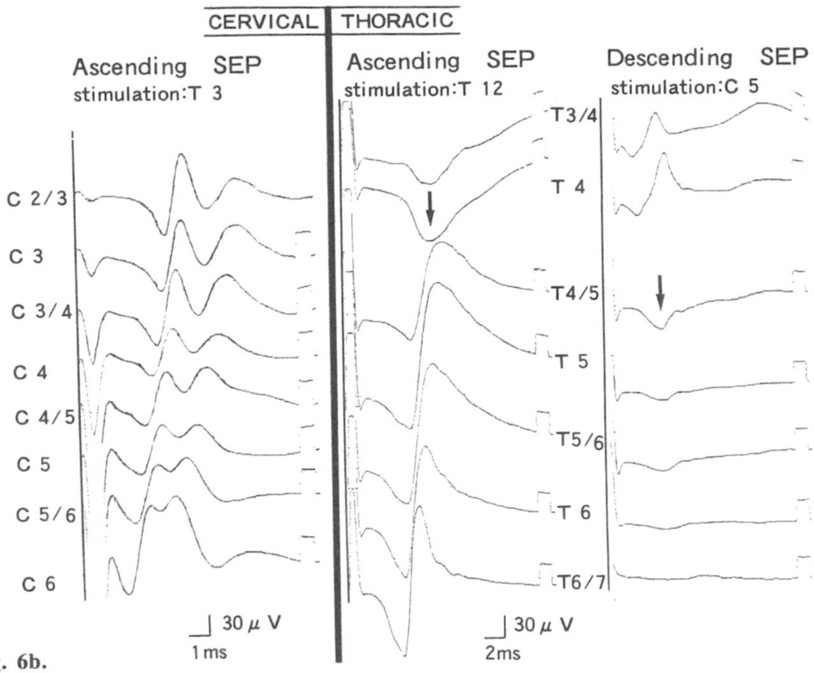

Fig. 6b.

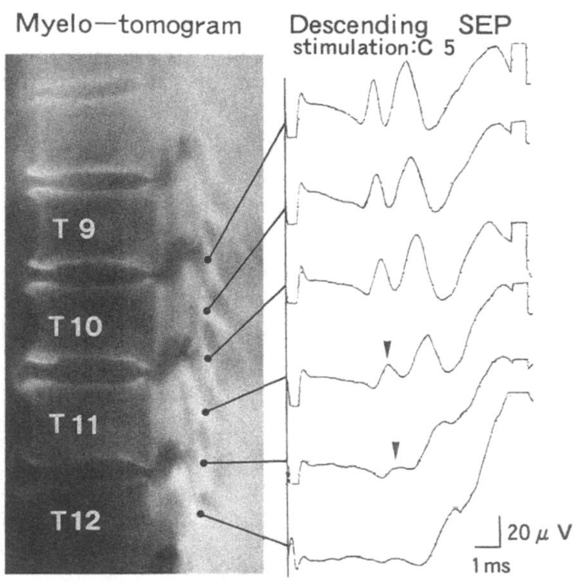

Fig. 7. Case 10. OYL from T7–8 to T11–12. The myelotomogram showed posterior indentations at T8–9, T9–10, and T10–11, and incomplete block at T11–12. On descending SEP, the amplitude decreased gradually below T10–11 (*black arrowhead*)

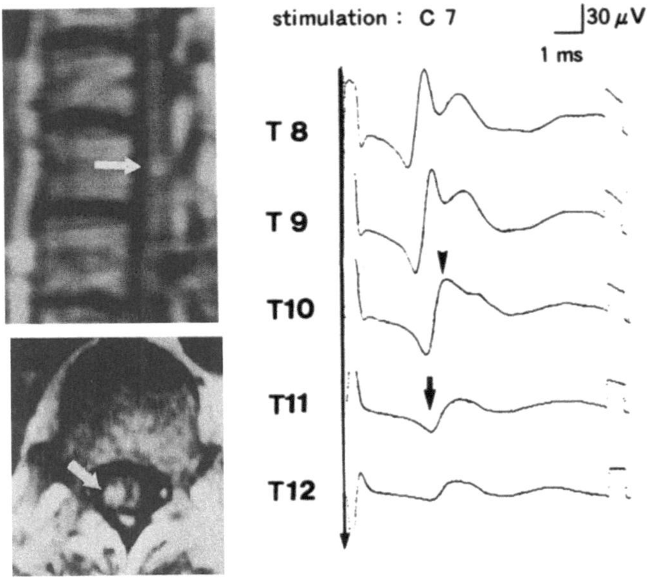

Fig. 8. Case 16. Hematomyelia. A high signal area (*white arrow*) was detected at T11 on MRI (SE 500,20). Descending SEP showed positive-going potentials at T11 (*black arrow*)

Each of these is useful in the diagnosis of cervical myelopathy. These tests, except for conducted SEP, however, appear to be less useful in the level-specific diagnosis of thoracic myelopathy, because they can make only a rough diagnosis to eliminate cervical myelopathy. We have utilized conducted SEPs in thoracic myelopathy patients in order not only to determine the spinal level of the main lesion but also to evaluate the severity of myelopathy.

One of the advantages is that both the lower and upper margins of the damage in the upper or middle thoracic spine can be diagnosed using both ascending and descending SEPs, such as in cases 3 and 17. However, the lower margin of the damage in the lower thoracic spine, including conus medullaris, should be determined with the use of segmental SEPs elicited by stimulation of the sciatic nerve, not by conducted SEP [9]. In multilevel compressive lesions, such as cases 3, 17, and 18, the main lesion could be identified, simplifying the determination of the surgical procedure and the range of the required decompression. Utilizing these SEP data, we performed decompression surgery at least over the range where marked SEP abnormalities were seen. When no abnormality is detected by myelography or other radiographical techniques, such as in cases of hematomyelia or spinal cord infarction, it is possible to identify the localization of the damaged spinal cord only by using the SEP method. In our study, it can be asserted that the conducted SEP was very useful for the level-specific diagnosis of thoracic myelopathy in all cases.

There is some tendency for the waveform patterns of SEP to change in proportion to the compression of the spinal cord as shown on the myelograms. Most of positive-going potentials appeared at the levels of complete block, while only one of marked amplitude decrease occurred there. Furthermore, there is a greater tendency for positive-going potentials to be elicited in patients with more severe myelopathy, than for decrease of the SEP amplitude. The waveform patterns of the amplitude decrease, including the disappearance of potentials, are recorded gradually along several segments of the moderately damaged spinal cord without clear delimitation, such as in the cases of OYL. On the other hand, the positive-going potentials occur abruptly at the involved segment of the severe damaged spinal cord with well-defined delimitation, such as in the cases of OPLL with OYL and hematomyelia [7].

It was not possible to evaluate the prognosis of the postoperative neurological symptoms using only the preoperative SEP waveform. Positive-going potentials represent severely damaged spinal cord function, but the prognosis of a patient with positive-going potentials is not always poor.

We conclude that conducted SEPs as an electrodiagnostic method have high sensitivity and accuracy in the level-specific diagnosis of thoracic myelopathy.

References

1. Baba H, Shima I, Tomita K et al. (1985) Clinical usefulness of spinal cord evoked potentials. In: Schramm J, Jones SJ (eds) Spinal cord monitoring. Springer, Berlin Heidelberg New York, pp. 245–249
2. Baba H, Shima I, Tomita K et al. (1985) Somatosensory evoked potentials in cervical myelopathy. In: Schramm J, Jones SJ (eds) Spinal cord monitoring. Springer, Berlin Heidelberg New York, pp 284–289
3. Baba H, Tomita K, Umeda S et al. (1988) Clinical study of spinal cord evoked potentials. In: Ducker TB, Brown RH (eds) Neurophysiology and standards of spinal cord monitoring. Springer, Berlin Heidelberg New York Tokyo, pp 216–221
4. Baba H, Tomita K, Umeda S et al. (1988) Somatosensory evoked potentials in the cervical OPLL myelopathy. Clin Orthop Surg 23:1409–1416
5. Baba H, Tomita K, Umeda S et al. (1987) Somatosensory evoked potentials in the upper cervical spine disorder. Orthop Surg Traumatol 30:1075–1082
6. Okuma T (1987) Level diagnosis of cervical myelopathy using evoked spinal action potentials. J Jpn Orthop Ass 61:477–489
7. Kurokawa T (1979) Clinical application of the evoked electrospinogram. Adv Neurol Sci 23:409–420
8. Nagata S (1988) Experimental studies on the spinal cord evoked potentials in compression injury of the cervical spinal cord. J Juzen Med Soc 97:688–703
9. Schramm J, Shigeno T, Brock M (1983) Clinical signs and evoked response alternations associated with chronic experimental cord compression. J Neurosurg 58:734–741
10. Shimomiya K (1985) Clinical study of cervical spondylotic myelopathy using evoked spinal cord potentials. In: Schramm J, Jones SJ (eds) Spinal cord monitoring. Springer, Berlin Heidelberg New York, pp 290–301
11. Tomita K, Baba H, Takahashi K (1989) Total (anterior and posterior) decompression of the spinal cord. J Jpn Orthop Ass 63:501–506

Peripheral Versus Central Mechanisms for Integration

J. Kimura[1], C.K. Suh, K. Orii, and T. Yamada

Introduction

Analyses of sensory nerve action potentials (SNAP) and somatosensory evoked potentials (SSEPs) provide an important physiologic measure of the sensory system because electrophysiologic abnormalities depend on the kind and degree of damage in individual fibers within the nerve. In segmental demyelination, for example, failure to activate the next node of Ranvier in some fibers results in diminution of SNAP. The size of the recorded response, however, does not necessarily change in proportion to the number of functional axons; a major reduction in amplitude or area can result from physiologic desynchronization of the axonal volleys [1, 6, 8].

Conduction block, therefore, is much more difficult to document than previously was thought [7, 9–10]. In particular, one cannot equate the degree of amplitude reduction with the severity of conduction block in analyzing complicated patterns of SSEPs recorded over the scalp [2–5]. For example, the waveform analysis may reveal relative preservation of SSEPs despite major abnormalities of sensory conduction of a demyelinative peripheral neuropathy.

We have previously postulated that duration-dependent phase cancellation dictates the relationship between the size of SNAPs and the length of the nerve segment under study [6]. A paired shock technique used in the present experiment simulates the effect of desynchronized input. It serves as a model to evaluate normal and abnormal temporal dispersion that could substantially alter the waveform in peripheral versus central sensory processing.

Materials and Methods

In ten healthy subjects, stimulation of the second (S1) or third (S2) digit elicited median SNAP at the wrist and SSEP over the scalp. A concomi-

[1] Department of Neurology, Kyoto University Hospital, Sakyo-ku, Kyoto, 606 Japan

K. Shimoji et al. (Eds.)
Spinal Cord Monitoring and Electrodiagnosis

tant application of S1 and S2, both 0.1 ms in duration and supramaximal in intensity, simulated the effect of desynchronizated inputs at varying interstimulus intervals. Both stimuli, delivered singly or in combination, evoked a reproducible response on repeated trials. The interstimulus intervals between S1 and S2 ranged from 0 to 2.0 ms in increments of 0.1 or 0.2 ms and from 2 to 10 ms in increments in 0.5 ms.

A multichannel averager allowed simultaneous recording of SNAP at the wrist and SSEP from the scalp. The two pairs of electrodes for peripheral potentials consisted of a common distal lead, G1, at the distal wrist crease, and two proximal leads, G2, located 2 cm and 6 cm from G1 (Fig. 1). For SSEP, a pair of surface electrodes was placed over C3 or C4 and A1 and A2, according to the international derivation. Each test set comprised an average of 1000 summated potentials with a frequency response of 10 Hz–10 kHz (3 dB down).

Two tracings obtained for each electrode derivation confirmed consistency. For SNAPs, we defined the amplitude from the baseline to the negative peak, the duration from the onset of the final return to the baseline,

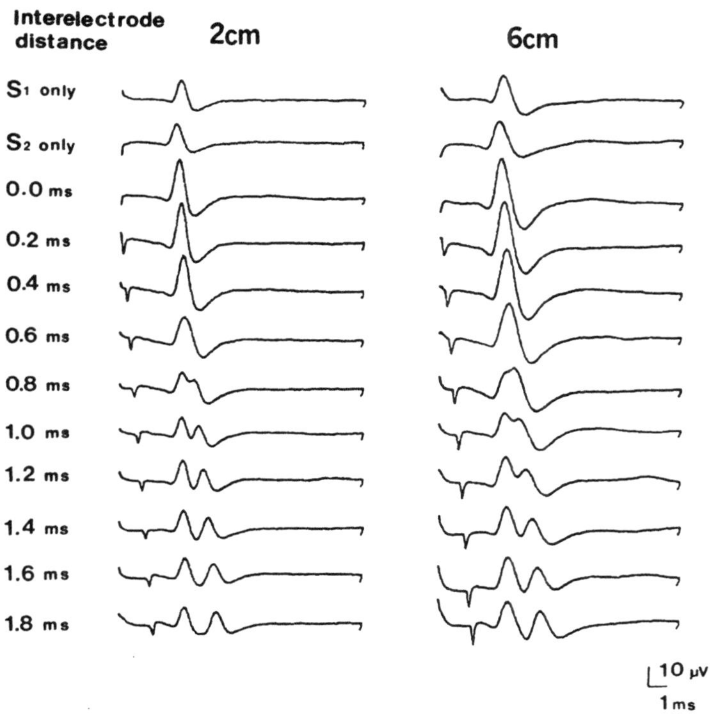

Fig. 1. Orthodromic sensory potentials at the wrist elicited by S1 or S2 only (top two tracings), or by both S1 and S2 at interstimulus intervals ranging from 0 to 1.8 ms. The area under the negative peak reached a minimal value at 0.8 ms with the pick-up electrodes placed 2 cm apart and at 1.0 ms with a longer interelectrode distance of 6 cm

and the area under the initial negative phase by integration. For SNAPs, measurement included peak to peak amplitude and peak to peak duration of N20, N30, and N60 for SSEP. Additionally, we compared the recorded response at each interestimulus interval to the algebraic sum of the two waveforms elicited individually by S1 and S2. The difference between the actually elicited potential and the corresponding computer summation revealed the degree of discrepancy between the theoretical and actual waveform at the respective interval.

Results

When recorded with G1 and G2 separated by 2 cm, SNAPs at the wrist showed a maximal area of 6.5 ± 2.7 μV × ms (Fig. 1, Table 1). At increasing interstimulus intervals, the response became progressively longer in duration, lower in amplitude, and smaller in area, reaching a minimal size of 4.5 ± 1.7 μV × ms at 0.6 ms. Further separation of S1 and S2 resulted in division of the recorded response into two discrete components, with gradual return of the total area to 6.0 ± 2.0 μV × ms at 1.4 ms.

When recorded with G1 and G2 separated by 6 cm (Fig. 1, Table 2), the area under the waveform changed from a maximal value of 13.1 ± 4.6 μV × ms to a minimal value of 9.3 ± 3.5 μV × ms at 0.9 ms with a partial recovery to 10.5 ± 3.3 μV × ms at 1.4 ms. Thus, the wider the separation between G1 and G2, the longer the duration of recorded responses and the greater the

Table 1. Orthodromic sensory nerve action potentials with electrodes separated by 2 cm

	Duration (ms)		Amplitude (μV)		Area (μV × ms)	
	Recorded response	Algebraic sum	Recorded response	Algebraic sum	Recorded response	Algebraic sum
S1 only	1.7 ± 0.3		8.0 ± 3.1		3.1 ± 1.4	
S2 only	1.8 ± 0.3		6.9 ± 3.0		3.1 ± 1.5	
0.0 ms	1.8 ± 0.2	1.8 ± 0.3	15.3 ± 5.8	14.6 ± 6.0	6.5 ± 2.7	6.1 ± 2.8
0.1	1.9 ± 0.2	1.9 ± 0.3	14.5 ± 6.0	13.3 ± 5.8	6.3 ± 2.7	5.9 ± 2.8
0.2	2.0 ± 0.3	1.9 ± 0.3	12.3 ± 5.5	11.4 ± 5.1	6.0 ± 2.7	5.4 ± 2.7
0.3	2.1 ± 0.3	2.1 ± 0.3	10.1 ± 4.2	9.1 ± 4.0	5.4 ± 2.5	4.9 ± 2.5
0.4	2.2 ± 0.2	2.2 ± 0.3	8.4 ± 3.0	7.4 ± 2.9	5.0 ± 2.4	4.5 ± 2.3
0.5	2.3 ± 0.2	2.3 ± 0.3	8.0 ± 2.5	7.1 ± 2.6	4.8 ± 2.2	4.3 ± 2.2
0.6	2.4 ± 0.3	2.4 ± 0.3	8.1 ± 2.5	7.2 ± 2.7	4.5 ± 1.7	4.3 ± 2.1
0.7	2.5 ± 0.3	2.5 ± 0.3	8.5 ± 2.7	7.4 ± 2.8	5.0 ± 1.8	4.5 ± 1.9
0.8	2.6 ± 0.2	2.5 ± 0.3	8.8 ± 2.8	7.5 ± 2.8	5.4 ± 1.7	4.8 ± 1.9
0.9	2.7 ± 0.3	2.6 ± 0.2	8.9 ± 3.0	7.6 ± 2.8	5.7 ± 1.7	5.1 ± 1.9
1.0	2.8 ± 0.3	2.7 ± 0.3	8.8 ± 2.7	7.6 ± 2.7	6.1 ± 1.7	5.5 ± 2.1
1.1	2.9 ± 0.3	2.8 ± 0.3	8.7 ± 2.8	7.5 ± 2.6	6.3 ± 1.9	5.7 ± 2.3
1.2	3.0 ± 0.3	2.9 ± 0.3	8.5 ± 2.8	7.4 ± 2.6	6.4 ± 1.9	5.9 ± 2.4
1.3	3.1 ± 0.3	3.0 ± 0.2	8.3 ± 2.6	7.2 ± 2.6	6.4 ± 1.9	6.3 ± 2.6
1.4	3.2 ± 0.3	3.1 ± 0.3	8.1 ± 2.7	7.1 ± 2.6	6.9 ± 2.0	6.6 ± 2.7

Table 2. Orthodromic sensory nerve action potentials with electrodes separated by 6 cm

	Duration (ms)		Amplitude (μV)		Area (μV × ms)	
	Recorded response	Algebraic sum	Recorded response	Algebraic sum	Recorded response	Algebraic sum
S1 only	2.2 ± 0.4		10.7 ± 3.9		6.2 ± 2.1	
S2 only	2.4 ± 0.4		10.4 ± 3.9		6.3 ± 2.3	
0.0 ms	2.3 ± 0.3	2.4 ± 0.3	21.9 ± 8.1	21.0 ± 7.8	13.1 ± 4.6	12.1 ± 4.1
0.1	2.3 ± 0.2	2.4 ± 0.4	21.5 ± 7.9	20.3 ± 7.6	12.9 ± 4.5	12.0 ± 4.3
0.2	2.4 ± 0.3	2.5 ± 0.3	19.8 ± 7.3	18.5 ± 6.7	12.5 ± 4.5	11.6 ± 4.1
0.3	2.6 ± 0.3	2.6 ± 0.3	17.4 ± 6.1	16.3 ± 5.7	11.9 ± 4.0	11.0 ± 3.8
0.4	2.7 ± 0.2	2.7 ± 0.3	14.7 ± 5.0	13.8 ± 4.6	11.0 ± 3.8	10.4 ± 3.5
0.5	2.8 ± 0.2	2.8 ± 0.4	12.3 ± 4.4	11.3 ± 3.6	10.7 ± 3.7	9.8 ± 3.3
0.6	2.9 ± 0.2	2.9 ± 0.4	10.2 ± 3.3	9.8 ± 3.2	9.9 ± 3.4	9.3 ± 3.1
0.7	3.0 ± 0.3	2.9 ± 0.5	10.4 ± 3.4	9.6 ± 3.2	10.0 ± 3.3	9.0 ± 3.0
0.8	3.1 ± 0.2	3.1 ± 0.4	11.0 ± 4.3	9.9 ± 3.4	9.8 ± 4.0	8.7 ± 2.7
0.9	3.2 ± 0.2	3.2 ± 0.4	11.3 ± 4.0	10.1 ± 3.4	9.3 ± 3.5	8.2 ± 2.7
1.0	3.4 ± 0.1	3.3 ± 0.3	11.2 ± 3.3	10.2 ± 3.4	9.8 ± 3.7	8.2 ± 2.7
1.1	3.4 ± 0.2	3.4 ± 0.3	10.8 ± 3.2	10.3 ± 3.2	9.6 ± 3.4	8.4 ± 2.6
1.2	3.6 ± 0.2	3.5 ± 0.4	10.9 ± 3.1	10.2 ± 3.0	9.7 ± 3.1	8.8 ± 2.6
1.3	3.6 ± 0.3	3.6 ± 0.4	10.6 ± 2.9	10.3 ± 2.9	9.8 ± 3.1	9.0 ± 2.6
1.4	3.7 ± 0.2	3.7 ± 0.4	10.4 ± 2.9	10.1 ± 2.7	10.5 ± 3.3	9.4 ± 2.6

interstimulus interval between S1 and S2 that gives rise to the maximal reduction in the size of the potential ($p < 0.01$). The SNAP actually elicited by combined stimulation of S1 and S2 slightly exceeded the computer assisted algebraic sum of the two sensory potentials individually elicited by S1 and S2 (Tables 1 and 2). Statistical analyses, however, showed no significant difference ($p > 0.05$) between the two either in amplitude or time course.

In contrast to SNAPs, which showed a substantial change in size as a function of interstimulus interval between S1 and S2, SSEPs elicited by the same paired stimuli remained nearly constant up to 1.0 ms. At longer intervals, SSEPs showed very complex waveform changes, that varied substantially from one subject to the next (Figs. 2 and 3). The computer assisted algebraic sum of the two waveforms obtained individually by S1 or S2 always exceeded the corresponding SSEP actually evoked by stimulation of both S1 and S2. On average, SSEP was reduced in amplitude by 15%–20%, compared with the calculated value. Irrespective of the interstimulus interval, the subtracted potential nearly, though not exactly, matched the waveform elicited by S2 alone in latency and amplitude.

Discussion

The impulses of slow fibers lag increasingly behind those of fast fibers as they propagate along the conduction path. As a consequence, potentials recorded at a distance become lower in amplitude, longer in duration, and smaller in

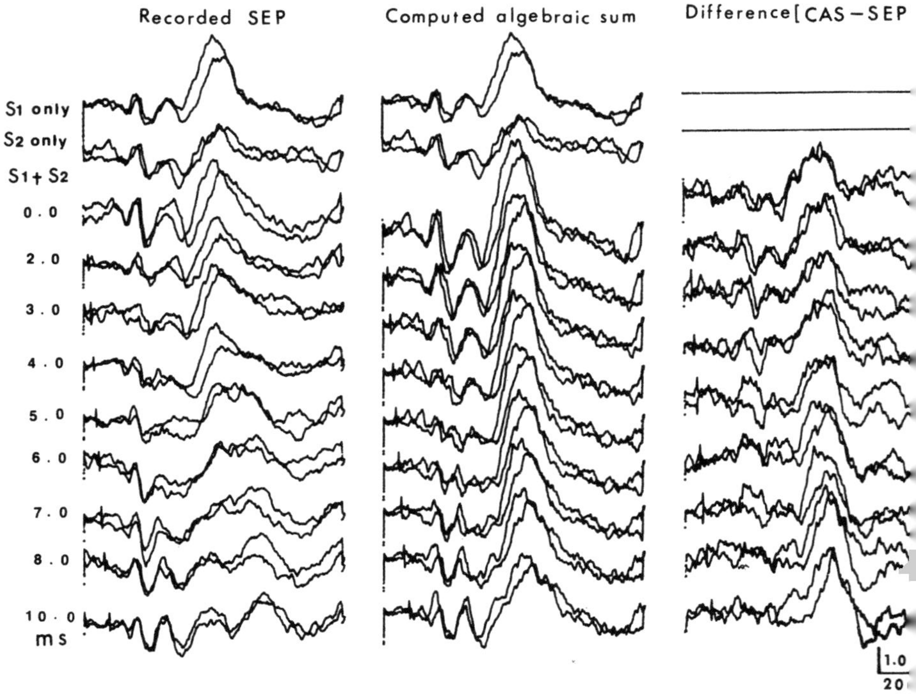

Fig. 2. SSEP at C3 elicited by S1 or S2 only (top two tracings), or by both S1 and S2 at interstimulus intervals ranging from 0 to 10 ms. The algebraic sum of their separate responses (*middle column*) always exceeded the corresponding recorded response (*left*) with the difference showing a relatively constant waveform (*right*)

area, even though the same number of sensory fibers contribute to the action potential. We have previously documented a linear relationship between the size of the response and length of the nerve segment. Such physiologic changes in waveform result from duration-dependent phase cancellation between fast and slow conducting impulses.

In the present study, we have tested the effect of interelectrode distance on the time course of phase cancellation between "fast" and "slow" components. The distance between G1 and G2 determines the duration of unit discharges, which in turn dictates the degree of phase cancellation. A greater interelectrode distance results in a longer duration of individual responses compared to the shorter duration of responses recorded by more closely placed electrodes. The cancellation effect, then, appears at a longer interstimulus interval between S1 and S2. This observation again supports the view that the duration of unit discharge dictates the effect of temporal dispersion on amplitude. A latency difference slightly less than one half of the total duration of unit discharge maximizes the phase cancellation between

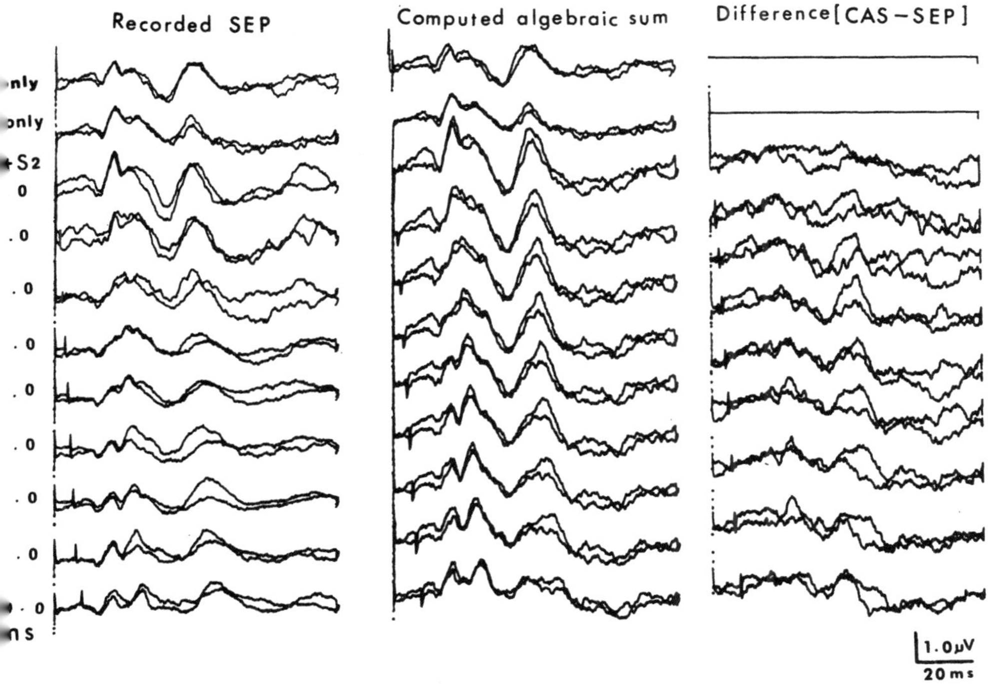

Fig. 3. The same arrangement as in Fig. 2 to show individual variability in the degree of difference (*right column*) between the computed algebraic sum (*middle*) and actually recorded SSEP (*left*)

two components. Consequently, this degree of temporal dispersion gives rise to maximal loss of the area under the recorded waveform.

The algebraic sum of two desynchronized potentials approximately equals the size of the SNAP actually recorded following the delivery of the two stimuli. This finding supports the hypothesis that diminution of the surface recorded SNAP results solely from interaction between the two unit discharges. In our model, the minimal response showed an average reduction in amplitude or areas of 30%–40%. This is nearly the same amount of change as induced by physiologic temporal dispersion of sensory potentials previously documented for median, ulnar, and radial nerves. Further, a 0.5–1.0 ms interstimulus interval required to produce the maximal change in sensory potential equals approximately the estimated range of latency difference between the fast and slow sensory fibers [1–2].

The present study has revealed that a slight shift in latency, on the order of 1 ms, between the "fast" and "slow" conducting fibers results in a major reduction of sensory action potential. When the interstimulus interval exceeds 1 ms, the area gradually returns to normal because, with such an extreme dispersion, the two discharges no longer superimpose. When the two potentials are separated by slightly less than one half the duration of the single unit, the combined response will have a relatively normal waveform but a reduced area. In the absence of conduction block, the electrode placement affects the degree of duration-dependent phase cancellation. A greater separation of the recording electrodes increases the duration of the unit discharge, thereby causing the maximal cancellation at a greater degree of desynchronization.

In contrast to SNAP, SSEP follows no predictable time course of phase cancellation, the degree of SSEP deviation varying substantially from one subject to the next. Subtraction of an actually recorded potential from the corresponding computer summation, however, yields a relatively constant waveform individually, at the interstimulus interval up to a few milliseconds. Over this range, the measured amplitude of SSEPs elicited by two stimuli always falls short of the algebraic sum of their separate amplitude. Thus, desynchronized inputs have entirely different effects on SSEPs and SNAPs. We postulate that the passage of an initial impulse not only occludes but also inhibits a subsequent impulse at central synapses, where the volleys from two sources converge. Thus, SSEP undergoes changes from the interaction of two inputs in addition to a temporal dispersion based on phase cancellation.

The loss of area under the temporally dispersed waveform results from a duration-dependent phase cancellation in SNAP and more complicated synaptic modulation in SSEP. Of course, one cannot extrapolate from a simple model involving only two unitary response what happens physiologically with thousands of nerve fibers conducting at various velocities. The basic principle of the phenomenon described herein, however, applies to any waveform analysis. As one of the distinguishing features, we predict that multiple unitary discharges tend to smooth the shape of the composite potential, making the identification of temporally dispersed waveforms more difficult than depicted in our model.

All these factors affect the interpretation of nerve conduction studies in the clinical domain. In SSEPs, fast impulses that reach the synapse first activate the common pathway shared by slow impulses arriving at the same region subsequently. This mechanism would favor the inputs from the fast conducting fibers over those from the slow conducting fibers. In other words, occlusion and inhibition of the delayed components at the point of convergence, in effect, resynchronize the desynchronized inputs giving rise to a relatively normal waveform. In addition, a small number of afferent inputs suffice to evoke a surprisingly large SSEP based on divergence at each synapse. On the other hand, an absent or nearly absent SNAP results from phase cancellation of temporally dispersed impulses even in the presence of a sufficient number of surviving axons. Thus, stimulation of the peripheral nerve in severe neuropathy, can evoke a surprisingly normal SSEP despite a very abnormal SNAP recorded simultaneously.

Summary

Stimulation of the second (S1) or third (S2) digit elicits a median SNAP at the wrist and SSEP over the scalp. Hence, a concomitant application of S1 and S2 with varying interstimulus intervals simulates the effect of desynchronized inputs on peripheral versus central processing of sensory impulses.

In 20 hands, a shift in latency on the order of 1 ms between S1 and S2 caused a major reduction in SNAP by as much as 30%–40%. A latency difference slightly less than one half of the total duration of unit discharge maximized the phase cancellation between the two components and consequently the loss of area under the waveform. In contrast, the combined use of S1 and S2 elicited a smaller SSEP than the algebraic sum of their separate responses, by 15%–20% on average, at interstimulus intervals up to a few milliseconds.

Thus, unlike SNAP, that shows only a temporal dispersion based on phase cancellation, SSEP undergoes additional changes from interaction of two inputs. We postulate that the passage of an initial impulse not only occludes but also inhibits a subsequent impulse at central synapses, where the volleys from two sources converge.

References

1. Dorfman LJ (1984) The distribution of conduction velocities (DVC) in peripheral nerves: a review. Muscle Nerve 7:2–11
2. Eisen A, Purves S, Horich M (1982) Central nervious system amplification: its potential in the diagnosis of early multiple sclerosis. Neurology 32:359–364
3. Gandevia SC, Burke D, McKeon BB (1983) Convergence in the somatosensory pathway between cutaneous afferents from the index and middle fingers in man. Exp Brain Res 50:415–425
4. Kakigi R (1986) Ipsilateral and contralateral SEP components following median nerve stimulation: effects of interfering stimuli applied to the contralateral hand. Electroencephalogr Clin Neurophysiol 64:246–259
5. Kang R, Hermann D, MacGilla M, Zarzecki P (1985) Convergence of sensory inputs in somatosensory cortex: interactions from separate afferent sources. Exp Brain Res 57:271–278
6. Kimura J, Machida M, Ishida T, Yamada T, Rodnitzky RL, Kudo Y, Suzuki S (1986) Relation between size of compound sensory or muscle action potentials, and length of nerve segment. Neurology 36:647–652
7. Kimura J, Sakimura Y, Machida M, Fuchigami Y, Ishida T, Claus D, Kameyama S, Nakazumi Y, Wang J, Yamada T Effect of desynchronized inputs on compound sensory and muscle action potentials. Muscle Nerve
8. Lee R, Ashby P, White D, Aguayo A (1975) Analysis of motor conduction velocity in human median nerve by computer simulation of compound action potentials. Electroencephalogr Clin Neurophysiol 39:225–37
9. Lewis RA, Sumner AJ, Brown MJ, Asbury AK (1982) Multifocal demyelinating neuropathy with persistent conduction block. Neurology (NY) 32:958–64
10. Olney RK, Miller RG (1984) Conduction block in compression neuropathy: recognition and quantification. Muscle Nerve 7:662–667

Central Somatosensory Conduction Time

B.R. CANT

Introduction

Of the things which can be measured in clinical neurophysiology those that are time-dependent, frequency, latency, and conduction velocity, generally have the narrowest normal range and are of greatest diagnostic utility. The velocity of an evoked potential within any section of the central nervous system is more narrowly distributed than its amplitude and because of this minor abnormalities of conduction velocity should be more readily identified than minor abnormalities of amplitude. Unfortunately, the measurement of conduction velocities within segments of the central nervous system using surface electrodes requires assumptions concerning the lengths of pathways and synaptic delays which cannot be verified. The latency of evoked potentials at various sites can be measured quite accurately, however, and the only assumption which needs to be made concerning these is the origin of the potentials recorded at these sites.

In studying somatosensory function within the central nervous system, it is usually most convenient to stimulate peripheral nerves. For that reason, measuring centrally conduction by using latencies requires exclusion of the variablity due to the effect of physiological and pathological factors on peripheral nerve function. During the past decade, this has been attempted by measuring the interval between potentials evoked from the spinal cord and cerebral cortex after stimulation of the median nerve at the wrist. This has been called the central conduction time (CCT) [3, 8]. In this paper the work done by Anne Hume, Nigel Shaw, James Judson, and myself in this field is briefly reviewed and also the work of others is referred to.

The N13 wave recorded from the cervical cord is a convenient marker for the entry of somatosensory impulses into the cord. This potential, which arises primarily from interneurons in the dorsal horn, is most clearly recorded by using an electrode pair placed anteriorly and posteriorly to the neck. When one electrode is placed over the posterior cervical area and the other

North Shore Medical Centre, 66 Pacific Highway, St. Leonards NSW 2065, Australia

K. Shimoji et al. (Eds.)
Spinal Cord Monitoring and Electrodiagnosis

over the midfrontal area of the scalp, the recorded potential is a more complex amalgam than was thought when this montage was originally adopted. However, this does not significantly degrade the information obtained and as this montage has been used in most of the studies carried out, it is the one to which is here referred. A pair of electrodes located at the midfrontal area and the post-central area recorded the N20 potential evoked by the arrival of somatosensory impulses at the cerebral cortex.

To reduce the number of references many reports by various authors cited in a previous review are indicated by reference to that review [3].

Physiological Factors

Various physiological factors have little or no effect upon the CCT. Age has no significant effect between the second and fifth decades [10]. There is no good evidence that sex has a significant effect [3]. Most importantly, although height and arm length have a marked effect on the latencies of the cervical and scalp responses, they have no effect upon the CCT [8]. The CCT increases exponentially with decreasing central body temperature, but the effect of minor variations in temperature is small and there is no evidence that the temperature of the upper limbs has any influence on the CCT [11, 3]. Variations in the partial pressure of blood gasses and pH do not appear to affect the CCT [9]. With a stimulus above threshold, the CCT is not influenced by intensity [8]. Within the range normally used, stimulus frequency has no effect upon CCT. If very high frequencies have the same effects upon the somatosensory evoked potentials recorded from humans as they have upon those recorded from animals, it is likely that they would prolong the CCT, but as these stimulus frequencies are not tolerated by conscious humans, this is of little clinical relevance [17].

Pharmacological Effects

Of the various pharmacological effects which have been demonstrated the most important are those due to anaesthetic agents. These are variable. Halothane has an effect upon the CCT as does isoflurane, while fentanyl and sodium thiopental have no effect [18, 11, 7]. Phenobarbital has a small effect upon the CCT and in animals pentobarbital has been demonstrated to have a small effect [9, 16]. Carbamazepine and phenytoin have a small effect upon the CCT, while sodium valproate does not [13]. None of these pharmacological effects is marked and, with the exception of those due to some anaesthetic agents, they are of doubtful significance in the clinical investigation of individual patients.

Cerebral Trauma

Our major interest has been the measurement of the CCT in cerebral trauma and its use as a method of investigating the prognosis and pathophysiology in that disorder [4, 9, 12]. We have classified somatosensory evoked potentials recorded from patients with head injuries into three broad groups. Category 1 are evoked potentials demonstrating normal and symmetrical CCTs. Category 2 are evoked potentials in which there is an increase in the CCT on one or both sides or asymmetrical CCTs. In category 3, the cortical evoked potential is absent on one or both sides. It is important to note that this classifiication takes into account only the CCT and the presence or absence of N20. The amplitude of N20 if it is present is immaterial. We have found the classification of evoked potentials in this way provides a more accurate prognosis of the outcome from cerebral trauma during the acute period than any other clinical or investigative finding. The prognostic value of these findings remains the same within the first 4 days. In patients with a Glascow Coma Score of eight or less, 87% of patients with category 1 evoked potentials recorded within the first 48 h have a favourable outcome (good recovery or moderate disability), while 92% of those patients with category 3 evoked potentials have an unfavourable outcome (severe disability, persistent vegetative state, or death). Of those patients with category 2 evoked potentials 73% have a favourable outcome [12].

This type of analysis conceals the more direct relationship of CCT to the severity of the cerebral trauma. The CCT is correlated with the severity of the injury as measured by the outcome. This correlation persists over many months and, as the patients recover, the parallel reduction in the CCT demonstrated by patients with a good outcome and those remaining disabled suggests that this parameter is a measurement of some basic pathophysiological disturbance which occurs with cerebral trauma [9]. The relative contributions of synaptic delay and delayed axonal conduction due to demyelination remain unclear because of the paucity of pathological studies performed in the postacute period.

Other Clinical Applications

Central conduction can be prolonged in a number of disease states. In multiple sclerosis, the CCT can be prolonged in the presence or absence of sensory symptoms in the limbs studied [3]. This may or may not be accompanied by an alteration in the amplitude of the evoked potentials. Central conduction time has been shown to be abnormal in acute and subacute cerebral ischaemia [3]. A correlation between CCT and regional cerebral blood flow has been demonstrated experimentally [3]. With cerebral infarction the CCT is prolonged and this has been used to differentiate multi-infarct dementia from Alzheimer's disease [1].

In various inherited ataxias the CCT is prolonged. This may occur in Friedreich's ataxia, and a similar measurement, from N9 to N20, has been used to assist the clinical classification of the inherited ataxias [3]. A prolonged CCT has been found in a minority of patients with Charcot-Marie-Tooth disease suggesting that there is a variant of that disorder with central involvement [3]. In metachromatic leukodystrophy, the CCT may be prolonged [20]. A prolonged CCT has been reported in a minority of patients with myotonic dystrophy [2]. CCT is also prolonged in some patients with vitamin B_{12} deficiency [21]. Similarly, the Guillain-Barré syndrome is rarely associated with an increase in the CCT [15]. Of patients with cervical myelopathy due to degenerative disease of the cervical spine 25% have been shown to have an increased CCT [19]. There is controversy concerning the CCT in patients with motor neuron disease [5, 6, 14]. Just as there is uncertainty as to the significance of abnormal sensory findings in a minority of patients with motor neuron disease, so there have been conflicting opinions as to the significance of abnormalities of the CCT found in some of them.

Conclusion

In summary, by recording evoked potentials from the neck and the scalp following median nerve stimulation at the wrist, it is possible to measure the time taken for the conduction of somatosensory impulses from the cervical cord to the cerebral cortex. This is a robust measurement which has been replicated by many groups. It is little influenced by a variety of physiological and pharmacological factors, but it is altered by various types of pathology. Because of this it can be used to demonstrate the presence of pathology and to assess its severity. In one particular disorder, cerebral trauma, measurement of the CCT allows the early and accurate assessment of damage to the central nervous system.

References

1. Abbruzzese G, Reni L, Cocitio L, Ratto S, Abbruzzese M, Favale E (1984) Short-latency somatosensory evoked potentials in degenerative and vascular dementia. J Neurol Neurosurg Psychiatry 47:1034–1037
2. Bartel PR, Lotz BP, Van Der Meyden CH (1984) Short-latency somatosensory evoked potentials in dystrophia myotonica. J Neurol Neurosurg Psychiatry 47:524–529
3. Cant BR, Shaw NA (1986) Central somatosensory conduction time: method and clinical applications. In: Cracco RQ, Bodis-Wollner I (eds) Evoked potentials. Frontiers of clinical neuroscience, vol 3. Liss, New York, pp 58–67

4. Cant BR, Hume AL, Judson JA, Shaw NA (1986) The assessment of severe head injury by short-latency somatosensory and brainstem evoked potentials. Electroencephalogr Clin Neurophysiol 65:188–195
5. Cascino GD, Ring SR, King PJL, Brown RH, Chiappa KH (1988) Evoked potentials in motor system diseases. Neurology 38:231–238
6. Cosi V, Poloni M, Mazzini L, Callieco R (1984) Somatosensory evoked potentials in amyotrophic lateral sclerosis. J Neurol Neurosurg Psychiatry 47:857–861
7. Ganes T, Lundar T (1983) The effect of thiopentone on somatosensory evoked responses and EEGs in comatose patients. J Neurol Neurosurg Psychiatry 46:509–514
8. Hume AL, Cant BR (1978) Conduction time in central somatosensory pathways in man. Electroencephalogr Clin Neurophysiol 45:361–375
9. Hume AL, Cant BR (1981) Central somatosensory conduction after head injury. Ann Neurol 10:411–419
10. Hume AL, Cant BR, Shaw NA, Cowan JC (1982) Central somatosensory conduction time from 10 to 79 years. Electroencephalogr Clin Neurophysiol 54:49–54
11. Hume AL, Durkin MA (1986) Central and spinal somatosensory conduction times during hypothermic cardiopulmonary bypass and some observations on the effects of fentanyl and isoflurane anesthesia. Electroencephalogr Clin Neurophysiol 65:46–58
12. Judson JA, Cant BR, Shaw NA (1990) The early prediction of outcome from cerebral trauma by somatosensory evoked potentials. Crit Care Med 18:363–368
13. Mervaala E, Keranen T, Tiihonen P, Riekkinen P (1987) The effects of carbemazapine and sodium valproate on SEPs and BAEPs. Electroencephalogr Clin Neurophysiol 68:475–478
14. Radtke RA, Erwin A, Erwin CW (1986) Abnormal sensory evoked potentials in amyotrophic lateral sclerosis. Neurology 36:796–801
15. Ropper AH, Chiappa KH (1986) Evoked potentials in Guillain-Barré syndrome. Neurology 36:587–590
16. Shaw NA, Cant BR (1981) The effects of pentobarbital on central somatosensory conduction time in the rat. Electroencephalogr Clin Neurophysiol 51:674–677
17. Shaw NA, Cant BR (1984) The effect of experimental concussion on somatosensory evoked potentials. Aust J Exp Biol Med Sci 62:361–371
18. Symon L, Wang AD, Costa E, Silva IE, Gentile F (1984) Perioperative use of somatosensory evoked responses in aneurysm surgery. J Neurosurg 60:269–275
19. Veilleux M, Daube JR (1987) The value of ulnar somatosensory evoked potentials (SEPs) in cervical myelopathy. Electroencephalogr Clin Neurophysiol 68:415–423
20. Wulff CH, Trojaborg W (1985) Adult metachromatic leukodystrophy: neurophysiologic findings. Neurology 35:1776–1778
21. Zegers de Beyl D, Delecluse F, Verbanck P, Borenstein S, Capel P (1988) Somatosensory conduction in vitamin B12 deficiency. Electroencephalogr Clin Neurophysiol 69:313–318

Somatosensory Evoked Potentials After Posterior Tibial Nerve Stimulation in Children with Neurological Disorders

K. Mutoh[1], T. Okuno, M. Ito, T. Fujii, and H. Mikawa

Somatosensory evoked potentials after posterior tibial nerve stimulation (PTN-SSEPs) have been studied extensively in adults, and have been reported to be clinically useful [9]. However, relatively few SSEP studies have been conducted with lower limb stimulation in infants and children [1–3]. SSEPs after lower limb stimulation seem to be more informative than those after upper limb stimulation, since they can afford objective estimations about sensory afferent conduction over the peripheral nerve through the spinal cord to the cerebral cortex. Recently, we studied PTN-SSEPs in normal infants and children, and reported that maturation of afferent conduction in the central nervous system is completed in humans over 12 years of age [6, 7]. In the present study, results of PTN-SSEPs in 35 patients with various neurological disorders were reviewed in order to determine the clinical usefulness of PTN-SSEPs.

Methods and Materials

PTN-SSEPs were recorded as reported previously [6, 7], and were analyzed using a Nicolet CA 1,000 following unilateral percutaneous stimulation of the PTN at the ankle. The square-wave stimulus pulses were 200 μs in duration, just above the motor threshold in intensity, and were delivered at a rate of 5.1–7.1/s. An analysis time of 50 ms and a bandwidth of 30–1500 Hz were employed. We averaged 2000–4000 responses. The room temperature was maintained between 24°C and 26°C, and at least two responses were obtained to assure consistency.

Silver-silver chloride disk electrodes were applied with electrode paste, and the impedance was maintained at less than 5 kΩ. The recording electrodes were attached at the popliteal fossa ipsilateral to the stimulated nerve (PF), the 3rd lumbar and 7th cervical spines (L3S and C7S), 2 cm

[1] Department of Pediatrics, Faculty of Medicine, Kyoto University, 54 Kawahara-cho, Shogoin, Sakyo-ku, Kyoto, 606 Japan

K. Shimoji et al. (Eds.)
Spinal Cord Monitoring and Electrodiagnosis

rostral to them (L3S+ and C7S+), Fpz, Cz′ (the midpoint between Cz and Pz of the 10–20 international system), the contralateral knee (cKn), shoulder (cSh), and earlobe (cEL). Two recording sessions were conducted, one mainly for spinal evoked potentials with bipolar derivations, and another for short-latency SSEPs with a noncephalic reference. The former included the following four derivations: PF–cKn, L3S–L3S+, C7S–C7S+, and Cz′–Fpz. The latter comprised the following four derivations: Cz′–cKn, cEL–cKn, cSh–cKn, and Fpz (or L3S)–cKn. The results of each recording session were stored on a floppy disk, and the SSEPs with Cz′–cEL and Cz′–cSh derivations were later obtained by subtraction.

The results of PTN-SSEPs were estimated as reported previously [6, 7], and were compared with the normative data of age-matched controls in our laboratory. The obtained values were judged as abnormal if they differed by 2 SD of the mean for age-matched controls.

PTN-SSEPs were recorded in 35 patients, between the ages of 0 months and 24 years, with various neurological disorders of childhood-onset. In some cases, repeat examinations could be conducted, and a total of 40 recordings were obtained. The 35 subjects were divided into three groups according to the nature of their lesion. Group I consisted of 14 patients who were

Table 1. Clinical features and results of PTN-SSEPs of 14 patients in Group I

Patient/ sex	Age when SSEP recorded (y:m)	Initial symptom	Neurological findings on SSEP examination		
			Motor	Sensory	Others
1/F	9:7	Hemiplegia	+	+	MRI: normal cervical cord
2/F	2:6	Paraplegia	+	+	MRI: normal lumbar cord
3/F	3:3	Paraplegia	+	–	MCV lowered
4/F	10:1	Paraplegia	+	+	Increased CSF protein
5/M	10:3	Monoplegia	+	–	MCV lowered
6/F	6:9	Paresthesia	–	+	AMoL
7/F	0:2	Paraplegia	+	+	MRI abnormal (T1–T4)
8/M	7:10	Paraplegia	+	+	Myelography: block at T10
9/F	3:5	Hemiplegia	+	–	Head CT: LDA in midbrain
10/F	1:4	Hemiplegia	+	–	Head CT: infarction
11/F	14:5	Visual disturbance	–	–	Optic neuritis
12/F	13:3	Paresthesia	–	+	Headache
13/F	12:11	Quadriplegia	–	–	MRI: normal cervical cord
14/M	7:5	Paraplegia	–	–	MRI: abnormal in cerebrum

F, female; *M*, male; *y*, year; *m*, month; *MRI*, magnetic resonance imaging; *MCV*, motor nerve conduction velocity; *CSF*, cerebrospinal fluid; *LDA*, low density area; *AMoL*, acute monocytic leukemia; *Ab*, abnormal; *N*, normal; *n.j.*, not judged; *GBS*, Guillain-Barré syndrome; *MS*, multiple sclerosis; *VCR*, vincristine; *ADEM*, acute disseminated encephalomyelitis

suspected of having a localized lesion in the peripheral and/or central nervous system (Table 1). Group II included 16 cases who mainly suffered from degenerative or metabolic disorders, and were investigated with PTN-SSEPs in order to ascertain the extent of the pathological process in the nervous system (Table 2). Group III included five patients with myelodysplasia (Table 3). The results of PTN-SSEPs are described separately for these three groups.

Results

Group I

Fourteen patients were suspected of having a localized lesion in the nervous system (Table 1), and all the patients in this group became acutely ill. Of the 14 patients in Group I, PTN-SSEPs were recorded in the acute stage in Patients 1–12, while they were recorded only during convalescence in Patients 13 and 14. Of the 12 subjects who were examined in the acute period, PTN-SSEP was abnormal in eight patients. Of these eight patients, afferent

Results of PTN-SSEPs	State of afferent conduction					Final diagnosis
	Distal PTN	Proximal PTN	Spinal cord	Brain stem	Cerebrum	
Ab	N	N	Ab	n.j.	N	Myelitis
Ab	N	N	Ab	n.j.	n.j.	Myelitis
Ab	Ab	Ab	n.j.	n.j.	n.j.	Polyneuritis
Ab	Ab	Ab	n.j.	n.j.	n.j.	GBS
Ab	Ab	N	N	N	N	Polyneuritis
Ab	Ab	Ab	N	N	N	VCR neuropathy
Ab	N	N	Ab	n.j.	n.j.	Spinal cord hemorrhage
Ab	N	N	Ab	n.j.	n.j.	Malignant lymphoma
N	N	N	N	N	N	Brainstem encephalitis
N	N	N	N	N	N	Moyamoya disease
N	N	N	N	N	N	Possible MS
N	N	N	N	N	N	Hysteria
N	N	N	N	N	N	Myelitis
N	N	N	N	N	N	ADEM

Table 2. Clinical features and results of PTN-SSEPs of 16 patients in Group II

Patient/ sex	Age when SSEP recorded (y:m)	Initial symptom	Neurological findings on SSEP examination		
			Motor	Sensory	Others
15/M	6:5	Gait disturbance	+	−	Mental regression
16/F	3:0	Ataxia	+	−	Psychomotor regression
17/F	1:6	Delay of development	+	−	Hypotonia
18/M	11:4	Spasticity	+	−	CT: calcification in basal ganglia
19/M	6:1	Psychomotor regression	+	+	Optic nerve atrophy
20/M	24:11	Spastic paraplegia	+	+	Optic nerve atrophy
21/F	5:11	Psychomotor regression	+	−	Convulsion
22/F	1:2	Psychomotor regression	+	−	Convulsion
23/M	18:7	External ophthalmoplegia	+	−	Hearing loss
24/M	14:10	Ataxia	+	+	Vitamin E level decreased
25/M	12:5	Paresthesia	−	+	MCV mildly lowered
26/F	4:11	Delay of development	+	−	Mentally retarded
27/M	9:9	Delay of development	+	−	Mentally retarded Scoliosis
28/M	16:11	Paresthesia	−	+	CSF: HTLV-I antibody (+)
29/M	1:8	Convulsion (intractable)	+	−	Severe delay of development
30/F	1:3	Facial diplegia	+	−	Mentally normal Motor delay

F, female; *M*, male; *y*, year; *m*, month; *CT*, computed tomography; *MCV*, motor nerve conduction velocity; *CSF*, cerebrospinal fluid; *N*, normal; *Ab*, abnormal; *n.j.*, not judged

conduction was disturbed in the peripheral nerve in four, and in the spinal cord in four. In all four cases with disturbed conduction in the spinal cord (Patients 1, 2, 7, and 8), the affected segment could be determined by recording spinal SSEPs from adjacently attached electrodes. Both the patients who were examined during convalescence showed normal PTN-SSEPs. The PTN-SSEPs of Patient 1 are shown in Fig. 1.

Group II

Patients in Group II had rather insidious onsets and took a more chronic course (Table 2). PTN-SSEPs were abnormal in 11 of the 16 subjects (68.8%) in Group II.

Results of PTN-SSEPs	State of afferent conduction					Final diagnosis
	Distal PTN	Proximal PTN	Spinal cord	Brain stem	Cerebrum	
N	N	N	N	N	N	Adrenoleukodystrophy
Ab	N	N	N	N	Ab	Leukodystrophy
N	N	N	N	N	N	Leukodystrophy
Ab	Ab	Ab	N	N	N	Cockayne syndrome
Ab	Ab	N	Ab	n.j.	n.j.	Neuroaxonal dystrophy
Ab	N	Ab	Ab	n.j.	n.j.	Hallervorden-Spatz syndrome
Ab	Ab	N	N	N	Ab	Leigh encephalopathy
Ab	Ab	N	N	N	Ab	Leigh encephalopathy
N	N	N	N	N	N	External ophthalmoplegia
Ab	N	N	Ab	n.j.	n.j.	Vitamin E deficiency
N	N	N	N	N	N	Hypoparathyroidism
N	N	N	N	N	N	Cerebral palsy
Ab	N	N	Ab	n.j.	n.j.	Cerebral palsy
Ab	N	Ab	n.j.	n.j.	n.j.	Cisplatin neuropathy
Ab	Ab	Ab	Ab	Ab	Ab	Phenytoin intoxication
Ab	Ab	Ab	n.j.	n.j.	n.j.	Möbius' syndrome

Of three cases with leukodystrophy (Patients 15–17), PTN-SSEPs were abnormal only in Patient 16, in whom the etiology was unknown. Patient 15 had adrenoleukodystrophy (ALD), and when he was examined 2 months after an onset of the disease with convulsions and gait disturbance, the PTN-SSEPs were normal. In the three patients whose disturbances pathologically involved spheroids (Patients 19, 20, and 24), a similar abnormality was observed, i.e., disturbed afferent conduction in the spinal cord. In Patients 21 and 22, in whom Leigh encephalopathy was suspected, similar PTN-SSEP abnormalities were observed in the distal ends of the peripheral and central nervous systems (Fig. 2). PTN-SSEPs were normal in a patient with chronic progressive external ophthalmoplegia (CPEO, Patient 23), and in a patient with primary hypoparathyroidism (Patient 25).

Table 3. Clinical features and results of PTN-SSEPs of five patients in Group III

Patient/ sex	Age when SSEP recorded (y:m)	Initial symptom	Neurological findings on SSEP examination		
			Motor	Sensory	Others
31/F	1:0	Myelomeningocele	+	+	Postoperative VP shunt (+)
32/M	0:0	Myelomeningocele	+	+	Postoperative VP shunt (+)
33/M	0:0	Myelomeningocele	+	+	Postoperative VP shunt (+)
34/M	0:0	Lumbosacral tumor	–	–	Preoperative Lipoma
35/F	2:5	Lumbosacral tumor	–	–	Preoperative Lipoma

VP, ventriculoperitoneal; *y*, year; *m*, month; *F*, female; *M*, male; *N*, normal; *Ab*, abnormal; *n.j.*, not judged

In the two patients with cerebral palsy (CP), PTN-SSEPs were abnormal in Patient 27, who had severe scoliosis, while they were normal in Patient 26, who did not have scoliosis. Patient 28 had acute myelocytic leukemia (AML), and was suspected of having human T-cell lymphotropic virus type I (HTLV-I) associated myelopathy (HAM) following a blood transfusion. He showed disturbed afferent conduction in the proximal segment of peripheral nerves, which was probably due to cisplatin administration. Patient 29 had Rasmussen's encephalitis, and afferent conduction was severely disturbed in both the peripheral and central nervous system. This was considered to be due to a high serum level of phenytoin (PHT). Patient 30 suffered from Möbius' syndrome, and showed disturbed afferent conduction at the dorsal roots (absent L3S potential).

Group III

PTN-SSEPs were examined in the five patients with myelodysplasia (Table 3). In Patients 34 and 35 who had spina bifida cystica with lipoma in the lumbosacral area, afferent conduction was preserved through the cauda equina up to the cerebral cortex. In Patients 31–33 who had myelomeningoceles, SSEPs were identified only up to the cauda equina.

Discussion

Although SSEPs after lower limb stimulation have been utilized clinically in adults [9], few PTN-SSEP studies on pathological cases have been reported in

Results of PTN-SSEPs	State of afferent conduction					Final diagnosis
	Distal PTN	Proximal PTN	Spinal cord	Brain stem	Cerebrum	
Ab	N	N	Ab	n.j.	n.j.	Myelomeningocele
Ab	N	N	Ab	n.j.	n.j.	Myelomeningocele
Ab	N	N	Ab	n.j.	n.j.	Myelomeningocele
N	N	N	N	N	N	Spina bifida cystica
N	N	N	N	N	N	Spina bifida cystica

infants and children [1–3, 5, 8, 10]. In the present study, the 35 subjects were divided into three groups according to the nature of their lesion. Fourteen cases suspected of having a localized lesion along the afferent pathway were included in Group I, and eight of the 14 subjects (57.1%) showed abnormal PTN-SSEPs. Four of the eight cases with abnormal PTN-SSEPs in Group I later proved to have a focal spinal cord lesion, and the level of spinal cord involvement could be determined as shown in Fig. 1. Recording PTN-SSEPs was very useful in these patients with a focal lesion in the spinal cord [10], especially when MRI failed to reveal any abnormality. Of the four patients with peripheral neuropathy, two patients showed no complex nerve action potentials at the PF, and afferent conduction in the more rostral segments could not be estimated.

Group II included 11 patients with metabolic or degenerative disorders, and seven of the 11 cases showed abnormal PTN-SSEPs. Three of these seven cases showed disturbed afferent conduction in the spinal cord, which was probably due to dorsal column involvement of spheroids. The other four cases with abnormal PTN-SSEPs (Patients 16, 18, 21, and 22) showed disturbed conduction in the distal end of the peripheral and/or central nervous system, and slowing of conduction velocity in the spinal cord was never encountered. This was in line with the results by Cracco et al. [2, 3]. In a patient with Möbius' syndrome, afferent conduction from the lower limb was disturbed at the spinal entry, and generalized anomalous development of nerve roots was suggested as an etiology of facial diplegia in this case.

Group III included five cases of myelodysplasia in the lumbar area, and afferent conduction through the dysraphic cord was judged to be preserved in the two patients with spina bifida cystica. In contrast, PTN-SSEPs were absent over the cervical cord and scalp in the three patients with myelomeningoceles. Since PTN stimulation of the ankle transmits afferent volleys via

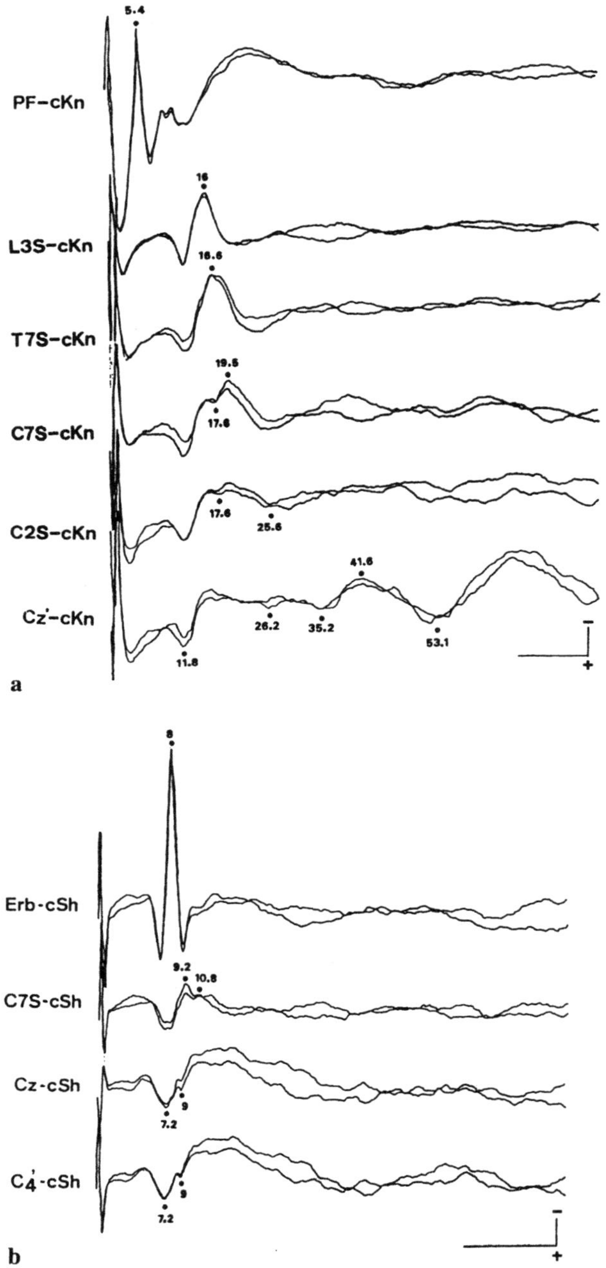

Fig. 1a,b. PTN-SSEPs in a 9-year-old girl with acute myelitis of the cervical cord (Patient 1). The horizontal line represents 10 ms, while the vertical line denotes 1 μV. **a** Two short-latency SSEPs, PI and PII, were identified at C7S (latency of PI, 11.8 ms and of PII, 17.6 ms). Although cortical responses as well as short-latency SSEPs, which normally follow PI and PII, were not clearly observed at Cz′, a low-amplitude positive wave which probably corresponded to P1 (35.2 ms) was observed at Cz′. **b** SSEPs after left median nerve stimulation showing disturbed afferent conduction through the cervical cord. SSEPs after right median nerve stimulation were normal

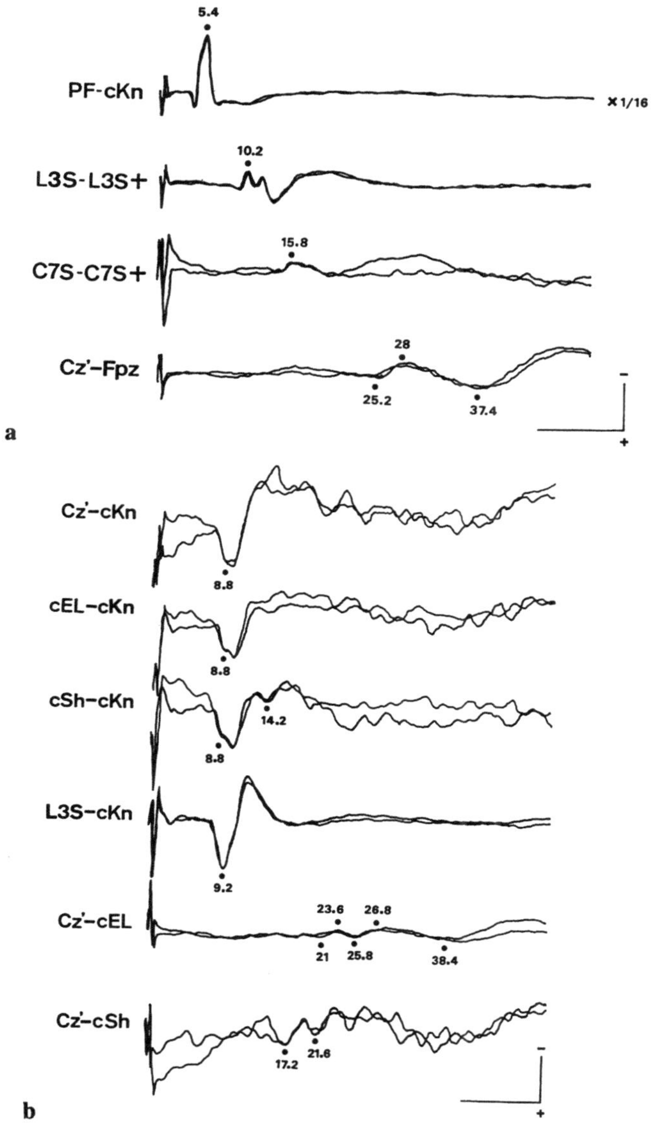

Fig. 2a,b. PTN-SSEPs recorded in a 5-year-old girl with Leigh encephalopathy (Patient 21). The horizontal line represents 10 ms and the *vertical line* denotes 1 μV. **a** Although the main positive wave with Cz′–Fpz derivation was considered to correspond to P1, its latency was delayed (37.4 ms). Peripheral nerve conduction velocity between ankle and the PF was slow (27.8 m/s). Note that the display gain of the top trace is 1/16 of the other three traces. **b** Four positive short-latency PTN-SSEPs (PI, PII, PIII, and PIV) were observed with normal latencies (PI, 8.8 ms; PII, 14.2 ms; PIII, 17.2 ms; PIV, 21.6 ms). With cEL and cSh as references, a positive wave which corresponded to P1 was identified at Cz′, again with a delayed latency (38.4 ms)

multiple dorsal roots (L4, L5, and S1), the level of sensory deficits cannot be precisely determined by eliciting PTN-SSEPs, and further technical refinement is necessary.

As a whole, PTN-SSEPs were abnormal in 22 of the 35 patients (62.9%) investigated in the present study, and were also considered to be useful in the examination of the pediatric age group. Recording PTN-SSEPs was most useful in patients with acute neurological disorders, especially in those with focal spinal cord lesions. Of the eight Group I cases with abnormal PTN-SSEPs, half showed abnormalities in the peripheral nerves. Such abnormalities in the peripheral nerves were sometimes so severe that afferent conduction in the more rostral segments could not be estimated (Patients 3, 4, 28, and 30). This is one of the shortcomings of SSEPs elicited via peripheral nerve stimulation, and was also true in patients with complete lesions in the spinal cord, that were either localized (Patients 2, 7, 8, 31, 32, and 33) or diffuse (Patients 19, 20, 24, and 27). On these occasions, SSEPs recorded after median nerve stimulation was examined to elucidate the state of afferent conduction in the rostral segments, which was successful in most cases other than those with severe generalized demyelination.

References

1. Bell HJ, Dykstra DD (1985) Somatosensory evoked potentials as an adjunct to diagnosis of neonatal spinal cord injury. J Pediatr 106:298–301
2. Cracco JB, Bosch VV, Cracco RQ (1980) Cerebral and spinal somatosensory evoked potentials in children with CNS degenerative disease. Electroencephalogr Clin Neurophysiol 49:437–445
3. Cracco JB, Cracco RQ (1986) Spinal, brainstem, and cerebral SEP in the pediatric age group. In: Cracco RQ, Bodis-Wollner I (eds) Evoked potentials. Liss, New York, pp 471–482
4. Gilmore RL, Bass NH, Wright EA, Greathouse D, Stanback K, Norvell E (1987) Developmental assessment of spinal cord and cortical evoked potentials after posterior tibial nerve stimulation: Effects of age and stature on normative data during childhood. Electroencephalogr Clin Neurophysiol 62:241–251
5. Kamimura N, Shichida K, Tomita Y, Takashima S, Takeshita K (1988) Spinal somatosensory evoked potentials in infants and children with spinal cord lesions. Brain Dev 10:355–359
6. Mutoh K, Okuno T, Mikawa H, Hojo H (1988) Maturation of somatosensory evoked potentials upon posterior tibial nerve stimulation. Pediatr Neurol 4:342–9
7. Mutoh K, Hojo H, Mikawa H (1989) Maturational study of short-latency somatosensory evoked potentials after posterior tibial nerve stimulation in infants and children. Clin Electroencephalogr 20:91–102
8. Mutoh K, Ito M, Okuno T, Mikawa H, Minami S, Asato R, Yuge M (1989) Nontraumatic spinal intramedullary hemorrhage in an infant. Pediatr Neurol 5:53–56
9. Oken BS, Chiappa KH (1986) Somatosensory evoked potentials in neurological diagnosis. In: Cracco RQ, Bodis-Wollner I (eds) Evoked potentials. Liss, New York, pp 379–389
10. Schiff JA, Cracco RQ, Rossini PM, Cracco JB (1984) Spine and scalp somatosensory evoked potentials in normal subjects and patients with spinal cord disease: evaluation of afferent transmission. Electroencephalogr Clin Neurophysiol 59:374–387

Lumbosacral Evoked Potentials and Nerve Conduction Study to Posterior Tibial Nerve Stimulation in Patients with Long-Term Anticonvulsant Therapy

T. FUJIWARA[1] and K. SEKO

Noninvasive surface electrode recording would have many advantages for recording evoked potentials in the lumbosacral region, if the technique were stable and without hazards. Our aims were (a) to establish a method of recording evoked potentials (EPs) in the lumbosacral region in normal subjects, and (b) to apply this to the diagnosis of spinal cord lesions and peripheral nerve dysfunction. Determination of a reasonable level for obtaining lumbosacral EPs could be also important in determining the lowest possible level for conduction velocity measurement in the spinal cord.

We examined 27 normal subjects and 37 epileptic patients on anticonvulsant therapy with suspected peripheral nerve dysfunction. We also investigated the utility of entire nerve conduction velocity (ENCV) measurement in the posterior tibial nerve.

Subjects and Methods

Twenty-seven healthy young volunteers were studied (19 men and eight women; mean age, 23 years; range, 17–36 years; mean height, 166 cm; range, 163–182 cm). Thirty-seven epileptic patients (21 men and 16 women) were also examined. The mean age of the patients was 34.9 years, ranging from 18 to 58 years, and the mean height was 166 cm, ranging from 144 to 177 cm. The patients were divided into four groups by seizure frequency (good or poor response to drug therapy) and duration of medication (less than or over 13 years): Group IA consisted of well-controlled patients with no or few seizures, who had been on anticonvulsant therapy for less than 13 years. Group IB consisted of well-controlled patients who had been on medication for over 13 years. Group IIA consisted of poorly-controlled patients with frequent seizures but who had been on medication for less than 13 years.

[1] College of Medical Technology, Kyoto University, 53 Kawahara-cho, Shogoin, Sakyo-ku, Kyoto, 606 Japan

K. Shimoji et al. (Eds.)
Spinal Cord Monitoring and Electrodiagnosis

Group IIB consisted of poorly-controlled patients who had been on medication for over 13 years. The mean duration of treatment of the patients was 14.0 years, ranging from 3 to 32 years.

During examination of the ENCV, the subjects lay comfortably supine in a semidarkened room. A stimulus of 0.2 ms duration and a current strength sufficient to produce mild or moderate twitching of the abductor hallucis muscle was applied to the skin over the posterior tibial nerve at the ankle. Averaging was made between 512 and 1,024 using a total analysis time of 50 ms.

Seven disk-shaped surface electrodes were placed over the skin above the vertebral column at the following sites: 20 cm, 15 cm, 10 cm, and 5 cm above Jacoby's line, at Jacoby's line (L4–5 vertebral level), and 5 cm and 10 cm below the line. These electrodes were referred to the contralateral knee on the side of stimulation for monopolar recording, and adjacent electrodes were successively connected longitudinally for bipolar recording. The level generating EPs with the maximum amplitude was determined by both the findings of the monopolar recording and the phase reversal of the main negative peaks in the bipolar recording. The ENCV was calculated by dividing the distance

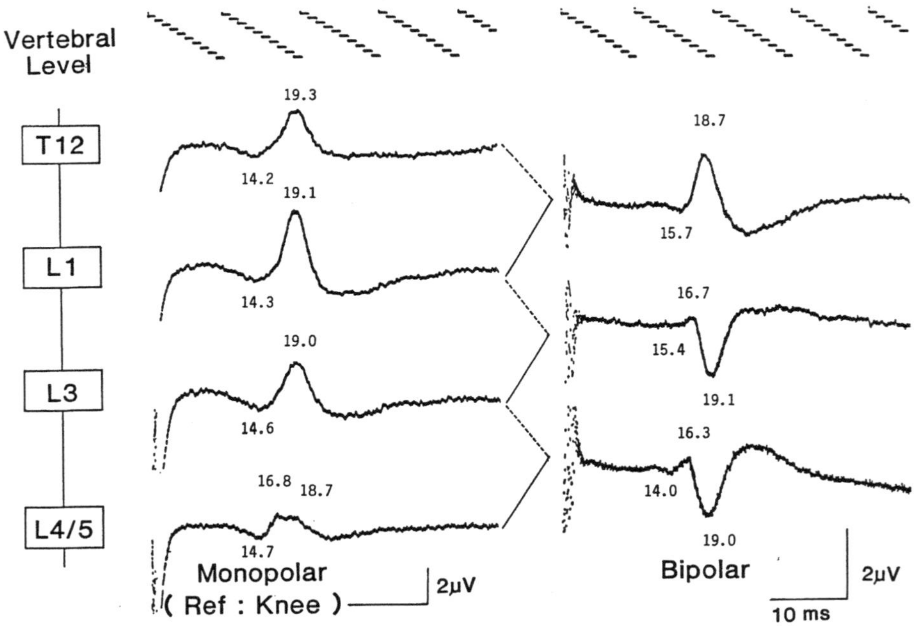

Fig. 1. Lumbosacral evoked potentials recorded with surface electrodes in a normal subject. The four waves in the *left column* are the monopolar recording and the three waves in the *right column* are the bipolar recording. It is clearly seen that the EP with the greatest amplitude originates at the upper lumbar or 12th thoracic vertebral level with both recording methods

Table 1. Entire nerve conduction velocity (ENCV) values for the posterior tibial nerve in normal subjects

Recording level (cm)	Type of recording Monopolar *n*	Mean	SD	Bipolar *n*	Mean	SD
L + 20	3	51.0	3.04		–	
L + 15	13	52.0	1.88	3	51.0	1.65
L + 10	27	50.3	1.83	12	52.2	2.70
L + 5	27	48.4	2.03	27	54.2	3.09
Lumbar (Jacoby's line)	24	46.7	2.38	25	54.3	2.37
L − 5	7	46.0	2.49	7	54.3	2.24
L − 10	1	52.5	0	1	56.8	0
Maximum amplitude level[a]	27	49.8	1.97	14	51.9	2.48

L, Jacoby's line; *L*+, above Jacoby's line; *L*−, below Jacoby's line
[a] mean ENCV values obtained by calculating only the N21 latency at the greatest amplitude in each subject

from the ankle to the recording site at each lumbar level by the peak latency of the evoked potential N21.

In this study, the motor nerve conduction velocity (MCV, popliteal–ankle), F wave conduction velocity (FWCV, lumbar-ankle, lumbar-popliteal, popliteal–ankle) and sensory nerve conduction velocity (SCV, ankle–popliteal) of the posterior tibial nerve as well as the SCV of the sural nerve (ankle–soleus), were measured simultaneously.

Results

Normal Control Group

Figure 1 shows the actual wave forms of the lumbosacral EPs recorded from one normal subject. The four waves in the left column are the ones recorded from the 5th lumbar to the 12th thoracic vertebral levels by monopolar recording using a reference electrode on the contralateral knee. The three waves in the right column are from bipolar recording. The potential with maximum amplitude at the first lumbar vertebral level in the monopolar recording and phase reversal of the main peak in the bipolar recording between two successive records occurs at the same level.

The latencies of the large negative peak (N21) were measured at every lumbosacral level, and the level of maximum amplitude was determined using both monopolar and bipolar recording methods in every case. The ENCVs for the posterior tibial nerve were calculated in the 27 normal subjects.

Table 1 shows the mean values of the ENCV at every recording level from 20 cm above to 10 cm below Jacoby's line, which was used as the point of reference. The maximum amplitude level shown in the lowest row indicates the mean ENCV values obtained by calculating only the N21 latency at maximum amplitude for each subject.

The mean ENCV calculated from the EPs at maximum amplitude was 49.8 m/s for monopolar recording, and 51.9 m/s for bipolar recording. These values most closely approximated to the ones obtained 10 cm above Jacoby's line. These mean values were very close to the FWCV, MCV, and SCV values for the posterior tibial nerve which were determined at the same time.

The lumbosacral EPs recorded below Jacoby's line had a configuration consisting of two negative peaks in many cases. On the other hand, those at the level above the line had a single peak in all cases. A shift in the negative peak of the EP was observed to some extent at recording sites from the lower to the upper lumbar levels above the line in both the monopolar and bipolar recordings.

Epileptic Patient Group

The 37 epileptic patients mainly suffered from grand mal seizures and all were taking phenytoin, which was administered in combination with phenobarbital or, in a few patients, with valproic acid.

Monopolar Recording

The mean ENCVs were also calculated using the N21 latency of the EP at every level including the maximum amplitude level for monopolar recording. Statistical comparisons were made among the four groups (Groups I and II determined by the effect of drug therapy, Groups A and B determined by the duration of medication) (Table 2).

Table 2. Mean ENCV (and SD) values obtained by monopolar recording in the 37 epileptic patients

Duration of medication	Group I				II			Total		
	n	mean	SD		*n*	mean	SD	*n*	mean	SD
A (< 13 years)	9	47.6	2.10	→	6	46.0	1.77	15	46.9	2.13
B (> 13 years)	6	46.6	2.85	↘	11	43.6	1.95	17	↓ 44.7	2.72
Total	15	47.2	2.48	➞	17	44.4	2.20	32	45.7	2.72

→, $p < 0.05$; ➞, $p < 0.001$

The mean ENCV (43.6 m/s) calculated from the maximum amplitude EP was significantly reduced in Group IIB ($p < 0.001$) compared with that in Group IA (47.6 m/s), and the mean value for Group B as a whole (44.7 m/s) was significantly reduced compared to Group A as a whole (46.9 m/s, $p < 0.001$). A significant fall in the mean ENCV between Groups I and II as a whole was also observed ($p < 0.001$).

Bipolar Recording

The negative peak of the lumbosacral EP (N21) at the maximum amplitude level with bipolar recording was also used for calculation of the ENCV with similar results. The mean ENCV in Group IIB (46.6 m/s) was significantly lower compared with Group IA (50.4 m/s), and the mean value for Group B as a whole (47.6 m/s) was significantly lower than for Group A as a whole (49.7 m/s). However, the statistical significance of difference between these groups was less than with the results from monopolar recording ($p < 0.01$) (Table 3).

Discussion

Recording of the evoked potentials from the spinal cord with surface electrodes was first successfully reported by Liberson et al. (1966), and more recently the technique has been developed and applied clinically as reported by many authors [1–3, 6, 7]. A noninvasive and simple recording technique using surface electrodes would be of great value to neurologists, if neurophysiological analysis of the lumbosacral EP wave form could detect lesions in the spinal cord and peripheral nerves.

Wave Forms of the Lumbosacral EPs and Maximum Amplitude Level

It is well known that the lumbosacral EP has a different wave form at different recording levels: it has a bimodal (two negative peaks) configuration in the sacral region (cauda equina potential), and a unimodal (single negative peak) configuration in the lumbar or lower thoracic region (lumbar potential). These two negative waves observed from the sacral to lower lumbar levels (i.e., in the cauda equina region) are designated the DR-potentials (for dorsal root) and the VRr-potential (for ventral root, reflexly elicited) [2], or the R-potential (for root) and the A-potential (for anterior root) [3], for the first and second negative peaks, respectively. The first peak of the wave configuration is supposed to be the potential corresponding to activation of

Table 3. Mean ENCV (and SD) values from bipolar recording in the 37 epileptic patients

Duration of medication	Group I *n*	mean	SD	II *n*	mean	SD	Total *n*	mean	SD
A (< 13 years)	9	50.4	2.88 ↘	6	48.5	2.77	15	49.7 ↓	2.99
B (> 13 years)	5	47.6	3.75	7	46.6	2.99	12	47.6	3.20
Total	14	49.4	3.50 ⟶	13	47.2	2.88	27	48.7	3.25

→, $p < 0.05$; ⟶, $p < 0.01$

the dorsal root and the second peak to reflect the efferent motor activity in the anterior root that occurs when reflex activity is evoked. On the other hand, the greatest negative peak is observed at the upper lumbar or 12th thoracic level. This level corresponds to the lower end of the spinal cord, but the exact origin of the EP is a matter of controversy: activation in the dorsal horn [3] and origin as a postsynaptic cord potentials [6] have both been proposed.

It has been reported that the amplitude of the EP is usually greatest at the upper lumbar or 12th thoracic vertebral level. In this study, the maximum amplitude level was exactly determined using both monopolar and bipolar recordings. The reason for determination of this level is that it can be attributed to the final entrance zone of peripheral nerves into the spinal cord, and also that it is very important to decide exactly the lowermost point for measuring spinal cord conduction velocity.

It appeared to be an important finding that phase reversal on bipolar recording was observed above and below the level at which the maximum amplitude EP was recorded on monopolar recording.

Meaning of ENCV Measurement and Reduction of ENCV in Epilepsy

The ENCV is thought to correspond to the entire nerve conduction time from the ankle to the maximum amplitude level in the lumbar region where the posterior tibial nerve terminates. Accordingly, the ENCV can be regarded as the averaged sensory conduction velocity of the entire nerve. In this study, we investigated the electrophysiological meaning of the ENCV in epileptic patients who were thought to have peripheral nerve abnormalities.

Electrophysiological abnormalities of the peripheral nerves have been found more often than clinical manifestations in epileptic patients with long-term anticonvulsant therapy. These findings have generally consisted of a slight reduction in motor or sensory conduction and reduced or absent sensory action potentials [8]. Fujiwara et al. (1979) reported that electrophysiological abnormalities suggestive of peripheral nerve dysfunction were related to the duration of anticonvulsant therapy and the severity of epilepsy (i.e., good or poor seizure control with therapy). In the present study, these results were confirmed again: the ENCV of the posterior tibial nerve decreased in epileptic patients on anticonvulsant therapy for over 13 years, especially in poorly controlled cases. The difference was more significant with monopolar than with bipolar recording, and analysis of the EP at the level of maximum amplitude was best for detecting peripheral nerve dysfunction.

Summary and Conclusion

1. The lumbosacral EPs evoked by stimulation of the posterior tibial nerve at the ankle were recorded with surface electrodes in 27 normal subjects and 37 epileptic patients, and the maximum amplitude level of the EP was determined using both monopolar and bipolar recording methods.
2. The entire nerve conduction velocity along the posterior tibial nerve was calculated using the N21 latency of the EP at the maximum amplitude level. The mean value of the ENCV in normal subjects was 49.8 m/s with monopolar recording, and 51.9 m/s with bipolar recording. These values approximated closely to the FWCV, MCV, and SCV values for the posterior tibial nerve which were determined at the same time.
3. The mean ENCV in poorly-controlled epileptic patients who had been on medication for over 13 years (group IIB) was significantly decreased compared to well-controlled cases who had been treated for less than 13 years (group IA).
4. Monopolar recording was more sensitive for detecting lowered nerve conduction in the epileptic patients than was bipolar recording.
5. This study suggests that the nerve conduction studies using lumbosacral EPs are useful for detecting peripheral nerve dysfunction, especially if the negative peak of the EP at the maximum amplitude level seen with monopolar recording is used.
6. It appears reasonable that the level where the maximum amplitude EP originates should be used as the lowermost point in calculating spinal cord conduction velocity.

References

1. Cracco JB, Cracco RQ, Graziani LJ (1975) The spinal evoked responses in infants and children. Neurology (Minneapolis) 25:31–36

2. Delbeke J, McComas AJ, Kopec SJ (1978) Analysis of evoked lumbosacral potentials in man. J Neurol Neurosurg Psychiatry 41:293–302
3. Delwaide PJ, Schoenen J, De Pasqua V (1985) Lumbosacral spinal evoked potentials in patients with multiple sclerosis. Neurology 35:174–179
4. Fujiwara T, Seko K, Akiguchi I et al. (1979) Peripheral nerve function in patients with long-term anticonvulsant therapy. Clin Neurol (Japan) 19:735–743
5. Liberson WT, Gratzer M, Zalis A et al. (1966) Comparison of conduction velocities of motor and sensory fibers determined by different methods. Arch Phys Med Rehab 47:17–23
6. Lueders H, Dinner DS, Lesser RP, Klem G (1983) Origin of far-field subcortical evoked potentials to posterior tibial and median nerve stimulation. Arch Neurol (Chic) 40:93–97
7. Phillips II LH, Daube JR (1980) Lumbosacral spinal evoked potentials in humans. Neurology 130:1175–1183
8. Trimble MR, Reynolds EH (1984) Neuropsychiatric toxicity of anticonvulsant therapy. In: Mattews WB, Glaser GH (eds) Recent advances in clinical neurology, No 4. Churchill Livingstone, Edinburgh London, pp 261–280

Clinical Assessment of the Prognosis and Severity of Spinal Cord Injury Using Corticospinal Motor Evoked Potentials

T. Hirayama[1], T. Tsubokawa, S. Maejima, T. Yamamoto, and Y. Katayama

Abstract

In nine patients with acute cervical spinal cord injury, corticospinal motor evoked potentials (MEPs) elicited by either epidural stimulation at the motor cortex or transcranial stimulation were recorded during acute periods in order to obtain information concerning motor function, in comparison with other evoked potentials (SSEPs, somatosensory evoked potentials and SESPs, spinally evoked spinal cord potentials). The courses of acute cervical spinal cord injury were classified into four patterns based on the results for the multimodality evoked potentials. The first pattern, which failed to show any activity in SSEPs, SESPs, or corticospinal MEPs throughout the course of sequential recordings for 4 days, was associated with a very poor prognosis. Patients with the second pattern, which showed recovery of detectable activity only in SESPs, did not undergo improvement of paraplegia. Those with the third pattern, which showed recovery of detectable activity in corticospinal MEPs as well as SESPs, eventually acquired a capacity to walk. Those with the fourth pattern, which showed detectable activity in SSEPs, SESPs, and corticospinal MEPs, had a good prognosis and their paraplegia improved within a week. These results suggest that recordings of multimodality evoked potentials can be useful for determining the prognosis and severity of acute cervical spinal cord injury.

Introduction

At the acute stage of spinal cord injury, it is usually difficult to evaluate the degree of injury and to assess the prognosis from the neurological symptoms and diagnostic imaging alone. Attempts have been made to use evoked potentials

[1] Department of Neurological Surgery, Nihon University School of Medicine, 30-1 Ohyaguchi-Kamimachi, Itabashi-ku, Tokyo 173, Japan

K. Shimoji et al. (Eds.)
Spinal Cord Monitoring and Electrodiagnosis

for these purposes but none of the results have proved sufficient [6, 7]. The main reason is that the traditional evoked potentials, such as the somatosensory evoked potentials (SSEPs) and spinally evoked spinal cord potentials (SESPs), although suitable for objectively evaluating the function of the afferent pathways, do not directly reflect the function of the efferent pathways, which tend to be clinically more important, from the brain to the spinal cord. We experimentally investigated corticospinal motor evoked potentials (MEPs), which are considered to reflect pyramidal tract function. The motor area of the cerebral cortex was stimulated and the response wave was recorded in the epidural space of the spinal cord [2, 10]. We clinically applied this technique for evaluating the grade and prognosis of spinal cord injury. Its clinical usefulness was confirmed on the basis of comparative studies with other types of evoked potentials [3, 8, 9].

We undertook a comparison of conventional epidural stimulation and transcranial stimulation (a noninvasive stimulation) by clinically applying epidural recordings and muscle responses. This paper reports our findings.

Materials and Methods

The present study involved nine patients with acute cervical spinal cord injury. Corticospinal MEPs and muscle responses were recorded in addition to conventional SSEPs and SESPs. The methods of stimulation for the corticospinal MEPs included epidural stimulation and transcranial stimulation. In epidural stimulation, a RESUME electrode was inserted through a burr hole near the motor area of the cerebral cortex. Transcranial stimulation was performed according to the method of Merton and Morton [4]. A pulse of 2–8 mA for 500 μs was employed for epidural stimulation, and a pulse of 30–100 mA for 200 μs was used for transcranial stimulation. The potential wave was monitored through two flexible wire electrodes inserted into the epidural spaces of the rostral and caudal sides of the injured site of the cervical spinal cord. The reference electrode was placed at the paravertebral muscle. The muscle responses to transcranial stimulation were recorded simultaneously. The electrodes were left in place for 4 days and the responses were repeatedly recorded.

Results

Corticospinal MEPs in humans were obtained only by the stimulation of the motor area of the cerebral cortex. The conduction velocity was 50–75 m/s, as calculated from the recorded distance and the difference in latency. This potential responded well to stimulation with high frequencies of over 500 Hz

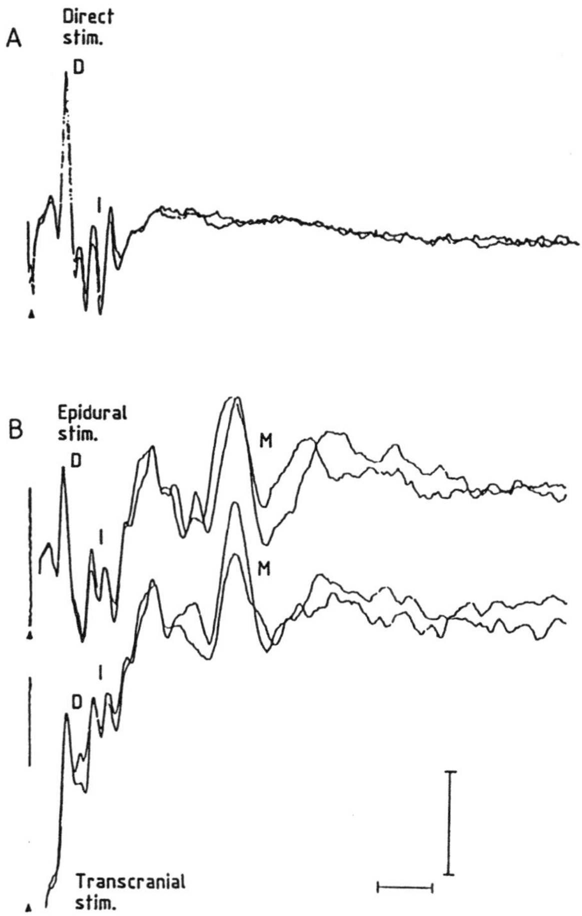

Fig. 1A,B. Normal pattern of corticospinal MEPs during **A** intracranial surgery and **B** the postoperative alert stage **A**. The corticospinal MEPs were observed as an initial negative spike (*D*) followed by multi-spikes of indirect responses (*I*) through interneurons. **B** The upper wave was triggered by epidural stimulation and the lower wave by transcranial stimulation. With both methods of stimulation, corticospinal MEPs were recorded at almost the same latency, and then multi-spike indirect responses and muscle responses (*M*) (calibrations: 5 ms, 2 μV)

and resisted the effect of thiopental sodium [3, 8, 9]. Based on these findings, the potential was considered to be the same as the pyramidal direct response reported by Patton and Amassian [5]. Its velocity matched the reported conduction velocity of the fast pyramidal tract neurons. It also shared several features in detail with the corticospinal MEPs, which we studied experimentally.

Representative corticospinal MEPs recorded during intracranial surgery are shown in Fig. 1A. The corticospinal MEPs were observed as an initial negative spike followed by multi-spikes of indirect responses through interneurons. The waves in Fig. 1B were recorded during the postoperative alert stage. The upper wave was triggered by epidural stimulation and the lower wave by transcranial stimulation. With both methods of stimulation, corticospinal MEPs were recorded at almost the same latency, and then multi-spike indirect responses and muscle responses from the reference electrode placed in the paravertebral muscle were recorded. The epidural recording of spinal potentials by transcranial stimulation in humans thus corresponded to the corticospinal MEPs, and clearly differed from the transcranially evoked spinal cord potentials (TESPs), which had been investigated experimentally in adult cats.

The sequential changes of various evoked potentials observed in the nine patients with acute injury of the cervical spinal cord and their correlations with prognosis are shown in Table 1. The outcome was either death or paraplegia in those patients (Cases 1–5) in whom no SSEPs, SESPs, or corticospinal MEPs were recorded and in one patient (Case 6) in whom no corticospinal MEPs were recorded during the course, even though SESPs were restored. An extremely poor prognosis for motor function was observed in these cases, while in those patients in whom SESPs were preserved, even if highly attenuated, or restored during the course (cases 6–8), there were some patients (Cases 7 and 8) in which corticospinal MEPs were restored after the appearance of SESPs. The prognosis for motor function was good in these cases. Improvement of clinical symptoms was noted at a relatively early stage in one patient (case 9) in whom corticospinal MEPs were preserved, although the wave pattern was very abnormal.

In cases in which corticospinal MEPs were observed on the caudal side of the injury during the acute stage of 4 days, corticospinal MEPs were noted after the restoration of SESPs. Corticospinal MEPs preceded neurological improvement of motor function. In one such patient (Case 7), the muscle

Table 1. Summary of changes in evoked potentials and outcome

Case	SSEP	SESP	MEP	Outcome
1	−	−	−	died
2	−	−	−	died
3	−	−	−	paraplegia
4	−	−	−	paraplegia
5	−	−	−	paraplegia
6	−	− → +	−	paraplegia
7	− → +	− → +[a]	− → +[b]	capable of walking[c]
8	− → +	+	− → +	capable of walking
9	+	+	+	capable of walking

−, no activity; +, detectable activity
[a] 2 days after injury
[b] 4 days after injury
[c] with assistance

response elicited by transcranial stimulation was monitored simultaneously (Fig. 2). This patient was diagnosed as having motor and sensory disturbance below the level of C6 on initial examination. A roentgen study of the cervical vertebrae revealed anterior dislocation of C5 and fracture of its spinous process. These findings were considered to be a result of hyperflexion injury. Since studies with multimodality evoked potentials failed to detect SSEPs, SESPs, and corticospinal MEPs at the early stage after admission, this patient was first regarded as having a transectional injury below the level of C6. However, corticospinal MEPs began to appear on the caudal side of the injury following restoration of SESPs. In spite of the recovery of the corticospinal MEPs, although with a reduction in conduction velocity to 38 m/s, no muscle responses were noted in the quadriceps femoris. When muscle responses began to become detectable again, the neurological symptoms of motor function also started to improve. Thus, a diagnosis of central injury was possible in this case at an early stage.

At present we are evaluating magnetic stimulation, which is another method of transcranial stimulation. Although the area of stimulation cannot be defined as precisely as in electric stimulation, magnetic stimulation has the advantage of being less painful and less influenced by the cranium. Further studies are scheduled using this stimulation method (Fig. 3).

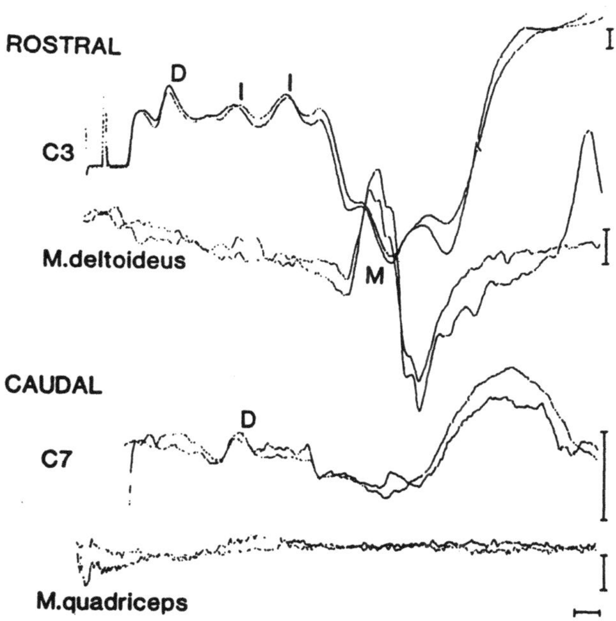

Fig. 2. Changes in corticospinal MEPs and muscle responses recorded on the rostral and caudal sides of the lesion (C6) at the acute phase of spinal cord injury. In spite of the recovery of corticospinal MEPs, although with a reduction in conduction velocity to 38 m/s, no muscle responses were observed in the quadriceps femoris (calibrations: 2 ms, 2 μV)

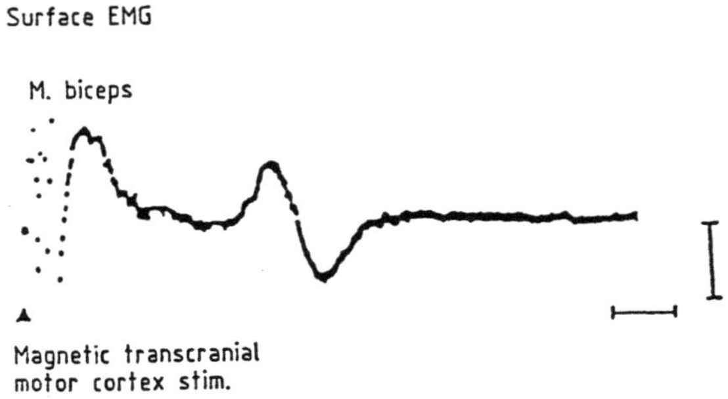

Fig. 3. Muscle response elicited by transcranial magnetic stimulation (calibrations: 5 ms, 200 μV)

Discussion

The present clinical study, based on basic experimental research, revealed several new findings.

Firstly, TESPs in human beings were found to be the same as corticospinal MEPs, while TESPs in adult cats, although efferent spinal evoked potentials, did not reflect the function of the pyramidal tract [2]. The reason for this difference, although it remains to be investigated, appears to be related to differences in shape or intracranial content. It is possible to stimulate the motor area of the cerebral cortex by the noninvasive transcranial stimulation method. Since this examination technique requires less stress, the clinical significance of corticospinal MEPs will be enhanced.

Secondly, when multimodality evoked potentials were investigated, cases which had been considered to have transectional injury based on their neurological symptoms on admission could be classified into four types: (a) an outcome of transectional injury was demonstrated where no SSEPs, SESPs, or corticospinal MEPs were present; (b) although SESPs were restored during the course, there were no corticospinal MEPs and the prognosis for the motor function was poor; (c) SESPs were preserved even at the early stage or restored during the course, and were followed by corticospinal MEPs with a good prognosis for motor function and final outcome of possible central injury; and (d) SSEPs, SESPs, and corticospinal MEPs were preserved, and clinical improvement was observed at a relatively early stage of the course. In each case, the corticospinal MEPs were most correlated with the prognosis for motor function, and repeated recording of this response was considered to be useful for early evaluation of the prognosis of motor function and for grading of the spinal cord injury.

Thirdly, the experimental study in adult cats revealed that corticospinal MEPs, corresponding to the pyramidal tract, ran through the deep layer of the dorsolateral funiculus, while the N1 element of SESPs was found to reflect to the function of the tract running through the superficial layer of the dorsolateral funiculus. Central injury was thus recognized as a dissociation of the corticospinal MEPs and the N1 element of SESPs. The results of the present clinical study were consistent with the findings of the above-mentioned experimental studies [2]. Corticospinal MEPs, which had been abolished at the acute phase of central injury, were restored before the neurological symptoms began to improve. This clearly demonstrated the importance of multimodality evoked potentials, including corticospinal MEPs, for improving our understanding of the pathophysiology of acute phase spinal cord injury and for grading its severity.

Fourthly, the muscle response elicited by transcranial stimulation was recorded. Since this response was mediated through the alpha motoneurons of the anterior horn of the spinal cord, it depended on the excitability of these neurons [1]. Dissociation from the corticospinal MEPs was considered to be caused by a lowered excitability of the alpha motoneurons at the acute phase of spinal cord injury. Thus, the value of the muscle response was limited at the acute phase of spinal cord injury.

References

1. Katayama Y, Becker DP, Hayes RL (1985) Depression of afferent-induced primary afferent depolarization at the lumbar spinal cord following concussive head injury. Brain Res 335:392–395
2. Katayama Y, Tsubokawa T, Sugitani M, Maejima S, Hirayama T, Yamamoto T (1986) Assessment of spinal cord injury with multimodality evoked potentials. Part 1: Localization of lesions in experimental spinal cord injury. Neuro-orthopedics 1:130–141
3. Katayama Y, Tsubokawa T, Maejima S, Hirayama T, Yamamoto T (1988) Corticospinal direct response recorded from the spinal epidural space to stimulation of the exposed motor cortex in humans. In: Ducker TB, Brown RH (eds) Neurophysiology and standards of spinal cord monitoring. Springer, Berlin Heidelberg New York Tokyo, pp 101–105
4. Merton PA, Morton HB (1980) Stimulation of the cerebral cortex in the intact human subject. Nature 285:227
5. Patton HD, Amassian VE (1954) Single and multiple analysis of cortical stage of pyramidal tract activation. J Neurophysiol 17:345–357
6. Tsubokawa T, Ramsay RE (1982) Evoked responses: Use in a neurological intensive care unit. In: Green BA, Marshall LF, Gallagher TJ (eds) Intensive care for neurological trauma and disease. Academic Press, New York, pp 206–216
7. Tsubokawa T (1987) Clinical value of multimodality spinal cord evoked potentials in prognosis of spinal cord injury. In: Vigouroux RP, Harris PH (eds) Thoracic and lumbar spine and spinal cord injuries. Springer, Wien New York, pp 65–92
8. Tsubokawa T, Katayama Y, Maejima S, Hirayama T, Yamamoto T (1987) Assessment of spinal cord injury with multimodality evoked spinal cord potentials. Part 2: Correlation with neurological outcome in clinical spinal cord injury. Neuro-orthopedics 3:82–89
9. Tsubokawa T, Katayama Y, Maejima S, Hirayama T, Yamamoto T (1988) Usefulness of the motor evoked potential (pyramidal D response) for the assessment of the spinal cord injury in humans. In: Ducker TB, Brown RH (eds) Neurophysiology and

standards of spinal cord monitoring. Springer, Berlin Heidelberg New York Tokyo, pp 106–113
10. Yamamoto T, Tsubokawa T, Hirayama T, Maejima S, Katayama Y (1985) Non-invasive monitoring of cortico-spinal evoked responses from the spinal epidural space in the cat. Nihon Univ J Med 27:17–21

Killed-End Corticospinal Motor Evoked Potential (MEP) in Patients with Spinal Cord Injury

Y. KATAYAMA[1], T. TSUBOKAWA, S. MAEJIMA, T. HIRAYAMA, and T. YAMAMOTO

Introduction

A monophasic positive potential is recorded when a nerve impulse approaches but never passes beyond the recording site. This type of potential has been termed the killed end potential (KP). Studies in experimental animals have repeatedly suggested a clinical value for the KP in localizing the level of spinal cord lesions [3, 4, 9]. The KP has been recorded intraoperatively in patients with spinal cord lesions [7, 10, 11]. We have also demonstrated that the KP of spinospinal responses can be readily recorded preoperatively in patients with eletrodes inserted percutaneously into the spinal epidural space and indeed has a value for localizing lesions [5].

All of these studies, however, have employed spinospinal or periphero-spinal responses for recording the KP. We describe here the KP of a cortico-spinal response (corticospinal motor evoked potentials, MEP) [2, 6, 8, 11] recorded percutaneously in patients. Unlike spinospinal or peripherospinal responses, the corticospinal MEP directly reflects conduction disturbances along corticospinal tract axons. The KP of the corticospinal MEP may thus provide us with information that is clinically more important.

The KP is usually higher in amplitude than normal evoked potentials. This is accounted for by injury currents due to a low transmembrane resistance at the damaged site of axons [1]. Thus, changes in the amplitude of KP may reflect changes in the transmembrane resistance and the injury currents [8]. Injury currents have been implicated in secondary processes of axonal degeneration mainly through the influx of calcium ion [1]. Furthermore, inhibition of axonal degeneration by an artificially applied bucking voltage has been demontrated experimentally and suggested to have a clinical value [1]. Characterization of the time courses of changes in KP in patients with spinal cord injury may therefore yield useful information for establishing a new therapeutic approach to minimize secondary process of axonal degeneration.

[1] Department of Neurological Surgery, Nihon University School of Medicine, 30-1 Ohyaguchi-Kamimachi, Itabashi-ku, Tokyo, 173 Japan

K. Shimoji et al. (Eds.)
Spinal Cord Monitoring and Electrodiagnosis

Materials and Methods

Data were obtained from four patients with cervical spinal cord injury. All patients gave informed consent for the monitoring of their spinospinal and corticospinal responses. Under fluoroscopic control, a pair of flexible platinum electrodes (Medtronic M8483, Medtronic Inc., Minneapolis, MN) were inserted into the epidural space at the lower cervical or upper thoracic level and advanced rostrally, one above and the other below the presumed level of the lesion. A detailed description of the technique has been given elsewhere [4, 5]. The corticospinal MEP was recorded monopolarly from the epidural electrodes with a reference electrode placed at the paravertebral muscles, in response to motor cortex stimulation. The motor cortex was stimulated transcranially [2, 8] or epidurally with small electrodes screwed into the skull [6, 11]. The stimuli were applied with monophasic square-wave pulses of 0.5 ms duration delivered at 4 Hz. Spinospinal responses were also recorded from the epidural electrode placed above the presumed level of the lesion, in response to stimulation with the epidural electrode placed below the presumed level of the lesion. Signals from the electrodes were fed into an amplifier with a bandpass range of 5 Hz–5 kHz and were averaged for 32–46 sweeps using a signal processor. Multiple recordings were performed with stepwise withdrawal of the rostrally placed epidural electrodes. In two patients, the epidural electrode was left in place for 3 days at the level where the KP was best recorded.

Results

The electrode was successfully advanced rostrally beyond the level of the lesion in three patients. In the region rostral to the lesion, a normal corticospinal MEP was clearly recorded: an initial biphasic positive-negative response followed by three additional biphasic positive-negative responses was observed. These responses, termed direct (D) and indirect (I) waves, respectively, were followed by a muscle response originating from the paravertebral muscle where the reference electrode was placed. With stepwise withdrawal of the epidural electrode, the latencies of the D and I waves became progressively prolonged. The physiological characteristics of these waves recorded with the same technique have been described elsewhere [5].

At the presumed level of the lesion, however, both the D and I waves suddenly changed to a monophasic positive response (Fig. 1). It was also noted that the monophasic positive I waves were superimposed on a progressive positive potential shift which occurred following the monophasic positive D wave (Fig. 1). The positive D and I waves, and the positive potential shift were always observed at the same time. It appeared that the positive D and I waves which occurred sequentially before the preceding positivity returned to the resting level formed the progressive potential shift. Since the positive

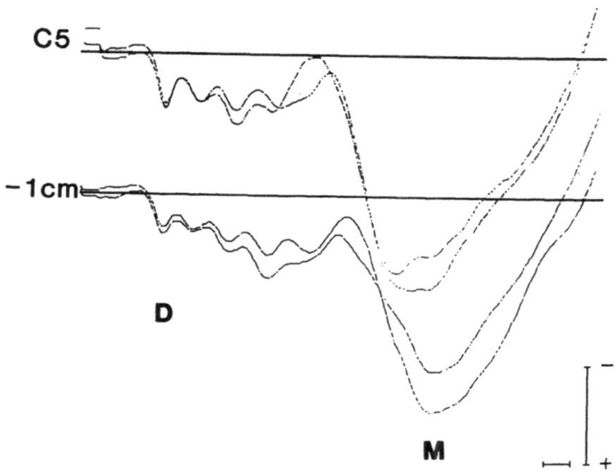

Fig. 1. Representative examples of killed end recordings of the corticospinal MEP in a patient with cervical spinal cord injury while the electrode was withdrawn in a stepwise fashion. At the level of C5, the presumed level of the lesion, monophasic positive D (*D*) and three to four I waves superimposed on a progressive positive potential shift were maximally observed. This positive wave complex was recorded at a level up to 3 cm more caudal, although the amplitude decreased progressively. The latencies remained constant. These responses were followed by a muscle response (*M*) originating from the paravertebral muscle where the reference electrode was placed. *Calibrations*, 1 ms, 10 μV. Upwards corresponds to negativity

potential shift was interrupted by a muscle response picked up by the reference electrode, it was uncertain how long this positive potential shift was sustained.

Sequential recordings during the stepwise withdrawal demonstrated that the level associated with the largest positive wave complex was located near the site where the D and I waves changed from biphasic positive-negative responses to a monophasic positive response. The amplitude decreased progressively with further withdrawal of the electrode, whereas the latencies remained constant (Fig. 1). Positivity was detected at a level up to 3 cm caudal to the level where the largest positive wave complex was observed. Recordings at more caudal levels showed no responses besides the muscle response recorded from the reference electrode. It was not possible to advance the electrode rostrally beyond the lesion in one patient. However, a positive wave complex identical to that observed in others was recorded within an area 2–3 cm caudal to the presumed level of the lesion.

Discussion

The characteristics of the positive wave complex observed in the present study were consistent with those of the KP recorded in a volume conductor. Thus, the KP of the D and I waves were each detected as a monophasic positive response. The sequential occurrence of positivity appeared to cause the progressive positive potential shift in the region caudal to the level of injury. The corticospinal MEP in experimental animals was originally described as a large KP. Injury to the axons provides a low-resistance pathway for current flows at the damaged site. Such injury currents result in distally negative voltage gradients, the classical demarcation potential [1]. The higher amplitude of KP as compared with the normal evoked potential has been explained by this demarcation potential [8]. When the impulse approaches the site of injury, the current flows from the approaching volley are opposite in direction to the current flows for the demarcation potential. Thus, the volume-recorded positivity of the approaching volley appears more positive. This suggests that the amplitude of the KP reflects the density of the injury currents.

Experimental transection of the spinal cord has revealed that enormous currents enter the cut surface, which are associated with influx of calcium ion. In addition, influx of sodium ion, the largest component of the injury currents, is known to induce mitchondria to release sequestered calcium ion. These processes give rise to a very high intracellular concentration of calcium ion which can be expected to have severe effects on the axonal cytoskeleton as well as the neurofilaments. It was recently demonstrated that degenerative processes of damaged axons can be inhibited by an artificially applied bucking voltage which counteracts the injury currents [1]. The injury currents thus appear to play an important role in the processes of axonal degeneration in spinal cord injury, presumably through an increase in the intracellular concentration of calcium ion.

The injury currents decline steeply within hours after experimental transection of the spinal cord and a relatively stable density of current entering the cut surface of the cord persists for the next 4–5 days [1]. Although confirmation in experimental animals is required, accumulation of clinical data from future studies regarding the relationships of time courses of the KP of the corticospinal MEP to the final outcome of the motor deficits may confirm the suggested role of injury currents in the degeneration processes in clinical spinal cord injury.

Acknowledgment. The authors wish to thank Dr. Kaoru Sakatani, Department of Neurological Surgery, New York University, for helpful discussions.

References

1. Borgens RB (1988) Voltage gradients and ionic currents in injured and regenerating axons. In: SG Waxman (ed) Advances in Neurology, vol 47: Functional Recovery in Neurological Disease, pp 51–66
2. Boyd S, Rothwell JC, Cowan JMA, Webb PJ, Morley T, Asselman P, Marsden CD (1986) A method of monitoring function in corticospinal pathways during scoliosis surgery with a note on motor conduction velocities. J Neurol Neurosurg Psychiatry 49:251–257

3. Cracco RQ, Evans B (1978) Spinal evoked potential in the cat: effects of asphyxia, strychnine, cord section and compression. Electroencephalogr Clin Neurophysiol 44:187–201
4. Deeke L, Tator CH (1973) Neurophysiological assessment of afferent and efferent conduction of the injured spinal cord of monkeys. J Neurosurg 39:65–74
5. Katayama Y, Tsubokawa T, Yamamoto T, Hirayama T, Maejima S (1988) Preoperative determination of the level of spinal cord lesions from the killed end potential. Surg Neurol 29:91–94
6. Katayama Y, Tsubokawa T, Maejima S, Hirayama T, Yamamoto T (1988) Corticospinal direct response in humans: identification of the motor cortex during intracranial surgery under general anesthesia. J Neurol Neurosurg Psychiatry 51:50–59
7. Kurokawa T (1978) Electrospinogram. Brain and Nerve 30:467–484 (in Japanese)
8. Levy WJ, York DH, McCaffrey M, Tanzer F (1984) Motor evoked potentials from transcranial stimulation of the motor cortex in humans. Neurosurgery 15:287–302
9. Schramm J, Krause R, Shigeno T, Brock M (1983) Experimental investigation on the spinal cord evoked injury potential. J Neurosurg 59:485–492
10. Tamaki T, Yamane T (1975) Clinical utilization of the evoked spinal cord action potential in spine and spinal cord surgery. Electroencephalogr Clin Neurophysiol 39:539 (abstract)
11. Tsubokawa T, Katayama Y, Maejima S, Hirayama T, Yamamoto, T (1987) Assessment of spinal cord injury. Neuro-orthopedics 3-82–89

Correlation of Somatosensory Evoked Potential Abnormalities with Brain and Cervical Cord Magnetic Resonance Imaging in Multiple Sclerosis

S.J. JONES[1], G. TURANO, D.H. MILLER, G.H. DU BOULAY, and R. KAKIGI

In patients with multiple sclerosis (MS), abnormalities of somatosensory evoked potentials (SSEP) can frequently be detected following upper or lower limb stimulation [8, 9]. In a very large proportion of cases, magnetic resonance imaging (MRI) of the brain reveals disseminated lesions, mostly in the subcortical white matter but also in the cerebral and cerebellar cortex, the midbrain, brainstem, spinal cord and optic nerves [6]. Although many lesions seen on MRI are "clinically silent", it is usually assumed that they correspond to demyelinated plaques seen at autopsy. It is also assumed that SSEP abnormalities are due to disturbed conduction through plaques in the sensory pathway, but there is little direct evidence for this [5]. In MS patients examined with both SSEP and MRI, the opportunity exists to compare the incidence of clinical, electrophysiological and MRI abnormalities and to see to what degree and with what specificity the abnormalities seen on MRI are reflected by electrical defects of central sensory nerve conduction.

Material and Methods

Thirty-one patients were examined, 21 women and 10 men, aged 22 to 51 years (mean 35.4 years). The SSEP control group also comprised 31 subjects, 15 women and 16 men, aged 18 to 43 years (mean 29.9 years). The patients were selected on the basis of a cervical cord syndrome (duration 1–139 months, mean 32 months) which was relapsing and remitting in 21, progressive in 7 and acute in 3. There was no evidence of traumatic, vascular, compressive or degenerative pathology, and the lesion was therefore considered compatible with demyelination. According to the criteria of Poser et al. [7], three patients were diagnosed as clinically definite MS, 11 as laboratory supported definite MS, 13 as clinically probable MS and 4 as isolated cord lesions.

[1] Medical Research Council External Staff, The National Hospital for Nervous Diseases, Queen Square, London WC1N 3BG, UK

K. Shimoji et al. (Eds.)
Spinal Cord Monitoring and Electrodiagnosis

Median nerve SSEP were recorded from eight sites on the head and neck (principally the ipsilateral clavicle, seventh and second cervical vertebrae and the scalp overlying the contralateral sensory cortical hand area) with reference to Fz. Noncephalic reference SSEP were derived off-line by subtracting the appropriate Fz reference waveforms (contralateral clavicle and supraglottal) from each trace. Tibial nerve SSEP were recorded from sites overlying the sensory cortical foot area and the left and right hand areas. Square wave impulses 200 μs in duration were delivered to the median nerve at either wrist and the posterior tibial nerve at either ankle, with a frequency of 3.2/s and an intensity sufficient to produce a moderate muscle twitch. The responses were amplified with a band pass of 1 Hz–5 kHz (less than 3 dB down), digitized at a frequency of 8 kHz (median) or 4 kHz (tibial), and averaged in groups of 400 with an epoch of 32 ms (median) or 64 ms (tibial) starting 3 or 13 ms after the stimulus. Normal limits of SSEP latency, interpeak latency and left-right latency difference were defined by the control mean plus 2.5 standard deviations (males and females separately). Normal amplitude limits were defined by the mean minus 2.5 standard deviations of the logarithm of control values.

MRI of the brain and cervical cord was provided in sagittal and transverse view with slices 5 mm deep, using a Picker 0.5 Tesla superconducting scanner. A T2 (SE2000/60) weighted sequence was used in every case, and more heavily T2 or T1 weighted sequences in some cases where changes appeared equivocal. Cord and brain MRI were interpreted by one author (DM) who was blind to the electrophysiological findings. Brain MRI was interpreted independently by a second author (GHdB), also blind to the electrophysiological findings, in order to be sure that only unequivocal lesions were recorded and to classify them according to whether or not they might involve the sensory pathway.

Results

Abnormal SSEP were recorded in 21 patients (68%). In median nerve responses with Fz reference, the clavicular N9 was always normal but N13 (Cv7 and/or Cv2) and N20 (cortical hand area) were each abnormal in 52% of patients (Fig. 1a). The cortical P40 following posterior tibial nerve stimulation was abnormal in 32% of patients, but was often relatively preserved in comparison to upper limb responses (e.g., Fig. 1a). In noncephalic reference waveforms, P14 (hand area-clavicle) was abnormal in 45% (sometimes not interpretable) and N13 (Cv7-supraglottal) was abnormal in 32%.

MRI of the brain and/or the cervical cord was abnormal in 28 patients (90%). Cord abnormalities, seen in 77%, were usually manifested as areas of increased signal (Fig. 1b) although atrophy was seen in 10% and swelling in 16%. Abnormalities were classified according to the locus and number

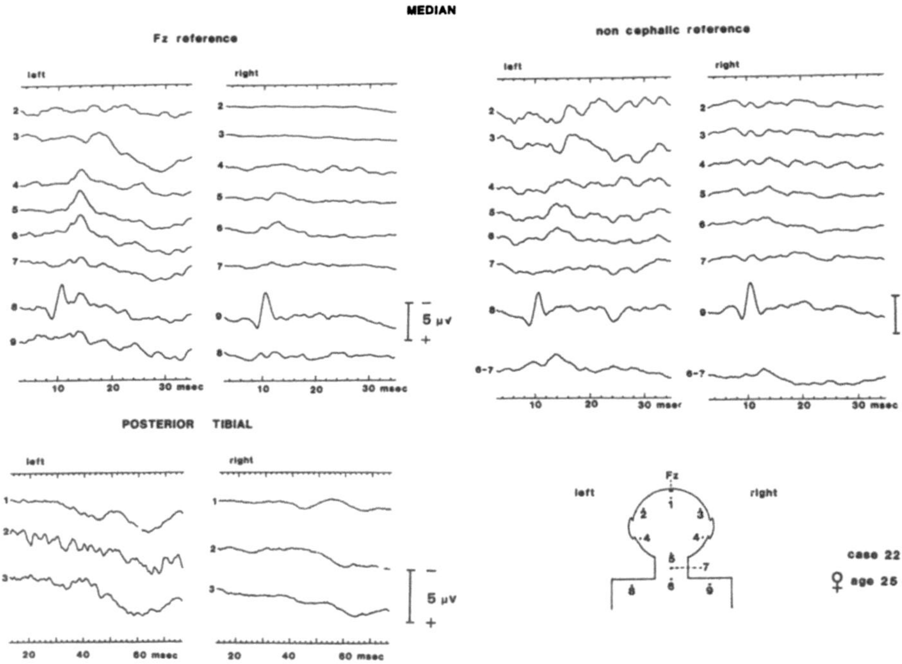
MEDIAN
Fz reference
non cephalic reference
left
right
5 µv
10
20
30 msec
6-7
POSTERIOR TIBIAL
60 msec
40
Fz
case 22
♀ age 25

a

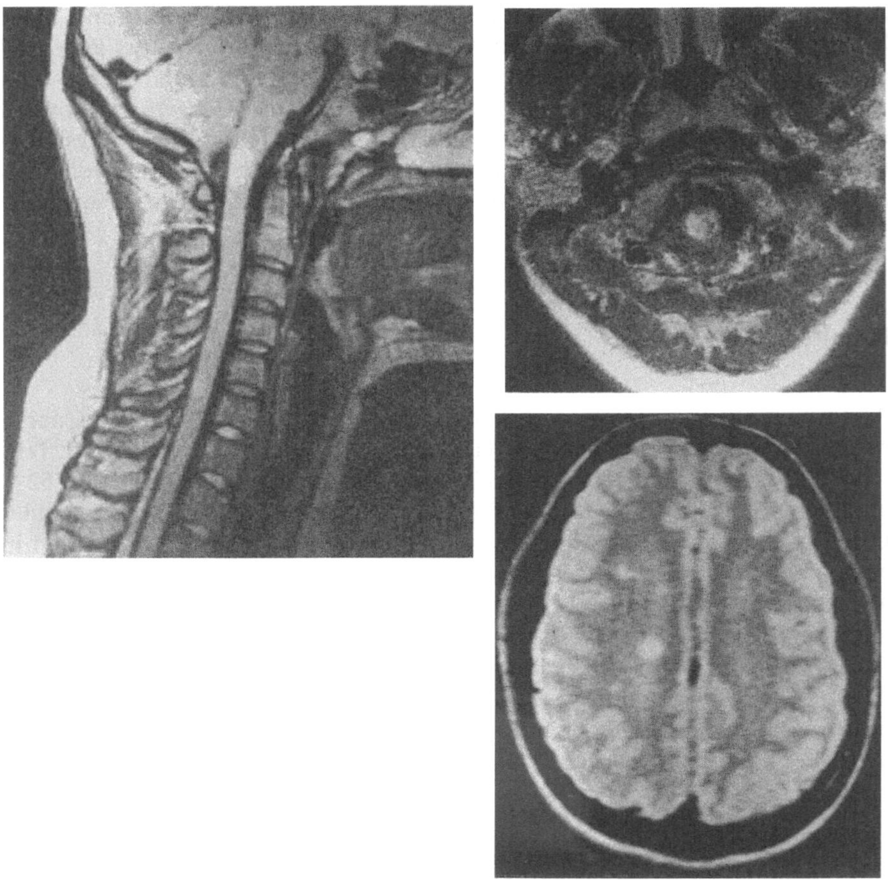

b

of segmental levels affected and, in transverse views, whether the lesion appeared to involve anterior and/or posterior structures on the left and/or right. Brain MRI abnormalities possibly involving the sensory pathways were seen in 68% of patients, at the level of the medulla (16%), pons (16%), internal capsule (10%), sensory radiations (55%) and cortex (3%).

Correlation Between SSEP and MRI Abnormalities

The significance of correlations was examined using the Fisher exact probabilities test (Table 1). In median nerve SSEP with Fz reference, abnormalities of N13 and/or N20 were significantly correlated with the presence of MRI abnormalities in the cervical cord ($p = 0.012$). A significant correlation was obtained with MRI abnormalities involving the posterior ($p = 0.018$) but not the anterior side of the cord. For each arm, median nerve SSEP abnormalities were significantly correlated with cervical MRI abnormalities affecting the ipsilateral (left, $p = 0.0002$; right, $p = 0.018$), but not the contralateral side. No significant correlations could be demonstrated between cervical MRI abnormalities and abnormalities of posterior tibial or noncephalic reference median nerve SSEP.

Although in the overall group of 31 patients SSEP and cervical MRI abnormalities were closely correlated, partial or complete disagreement occurred in 13. There were nine in whom SSEP from one (3) or both (6) arms were normal in spite of MRI abnormalities apparently involving the posterior ipsilateral sector of the cord (Fig. 2a,b). The converse pattern, of abnormal SSEP without appropriate MRI abnormalities in the cord, was seen unilaterally in four patients, one of whom had a delay of N20 alone which could have been due to an MRI abnormality in the sensory radiation on the side contralateral to the stimulus.

In relation to MRI of the brain, the incidence of SSEP abnormalities following upper or lower limb stimulation was not significantly influenced by the presence or absence of lesions possibly involving the sensory pathways at the level of the medulla, pons, internal capsule, sensory radiations or sensory cortex on the appropriate side.

Concerning the level of the cord lesion, in supraglottal reference median

Fig. 1. **a** Normal SSEP following left median nerve stimulation but abnormal responses on the *right* in a patient presenting with an acute right-sided cervical cord syndrome ("Useless hand of Oppenheim"), diagnosed as clinically probable MS. The cervical N13 was of reduced amplitude and the cortical N20 absent. Abnormalities were less clearly defined in non-cephalic reference median nerve SSEP, and posterior tibial nerve responses were within normal limits. **b** MRI of the brain and cervical cord were interpreted without knowledge of SSEP findings. The sagittal view (*left*) shows a region of high signal at the cervicomedullary junction. The transverse view (*upper right*) at this level shows a high signal area mainly involving the right posterior quadrant. Transverse MRI of the brain (*lower right*) revealed at least two right-sided abnormalities in the subcortical white matter, plus a central lesion of the midpons (not shown)

Table 1. Correlation between SSEP and MRI abnormalities (Fisher exact probabilities test)

SSEP	Median						Tibial		
Reference	Fz			Noncephalic			Fz		
MRI	L/R	L	R	L/R	L	R	L/R	L	R
Cervical cord	0.012			> 0.1			> 0.2		
Left		0.002	> 0.3		> 0.3	–		> 0.1	–
Right		> 0.1	0.018		–	> 0.1		–	> 0.1
Posterior	0.013			> 0.3			> 0.1		
Anterior	> 0.2			> 0.3			> 0.1		
Brain sensory pathways	> 0.1			> 0.3			> 0.3		
Left		> 0.3	> 0.2		–	–		–	–
Right		> 0.2	> 0.1		–	–		–	–

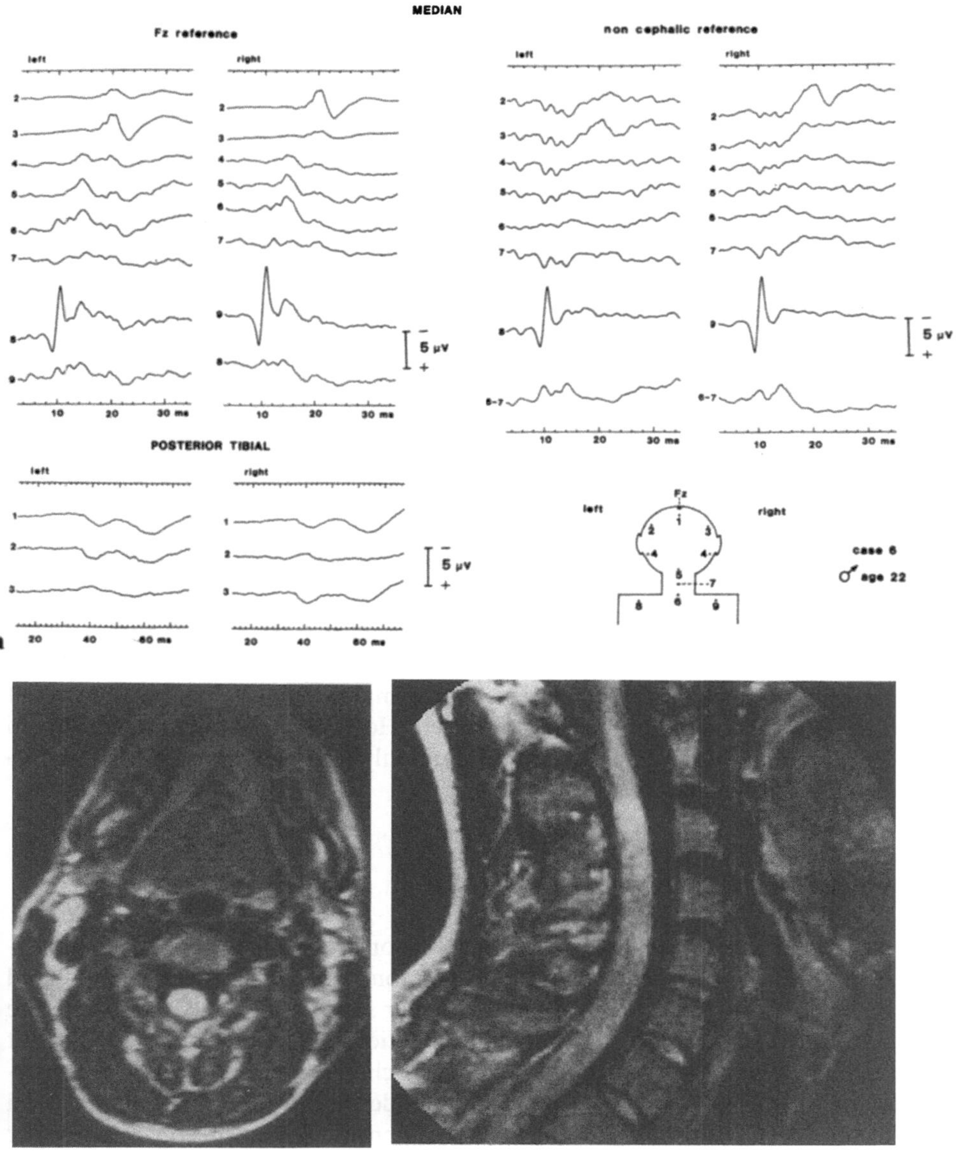

Fig. 2. a Normal SSEP in a patient presenting with a relapsing and remitting cervical cord syndrome, manifested by Lhermitte's symptom, mild sensory symptoms and signs in the limbs and urinary urgency, diagnosed as laboratory supported definite MS. **b** Sagittal MRI (*right*) of the cervical cord shows a region of high signal at Cv 3. The transverse view (*left*) at this level shows widespread involvement throughout the cord. Transverse MRI of the brain (not illustrated) revealed abnormalities in the right internal capsule and the left and right sensory radiations

Table 2. Correlation of N13 and P14 abnormalities with level of cervical MRI abnormality (Fisher exact probabilities test)

Component Electrode locations		N13 Cv7-Fz	N13 Cv7-supraglottal	P14 Hand area-clavicle
MRI levels (*n*)				
Cv1-CMJ	(6)	0.01	> 0.4	0.017
Cv3-Cv2	(17)	> 0.2	> 0.3	> 0.3
Cv5-Cv4	(8)	> 0.1	> 0.2	> 0.2
Cv7-Cv6	(4)	0.06	0.029	> 0.2

CMJ, cervicomedullary junction

nerve SSEP the incidence of N13 abnormalities was highest (100%) in patients with cervical MRI abnormalities at Cv6–Cv7 ($p = 0.029$), diminishing progressively at more rostral levels (see Table 2). In clavicle reference recordings, P14 abnormalities were most strongly associated ($p = 0.017$) with MRI lesions at C1 or the cervicomedullary junction (CMJ). In Fz reference recordings, abnormalities of N13 were strongly associated with MRI lesions at C1 or the CMJ ($p = 0.01$) and weakly with abnormalities at Cv6–Cv7 ($p = 0.06$).

The interpeak latency N9–N20 and the absolute latency of P40 were both significantly correlated with the number of levels affected by MRI abnormalities on the ipsilateral side of the cord (left, $r = 0.57$ and 0.58, $p < 0.01$; right, $r = 0.38$ and 0.41, $p < 0.02$). The relationship between P40 latency and the number of levels affected was approximately linear (Fig. 3), equivalent to a delay of about 2 ms per level affected. However, too much weight should be placed on this finding since there were only four patients with MRI abnormalities longer than two levels.

Clinical Correlations

No significant correlations could be demonstrated between SSEP abnormalities and the presence of sensory symptoms and/or signs in the stimulated limb, nor with impairment of any particular sensory modality. Similarly, the presence and location of MRI abnormalities in the cervical cord did not correlate significantly with any pattern of clinical sensory or motor impairment on the ipsilateral or contralateral side of the body. The upper level of the MRI abnormality in the cord was consistently at or above the upper level of the lesion as indicated by the segmental distribution of sensory impairment.

A relatively close correspondence between clinical, SEP and MRI abnormalities was seen in four patients presenting with a "useless hand of Oppenheim", characterised by a profound loss of joint-position sense. In all four the SSEP from the affected arm were abnormal (Fig. 1) while tibial SSEP were abnormal in two. In each case, MRI of the cord showed a region of abnormal signal involving the side ipsilateral to the clinical deficit; the

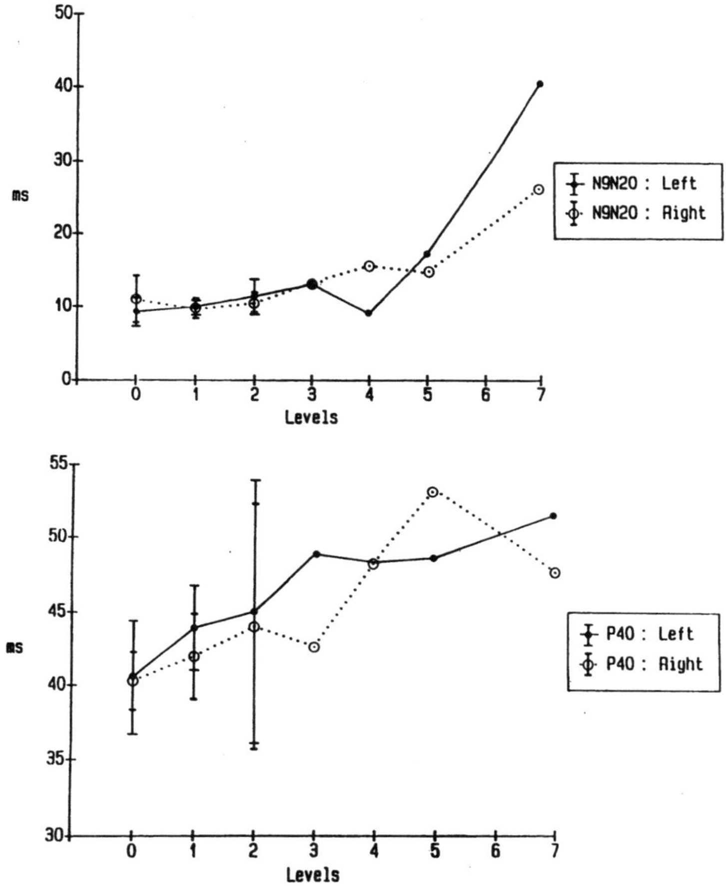

Fig. 3. Correlation of N9–N20 interpeak latency (median SSEP) and absolute latency of P40 (posterior tibial SSEP) with rostrocaudal extent (number of levels involved) of MRI abnormalities on the ipsilateral side of the cervical cord. All correlation coefficients were significant ($p < 0.02$ or $p < 0.01$). Standard deviations are shown for 0, 1 and 2 levels affected. Longer abnormalities affecting 3, 4, 5 and 7 levels were seen in only one patient each

abnormality involved the posterior part of the cord in three patients and the anterior part in one. The level of the MRI abnormality in these four ranged from C5 to the CMJ.

Discussion

The findings of the study indicate that, in patients with demyelinating cervical cord disease, MRI abnormalities in the cord are associated to a significant degree with SSEP abnormalities from the upper limbs. When the MRI was classified according to involvement of the left, right, anterior or posterior sides of the cord, significant SSEP correlations were demonstrated only for ipsilateral and posterior abnormalities. This is not a surprising finding, but provides the first direct confirmation that MRI abnormalities in the cord are associated with nerve conduction defects in appropriate tracts.

The lack of correlation between cervical MRI and lower limb SSEP abnormalities is more surprising. This was not due to the presence of lesions at lower levels of the cord which were detected by tibial SSEP but not by MRI, since tibial SSEP showed a much lower incidence of abnormality than median in this group of patients. Clearly, myelin damage in the cord was often highly localized, affecting sensory fibres from the upper limb while sparing those from the lower limb.

The correlation between MRI abnormalities and noncephalic reference median nerve SSEP was also poor, perhaps due to the lower signal/noise ratio and hence occasional noninterpretability of these waveforms. However, the association between abnormalities of supraglottal reference N13 and MRI lesions at C6–C7, and between clavicle reference P14 and MRI lesions at C1 and the CMJ, is consistent with current views regarding the origin of these components. In noncephalic reference SSEP it is probable that N13 is generated largely in the dorsal horn at low cervical level and P14 at the rostral end of the cord or the lower brainstem. The N13 recorded with Fz reference represents a fusion of these components [1–4].

The correlation between SSEP and brain MRI abnormalities possibly involving the sensory pathways was poor. Apart from the lack of significance in the Fisher test, the fact that the Fz reference N13 component was usually affected in addition to N20 indicates that MRI lesions of the internal capsule, sensory radiations and cortex, also seen in a large proportion, cannot have been the major factor responsible. In MRI it is currently impossible to identify the course of sensory pathways with certainty, and it may also be the case that the fibres of the sensory radiation are sufficiently dispersed that many can survive the presence of small, disseminated lesions, to the extent that the cortical SSEP remain within normal limits.

The correlation between SSEP abnormalities and the pattern of clinical involvement was poor, perhaps because the SSEP reflects only sensory conduction in pathways responsible for the "large fibre" modalities of touch and proprioception. The correlation between the extent of MRI abnormalities in transverse views and patterns of clinical impairment was also poor. This, plus the fact that normal SSEP in the presence of MRI abnormalities apparently involving the appropriate sector of the cord were more common than the converse finding, may be due to the fact that areas of abnormal MRI signal chiefly reflect an increase in the water content of the tissue. Although this is likely to be centred on the region of myelin damage, during the acute stage where demyelination is accompanied by oedema it may spread to adjacent areas where axonal conduction is unimpaired. Even in chronic lesions the degree of signal abnormality, due partly to astrocytosis, is unlikely to be closely related to the extent of demyelination or axon loss.

MRI is a sensitive technique for demonstrating the presence of disseminated lesions in the central nervous system. In patients with multiple sclerosis, MRI abnormalities in the cervical cord appear to be closely related to the locus of demyelinated plaques causing clinical symptoms and abnormal SSEP, but probably extend beyond the immediate area of myelin damage. In conjunction with clinical and electrophysiological data, the manner in which MRI abnormalities evolve over time may provide new insights into the pathophysiology of the disease.

References

1. Desmedt JE, Cheron G (1981) Prevertebral (oesophageal) recording of subcortical somatosensory evoked potential in man: the spinal N13 component and the dual nature of the spinal generators. Electroencephalogr Clin Neurophysiol 52:257–275
2. Jacobson GP, Tew JM (1988) The origin of the scalp recorded P14 following electrical stimulation of the median nerve: intraoperative observations. Electroencephalogr Clin Neurophysiol 71:73–76
3. Jones SJ (1977) Short latency potentials recorded from the neck and scalp following median nerve stimulation in man. Electroencephalogr Clin Neurophysiol 43:853–863
4. Lueders H, Lesser R, Hahn J, Little J, Klem G (1983) Subcortical somatosensory evoked potentials to median nerve stimulation. Brain 106:341–372
5. Matthews WB, Esiri M (1979) Multiple sclerosis plaque related to abnormal somatosensory evoked potentials. J Neurol Neurosurg Psychiatry 42:940–942
6. Ormerod IEC, Miller DH, McDonald WI, Boulay EPGH du, Rudge P, Kendall BE, Moseley IF, Johnson G, Tofts PS, Halliday AM, Bronstein AM, Scaravili F, Harding AE, Barnes D, Zilkha KJ (1987) The role of NMR imaging in the assessment of multiple sclerosis and isolated neurological lesions: a quantitative study. Brain 110:1579–1616
7. Poser CM, Paty DW, Scheinberg L, McDonald WI, Davis F, Ebers GC, Johnson KP et al. (1983). New diagnostic criteria for multiple sclerosis: guidelines for research protocol. Ann Neurol 13:227–231
8. Small DG, Matthews WB, Small M (1978) The cervical somatosensory evoked potential in the diagnosis of multiple sclerosis. J Neurol Sci 35:211–224
9. Trojaborg W, Petersen E (1979) Visual and somatosensory evoked potentials in multiple sclerosis. J Neurol Neurosurg Psychiatry 42:323–330

Comparison Between Magnetic Resonance Imaging and Somatosensory Evoked Potentials in Chronic Spinal Cord Injury Patients

S. Nagata[1], E. Transfeldt, B. Roche, F. Torres, and K. Heithoff

Summary

Comparison between magnetic resonance imaging (MRI) and somatosensory evoked potentials (SSEP) was done in 30 chronic spinal cord injury patients. A high incidence of syringomyelia (42%) as a posttraumatic later complication was found from the result of the MRI study.

Positive correlation was seen between MRI and SEEP (88%), and neurological evaluation and SSEP (81%). SSEP were useful in detecting spinal cord lesions at levels that are clinically normal. But a false rate of posterior tibial nerve stimulation in the SSEP was relatively high (28%) compared with median (5%) and ulnar nerve stimulation (6%).

Introduction

Prognosis and further treatment are vitally dependent on the completeness of a spinal cord injury. Evaluation of severe spinal injuries continues to remain a difficult process. Many patients who initially do not appear to have any neurological function may show some recovery. There is considerable concern that many patients being labeled as complete injuries are in fact anatomically incomplete and may even have electrophysiological pathways which may allow conductivity across the area of damage. (It is conceivable that many of these patients may in fact benefit from functional electrical stimulation or even patterned electrical stimulation in being able to regain some recovery.) Patients with compressive lesions of the spinal cord who show electrophysiological or clinical conductivity across the area of spinal cord injury are better candidates for early spinal decompression. Somatosensory and motor evoked potentials do have the ability to measure the functional

[1] Department of Orthopedic Surgery, Kanazawa University School of Medicine, 13-1 Takara-machi, Kanazawa, 920 Japan

K. Shimoji et al. (Eds.)
Spinal Cord Monitoring and Electrodiagnosis

integrity of the spinal cord at the level of injury. However, it is not quite clear to what extent the cord has to be damaged to be totally without somatosensory evoked potential (SSEP). A databank of information needs to be developed in order to make more concrete statements about the abnormalities recorded in the evoked potentials in patients with spinal cord injuries.

Correlation of these evoked potentials clearly needs to be done with the clinical function of the patient. In addition, modern (radiological) imaging techniques such as magnetic resonance imaging (MRI) provide the potential to give more accurate anatomical information about the spinal cord tissue at the level of the injury. High resolution MRI is able to provide useful information about normal and abnormal segments of the spinal cord.

We feel therefore that it would be of value to make a correlation of the evoked potentials, the MRI, as well as the clinical function in patients with acute and chronic spinal cord injuries. Both of these investigative techniques have been used independently in spinal cord injuries, but there has been no good study to correlate these two functions with each other and with clinical function.

Materials and Methods

Thirty chronic spinal cord injury patients were examined, including 24 men and 6 women. Ages were from 22 to 60 (average 34.5). The period from onset to examination ranged from 8 months to 22 years (average 8.2 years). There were 23 quadriplegics and 7 paraplegics, 17 complete and 13 incomplete patients.

Clinical evaluation was done about each patient in our spinal cord injury clinic concerning motor and sensory function, reflexes, muscle tone, etc. Short latency SSEP was recorded by Nicolet Pathfinder II at the EEG laboratory of the University of Minnesota. Median, ulnar and posterior tibial nerves were selected as stimulation sites. Montages were Fz-Erb (Ch. 1), Fz-C7 (Ch. 2), Fz-C2 (Ch. 2), and Fz-Cc′ (Ch. 4) in median and ulnar nerve stimulation, L3-L1 (Ch. 1), Fz-C7 (Ch. 2), FDz-Cz (Ch. 3), Ci'-Cc′ (Ch. 4) in posterior tibial nerve stimulation. Timebase was 40 ms in median nerve and ulnar nerve stimulation and was 80 ms in posterior tibial stimulation. Some 1024 repetitions were averaged. Band-pass filter was from 30–3000 Hz. Duration of stimulation was 200 μs. Stimulation rate was 5.1/s in median and ulnar nerve stimulation and 4.1/s in posterior tibial nerve stimulation. Maximum strength of stimulation was 20 mA in median and ulnar nerve stimulation and was 25 mA in posterior tibial nerve stimulation. Maximum tolerable strength of stimulation was used for each stimulation site.

MRI was performed at the level of spinal injury and additional levels imaged only if indicated by clinical condition or if pathological changes extended beyond the level of spine injury. T1 and T2 weighted images were performed and an attempt was made to get detail with respect to quantity and quality of normal spinal cord, scar tissue, and possible cyst (syrinx) formation.

The results of the three examinations were compiled in each patient for cross correlation.

Results

SSEP and Clinical Evaluation

The afferents of the median nerve are primarily C4, C5, and C6 spinal cord levels. And the afferents of ulnar nerve are mainly C7, C8, and T1 cord levels. We can estimate the spinal cord injury level using the technique of these SSEP in quadriplegia patients. Posterior tibial nerve stimulation is theoretically useful for determination whether the injury is complete or incomplete. The recording of lumbar and cervical potentials is also helpful for the estimation of the injury level.

The estimation of the cord injury level was done by SSEP recording of median and ulnar nerves. Between SSEP and clinical evaluations, a positive correlation concerning diagnosing levels of injury was seen in 22 patients (81%). Five patients showed poor correlation. In comparing complete and incomplete patients 13 had positive correlation and 1 showed poor correlation in complete injury patients.

The patient in whom results of SSEP and clinical evaluations showed poor correlation had a complete injury at the T5 level in clinical evaluation. But both sides of the SSEP by ulnar nerve stimulation showed severe abnormality. In the MRI examination, we found severe lower cervical cord atrophy and syringomyelia below the lesion.

MRI and Clinical Evaluation

All the patients who underwent the MRI evaluation showed spinal cord abnormality to some extent. The most frequent abnormality was myelomalacia (21 patients, 81%), 3 expansive and 18 focal myelomalacia. Secondly, the incidence of posttraumatic syrinx formation was documented in 11 (42%) patients. Six patients had extensive syringomyelia and five had focal syrinx at the lesion level.

Comparison between the results of the MRI and the clinical evaluation, positive correlation (complete neurological deficit and transverse necrosis or incomplete neurological deficit and partial cord damage) was seen in 20 patients (77%). Poor correlation was seen in six patients (23%). In the poor correlation patients, MRI findings showed the existence of pathology rostral

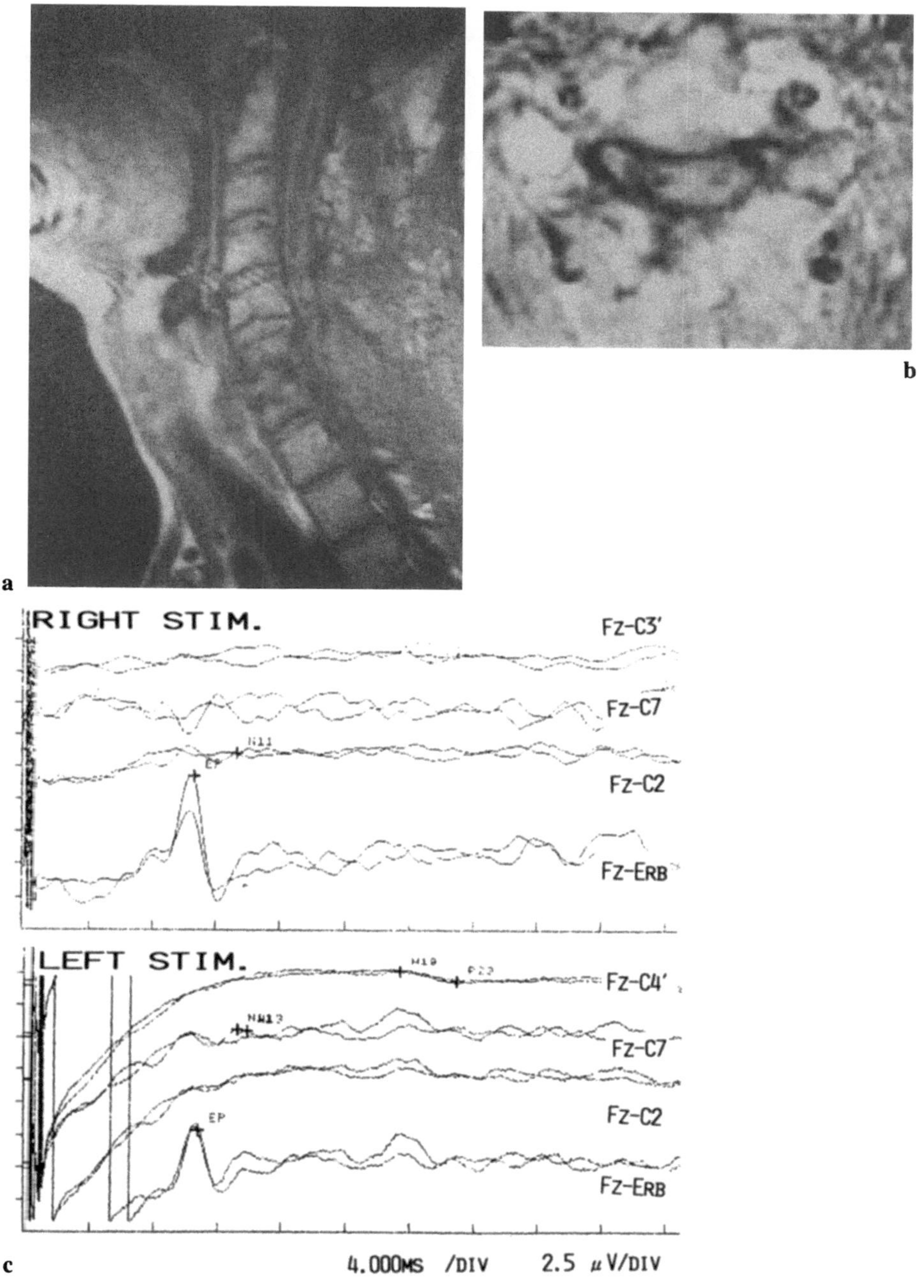

Fig. 1a–c. **a,b** The sagittal and axial cut in T1 weighted image of a 41-year-old man with C5–6 injury. A large syrinx above the primary injury level is demonstrated well. **c** The median nerve SSEP shows poor cortical response

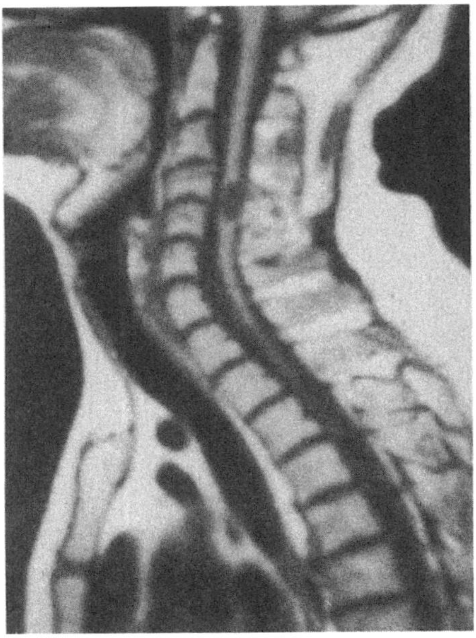

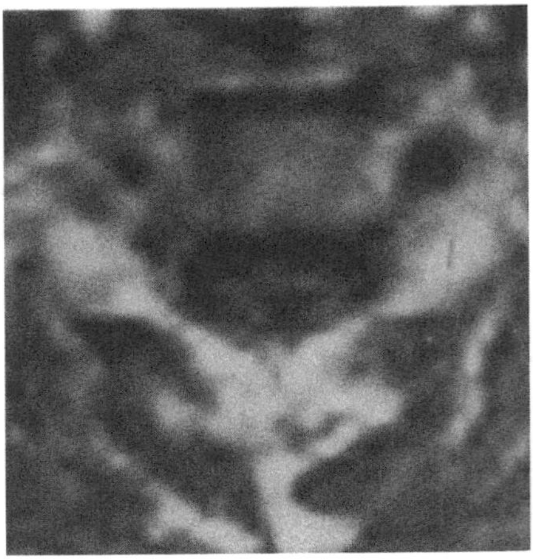

b

a

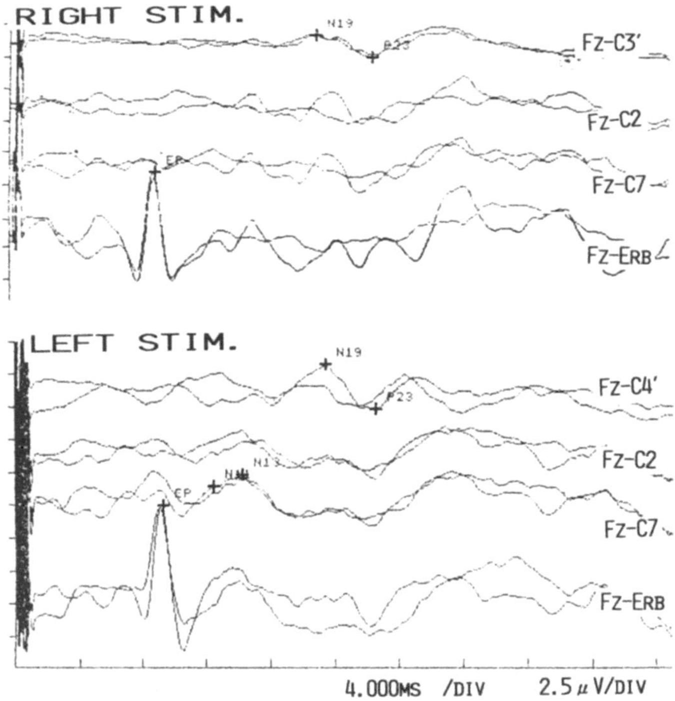

c

Fig. 2a–c. a,b The sagittal and axial cut in T1 weighted image of a 33-year-old man with C4 injury. Focal myelomalacia dorsal to C4–5 vertebra can be seen clearly. **c** The median nerve SSEP shows good cortical responses

to the primary lesion site: large syringomyelia in two patients; one with severe myelomalacia; and three with spinal cord atrophy.

MRI and SSEP

Positive correlation concerning spinal cord injury level was seen in 21 patients (88%). Four patients showed poor correlation: 2 complete and 2 incomplete.

False positive and false negative SSEP were seen in one patient each. The remaining two showed ascending pathology in MRI study and had positive correlation between SSEP and clinical evaluations.

Figure 1 demonstrates a positive correlation case. The patient is a 41-year-old man with a history of a C5–6 vertebral fracture. On clinical evaluation his injury level was C6. Usually this level of injury shows normal SSEP by median nerve stimulation, but his median SSEP showed no cortical response on right side stimulation and low amplitude prolonged cortical response on the left. MRI findings reveal a severe myelomalacia at the C5–6 vertebral level and a large syrinx extending above and below the primary injury level.

Poor correlation between MRI and SSEP is shown in Fig. 2. The patient is a 33-year-old man with a history of a C4 vertebral fracture. Clinical evaluation shows an incomplete injury at the C4–5 level. Median SSEP showed normal response, but his MRI results showed C4–5 level focal myelomalacia mainly involving the dorsal and dorsolateral column.

Discussion

For the chronic spinal cord injury patient, posttraumatic progressive myelopathy is a serious problem. Most often it is caused by late degenerative changes in the spinal cord such as severe myelomalacia, syringomyelia, and spinal cord atrophy, etc. Recent reports estimate this incidence at 2% or less in all post-traumatic spinal cord injury patients [1–4].

In our MRI study, all the patients had to some extent cord abnormality and six patients showed expansion of cord degeneration (23%). This incidence was relatively high compared with previous reports.

It is suggested that there are many chronic spinal cord injury patients who have remained neurologically stable with progressive degenerative changes beyond the primary lesion in their spinal cord. Therefore careful neurological follow-up of the chronic spinal cord injured patient is necessary.

MRI provides detailed anatomical information about the spinal cord. Distinction of pathological states is sometimes possible using MRI. However, MRI is limited in evaluation of the physiological function of the spinal cord.

On the other hand, SSEP supply valuable information about the residual spinal cord function in spinal cord injury patients. The extent of the lesion may also be detectable using sequential recording techniques [5].

Young reported that SSEP are reliable and sensitive indicators of residual neurological function in acute spinal injury, and incidence of discrepancies between SSEP and neurological function increased in chronic spinal cord injuries [7]. He also reported that these discrepancies were concerned with neural fatigue, habituation, and demyelination in animal models [6].

In our study, there were five poor correlations (19%) between SSEP and clinical evaluations. The causes of these discrepancies were thought to be (1) a demyelination change in SSEP false positive cases, and (2) the distribution of the residual tracts in SSEP false negative cases.

Between MRI and SSEP there was good correlation (84%), except in four patients. In two individuals, normal SSEP studies were obtained, while MRI clearly demonstrated myelopathy of the spinal cord. The second two showed normal SSEP and stable neurological examination, however, MRI revealed ascending myelopathy of the spinal cord.

Both SSEP and MRI studies have merits and faults. Trends in patient management toward cost effectiveness will support the use of noninvasive testing of SSEP over MRI study in monitoring chronic spinal cord abnormalities in limited patient situations. It is felt that SSEP may be useful in diagnosing abnormalities of the spinal cord prior to using more expensive investigations such as the MRI.

Concerning false negative and false positive rates of SSEP, posterior tibial stimulation SSEP showed a relatively high value compared with median and ulnar stimulation (28%), and all of them involved incomplete injury patients. Weak stimulation or technical problems may be the cause of this discrepancy. The high false rate indicates that the posterior tibial nerve stimulation is less reliable and sensitive in detecting spinal cord abnormalities.

Conclusion

1. This study documented a high prevalence of syringomyelia (42%) in chronic spinal cord injury.
2. There was a positive correlation of MRI and SSEP of 88% and a positive correlation of clinical evaluation and SSEP of 81%, specific to the spinal cord injury level.
3. In chronic spinal cord injury patients, monitoring with SSEP studies may prove invaluable in detecting abnormalities of the spinal cord prior to the patient becoming neurologically symptomatic.
4. Posterior tibial nerve stimulation is less reliable and sensitive in detecting chronic spinal cord abnormalities in the incomplete spinal cord injury patient.

References

1. Griffiths ER, McCormick CC (1981) Posttraumatic syringomyelia (cystic myelopathy). Paraplegia 19:96–97
2. Quencer RM, Sheldon JJ, Donovan Post MJ et al. (1986) MRI Chronically injured cervical spinal cord. AJR 147:125–132
3. Vernon JD, Chir B, Silver JR et al. (1982) Post-traumatic syringomyelia. Paraplegia 20:339–364
4. Williams B, Terry AF, Jones F et al. (1981) Syringomyelia as a sequel to traumatic paraplegia. Paraplegia 19:67–80
5. Yiannikas C (1983) Spinal Cord Trauma. In: Chiappa KH (eds) Evoked Potentials in Clinical Medicine. Raven Press, New York, pp 287–288
6. Young W, Mollin D (1988) Somatosensory evoked potentials in chronic spinal cord injury – an uptdate. In: Ducker TB, Brown RH (eds) Neurophysiology and standards of spinal cord monitoring. Springer Berlin Heidelberg New York Tokyo, pp 277–288
7. Young W (1985): Somatosensory evoked potentials (SSEPs) in spinal cord injury. In: Schramm J, Jones SJ (eds) Spinal Cord Monitoring. Springer, Heidelberg Berlin New York Tokyo, pp 127–142

Subject Index

H. M. Mayer, M. Brock, University of Berlin (Eds.)

Percutaneous Lumbar Discectomy

1989. XI, 218 pp. 134 figs. 44 tabs.
Hardcover DM 98,– ISBN 3-540-51032-X

Percutaneous lumbar discectomy is a new surgical method for treating lumbar disc diseases. The goal of the procedure is decompression of the spinal nerve root by percutaneous removal of the nucleus pulposus under local anesthesia. Probably 20% of all patients requiring lumbar disc surgery can be successfully treated by this method. During the past two years, percutaneous discectomy has spread rapidly, and it is now performed in most clinical departments engaged in spinal surgery.

The first International Symposium on Percutaneous Lumbar Discectomy, held in Berlin in August 1988, covered all current procedures known as "percutaneous discectomy" and the entire range of percutaneous techniques, both clinical and experimental. Its publication is important because of the recency of this new surgical procedure, the outstanding experience of the speakers – including the Japanese, American, and European "pioneers" of the technique – and last but not least the gaps in the knowledge of physicians concerning this topic. This procedure opens up new perspectives in the surgical treatment of degenerative diseases of the lumbar spine.

Springer-Verlag
Berlin
Heidelberg
New York
London
Paris
Tokyo
Hong Kong
Barcelona

MIX
Papier aus verantwortungsvollen Quellen
Paper from responsible sources
FSC® C105338

If you have any concerns about our products,
you can contact us on
ProductSafety@springernature.com

In case Publisher is established outside the EU,
the EU authorized representative is:
Springer Nature Customer Service Center GmbH
Europaplatz 3, 69115 Heidelberg, Germany

Printed by Libri Plureos GmbH
in Hamburg, Germany